D1345989

An understanding of the relationship between a pathogen and its host is essential for the development of effective disease control measures. This volume focuses on interactions at the molecular level, specifically between the proteins of the infectious agent and the proteins of the host that has been invaded. Both viral and bacterial systems are considered, with specific examples illustrating the rapid advances being made in defining the molecular mechanisms underlying infection.

MOLECULAR ASPECTS OF HOST–PATHOGEN INTERACTION

SYMPOSIA OF THE SOCIETY FOR GENERAL MICROBIOLOGY

Series editor (1996–2001): Dr D. Roberts, Zoology Department, The Natural History Museum, London

Volumes currently available:

28 STRUCTURE AND FUNCTION OF PROKARYOTIC MEMBRANES
29 MICROBIAL TECHNOLOGY: CURRENT STATE, FUTURE PROSPECTS
30 THE EUKARYOTIC MICROBIAL CELL
32 MOLECULAR AND CELLULAR ASPECTS OF MICROBIAL EVOLUTION
33 VIRUS PERSISTENCE
35 INTERFERONS FROM MOLECULAR BIOLOGY TO CLINICAL APPLICATIONS
36 THE MICROBE 1984 PART I: VIRUSES
PART II: PROKARYOTES AND EUKARYOTES
37 VIRUSES AND CANCER
38 THE SCIENTIFIC BASIS OF ANTIMICROBIAL CHEMOTHERAPY
39 REGULATION OF GENE EXPRESSION – 25 YEARS ON
41 ECOLOGY OF MICROBIAL COMMUNITIES
42 THE NITROGEN AND SULPHUR CYCLES
43 TRANSPOSITION
44 MICROBIAL PRODUCTS: NEW APPROACHES
45 CONTROL OF VIRUS DISEASES
46 THE BIOLOGY OF THE CHEMOTACTIC RESPONSE
47 PROKARYOTIC STRUCTURE AND FUNCTION: A NEW PERSPECTIVE
49 MOLECULAR BIOLOGY OF BACTERIAL INFECTION: CURRENT STATUS AND FUTURE PERSPECTIVES
50 THE EUKARYOTIC GENOME: ORGANISATION AND REGULATION
51 VIRUSES AND CANCER
52 POPULATION GENETICS OF BACTERIA
53 FIFTY YEARS OF ANTIMICROBIALS: PAST PERSPECTIVES AND FUTURE TRENDS
54 EVOLUTION OF MICROBIAL LIFE

MOLECULAR ASPECTS OF HOST–PATHOGEN INTERACTION

EDITED BY

M. A. McCRAE, J. R. SAUNDERS, C. J. SMYTH AND N. D. STOW

FIFTY-FIFTH SYMPOSIUM OF THE SOCIETY FOR GENERAL MICROBIOLOGY HELD AT HERIOT-WATT UNIVERSITY, EDINBURGH MARCH 1997

Published for the Society for General Microbiology

PUBLISHED BY THE PRESS SYNDICATE OF THE UNIVERSITY OF CAMBRIDGE
The Pitt Building, Trumpington Street, Cambridge CB2 1RP, United Kingdom

CAMBRIDGE UNIVERSITY PRESS
The Edinburgh Building, Cambridge CB2 2RU, United Kingdom
40 West 20th Street, New York, NY 10011-4211, USA
10 Stamford Road, Oakleigh, Melbourne 3166, Australia

First published 1997

Printed in the United Kingdom at the University Press, Cambridge

Typeset in Monotype Times 10/12pt

A catalogue record for this book is available from the British Library

ISBN 0 521 59215 1 hardback

CONTENTS

	page
Contributors	ix
C. Chothia Protein–protein and protein–carbohydrate recognition	1
D. J. Evans Picornavirus receptors, tropism and pathogenesis	23
G. R. Cornelis Cross-talk between *Yersinia* and eukaryotic cells	45
T. J. Foster, O. Hartford and D. O'Donnell Host–pathogen protein–protein interactions in *Staphylococcus*	67
M. Virji Mechanisms of microbial adhesion; the paradigm of *Neisseriae*	95
R. D. Everett Virus–host interactions in the control of the gene expression of nuclear replicating DNA viruses	111
J. Karn, N. J. Keen, M. J. Churcher, F. Aboul-ela, G. Varani and M. J. Gait Regulation of human immunodeficiency virus gene expression by Tat	129
M. E. Ward *Chlamydia* host and host cell interactions	153
C. J. Dorman, N. C. Nolan and S. G. J. Smith Control of type 1 fimbrial expression by a random genetic switch in *Escherichia coli*	191
M. B. Mathews Viruses and the protein synthesis machinery of the cell: offence, defence and dependence	211
G. Tran Van Nhieu, T. Adams, C. Dehio, R. Ménard, A. Skoudy, J. Mounier, R. Hellio, P. Gounon and P. Sansonetti *Shigella*-induced cytoskeletal reorganization during host cell invasion	237
R. Lartey, S. Ghoshroy, J. Sheng and V. Citovsky Transport through plasmodesmata and nuclear pores: cell-to-cell movement of plant viruses and nuclear import of *Agrobacterium* T-DNA	253

M. G. Rossmann
Viral cell recognition and entry 281
B. B. Finlay
Enteropathogenic *E. coli* and *Salmonella* species exploitation of host cells 311
P. Palese, P. Wang, T. Wolff and R. E. O'Neill
Host–viral protein–protein interactions in influenza virus replication 327
B. Caughey
Transmissible spongiform encephalopathies and the formation of protease-resistant prion protein 341

Index 357

CONTRIBUTORS

ABOUL-ELA, F. MRC Laboratory of Molecular Biology, Hills Road, Cambridge CB2 2QH, UK

ADAMS, T. Institut fur Mikrobiologie und Hygene der Humboldt-Universitat Charite, Dorotheenstrasse 96, D-10117 Berlin, Germany

CAUGHEY, B. Laboratory of Persistent Viral Diseases, Rocky Mountain Laboratories, National Institute of Allergy and Infectious Diseases, NIH, Hamilton, Montana 59840, USA

CHOTHIA, C. MRC Laboratory of Molecular Biology, Hills Road, Cambridge CB2 2QH, UK

CHURCHER, M. J. MRC Laboratory of Molecular Biology, Hills Road, Cambridge, CB2 2QH, UK

CITOVSKY, V. Department of Biochemistry and Cell Biology, State University of New York, Stony Brook, NY 11794, 5215, USA

CORNELIS, G. R. Microbial Pathogenesis Unit, International Institute of Cellular and Molecular Pathology and Faculte de Medecine, Universite Catholique de Louvain, Brussels, Belgium

DEHIO, C. Max Planck Institut fur Biologie, Ateilung Infektionbiologie, Speemannstrasse 34, D-72076 Tubingen, Germany

DORMAN, C. J. Department of Microbiology, Moyne Institute of Preventive Medicine, University of Dublin, Trinity College, Dublin 2, Republic of Ireland

EVANS, D. J. AMS, The University of Reading, Whiteknights, PO Box 228, Reading, Berks RG6 5AJ, UK

EVERETT, R. D. MRC Virology Unit, Church Street, Glasgow G11 5JR, UK

FINLAY, B. B. Biotechnology Laboratory and the Departments of Biochemistry & Molecular Biology and Microbiology & Immunology, University of British Columbia, Vancouver, BC, Canada V6T 1Z3

FOSTER, T. J. Department of Microbiology, Moyne Institute of Preventive Medicine, University of Dublin, Trinity College, Dublin 2, Republic of Ireland

GAIT, M. J. MRC Laboratory of Molecular Biology, Hills Road, Cambridge CB2 2QH, UK

GHOSHROY, S. Department of Biochemistry and Cell Biology, State University of New York, Stony Brook, NY 11794-5215, USA

GOUNON, P. Pathogenie Microbienne Moleculaire, Institut Pasteur, 28, rue du Dr Roux, 75724 Paris, Cedex 15, France

HARTFORD, O. Department of Microbiology, Moyne Institute of Preventive Medicine, Trinity College, Dublin 2, Republic of Ireland

HELLIO, R. Pathogenie Microbienne Moleculaire, Institut Pasteur, 28, rue du Dr Roux, 75724 Paris, Cedex 15, France

KARN, J. MRC Laboratory of Molecular Biology, Hills Road, Cambridge CB2 2QH, UK

KEEN, N. J. MRC Laboratory of Molecular Biology, Hills Road, Cambridge CB2 2QH, UK

LARTEY, R. Department of Biochemistry and Cell Biology, State University of New York, Stony Brook, NY 11794-5215, USA

MATHEWS, M. B. Cold Spring Harbor Laboratory, PO Box 100, Cold Spring Harbor, New York, NY 11724, USA *Present address*: Department of Biochemistry and Molecular Biology, UMDNJ – New Jersey Medical School, 185 South Orange Avenue, Newark, NJ 07103-2714, USA

MÉNARD, R. Department of Pathology, Michael Heidelberg Division of Immunology, NYU Medical Center, 550 First Avenue, New York, NY 10016-6402, USA

MOUNIER, J. Pathogenie Microbienne Moleculaire, Institut Pasteur, 28, rue du Dr Roux, 75724 Paris, Cedex 15, France

NOLAN, N. C. Department of Microbiology, Moyne Institute of Preventive Medicine, University of Dublin, Trinity College, Dublin 2, Republic of Ireland

O'CONNELL, D. Department of Microbiology, Moyne Institute of Preventive Medicine, University of Dublin, Trinity College, Dublin 2, Republic of Ireland

O'NEILL, R. E. Department of Microbiology, Mount Sinai School of Medicine, 1 Gustave L.Levy Place, New York, NY 10029, USA

PALESE, P. Department of Microbiology, Mount Sinai School of Medicine, 1 Gustave L.Levy Place, New York, NY 10029, USA

ROSSMANN, M. G. Department of Biological Sciences, Purdue University, West Fayette, Indiana 47907-1392, USA

SANSONETTI, P. Pathogenie Microbienne Moleculaire, Institut Pasteur, 28, rue du Dr Roux, 75724 Paris, Cedex 15, France

SHENG, J. Department of Biochemistry and Cell Biology, State University of New York, Stony Brook, NY 11794-5215, USA

SKOUDY, A. Pathogenie Microbienne Moleculaire, Institut Pasteur, 28, rue du Dr Roux, 75724 Paris, Cedex 15, France

SMITH, S. G. C. Department of Microbiology, Moyne Institute of Preventive Medicine, University of Dublin, Trinity College, Dublin 2, Republic of Ireland

TRAN VAN NHIEU, G. Pathogenie Microbienne Moleculaire, Institut Pasteur, 28, rue du Dr Roux, 75724 Paris, Cedex 15, France

VARANI, G. MRC Laboratory of Molecular Biology, Hills Road, Cambridge CB2 2QH, UK

VIRJI, M. Department of Paediatrics, University of Oxford, John Radcliffe

Hospital, Oxford OX3 9DU, UK *Present address*: School of Animal and Microbial Sciences, The University of Reading, Reading RG6 6AJ, UK
WANG, P. Department of Microbiology, Mount Sinai School of Medicine, 1 Gustave L.Levy Place, New York, NY 10029, USA
WARD, M. E. Mailpoint 814, Molecular Microbiology Group, Southampton University Medical School, Southampton SO16 6YD, UK
WOLFF, T. Department of Microbiology, Mount Sinai School of Medicine, 1 Gustave L. Levy Place, New York, NY 10029, USA

PROTEIN–PROTEIN AND PROTEIN–CARBOHYDRATE RECOGNITION

C. CHOTHIA

MRC Laboratory of Molecular Biology, Hills Road, Cambridge CB2 2QH, UK

INTRODUCTION

Protein recognition is a central aspect of biology: there is hardly any biological process in which it does not play an important role. This has made protein recognition a field in which a great variety of research work has been carried out in the past and is continuing at present. This review concentrates on what has been learnt from the molecular structures of protein–protein and protein–carbohydrate complexes and from experiments carried out to investigate various aspects of these structures.

In a review of these aspects of protein–protein recognition published in 1990, data were available from the atomic structures of 15 different complexes (Janin & Chothia, 1990). Now there is more than three times that number (see Table 1). Although most of the new structures have recognition sites whose general features are similar to those seen previously, there are some that are quite novel. In addition to the new structural data, experiments using protein engineering have determined the relative importance of residues in recognition sites and in the association processes. In Table 2, the protein–carbohydrate complexes of known atomic structure are listed. This Table shows again that most of these structures have been determined in the last four years.

PROTEIN–PROTEIN RECOGNITION

Structure of simple protein–protein recognition sites

Contact areas

The size of the recognition complex in a protein–protein complex can be described conveniently by the area that is buried by the contacts that the proteins make with each other. For complexes of known structure, this is determined simply by calculating the surface area of the proteins accessible to water (i) in isolation and (ii) in the complex. The surface buried on complex formation is the difference between these two numbers (Chothia & Janin, 1975). Table 1 lists the buried areas in the recognition sites in protein–protein complexes of known structure: antibodies and antigens; proteases and their

Table 1. *Dimeric protein–protein recognition complexes of known structure*

Protein 1	Protein II	Total area buried in the recognition site (Å^2)	Reference
	Antibody–antigen complexes		
Antibody D11.15	Lysozyme	1250	Chitarra *et al.*, 1993
Antibody D44.1	Lysozyme	1400	Braden *et al.*, 1994
Antibody D1.3	Lysozyme	1400	Bhat *et al.*, 1994
Antibody HyHEL 10	Lysozyme	1600	Padlan *et al.*, 1989
Antibody F9.13.7	Lysozyme	1700	Lescar *et al.*, 1995
Antibody HyHEL 5	Lysozyme	1700	Sheriff *et al.*, 1987
Antibody MOPC21	Domain III of protein G	1350	Derrick & Wigley, 1994
Antibody NC10	Neuraminidase	1350	Malby *et al.*, 1994
Antibody JE14	Histidine containing protein	1350	Prasad *et al.*, 1993
Antibody N10	Staphyloccal nuclease	1750	Bossart-Whitaker *et al.*, 1995
Antibody 730.14	Antibody 409.5.3	1850	Ban *et al.*, 1994
Antibody NC41	Neuraminidase	1950	Tulip *et al.*, 1992
	Protease–inhibitor complexes		
Trypsin	Pancreatic trypsin inhibitor	1400	Huber *et al.*, 1974
Trypsinogen	Pancreatic trypsin inhibitor	1400	Bode *et al.*, 1978
Trypsin	Soybean trypsin inhibitor	1450	Sweet *et al.*, 1974
Trypsin	Bitter gourd inhibitor	1550	Huang, Liu & Tang, 1993
Trypsinogen	Pancreatic secretory inhibitor	1750	Bolognesi *et al.*, 1982
Chymotrypsin	Ovomucoid inhibitor	1450	Fujinaga *et al.*, 1987
Chymotrypsin	Eglin C	1550	Frigerio *et al.*, 1992
Elastase	Ovomucoid inhibitor	1300	Bode *et al.*, 1986
Kallikrien	Pancreatic trypsin inhibitor	1400	Chen & Bode, 1983

Protease B	Ovomucoid inhibitor	1250	Read *et al.*, 1983
Subtilisin	Eglin	1500	Bode *et al.*, 1986
Subtilisin	Subtilisin inhibitor	1500	Takeuchi *et al.*, 1991
Subtilisin	Chymotrypsin inhibitor 2	1600	McPhalen & James, 1987
Carboxypeptidase A	Potato inhibitor	1350	Rees & Lipscomb, 1982
Papain	Stefin	2100	Stubbs *et al.*, 1990
Thrombin	Hirudin	3300	Rydel *et al.*, 1991
	Other protein–protein complexes		
Cytochrome c	Cytochrome c peroxidase	1150	Pelletier & Kraut, 1992
SH3	Core of HIV-1 nef	1200	Lee *et al.*, 1996
Ser/Thr kinase C-Raf1	Rap1A	1250	Nassar *et al.*, 1995
Glycerol kinase	Glucose-specific factor III	1300	Hurley *et al.*, 1993
Fc fragment	Protein A fragment	1300	Deisenhofer, 1981
Barnase	Barstar	1600	Guillet *et al.*, 1993
Uracil-DNA glycosylase	Uracil glycosylase inhibitor	1850	Savva & Pearl, 1995
Acetylcholine esterase	Fasciculin	2000	Harel *et al.*, 1995
Actin	DNase 1	1800	Kabsch *et al.*, 1990
Actin	Gelsolin segment 1	2050	McLaughlin *et al.*, 1993
Actin	Profilin	2050	Schutt *et al.*, 1993
HLA-DR1	Enterotoxin b	1550	Jardetzky *et al.*, 1994
HLA-DR1	Toxic shock syndrome toxin	~2000	Kim *et al.*, 1994
CDK2 kinase	CksHs1	1300	Bourne *et al.*, 1996
CDK2 kinase	CyclinA	3550	Jeffrey *et al.*, 1995
	Macromolecular chelation complexes		
Thrombin	Rhodniin	1900	van de Locht *et al.*, 1995
Trypsin	Ecotin	2850	McGrath *et al.*, 1994
Ribonuclease A	Ribonuclease inhibitor	2550	Kobe & Deisenhofer, 1995
β-lactamase	BLIP	2650	Strynadka *et al.*, 1996

Table 2. *Lectin–carbohydrate recognition complexes of known structure*

Protein	References
	Animal lectins
Mannose-binding protein-A	Weis *et al.*, 1992
	Kolatkar & Weis, 1996
Manose-binding protein-C	Ng, Drigaber & Weis, 1996
Galectin-1	Liao *et al.*, 1994
	Bourne *et al.*, 1994*b*
Galectin-2	Lobsanov *et al.*, 1993
Serum amyloid P component	Emsley *et al.*, 1994
Basic fibroblast growth factor	Ornitz *et al.*, 1995
	Plant lectins
Concanvalin A	Derewenda *et al.*, 1989
	Naismith *et al.*, 1994
Pea lectin	Rini *et al.*, 1993
Lathyrus ochrus lectins	Bourne, Rougé & Cambillau, 1990*a*
	Bourne *et al.*, 1990*b*
	Bourne, Rougé & Cambillau, 1992
	Bourne *et al.*, 1994*a*
	Bourne *et al.*, 1994*c*
Griffonia simplicifolia lectin IV	Delbaere *et al.*, 1993
Erythrina corallodendron lectin	Shaanan *et al.*, 1991
Soybean agglutinin	Dessen *et al.*, 1995
Wheat germ agglutinin	Wright, 1984
	Wright, 1990
	Wright, 1992
	Wright & Jaeger, 1993
Ricin	Montfort *et al.*, 1987
	Rutenber & Robertus, 1991
Snowdrop lectin	Hester *et al.*, 1995
Jacalin	Sankaranarayanan *et al.*, 1996
	Bacterial lectins
Enterotoxin	Sixma *et al.*, 1992
	Merritt *et al.*, 1994*b*
Cholera toxin	Merritt *et al.*, 1994*a*
Pertussis toxin	Stein *et al.*, 1994*b*
	Viral lectins
Influenza virus haemagglutinin	Weis *et al.*, 1988
	Sauter *et al.*, 1992*a*
	Sauter *et al.*, 1992*b*
	Watowich *et al.*, 1994
Polyoma virus protein 1	Stehle *et al.*, 1994
	Stehle & Harrison, 1996
P22 tailspike protein	Steinbacher *et al.*, 1996

inhibitors; and a miscellaneous set that includes HLA–superantigen complexes.

In Table 1, the values are given for total buried area on complex formation, i.e. the sum of the areas lost by each of proteins in the complex. For five structures, the buried areas are in the range 2550 to 3550 Å^2. These are discussed in two sections below, and here we will consider the 42 'simple' cases where the total buried areas range between 1150 and 2100 Å^2. In most of these, individual proteins bury 600 to 1000 Å^2 on association. These areas are some 20% of the surface of small proteins such as the 58-residue pancreatic trypsin inhibitor but only about 5% of the surface of large proteins such as the 306-residue carboxypeptidase A.

Examination of the different types of complexes shows that they have recognition sites with a similar range of sizes and mean values. For the 12 antibody–antigen complexes, the recognition sites bury 1250 to 1950 Å^2 and have a mean value of 1550 Å^2. For 16 proteases–inhibitor complexes and 14 miscellaneous complexes, the ranges are 1250 to 1750 Å^2 and 1150 to 2050 Å^2, respectively, and the mean values 1450 and 1600 Å^2.

Thus, these data show that recognition sites involve only a small region of protein surfaces; the total surface buried varies between 1100 Å^2 and 2100 Å^2 but in most cases is close to 1500 Å^2, and, for the very different types of recognition sites, the ranges in the size of recognition sites are very similar.

Contact residues and their chemical character

The total number of residues making intermolecular contacts ranges between 22 and 40; although for three-quarters of them the range is much smaller: 30 $\pm$ 4 (references listed in Table 1; unpublished calculations). In protease–inhibitor complexes, the number of residues contributed to this total by the two proteins is systematically unequal: 10 to 15 residues on the inhibitor contacts with 17 to 29 residues on the proteases. The differences arise from the shapes of the surface involved: an extended loop in the inhibitors fills a long groove formed by the active site and specificity pocket of the proteases (Janin & Chothia, 1990).

No particular amino acid composition is found for residues in recognition sites except for those in antibodies. In antibodies, nearly half the residues making contacts with antigens are aromatic (Padlan, 1990).

The chemical character of the accessible surface of an average protein is 55% non-polar, 25% polar and 20% charged (Miller *et al.*, 1987). The surface buried in recognition sites have, on average, the same amount of non-polar character, though in individual complexes it varies between 49 and 70%. On the other hand, the average amount of charged surface in recognition sites, 10%, is smaller than that of the average protein surface, and the amount of polar surface, 35%, is higher. Again, in individual complexes, there are wide variations about these mean values.

Hydrogen bonds

A comparative examination of protease–inhibitor and antigen–antibody complexes showed that, in almost all cases, their recognition sites have between 8 and 14 hydrogen bonds and 10 on average (references listed in Table 1; unpublished calculations). Between a quarter and half involve one charged group. Hydrogen bonds between two charged groups are uncommon in all but a few complexes (see below).

In addition to direct hydrogen bonds, a number of polar groups are linked through the hydrogen bonds of bridging water molecules. Between 6 and 12 such water molecules are seen in complexes whose structures have been determined at high resolution (Bolognesi *et al.*, 1982; Bode, Capamokas & Musil, 1987; McPhalen & James, 1987; Bhat *et al.*, 1994).

Residue packing

Visual inspection of recognition sites shows that the two sides have good complementarity and that cavities occur rarely. Early quantitative calculations on protease–inhibitor complexes showed that, overall, their interfaces are close packed with the contact residues occupying volumes that are the same as those in crystals of amino acids (Chothia & Janin, 1975). More recent calculations on the two antibody–antigen complexes, D1.3-lysozyme and HyHEL5-lysozyme gave the same result (Harpaz, Gerstein & Chothia, 1994).

Recognition sites in complexes whose formation involves disorder to order transitions

The structures of many of the constituents of protein complexes listed in Table 1 have been determined in their unassociated forms. Comparison of these with the associated forms show that small conformational changes occur commonly on complex formation. The changes consist mostly of rotation of surface side chains: relative shifts of up to ~1.5 Å in close-packed segments of the protein, e.g. α-helices, and changes in the conformations of loops that are relatively free of tertiary structure constraints (see references given in Table 1). There is evidence to show that, in general, these sorts of changes cost little in terms of energy (Gerstein, Lesk & Chothia, 1994). On the other hand, there are complexes whose formation involve disorder-to-order transitions and these do involve large energies. This can be illustrated by two examples.

In trypsinogen the specificity pocket is partially disordered but, on proteolytic activation of the zymogen to trypsin, the region becomes fully ordered. The interaction of pancreatic trypsin inhibitor with trypsin involves packing a side chain in the specificity pocket. The inhibitor also binds to trypsinogen: it produces an ordered specificity pocket and hence an interface which has the same structure as that in the trypsin complex (Bode, Schwager & Huber, 1978). The ordered structure is necessary for the zymogen and inhibitor to close pack and for the buried polar groups to form hydrogen bonds. However, this complex is much less stable than that formed by the inhibitor and trypsin: the association constant (K_a) is decreased by a factor of 10^6. Thus, the cost of producing the ordered structure of the specificity pocket is 8 kcal.

Thrombin is a serine protease that cleaves fibrinogen in the process of blood clot formation. The enzyme is very specific: of the 376 Arg/Lys-X bonds in fibrinogen, it cleaves only four. This specificity is achieved by the enzyme having not only the normal specificity pockets but two additional specificity sites. One of these, the apolar site, is adjacent to the catalytic site and the other, the exosite, some 35 Å away. Hirudin, a toxin produced by leeches to prevent the formation of blood clots, is a strong inhibitor of thrombin. It is a small protein of 65 residues. In solution, residues 3 to 48 form an ordered globular domain whilst 49 to 65 are in disordered extended chain. On binding to thrombin, the N-terminal peptide packs into the apolar site and the globular domain packs over the active site. Residues 49 to 65 pack across the surface of thrombin with the last five forming a helix that packs into the exocite (Rydel *et al.*, 1991).

The extended conformation of the C-terminal residues allow the small inhibitor to bind to the distant exosite on the enzyme. In total, 3300 $Å^2$ of surface are buried in the thrombin-hirudin complex: twice that in most of the other complexes listed in Table 1. The affinity constant is very high, $K_a = 10^{-14}$. The barnase–barstar complex has the same affinity constant but is produced by a buried surface of only 1600 $Å^2$ (Table 1). In the thrombin–hirudin complex, the large buried surface both gives the complex high stability and pays for the loss of entropy that occurs when the C-terminal peptide has an ordered structure.

Recognition sites in a complex whose formation involves a large change in conformation

Cyclin-dependent kinases (CDK) are involved in the coordination of the eukaryotic cell cycle. They are inactive by themselves but gain basal activity and specificity by binding to cyclin, and full activity with subsequent phosphorylation. The CDKs have one lobe that is mainly formed by a β-sheet and one that is mainly formed by α-helices; the active site is in a deep cleft between the two. CyclinA binds to CDK2 in a region at one side of the active site (Jeffrey *et al.*, 1995). It is in contact with both lobes and buries 3550 $Å^2$ of surface in the recognition site; an area twice that found in simple recognition complexes. This interaction not only stabilizes complex formation: it also stabilizes large changes in the structure of CDK2.

Comparison of the structure of free CDK2 with that found in the complex shows that the interactions with cyclinA produce a reorganization of the active site (Jeffrey *et al.*, 1995). On complex formation, the active site cleft is opened by the two lobes rotating relative to each other by 14°. A helix within the cleft, which is involved in function, rotates and moves several ångstroms further into the more open cleft. The T loop, which blocks the active site in the free enzyme and contains the site phosphorylated in the fully active form, undergoes large changes in conformation and position and packs against three helices in cyclinA.

Macromolecular chelation complexes

The simple recognition complexes that have been discussed so far have involved one continuous area of interaction; though, in the case of thrombin–hirudin, a single rather extended area. There is a small group of recognition complexes that uses two separate distant areas of interaction. These have been called macromolecular chelation complexes (McGrath *et al.*, 1994).

The protease inhibitor ecotin is formed by two 142 residue subunits. The subunits, related by a twofold axis, form a molecule with a shallow U-shaped surface. On binding to trypsin, two arms of the U packs against two areas on the surface of trypsin; one arm packs against the active site and the other against a region some 45 Å away (McGrath *et al.*, 1994).

Two other complexes have the same kind of interaction. The ribonuclease inhibitor has the shape of a horseshoe. The inner surface is formed by a 17-strand β-sheet and the outer surface by 16 α-helices. The 458-residue inhibitor packs around the 126-residue ribonuclease A with one arm against the active site and the other against the back of the enzyme (Kobe & Deisenhofer, 1995). BLIP, the β-lactamase inhibitor, is built in two tandemly repeated domains that form a polar concave surface. In the complex, this surface packs around one of the two lobes of β-lactamase covering the active site (Strynadka *et al.*, 1996).

In these three complexes the total area buried in the recognition sites is 2550 to 2850 Å^2; an area 50 to 100% greater than that found in simple recognition complexes. There are no detailed descriptions of the packing at the three recognition sites, but they do have unusual features. BLIP and ecotin have high affinities but are less specific than the inhibitors in simple complexes. BLIP can bind to a variety of class A β-lactamases and has an extensive layer of water molecules trapped between the enzyme and inhibitor (Strynadka *et al.*, 1996). Ecotin can bind not only trypsin but also chymotrypsin and elastase. In places Met in the specificity pocket of trypsin rather than the Lys or Arg of more specific inhibitors. This suggests that the high affinities and broader range of these inhibitors are produced by having larger but less specific recognition sites than those found in simple complexes.

The thrombin–rhodinin chelation complex is closer in structure to simple recognition complexes. Rhodinin is a thrombin-specific inhibitor from the assassin bug. It consists of two Kazel-type domains, 48 and 49 residues, respectively, linked by a seven-residue peptide. In the complex with thrombin, the first domain packs into the active site cleft and the second domain packs into the separate fibrinogen exosite (van de Locht *et al.*, 1995). The total area buried in the two recognition sites is 1900 $Å^2$.

The relative importance of residues that form recognition sites

The relative importance of the residues that form recognition sites has been determined for two complexes using protein engineering. In these experiments, residues that form intermolecular contacts have been changed, usually to Ala, and the effect on binding energy measured.

The human growth hormone activates its receptor by producing a complex that consists of one hormone molecule and two receptor molecules (de Vos, Ultsch & Kossiakoff, 1992). Formation of the complex involves an initial interaction that brings 33 hormone residues into contact with, or close to, 31 residues on one receptor molecule. All but four of these side chains were replaced by alanine and their effects on the affinity measured. For the hormone, mutations at eight sites reduced binding energies by between 1.1 and 2.4 kcal. At the other 24 sites, the effects of mutation are smaller (Cunningham & Wells, 1993). For the receptor, mutations at 11 sites reduced binding energies by 1 to 4.5 kcal and those at 17 other sites give smaller reductions (Clackson & Wells, 1995). The size of the change in binding energy of a residue on mutation correlates neither with the number of intermolecular contacts it makes nor with the extent of surface it buries in the recognition site. It does correlate with the position of the residue in the recognition site. The eight sites in the hormone and the eleven in the receptor that produce large changes form contiguous patches at the centre of the contact region. In the complex, they pack together to form the central half of the recognition site with hydrophobic residues at the core flanked by polar and charged groups (Clackson & Wells, 1995).

The action of the extracellular ribonuclease barnase is inhibited within the bacterium by the inhibitor Barstar. The two proteins have very high affinity, the K_a is 10^{14}. At the recognition site, 15 residues on the enzyme make contacts with, and form 14 hydrogen bonds to, 15 residues on the inhibitor (Guillet *et al.*, 1993; Buckle, Schreiber & Fersht, 1994). The inhibitor binds to the active site of the enzyme. The function of the enzyme, to recognize and cleave RNA, means that the active site contains Arg, Lys and His residues and the inhibitor recognizes these using Asp and Glu residues. Having charge–charge interactions at the centre of the contact is the opposite of what is found in the growth hormone–receptor complex just described. Mutations to Ala at the individual sites of a Lys, a His and two Arg at the centre of the barnase recognition site, and a Glu and two Asp at the centre of the barstar site, reduce the binding energy by 5 to 7.7 kcal (Schreiber & Fersht, 1993, 1995). These values are larger than those in the hormone–receptor complex because the interlocking set of charged hydrogen bonds formed by these residues means that the mutations leave buried charged groups without partners. The mutation of a Phe in barstar reduces the affinity by 3.4 kcal; other changes have small effects.

The experiments on these two very different complexes show that the identity of the residues that form the central half of the binding site are of major importance for recognition. Changes to residues in the peripheral half can be accommodated and give only very small changes in affinity.

Protein association

Kinetic studies on protein–protein complexes give association rate constants, k_{on}, in the range 10^5–10^9 M^{-1} s^{-1}. Calculations on what would be expected from protein diffusion coefficients show that, for the higher rate constants, association must be assisted by long-range interactions. For molecules such as trypsin or lysozyme, the diffusion coefficients in water are $\sim 10^6$ cm^2 s^{-1}, which give rates of collision of $\sim 10^9$ M^{-1} s^{-1}. This means that, in the complexes with the highest association rates, most collisions lead to association and, in those with low rates, about one in 10^4. Recognition sites cover only 5–20% of the protein surface and therefore, very few collisions can bring together the recognition sites in a correct alignment. Thus, for associations to be fast, they must be assisted by the formation of a loose complex that lasts long enough for it to isomerize a correctly packed interface and hence a stable structure.

Protein engineering experiments on the complexes thrimbin–hirudin (Stone, Dennis & Hofsteenge, 1989); cytochrome c–cytochrome b_5 (Northrup *et al.*, 1993; Guillemette *et al.*, 1994) and barnase–barstar (Schreiber & Fersht, 1996) have shown that the low affinity, non-specific, complexes are held together by long-range electrostatic forces, and that this is likely to be generally true for other systems with high association rates. In the case of the barnase–barstar complex, the mutation to alanine of the residues responsible for the electrostatic attraction decreases the association rate from 10^9 M^{-1} s^{-1} to a basal value of 10^5 M^{-1} s^{-1} (Schreiber & Fersht, 1996).

The stability of protein–protein complexes

There is no simple correlation between the structure of recognition sites and stability. Thus, for example, the complexes formed by trypsin–pancreatic trypsin inhibitor, barnase–barstar, ribonuclease–ribonuclease inhibitor and thrombin–hirudin have very similar affinity constants K_a of 10^{-13} to 10^{-14} but the surface buried in their recognition sites are 1250, 1600, 2550 and 3300 $Å^2$, respectively.

Recognition sites have evolved by natural selection to achieve affinities sufficient for their proper function. This does not mean the sites have optimal designs. This is shown, for example, by the mutation experiments on the human growth hormone–receptor complex described above: ten of the mutations to Ala produced small increases in affinity (Cunningham & Wells, 1993; Clackson & Wells, 1995).

In some cases, 'efficient' sites have evolved in which factors favouring disassociation are small, so the extent of the interactions required for a sufficiently high affinity is also small. In other cases, sites are less efficient; a large number of unfavourable aspects are balanced by a larger number of interactions.

The major terms opposing association are the loss of rotational, transitional and internal degree of freedom. The major terms favouring association are hydrophobic energy; van der Waals' interactions from surfaces buried in the recognition sites, and electrostatic energy from intermolecular hydrogen bonds. Although the nature of the terms is generally agreed, their magnitude and relative importance are matters of considerable discussion and dispute. The reader can find references to this work and a detailed discussion of its implications for the nature of protein–protein recognition in the recent papers by Janin (1995, 1996*a*, *b*). Two general points will be made here, however.

Theoretical estimates of the loss of transitional and rotational entropy that occurs on dimer formation indicate that it is proportional to the logarithm of the molecular weights of the two components (Finkelstein & Janin, 1989). This implies that the extent of interactions required to stabilize a complex will not increase greatly with the size of components.

A rough calculation of the hydrophobic free energy gained from the surface buried in the recognition site of simple complexes suggested it, together with the energy from hydrogen bonds, could account for their stability (Janin & Chothia, 1990). A recent more exact calculation took into account the chemical nature of the buried surfaces. Horton and Lewis (1992) derived the relationship:

$$\Delta G_{\mathrm{d}} \approx \sum \sigma_{\mathrm{i}} B_{\mathrm{i}} - \Delta G_{(0)}$$

where ΔG_{d} is the free energy of dissociation; σ_{i} is weight given to the area, B_{i}, buried by one five atom types and $\Delta G_{(0)}$ is 6.2 kcal. They showed that, for 15 complexes, the relation gives ΔG_{d} values that differ from experimental values by between 0 and 2.2 kcal and by 1 kcal on average. In addition to the hydrophobic character of the different atom types, the weights must incorporate some of the effects of van der Waals' interactions and hydrogen bonding.

The relation breaks down if large conformational changes or disorder-to-order transitions are involved in complex formation. It would also not predict the results mutations of recognition site residues. In effect, the relationship shows the relation between the observed affinity, and the extent and chemical nature of the surface buried in recognition sites, in 'efficient' simple complexes where a close-packed recognition site is formed by surfaces whose structure in isolation is very similar to those in the complex.

PROTEIN–CARBOHYDRATE RECOGNITION

Examination of carbohydrate binding sites in proteins shows that they can be divided into two groups (Quiocho, Vyas & Spurline, 1991; Vyas 1991). In the first group the binding closes around the carbohydrates and it becomes very largely inaccessible to solvent. These binding sites have high affinities, typically $K_D \simeq 10^{-7}$. They are found in transport proteins and certain enzymes. In the second group, the binding sites are shallow groves and the carbohydrate is only partly buried. These sites have lower affinities: $K_D \simeq 10^{-3}$ is typical for a monosaccharide. Proteins in this class include pathogens that function through the recognition of carbohydrates on the cell surfaces. Both groups of proteins have similar kinds of interactions with carbohydrates; their affinity constants are different because the extent of the interactions is greater in the first group than in the second (Weis & Drickamer, 1996).

Protein recognition sites for carbohydrates

Table 2 lists the lectin–carbohydrate complexes whose atomic structures are currently known. Most of these were discussed in two excellent recent reviews (Weis & Drickamer, 1996; Rini, 1995) and these influenced strongly much of what is presented in this section.

Hydrogen bonding

Carbohydrates have the general formula $(CH_2O)n$. The hexoses, for example, have an $n = 6$ with five C and one O in a six-atom ring and the other five oxygens as ring substituent OH groups out into solution or making contacts with recognition proteins. These features make the surface of carbohydrates largely made up by polar groups and, as might be expected, the hydrogen bonds they make with protein polar groups play the major role in recognition.

The structure of a sugar binding site in a particular protein can be seen as a particular selection and arrangement of a set of general types of interactions (Weis & Drickamer, 1996). Carbohydrate OH groups make up to three hydrogen bonds with protein groups. If *OH* is the hydroxyl of carbohydrate, these hydrogen bonds have the form:

$$(NH)_n\text{- - - -}OH\text{- - - -}O{=}C \qquad \text{where } n = 1 \text{ or } 2$$

The protein NH groups come from the main chain or the side chains of Asn and Gln, mainly, but those in Arg or Lys side chains can also be used. The protein O=C group can be a carbonyl group from the main chain or from side chains of Asn and Gln; it can also be a carboxylate group of Asp or Glu. Hydrogen bonds using the side chains of Ser, Tyr or Thr are less common (Weis & Drickamer, 1996).

Neighbouring OH groups on sugars can hydrogen bond to two polar groups in a single amino acid such as the amide in Asn and Gln. The bifurcated hydrogen bonds, made by both the protein and sugar polar groups, form an interlocking network of interactions that is largely buried in the sugar–protein interface.

Hydrophobic interactions

Although carbohydrates are highly polar, they do contain hydrophobic groups. In galactose sugars, for example, the carbon atoms at five epimeric positions in the ring and one at an adjacent exocylic position form a continuous hydrophobic region. With very few exceptions, this region is found packed against aromatic side chains in protein–carbohydrate complexes of known structure (Rini, 1995; Weis & Drickamer, 1996). For other carbohydrates, non-polar recognition is more varied. The shielding of the hydrophobic surface from contact with water will make a significant contribution to the stability of the complex.

Specificity of sugar recognition

Discrimination between different carbohydrates can involve rather subtle differences in the structure of binding sites. The differences responsible for specificity can, however, be roughly divided into two types. First, the general shape of binding site can be such as to exclude ligands that have shapes different to that of the proper ligand. Secondly, the polar groups in the protein that become buried by the interaction are complementary to those on a particular carbohydrate, so buried polar groups are fully hydrogen bonded only by the appropriate ligand (Rini, 1995; Weis & Drickamer, 1996).

Multivalency

Pathogens of known structure that bind carbohydrates are in group II where binding sites are shallow and have low affinities for monosaccharides. High affinities are produced in some members of this group by having an extended binding site that recognizes a second saccharide linked to the primary determinant. This has been called subsite multivalency (Rini, 1995). The interactions made by the subsidiary site are smaller than those made by the primary site but, together, they produce high affinity complexes: $K_D \simeq 10^{-7}$ rather than $K_D \simeq 10^{-3}$; as, for example, in the *Lathyrus ochrus* lectin (Bourne *et al.*, 1990*b*).

Note that subsite multivalency also increases specificity. Having an intrinsic low affinity for monosaccharide prevents interference from competing small sugar ligands.

The second way that the affinity of proteins in this group is increased is by having several subunits or domains each with its separate binding sites, i.e. subunit or domain multivalency. This can give very large increases in affinity with $K_D \simeq 10^{-9}$. Commonly, the subunits or domains have been produced by gene duplications. Striking examples or this kind of multivalency are found in the several pathogens of known structure.

Multivalency in pathogens with carbohydrate binding sites

Multivalency is found in all pathogens which have carbohydrate binding sites and for which structures are known. One major group of bacterial toxins, called the AB_5 toxins, has five B subunits, each of which binds carbohydrates on the cell surface as the first step in the translocation of the toxic A subunit into the interior of the cell. The AB_5 group includes the cholera and Shiga toxin families and pertussis toxin. Structures have been determined for all, or part, of two members of the cholera family (Sixma *et al.*, 1991; Merritt *et al.*, 1994*a*), two members of the Shiga family (Stein *et al.*, 1992; Fraser *et al.*, 1994) and pertussis toxin (Stein *et al.*, 1994*a*). In the individual members of the cholera and Shiga families, the B subunits are identical. The B subunits of different members of these families have high sequence identities. The B subunits in the cholera family have little or no sequence identity with those in the Shiga family but they do have folds and binding sites that are so similar that they indicate a distant evolutionary relationship. The B subunits of pertussis toxin differ in size and sequence. Three are very similar in size and structure to those in the cholera and Shiga families (Stein *et al.*, 1994*a*). The other two have this structure also but linked to a second domain that is homologous to the C-type lectins.

In all these toxins, the five B subunits are arranged so as to be related by a pure or pseudo-fivefold axis. This puts the five carbohydrate binding sites on one side of the molecule in an arrangement that is sufficiently widely spaced to allow rapid interaction with the large, nearly planar, array of oligosaccharides presented by cell surfaces.

Ricin, the caster plant toxin, is a heterodimer AB. The ricin A subunit is homologous to the toxic A subunit in the Shiga family. The B subunit has two domains each with the same fold and a carbohydrate binding site (Rutenber & Robertus, 1991).

Viruses also use multivalency to produce strong binding to cell surfaces. The outer shell of polymavirus contains 360 copies of the protein VP1 arranged in 72 pentamers. Each of these VP1 proteins has a carbohydrate binding site. For individual sites $K_D \simeq 10^{-3}$ but, when the virus binds to the cell, the co-operative of multiple interactions produces strong binding (Stehle *et al.*, 1994). The haemagglutinin of influenza virus binds to cell surfaces and then undergoes a series of transformation that result in the fusion of the virus and cell membranes. The molecule is a trimer, 135 Å long, and with one end attached to the virus. Each of the tree subunits has a domain at the distal end that binds cell surface carbohydrates (Weis *et al.*, 1988).

CONCLUSIONS

Simple protein–protein complexes bury approximately 1500 Å^2 of surfaces in a recognition site that is close packed and has a chemical character similar to the average protein surface. Small changes in structure can produce large changes in affinity. Complexes that involve disorder to order transitions, or large changes in conformation, involve much larger recognition sites.

Proteins involved in carbohydrate recognition have relatively shallow sites where the major interactions involve hydrogen bonds. Specificity can be increased by subsite multivalency. High affinities are produced in proteins by their having multiple copies of the carbohydrate binding domains. This is universal in the carbohydrate-binding part of the pathogens of known structure.

REFERENCES

Ban, N., Escobar, C., Garcia, R., Hasel, K., Day, J., Greenwood, A. & McPherson, A. (1994). Crystal structure of an idiotype–anti-idiotype Fab complex. *Proceedings of the National Academy of Sciences, USA*, **91**, 1604–8.

Bhat, T. N., Bentley, G. A., Boulot, Greene, M. I., Tello, D., Dall'Acquua, W., Souchon, Schwarz, F. P., Mariuzza, R. A. & Poljak, R. J. (1994). Bound water molecules and conformational stabilization help mediate an antigen–antibody association. *Proceedings of the National Academy of Sciences, USA*, **91**, 1089–93.

Bode, W., Schwager, P. & Huber, R. (1978). The transition of bovine trypsinogen to a trypsin-like state upon strong ligand binding. *Journal of Molecular Biology*, **118**, 99–112.

Bode, W., Wei, A. Z., Huber, R., Meyer, E., Travis, J. & Neumann, S. (1986). X-ray crystal structure of the complex of human leukocyte (Pmn elastase) and the third domain of the turkey ovomucoid inhibitor. *EMBO Journal*, **5**, 2453–58.

Bode, W., Papamokas, E. & Musil, D. (1978). The high resolution X-ray crystal structure of the complex formed between subtilisin Carlsberg and eglin C, an elastase inhibitor from the leech *Hirudo medicinalis*. *European Journal of Biochemistry*, **166**, 673–92.

Bolognesi, M., Gatti, G., Menegatti, E., Guarneri, M., Marquart, M., Papamokos, E. & Huber, R. (1982). Three dimensional structure of the complex between pancreatic secretory inhibitor (Kazel type) and trypsinogen at 1.8 Å resolution. *Journal of Molecular Biology*, **162**, 839–68.

Bossart-Whitaker, P., Chang, C. Y., Novotny, J., Benjamin, D. C. & Sheriff, S. (1995). The crystal structure of the antibody N10–*Staphylococcal nuclease* complex at 2.9 Å resolution. *Journal of Molecular Biology*, **253**, 559–75.

Bourne, Y., Rougé, P. & Cambillau, C. (1990*a*). X-ray structure of a (α-man(1-3)β-man(1-4)glcnac)–lectin complex at 2.1 Å resolution: the role of water in sugar–lectin. *Journal of Biological Chemistry*, **265**, 18161–65.

Bourne, Y., Roussel, A., Frey, M., Rougé, P., Fontecilla-Camps, J.-C. & Cambillau, C. (1990*b*). Three-dimensional structures of complexes of *Lathyrus ochrus* isolectin I. *Proteins*, **8**, 365–76.

Bourne, Y., Rougé, P. & Cambillau, C. (1992). X-ray structure of a biantennary octasaccharide–lectin complex refined at 2.3 Å resolution. *Journal of Biological Chemistry*, **267**, 197–203.

Bourne, Y., Ayouba, A., Rougé, P. & Cambillau, C. (1994*a*). Interaction of a legume lectin with two components of the bacterial cell wall. *Journal of Biological Chemistry*, **269**, 9429–39.

Bourne, Y., Bolgiano, B., Liao, D.-I., Strecker, G., Cantau, P., Hertzberg, O., Feizi, T. & Cambillau, C. (1994*b*). Cross-linking of mammalian lectin Galectin 1 by complex biantennary saccharides. *Nature Structural Biology*, **1**, 863–70.

Bourne, Y., Mazurierm, J., Legrand, D., Rougé, P., Montreuil, J., Spik, G. & Cambillau, C. (1994*c*). Structures of a legume lectin complexed with the human lactotransferrin N2 fragment and with an isolated biantennary glycopeptide: the role of the fucose moiety. *Structure*, **2**, 209–19.

Bourne, Y., Watson, M. H., Hickey, M. J., Holmes, W., Rocque, W., Reed, S. I. & Tainer, J. A. (1996). Crystal structure and mutation analysis of the human CDK2 kinase complex with cell cycle-regulatory protein CkdHs1. *Cell*, **84**, 863–74.

Braden, B. C., Souchon, H., Eiselé, J.-L., Bentley, G. A., Bhat, T. N., Navaza, J. N. & Poljak, R. J. (1994). Three dimensional structures of the free and antigen-complexed Fab from monoclonal anti-lysozyme antibody D44.1. *Journal of Molecular Biology*, **243**, 767–81.

Buckle, A. M., Schreiber, G. & Fersht, A. R. (1994). Protein–protein recognition: crystal structural analysis of a barnase–barstar complex at 2.0 Å resolution. *Biochemistry*, **33**, 8878–89.

Chen, Z. & Bode, W. (1983). Refined 2.5 Å X-ray crystal structure of the complex formed by porcine kallikrein A and the bovine pancreatic trypsin inhibitor. *Journal of Molecular Biology*, **164**, 283–311.

Chitarra, V., Alzari, P. M., Bentley, G. A., Bhat, T. N., Eisele, J. L., Houdusse, A., Lescar, J., Souchon, H. & Poljak, R. J. (1993). Three-dimensional structure of a heteroclitic antigen–antibody cross-reaction complex. *Proceedings of the National Academy of Sciences, USA*, **90**, 7711–15.

Chothia, C. & Janin, J. (1975). Principles of protein–protein recognition. *Nature*, **256**, 705–8.

Clackson, T. & Wells, J. A. (1995). A hot spot of binding energy in a hormone–receptor interface. *Science*, **267**, 383–6.

Cunningham, B. C. & Wells, J. A. (1993). Comparison of a structural and functional epitope. *Journal of Molecular Biology*, **234**, 554–63.

Deisenhofer, J. (1981). Crystallographic refinement and atonic models of a human Fe fragment and its complex with fragment B of protein A from *Staphylococcus aureus* at 2.9 and 2.8 Å resolution. *Biochemistry*, **20**, 2361–70.

Delbaere, L. T. J., Vandonselaar, M., Prasad, L., Quail, J. W., Wilson, K. S. & Dauter, Z. (1993). Structure of the lectin IV of *Griffonia simplicifolia* and its complex with the Lewis b human blood group determinant at 2.0 Å resolution. *Journal of Molecular Biology*, **230**, 950–65.

Derewenda, Z., Yariv, J., Helliwell, J. R., Kalb, A. J., Dodson, E. J., Papiz, M. Z., Wan, T. & Campbell, J. (1989). The structure of the saccharide-binding site of concanavalin. *EMBO Journal*, **8**, 2189–93.

Derrick, J. P. & Wigley, D. B. (1994). The third IgG-binding domain from streptococcal protein G. *Journal of Molecular Biology*, **243**, 906–18.

Dessen, A., Gupta, D., Sabesan, S., Brewer, C. F. & Sacchettini, J. C. (1995). X-ray crystal structure of the soybean agglutinin cross-linked with a biantennary analog of the blood group I carbohydrate antigen. *Biochemistry*, **34**, 4933–42.

Emsley, J., White, H. E., O'Hara, B. P., Oliva, G., Srinivasan, N., Tickle, I. J., Blundell, T. L., Pepys, M. B. & Wood, S. P. (1994). Structure of pentameric human serum amyloid P component. *Nature*, **367**, 338–45.

Finkelstein, A. V. & Janin, J. (1989). The price of lost freedom: entropy of biomolecular complex formation. *Protein Engineering*, **3**, 1–3.

Fraser, M. E., Chernaia, M. M., Kozlov, Y. V. & James, M. N. G. (1994). Crystal Structure of the holotoxin from *Shigella dysenteriae* at 2.5 Å resolution. *Nature Structural Biology*, **1**, 59–64.

Frigerio, F., Coda, A., Puglise, L., Lionetti, C., Menegatti, E., Amiconi, G., Schnebli, H. P., Ascenzi, P. & Bolognesi, M. (1992). Crystal and molecular structure of the bovine α-chymotrypsin–eglin C complex at 2.0 Å resolution. *Journal of Molecular Biology*, **225**, 107–23.

Fujinaga, M., Sielecki, R., Read, R. J., Ardelt, W., Laskowski, M. & James, M. N. J. (1987). Crystal and molecular structures of α-chymotrypsin with its inhibitor turkey ovomucoid third domain at 1.8 Å resolution. *Journal of Molecular Biology*, **195**, 397–418.

Gerstein, M., Leak, A. M. & Chothia, C. (1994). Structural mechanisms for domain movements in proteins. *Biochemistry*, **33**, 6739–49.

Guillemette, J. G., Barker, P. D., Eltis, L. D., Lo, T. P., Smith, M. & Brayer, G. D. (1994). Analysis of the biomolecular reduction of ferrocytochrome c by ferrocytochrome b_5 through mutagenesis and molecular modelling. *Biochimie*, **76**, 592–604.

Guillet, V., Lapthorn, A., Hartley, R. W. & Mauguen, Y. (1993). Recognition between a bacterial ribonuclease, barnase, and its natural inhibitor, barstar, *Structure*, **1**, 165–77.

Harel, M., Kleywegt, G. J., Ravelli, R. B. G., Silman, I. & Sussman, J. L. (1995). Crystal structure of an acetylcholinesterase–fasciclin complex: interaction of a three-fingered toxin from snake venom with its target. *Structure*, **3**, 1355–66.

Harpaz, Y., Gerstein, M. & Chothia, C. (1994). Volume changes on protein folding. *Structure*, **2**, 641–9.

Hester, G., Kaku, H., Goldstein, I. J. & Wright, C. S. (1995). Structure of mannose-specific snowdrop (*Galanthus nivalis*) lectin is representative of a new plant lectin family. *Nature Structural Biology*, **2**, 472–9.

Horton, N. & Lewis, M. (1992). Calculation of the free energy of association for protein complexes. *Protein Science*, **1**, 169–81.

Huang, Q., Liu, S. & Tang, Y. (1993). The refined 1.6 Å resolution crystal structure of the complex formed between porcine β-trypsin and MCTI-A, a trypsin inhibitor of the squash family. *Journal of Molecular Biology*, **229**, 1022–36.

Huber, R., Kukla, D., Bode, W., Schwager, P., Bartels, K., Deisenhofer, J. & Steigemann, W. (1974). Structure of the complex formed by bovine trypsin and bovine pancreatic trypsin inhibitor. II Crystallographic refinement at 1.9 Å resolution. *Journal of Molecular Biology*, **89**, 73–101.

Hurley, J. H., Faber, H. R., Worthylake, D., Meadow, N. D., Roseman, S., Pettigrew,

D. W. & Remington, S. J. (1993). Structure of the regulatory complex of *E. coli* III^{Gel} with glycerol kinase. *Science*, **259**, 673–7.

Janin, J. (1995). Elusive affinities. *Proteins*, **21**, 30–9.

Janin, J. (1996*a*). Quantifying biological specificity: the statistical mechanics of molecular recognition. *Proteins*, **25**, 438–45.

Janin, J. (1996*b*). Protein–protein recognition. *Progress in Biophysics and Molecular Biology*, **64**, 145–66.

Janin, J. & Chothia, C. (1990). The structure of protein–protein recognition sites. *Journal of Biological Chemistry*, **265**, 16027–30.

Jardetzky, T. S., Brown, J. H., Gorga, J. C., Stern, L. J., Urban, R. G., Chi, Y.-I., Stauffacher, C., Strominger, J. L. & Wiley, D. C. (1994). Three-dimensional structure of a human class II histocompatibility molecule complexed with superantigen. *Nature*, **368**, 711–18.

Jeffrey, P. D., Russo, A. A., Polyak, K., Gibbs, E., Hurwitz, J., Massagué, J. & Pavletich, N. P. (1995). Mechanism of CDK activation revealed by the structure of a cyclinA–CDK2 complex. *Nature*, **376**, 313–20.

Kabsch, W., Mannherz, H. G., Suck, D., Pai, E. F. & Holmes, K. C. (1990). Atomic structure of the actin:DNAse I complex. *Nature*, **347**, 37–44.

Kim, J., Urban, R. G., Strominger, J. L. & Wiley, D. C. (1994). Toxic shock syndrome toxin-1 complexed with a class II major histocompatibility molecule HLA-DR1. *Science*, **266**, 1870–7.

Kobe, B. & Deisenhofer, J. (1995). A structural basis of the interactions between leucine rich repeats and protein ligands. *Nature*, **374**, 183–6.

Kolatkar, A. & Weis, W. I. (1996). Structural basis of galactose recognition by C-type animal lectins. *Journal of Biological Chemistry*, **272**, 6679–85.

Lee, C.-H., Saksela, K., Mirza, U. A., Chait, B. T. & Kuriyan, J. (1996). Crystal structure of the conserved core of HIV-1 nef complexed with a Src family SH3 domain. *Cell*, **85**, 831–942.

Lescar, J., Pellegrini, M., Souchon, H., Tello, D., Poljak, R. J., Peterson, N., Greene, M. & Alzari, P. M. (1995). Crystal structure of a cross reaction complex between Fab F9.13.7 and guinea fowl lysozyme. *Journal of Biological Chemistry*, **270**, 18067–76.

Liao, D.-I., Kapadia, G., Ahmed, H., Vasta, G. R. & Herzberg, O. (1994). Structure of S-lectin, a developmentally regulated vertebrate β-galactoside binding protein. *Proceedings of the National Academy of Sciences, USA*, **91**, 1428–32.

Lobsanov, Y. D., Gitt, M. A., Leffler, H., Barondes, S. H. & Rini, J. M. (1993). X-ray crystal structure of the human dimeric s-lac lectin L-14-II. *Journal of Biological Chemistry*, **268**, 27034–8.

McGrath, M. E., Erpel, T., Bystroff, C. & Fletterick, R. J. (1994). Macromolecular chelation as an improved mechanism of protease inhibition: structure of the ecotin–trypsin complex. *EMBO Journal*, **13**, 1502–7.

McLaughlin, P. J., Gooch, J. T., Mannherz, H.-G. & Weeds, A. G. (1993). Structure of gelsolin segment 1-actin complex and the mechanism of filament severing. *Nature*, **364**, 685–92.

McPhalen, C. A. & James, M. N. G. (1987). Structural comparison of two serine proteinase–protein inhibitor complexes. Eglin C-subtilisin carlsberg and CI 2-subtilisin novo. *Biochemistry*, **27**, 6582–98.

Malby, R. L., Tulip, W. R., Harley, V. R., McKimm-Breschkin, J. L., Laver, W. G., Webster, R. G. & Colman, P. M. (1994). The structure of a complex between the NC10 antibody and influenza virus neuraminidase and comparison with the overlapping site of the NC41 antibody. *Structure*, **2**, 733–46.

Merritt, E. A., Sarfaty, S., Van den Akker, F., L'hoir, C., Martial, J. A. & Hol, W. G. J.

(1994*a*). Crystal structure of cholera toxin B-pentamer bound to receptor G_{M1} pentasaccharide. *Protein Science*, **3**, 166–75.

Merritt, E. A., Sixma, T. K., Kalk, K. H., van Zanten, B. A. M. & Hol, W. G. J. (1994*b*). Galactose binding site in *Escherichia coli* heat labile enterotoxin. *Molecular Microbiology*, **13**, 745–53.

Miller, S., Janin, J., Lesk, A. M. & Chothia, C. (1987). Interior and surface of monomeric proteins. *Journal of Molecular Biology*, **196**, 641–56.

Montfort, W., Villafranca, J. E., Monzingo, A. F., Ernsy, S. R., Katzin, B., Rutenber, E., Xuong, H. H., Hamlin, R. & Robertus, J. D. (1987). The three dimensional structure of ricin at 2.8 Å. *Journal of Biological Chemistry*, **262**, 5398–403.

Naismith, J. H., Emmerich, C., Habash, J., Harrop, S. J., Helliwell, J. R., Hunter, W. N., Raftery, J., Kalb, A. J. & Yariv, J. (1994). Refined structure of concanvalin A complexed with methyl α-D-mannopyranoside at 2.0 Å resolution and comparison with the saccharide-free structure. *Acta Crystallographica Sect. D*, **50**, 847–58.

Nasser, N., Horn, G., Herrmann, C., Scherer, A., McCormick, F. & Wittinghofer, A. (1995). The 2.2 Å crystal structure of the ras-binding domain of the serine/threonine kinase c-Raf1 in a complex with Rap1A and a GTP analogue. *Nature*, **375**, 554–60.

Ng, K. K. S., Drickamer, K. & Weis, W. I. (1996). Structural analysis of monosaccharide recognition by rat liver mannose-binding protein. *Journal of Biological Chemistry*, **271**, 663–74.

Northrup, S. H., Thomasson, K. A., Miller, C. M., Barker, P. D., Eltis, L. D., Guillemette, J. G., Inglis, S. C. & Mauk, A. G. (1993). Effects of charged amino acid mutations on the bimolecular kinetics of reduction of yeast iso-1-ferrocytochrome c by bovine ferrocytochrome b_5. *Biochemistry*, **32**, 6613–23.

Ornitz, D. M., Herr, A. B., Nilsson, M., Westman, J., Svahn, C. M. & Wakeman, G. (1995). FGF binding and FGF receptor activation by synthetic heparan-derived di- and trisaccharides. *Science*, **268**, 432–6.

Padlan, E. A. (1990). On the nature of antibody combining sites – unusual structural features that may confer on these sites an enhanced capacity for binding ligands. *Proteins*, **7**, 112–24.

Padlan, E. A., Silverton, E. W., Sheriff, S., Cohen, G. H., Smith-Gill, S. J. & Davis, D. R. (1989). Structure of an antibody–antigen complex: crystal structure of the HyHEL-10 Fab–lysozyme complex. *Proceedings of the National Academy of Sciences, USA*, **86**, 5938–52.

Pelletier, H. & Kraut, J. (1992). Crystal structure of a complex between electron transfer partners, cytochrome c peroxidase and cytochrome c. *Science*, **258**, 1748–55.

Prasad, L., Sharma, S., Vandonselaar, M., Quail, J. W., Lee, J. S., Waygood, E. B., Wilson, K. S., Dauter, Z. & Delbaere, L. T. J. (1993). Evaluation of mutagenesis for epitope mapping: structure of an antibody–protein antigen complex. *Journal of Biological Chemistry*, **268**, 10705–8.

Quiocho, F. A., Vyas, N. K. & Spurlino, J. C. (1991). Atomic interactions between proteins and carbohydrates. *Transactions of the American Crystallographic Association*, **25**, 23–35.

Read, R. J., Fujinaga, M., Sielecki, R. & James, M. N. G. (1983). Structure of the complex of *Streptomyces griseus* protease B and the third domain of the turkey ovomucoid inhibitor at 1.8 Å resolution. *Biochemistry*, **22**, 4420–33.

Rees, D. C. & Lipscomb, W. N. (1982). Refined crystal structure of potato inhibitor complex of carboxypeptidase A at 2.5 Å resolution. *Journal of Molecular Biology*, **160**, 475–98.

Rini, J. M. (1995). Lectin structure. *Annual Review of Biophysics and Biomolecular Structure*, **24**, 551–77.

Rini, J. M., Hardman, K. D., Einspahr, H., Suddath, F. L. & Carver, J. P. (1993). X-ray crystal structure of a pea lectin-trimannoside complex at 2.6 Å resolution. *Journal of Biological Chemistry*, **268**, 10126–32.

Rutenber, E. & Robertus, J. D. (1991). Structure of ricin B chain at 2.6 Å resolution. *Proteins*, **10**, 260–9.

Rydel, T. J., Tulinsky, A., Bode, W. & Huber, R. (1991). The refined crystal structure of the hirudin–thrombin complex. *Journal of Molecular Biology*, **221**, 583–601.

Sankaranarayanan, R., Sekar, K., Banerjee, R., Sharma, V., Surolia, A. & Vijayan, M. (1996). A novel mode of carbohydrate recognition in jacalin, a *Moraceae* plant with a β-prism fold. *Nature Structural Biology*, **3**, 596–603.

Sauter, N. K., Glick, G. D., Crowther, R. L., Park, S.-J., Eisen, M. B., Skehel, J. J., Knowles, J. R. & Wiley, D. C. (1992*a*). Crystallographic detection of a second binding site in influenza virus hemagglutinin. *Proceedings of the National Academy of Sciences, USA*, **89**, 324–28.

Sauter, N. K., Hanson, J. E., Glick, G. D., Brown, J. H., Crowther, Park, S.-J., Skehel, J. J. & Wiley, D. C. (1992*b*). Binding of influenza virus hemagglutinin to analogs of its cell-surface receptor, sialic acid: analysis by proton nuclear magnetic resonance spectroscopy and X-ray crystallography. *Biochemistry*, **31**, 9609–21.

Savva, R. & Pearl, L. H. (1995). Nucleotide mimicry in the crystal structure of the uracil-DNA glycosylase–uracil glycosylase inhibitor protein complex. *Nature Structural Biology*, **2**, 752–7.

Schreiber, G. & Fersht, A. R. (1993). Interaction of barnase with its polypeptide inhibitor barstar studied by protein engineering. *Biochemistry*, **32**, 5145–50.

Schreiber, G. & Fersht, A. R. (1995). Energetics of protein–protein interactions: analysis of the Barnase–Barstar interface by single mutations and double mutant cycles. *Journal of Molecular Biology*, **248**, 478–86.

Schreiber, G. & Fersht, A. R. (1996). Rapid, electrostatically assisted association of proteins. *Nature Structural Biology*, **3**, 427–31.

Schutt, C. E., Myslik, J. C., Rozycki, M. D., Goonesekere, N. C. W. & Lindberg, U. (1993). The structure of crystalline profilin-β-actin. *Nature*, **365**, 810–16.

Shaanan, B., Lis, H. & Sharon, N. (1991). Structure of a legume lectin with an ordered N-linked carbohydrate in complex with lactose. *Science*, **254**, 862–6.

Sheriff, S., Silverton, E. W., Padlan, E. A., Cohen, G. H., Smith-Gill, S. J., Finzel, B. C. & Davies, D. R. (1987). Three dimensional structure of an antibody–antigen complex. *Proceedings of the National Academy of Sciences, USA*, **84**, 8075–9.

Sixma, T. K., Pronk, S. E., Kalk, K. H., Wartna, E. S., Van Zanten, B. A. M., Witholt, B. & Hol, W. G. J. (1991). Crystal structure of cholera toxin-related heat-labile enterotoxin from *E. coli*. *Nature*, **355**, 561–4.

Sixma, T. K., Pronk, S. E., Kalk, K. H., Vanzanten, B. A. M., Berghuis, A. M. & Hol, W. G. J. (1992). Lactose binding to heat labile enterotoxin revealed by X-ray crystallography. *Nature*, **355**, 561–4.

Stehle, T., Yan, Y., Benjamin, T. L. & Harrison, S. C. (1994). Structure of murine polyomavirus complexed with an oligosaccharide receptor fragement. *Nature*, **369**, 160–3.

Stehle, T. & Harrison, S. C. (1996). Crystal structures of murine polyomavirus in complex with straight-chain and branched chain sialyloligosaccharide receptor fragements. *Structure*, **4**, 183–94.

Stein, P. E., Boodhool A., Tyrrell, G. J., Brunton, J. L. & Read, R. J. (1992). Crystal structure of the cell-binding B oligomer of verotoxin-1 from *E. coli*. *Nature*, **355**, 748–50.

Stein, P. E., Boodhoo, A., Armstrong, G. D., Cockle, S. A., Klein, M. H. & Read, R. J. (1994*a*). The crystal structure of pertussis toxin. *Structure*, **2**, 43–57.

Stein, P. E., Boodhoo, A., Armstrong, G. D., Heerze, L., Cockle, S. A., Klein, M. H. & Read, R. J. (1994*b*). Structure of a pertussis toxin–sugar complex as a model for receptor binding. *Nature Structural Biology*, **1**, 591–6.

Steinbacher, S., Baxa, U., Miller, S., Weintraub, Seckler, R. & Huber, R. (1996). Crystal structure of phage P22 tailspike protein complexed with *Salmonella* sp. O-antigen receptors. *Proceedings of the National Academy of Sciences, USA*, **93**, 10584–8.

Stone, R. S., Dennis, S. & Hofsteenge, J. (1989). Quantitative evaluation of the contribution of ionic interactions to the formation of the thrombin–hirudin complex. *Biochemistry*, **28**, 6857–63.

Strynadka, N. C. J., Jensen, S. E., Alzari, P. M. & James, M. N. G. (1996). A potent new mode of β-lactamase inhibition revealed by the 1.7 Å X-ray crystallographic structure of the TEM-1–BLIP complex. *Nature Structural Biology*, **3**, 290–7.

Stubbs, M. T., Laber, B., Bode, W., Huber, R., Jerala, R., Lenarcic, B. & Turk, V. (1990). The refined 2.4 Å X-ray crystal structure of recombinant human stefin B in complex with the cysteine proteinase papain: a novel type of proteinase inhibitor interaction. *EMBO Journal*, **9**, 1939–47.

Sweet, R. M., Wright, H. T., Janin, J., Chothia, C. & Blow, D. M. (1974). Crystal structure of the complex of porcine trypsin with soybean trypsin inhibitor (Kunitz) at 2.6 Å resolution. *Biochemistry*, **13**, 4212–18.

Takeuchi, Y., Satow, Y., Nakamura, K. T. & Mitsui, Y. (1991). Refined crystal structure of the complex subtilisin BPN and *Streptomyces* subtilisin inhibitor at 1.8 Å resolution. *Journal of Molecular Biology*, **221**, 309–25.

Tulip, W. R., Varghese, J. N., Laver, W. G., Webster, R. G. & Colman, P. M. (1992). Refined crystal structure of the influenza virus N9 neuraminidase-NC41 Fab complex. *Journal of Molecular Biology*, **227**, 122–48.

van de Locht, Lamba, D., Bauer, M., Huber, R., Friedrich, T., Kröger, B., Höffken, W. & Bode, W. (1995). Two heads are better than one: crystal structure of the insect derived double domain Kazal inhibitor rhodinin in complex with thrombin. *EMBO Journal*, **14**, 5149–57.

Vos, A. M. de, Ultsch, M. & Kossiakoff, A. A. (1992). Human growth hormone and extracellular domain of its receptor: crystal structure of the complex. *Science*, **255**, 306–12.

Vyas, N. K. (1991). Atomic features of protein–carbohydrate interactions. *Current Opinion in Structural Biology*, **1**, 732–40.

Watowich, S. J., Skehel, J. J. & Wiley, D. C. (1994). Crystal structures of influenza virus hemagglutinin in complex with high affinity receptor analogs. *Structure*, **2**, 719–31.

Weis, W. I., Brown, J., Cusack, S., Paulson, R. W. H., Skehel, J. J. & Wiley, D. C. (1988). The structure of the influenza virus haemagglutinin complexed to its receptor, sialic acid. *Nature*, **333**, 426–31.

Weis, W. I., Drickamer, K. & Hendrickson, W. A. (1992). Structure of a C-type mannose-binding protein complexed with an oligosaccharide. *Nature*, **360**, 127–34.

Weis, W. I. & Drickamer, K. (1996). Structural basis of lectin-carbohydrate recognition. *Annual Review of Biochemistry*, **65**, 441–73.

Wright, C. S. (1984). Structural comparison of the two distinct sugar binding sites in wheat germ agglutinin isolectin II. *Journal of Molecular Biology*, **178**, 91–104.

Wright, C. S. (1990). 2.2 Å resolution structure analysis of two refined N-acetylneuraminyl-lactose–wheat germ agglutinin isolectin complexes. *Journal of Molecular Biology*, **215**, 635–51.

Wright, C. S. (1992). Crystal structure of a wheat germ agglutinin/glycophorin–sialoglycopeptide receptor complex. *Journal of Biological Chemistry*, **267**, 14345–52.

Wright, C. S. & Jaeger, J. (1993). Crystallographic refinement and structure analysis of the complex of wheat germ agglutinin with a bivalent sialoglycopeptide from glycophorin. *Journal of Molecular Biology*, **232**, 620–38.

PICORNAVIRUS RECEPTORS, TROPISM AND PATHOGENESIS

D. J. EVANS

AMS, The University of Reading, Whiteknights, PO Box 228, Reading, Berks RG6 5AJ, UK

Virus tropism can be defined at two levels: host tropism being the range of host species that a virus can productively infect, and tissue tropism reflecting the particular tissues or organs within the host that are the site(s) of virus replication. The pathogenesis of virus infection may be a direct consequence of virus replication, resulting in cell destruction or localized immune responses directed against virus antigens, or may be an indirect consequence of virus infection, resulting from inflammation or cytokine responses.

What are the determinants of virus tropism? The primary determinant of virus tropism is the presence on the cell surface of a suitable receptor which the virus can utilize to gain entry to the cell. In the absence of such a receptor (or an alternative means of gaining entry, such as antibody-dependent enhancement in the flaviviruses) the cell is non-permissive, even though the internal milieu may be capable of supporting all the remaining stages of the virus replication cycle. Because of their fundamental roles in determining virus tropism, considerable effort has been devoted in the last few years to identifying cellular receptors and this has allowed the role these molecules have in virus tropism to be evaluated. Nevertheless, it is apparent that the tissue tropism (i.e. the tissues in which the virus replicates in an infected host) may be determined by factors in addition to the virus receptor. This chapter will focus on the identification and characterization of the cellular receptors for the picornaviruses and describe the contributions to tropism and pathogenesis of the molecular interactions between virus particle and host receptor.

THE PICORNAVIRUSES

The family *picornaviridae* includes a wide range of pathogens of humans and other animals. The family is divided into five genera: the aphthoviruses which cause foot-and-mouth disease (FMD) of cattle and other cloven footed animals; the rhinoviruses which are a major cause of the common cold; the enteroviruses which cause a wide range of clinical syndromes including poliomyelitis; the hepatoviruses which cause hepatitis; and the cardioviruses whose clinical manifestations include encephalitis and myocarditis in mice.

Members of the family are closely related both structurally and genetically, and are some of the best studied of all animal viruses. Poliovirus, the type-specific species of the enterovirus genus, and probably the best studied of all picornaviruses, possesses a 7.5 kb positive sense RNA genome consisting of a 750 nt 5′ untranslated region (UTR), a single open reading frame (ORF) encoding a 220 kDa polyprotein, a 3′-UTR of 72 nt and a 3′-terminal poly(A) tract (Kitamura *et al.*, 1981; Racaniello & Baltimore, 1981; Stanway *et al.*, 1983). The polyprotein is divided into three regions as a result of primary post-translational proteolytic processing by virus-encoded proteases. The P1 region encodes the capsid proteins VP1–4, the structures of which are described below (Hogle, Chow & Filman, 1985). The P2 and P3 regions encode proteins involved in protein maturation and RNA replication, and include two proteases, a polymerase (Kitamura *et al.*, 1981), and putative membrane binding (Giachetti & Semler, 1991; Giachetti, Hwang & Semler, 1992) and NTP-binding functions (Teterina *et al.*, 1992).

The 5′-UTR has a highly ordered secondary and tertiary structure; it can be divided into a 5′ proximal 'clover-leaf' module that plays a critical role in replication, and forms a binding site for the virus 3CD proteins and at least one cellular factor, and an internal ribosome entry site (IRES) of approximately 500 nt involved in translation (Pelletier & Sonenberg, 1988; Pelletier *et al.*, 1988). The 3′-UTR also has a defined secondary structure, although this varies significantly between genera within the picornaviruses. The precise role, if any, that this region plays in virus replication remains to be determined, particularly following the demonstration that it can be exchanged between representatives of different genera without abrogating virus viability (Rohll *et al.*, 1995).

The icosahedral capsid ($T = 3$ triangulation number) is composed of 60 copies of each of the four P1-derived capsid proteins, VP1–4. VP1–3 are quasi-equivalent in structure, each forming an eight stranded anti-parallel β-barrel, flanked by two short α-helical regions. The capsid mediates two interactions critical to the life cycle of the virus; it is involved in the escape from immune surveillance and in the specific interaction with the cellular receptor. In the best characterized representatives of the picornaviruses there are distinct structures associated with each of these roles. The rhinoviruses and enteroviruses present a number of surface protrusions that form the antigenic sites, and which vary in response to immune selection (Minor, 1990). In addition, these two genera possess a distinct cleft or canyon that surrounds each of the five-fold axes of symmetry of the virus particle which has been implicated in receptor binding. The visualization of the canyon on the original rhinovirus 14 structure led Rossmann to propose the 'canyon hypothesis' (Rossmann, 1989*a*, *b*) in which he suggested that the region of the virus that interacts with the cellular receptor is protected from host-antibody driven immune selection and variation as a consequence of its location within a cleft physically too small to allow antibody access. Subsequent studies have supported this hypothesis (Colonno *et al.*, 1988; Olson *et al.*, 1993), although other regions of certain picornaviruses that do possess a canyon have been implicated in receptor interactions (La Monica, Kupsky & Racaniello, 1987; Murray *et al.*, 1988; Colston & Racaniello, 1994), and certain picornaviruses (most notably FMDV, see below) lack a canyon and are known to utilize a different mechanism for receptor interaction. Recent studies of co-crystallised rhinovirus and neutralizing antibody have demonstrated that antibodies can penetrate the canyon, suggesting that the structure of this region of the capsid may be dictated by the receptor interaction, rather than functioning to conceal critical residues from the immune response (Smith *et al.*, 1996).

PICORNAVIRUS RECEPTORS

Kaplan (1955) demonstrated that human kidney and amnion, two tissues not normally infected in polio victims, become permissive for poliovirus infection when cultured *in vitro*. Permissiveness correlated with the expression on the cell surface of receptors that could mediate virus attachment and eclipse. A series of seminal studies by Holland and colleagues established that only primate cells permissive to poliovirus infection possessed specific receptors, whereas non-primate cells which were resistant to infection were devoid of such receptors (McLaren, Holland & Syverton, 1959; Holland, 1961, 1962; Holland & McLaren, 1961).

Further experiments demonstrated that not all picornaviruses utilize the same receptor. Trypsin treatment of cells destroyed the ability to bind poliovirus and rhinovirus, but had no effect upon echovirus, group B coxsackieviruses or cardioviruses (in contrast, binding of the latter two could be prevented by pre-treatment of cells with chymotrypsin). Similarly, neuraminidase prevented cardiovirus, bovine enterovirus (BEV), equine rhinovirus and human rhinovirus (HRV) 87 binding, demonstrating that sialic acid must form a necessary constituent of their receptor (Kaplan, 1955; Zajac & Crowell, 1965).

The use of somatic cell hybrids demonstrated that the receptors for poliovirus, coxsackie B viruses, echoviruses and the major group rhinoviruses mapped to chromosome 19. However, it was the application of competitive binding studies and the use of monoclonal antibodies (mAbs) directed against cell surface components and which prevent virus infection that have formed the basis of our current understanding of picornavirus receptor usage and have provided the necessary information for the molecular identification of several specific receptors (Table 1, Fig. 1).

Table 1. *Picornavirus receptors and co-factors implicated in cell infection*

Genus	Virus	Serotypes	Receptor	Accessory factors
Aphthovirus	FMDV	7	$\alpha_v\beta_3$ vitronectin receptor	heparin sulphate
Cardiovirus	EMCV	1	VCAM-1	
Enterovirus	Poliovirus	1–3	PVR	CD44
	Group A Coxsackievirus	A13, A15, A18, A21	ICAM-1	
	Group A Coxsackievirus	A9	$\alpha_v\beta_3$ vitronectin receptor	
	Group B Coxsackievirus	B1, B3, B5	DAF	$\alpha_v\beta_6$ integrin
	Echovirus	1 and 8	$\alpha_2\beta_1$ (VLA-2)	β-2 microglobulin
	Echovirus	3, 6, 6′, 6″, 7, 11, 12, 13, 20, 21, 24, 29, 33	DAF	β-2 microglobulin
	Echovirus	22	$\alpha_v\beta_3$ vitronectin receptor	
Hepatovirus	Hepatitis A virus	1	α2macroglobin?	
Rhinovirus	Human rhinovirus	All 91 major group	ICAM-1	
	Human rhinovirus	All 10 minor group	LDLR	
	Human rhinovirus	87	Sialic acid	

For references see text.

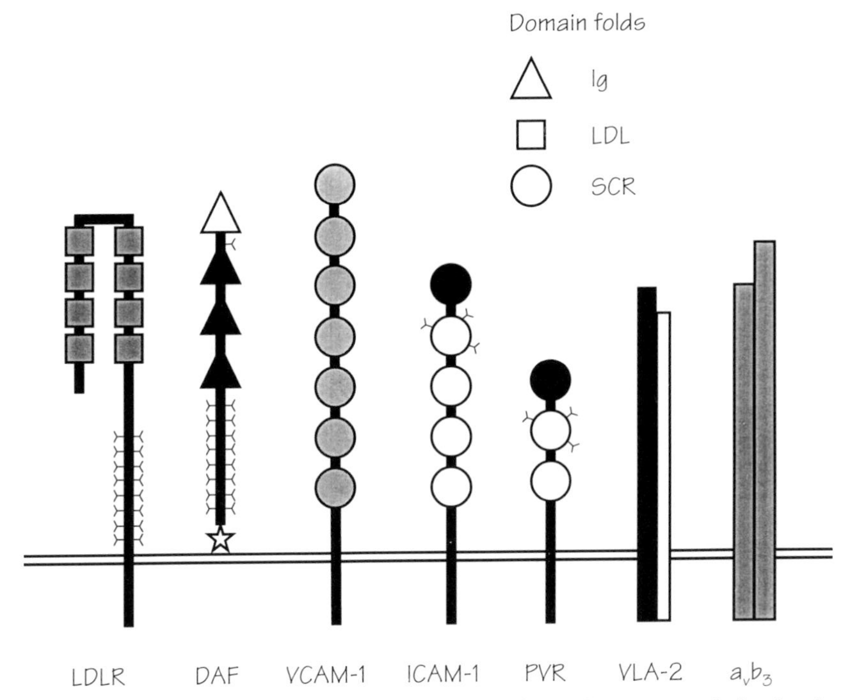

Fig. 1. Schematic representation of molecularly cloned picornavirus receptors indicating the domain structure and superfamily affiliations. Domains involved in virus binding are represented as symbols filled with black, those not involved are unfilled, and those for which information has yet to be obtained are mid-grey. References are listed in the text.

Enteroviruses

Poliovirus

The poliovirus receptor (PVR) was the first picornavirus receptor to be molecularly cloned and identified. This breakthrough depended upon the observation that murine L cells transfected with poliovirus RNA supported a single round of virus replication (Holland, McLaren & Syverton, 1959), and that transfection of L cells with a human genomic DNA library rendered some recipient cells permissive for poliovirus infection. The region of human DNA conferring the permissive phenotype was further characterized by a combination of rosetting and probing recovered cells with repetitive human DNA sequences, and a clone conferring the permissiveness was subsequently isolated from a cDNA library (Mendelsohn *et al.*, 1986; Mendelsohn, Wimmer & Racaniello, 1989; Koike *et al.*, 1990).

PVR is a type 1 membrane glycoprotein and a member of the immunoglobulin (Ig) superfamily. Differential splicing of the mRNA results in the synthesis of at least four different forms of the receptor, two of which are secreted (Koike *et al.*, 1990). The splicing affects the C-terminus of the encoded products and soluble forms of the receptor do not support virus infection. The cloning of a murine poliovirus receptor homologue (mph) that does not support virus binding or infection enabled the construction of hybrid receptors, and the identification of the residues critical for virus binding by homologue scanning mutagenesis. The virus binding domain was mapped to the first (i.e. most membrane distal) of the three Ig-like domains, although maximum poliovirus yields require a region of the receptor outside the first Ig-like domain (Morrison *et al.*, 1994; Bernhardt *et al.*, 1994). The C″ ridge of the first Ig-like domain is implicated in virus binding, which is analogous to the region of CD4 known to interact with HIV (Arthos *et al.*, 1989). However, certain mutations outside the binding site either delayed or prevented any cytopathic effect without preventing virus binding, suggesting that the interaction of the virus and receptor may be involved in events other than virus entry, such as triggering a signalling cascade that leads to cytolysis (Morrison *et al.*, 1994). The normal cellular role for PVR is not known. Mice transgenic for PVR are susceptible to poliovirus infection and develop symptoms that closely resemble human polio (Ren *et al.*, 1990; Koike *et al.*, 1991). These have been an important resource for our current understanding of poliovirus pathogenesis and neurovirulence (see below).

Echovirus

Three different proteins have been identified as echovirus receptors in the last five years. The integrin VLA-2 is the cellular receptor for echovirus serotypes 1 and 8, and was identified following demonstration that antibodies to the α-2 and β-1 subunits of the 145 kDa integrin heterodimer block virus binding (Bergelson *et al.*, 1992, 1993*a*). Human RD cells, which are deficient in α-2 expression, become permissive for virus binding and infection following transfection with an α-2-encoding cDNA. Further studies have demonstrated that echovirus 1 and collagen, one of the normal ligands for VLA-2, interact with adjacent or overlapping sites within domain I (residues 140–359) of the α-2 subunit of the receptor (Kamata, Puzon & Takada, 1994), although the mechanism by which the virus and natural extracellular matrix ligands interact with VLA-2 differ (Bergelson *et al.*, 1993*b*).

Many other echovirus serotypes, including 3, 6, 7, 11, 12, 20 and 21 use decay accelerating factor (DAF; also designated CD55), a 70 kDa component of the complement cascade (Bergelson *et al.*, 1994; Ward *et al.*, 1994). DAF is a GPI-linked member of the RCA (regulator of complement activity) family of proteins and possesses four extracellular domains that exhibit the characteristics of SCR (short consensus repeats) domains. Our laboratory identified DAF using a novel cloning technique (CELICS; cloning by immunocolourimetric screening) which allows the isolation (generally in a single round of screening) of a cDNA encoding a ligand to a mAb (Ward *et al.*, 1994). DAF was isolated from a HUVEC cDNA library using mAb854, raised to a HeLa cell membrane preparation (Minor *et al.*, 1984), and previously shown to block infection of RD cells by several echovirus serotypes. We have characterized the interaction of echovirus 7 and chimeric receptors constructed between DAF and its most closely related neighbour CD46 (also known as MCP; the measles virus receptor). Unlike the situation with PVR or ICAM-1 (see below), the first domain of DAF is not required for virus binding, and neither are the GPI-anchor or the serine/threonine-rich (S/T-rich) domain that separates the SCR-domains from the cell membrane (Clarkson *et al.*, 1995). In contrast, domains 2, 3 and 4 are all apparently required. It is not clear whether the integrity of the DAF molecule is dependent upon the presence of all three domains, only one of which directly contacts the virus particle, or if domains 2, 3 and 4 present a discontinuous binding site for the echovirus particle. The normal function of DAF, the regulation of complement activation, is also dependent upon the integrity of domains 2, 3 and 4, suggesting that the tertiary structure of the molecule may involve contributions from domains 2, 3 and 4 (Coyne *et al.*, 1992). However, it is interesting to speculate that the DAF-echovirus interaction may not involve docking of the receptor into a canyon on the virus surface, but instead involves an interaction of the three SCRs with contact residues surrounding a three-fold axis of virus symmetry.

Two lines of evidence suggest that Echovirus 22 (EV22) uses the $\alpha_v\beta_3$ vitronectin receptor for cell entry. The virus capsid protein VP1 contains an arginine-glycine-aspartic acid (RGD) motif, characteristic of certain integrin binding proteins and cell infection is blocked by RGD-containing peptides signifying a role for this sequence in cell entry (Hyypia *et al.*, 1992; Stanway *et al.*, 1994). Furthermore, EV22 competes for cell surface binding with the enterovirus coxsackievirus A9, which uses the vitronectin receptor for infection of GMK cells (Roivainen *et al.*, 1994).

Enterovirus 70

Enterovirus 70 (EV70), the aetiological agent of acute haemorrhagic conjunctivitis, is unusual among the human enterovirus in being able to infect a wide range of non-primate cell lines in culture (Yoshii, Natori & Kono, 1977). Karnauchow and colleagues have recently demonstrated that HeLa cell infection by EV70 involves decay accelerating factor, an observation supported by the gain in permissivity of murine NIH 3T3 cells transiently transfected with human DAF (Karnauchow *et al.*, 1996). The anti-DAF mAbs used in this study were specific for human DAF, and could not block infection of primate LLC-MK_2 or rodent CHO cell lines. Therefore, the receptor(s) used by EV70 on other cell lines remains to be determined; the limited homology between human and rodent DAF ($<50\%$) prevents recognition by mAbs specific for human DAF but also suggests that this virus may utilize molecules other than DAF for cell entry.

Coxsackievirus

Coxsackievirus A9 can utilize the $\alpha_v\beta_3$ vitronectin receptor for cell entry and possesses an RGD motif at the C-terminus of VP1. RGD-containing peptides block attachment of virus to AGM cells, unless the virus is pretreated with trypsin or human intestinal fluid. This suggests that the virus may exhibit dual receptor tropism, and be capable of exploiting one or more alternative receptors, which may be of importance in the pathogenesis of enteric infections (Roivainen *et al.*, 1991; Chang *et al.*, 1992).

Infection of HeLa cells by Coxsackievirus serotypes A13, A15, A18 and A21 is blocked by mAbs that also prevent infection by major group rhinoviruses, demonstrating that these serotypes use ICAM-1 as a cellular receptor (Colonno, Callahan & Long, 1986). Coxsackieviruses B1, B3 and B5 all use DAF as a cellular receptor (Shafren *et al.*, 1995), although the inability of murine cells transfected with DAF to support virus infection suggests that other factors are also required.

Rhinoviruses

Major group rhinoviruses

The rhinoviruses are divided into two groups, designated major and minor, dependent upon their receptor usage. The major receptor group, which constitute approximately 90% of the 100+ serotypes, use ICAM-1 (intracellular adhesion molecule type 1) as a cellular receptor (Staunton *et al.*, 1989; Greve *et al.*, 1989; Tomassini *et al.*, 1989). Like PVR, ICAM-1 is a type 1 membrane glycoprotein of the Ig-superfamily. The 90 kDa protein is a receptor for the leukocyte adhesion molecules LFA-1 and Mac-1, and possesses five Ig-like domains, the first of which has been implicated in rhinovirus binding by the construction of receptor chimeras and site directed mutagenesis (Staunton *et al.*, 1990; McClelland *et al.*, 1991). Within the first Ig-like domain of ICAM-1, at least seven regions could be mutated to block rhinovirus binding by the introduction of single residue substitutions that did not, in themselves, disrupt the overall integrity of the protein (Staunton *et al.*, 1990; Register *et al.*, 1991; McClelland *et al.*, 1991). These regions are distributed throughout the first Ig-like domain, but when mapped onto the predicted tertiary fold of the protein are generally localized on the turns between the βB-βC, βD-βE and βF-βG strands of the Ig-like structure. Direct support for Rossman's canyon hypothesis has been provided by cryo-electron microscopy of a complex between HRV16 and the first two domains of soluble ICAM-1. These elegant studies have allowed visualization of the receptor 'docked' into the canyon, with 15–20 Å buried below the virion surface. Superimposition of the ICAM-1 density on to the known structure of CD4 suggests that the regions of ICAM-1 implicated in virus binding by site directed mutagenesis are likely to be those most closely associated with the virus surface (Olson *et al.*, 1993). The recently determined structure of vascular cell adhesion molecule 1 (VCAM-1), which like ICAM-1 is both an integrin binding member of the Ig-superfamily and a picornavirus receptor (see below), may enable a better understanding of rhinovirus–receptor interactions (Jones *et al.*, 1995).

Minor group rhinoviruses

The minor receptor group rhinoviruses can use members of the low density lipoprotein receptor (LDLR) family to infect cells. A protein secreted by HeLa cells was purified that exhibited HRV2 binding activity (Hofer *et al.*, 1992), and was subsequently shown by tryptic digestion and N-terminal sequencing to be the LDL receptor. Antibodies to LDLR block virus binding and infection (Hodits *et al.*, 1995), and a solubilized avian homologue of LRLR blocks minor group rhinovirus infection in tissue culture (Gruenberger *et al.*, 1995). Surprisingly, LDLR-negative FH fibroblasts remain permissive to minor group rhinoviruses, an observation attributed to the expression by this cell line of the structurally related α2-macroglobulin/LDLR-related protein (Hofer *et al.*, 1994).

A single serotype of human rhinovirus (serotype 87) uses neither ICAM-1 or LDLR for cell entry but instead requires the presence of sialic acid on cellular receptors (Uncapher, DeWitt & Colonno, 1991).

Cardioviruses

Encephalomyocarditis virus (EMCV) binding and infection of vascular endothelial cells (VEC) is inhibited by an anti-VCAM-1 monoclonal antibody, which also blocks infection of CHO cells transfected with VCAM-1 (Huber, 1994). VCAM-1 (vascular cell adhesion molecule 1) is an integrin-binding member of the immunoglobulin superfamily with seven extracellular Ig-like domains, the expression of which is restricted to endothelial cells (Jones *et al.*, 1995). Infection of human erythrocytes by EMCV involves neuraminidase-sensitive components of glycophorin A (Allaway & Burness, 1986, 1987).

Aphthoviruses

The cellular receptor for foot and mouth disease virus (FMDV) serotype A12 was identified on the basis of homology between the highly conserved RGD-motif occupying the G-H loop of VP1 and the natural ligands for certain integrin binding proteins. The G-H loop (and C-terminus) of VP1 is cleaved by trypsin treatment resulting in significant reductions in infectivity, and high concentrations of RGD-containing peptides also block virus binding and infection of susceptible cells (Fox *et al.*, 1989). Competition experiments between FMDV type A12 and Coxsackievirus A9 identified the vitronectin receptor (integrin $\alpha_v\beta_3$) as one receptor utilized by FMDV, which was confirmed by blocking virus binding and/or infection using anti- α_v or anti-β_3 monoclonal antibodies.

In addition to the RGD-motif, the FMDV capsid also contains a putative heparin binding motif near the C-terminus of VP1. Pre-treatment of cells in tissue culture with heparinase, heparin or platelet factor 4 significantly inhibited virus binding to paraformaldehyde fixed cells, suggesting a role for heparan sulphate (HS) in virus binding. Furthermore, cell lines deficient in HS, pretreated with heparinase or heparin were either non-permissive, or exhibited significantly reduced permissivity, for type O strains of FMDV. This study suggests that the initial interaction between FMDV and the cell surface is via the non-integrin component, such as the charged HS proteoglycan, which subsequently leads to integrin-dependent internalization (Jackson *et al.*, 1996). Herpes simplex virus also uses a similar complex entry process that involves an initial interaction with HS proteoglycans at the cell surface before secondary receptors mediate the fusion of the virus envelope with the plasma membrane (Wudunn & Spear, 1989).

Hepatoviruses

The hepatitis A virus (HAV) receptor remains to be definitively identified. Zajac *et al.* have studied the parameters that influence the attachment of HAV to a wide range of cell lines, including BS-C-1, Hep G2, HeLa and L cells. Calcium ions enhanced both attachment and infectivity. In contrast, the addition of 2% serum, or the high M_r fraction following gel filtration, markedly inhibited attachment. The major protein component of the high M_r serum fraction co-migrated with α2 macroglobulin (α2M), which was shown to associate with virus by sedimentation and co-precipitation (Zajac *et al.*, 1991). The authors speculate that the normal cellular role of α2M, a plasma glycoprotein that binds and inhibits endopeptidases resulting in the complexes being cleared from the circulation, may have been exploited by the virus to facilitate access to the liver.

ARE VIRUS RECEPTORS THE SOLE DETERMINANTS OF TISSUE TROPISM?

The primary criteria by which virus receptors are defined, including many of those detailed above, is the ability of the molecule to permit virus binding. In some situations further supporting evidence is provided in that receptor expression in a cell line that supports replication, but not infection, is sufficient to allow a full replicative cycle to occur (e.g. expression of PVR in murine fibroblasts, or VLA-2 α-2 subunit in CHO cells). Two lines of evidence, from *in vivo* or *in vitro* (tissue culture) studies, suggest that this is a simplistic view of the molecular determinants of tissue tropism. The tissue distribution of the virus receptor and comparison with the known sites of virus replication suggests that receptor expression alone does not determine tropism. In many instances, receptor expression is more widespread than the tissue distribution of replicating virus; conversely, different viruses that use the same receptor can exhibit distinctly different tropisms. A number of viruses are known to require additional cellular factors for infection that do not directly mediate virus binding. Examples from outside the picornaviruses include a requirement for the vitronectin binding proteins $\alpha_v\beta_3$ and $\alpha_v\beta_5$ by adenovirus 12 (Bai, Campisi & Freimuth, 1994), and the LESTR or CC-CKR-5 β-chemokine receptors used by human immunodeficiency virus 1 as fusion cofactors (Feng *et al.*, 1996; Alkhatib *et al.*, 1996; Deng *et al.*, 1996). As the number of picornavirus receptors identified increases, there is evidence that some picornaviruses also require cell surface co-factors. Those identified at the molecular level are involved in cell infection, though there are also undoubtedly intracellular factors that function during virus replication but which are outside the scope of this chapter.

CD44

Lymphocyte homing receptor (CD44) has been implicated as an additional determinant of poliovirus tropism and was originally identified as the ligand for a mAb (AF3) that blocked poliovirus infection (preferentially blocking type 2 poliovirus) of HeLa cells. CD44 occurs in several heterogeneous isoforms, AF3 being specific for a 100 kD variant of the CD44H isoform which is the major species found in haematopoietic and non-haematopoietic cells. PVR negative murine L-cells expressing CD44H do not support poliovirus infection, or binding of radiolabelled virus, demonstrating that CD44H does not act as a receptor for the virus. However, AF3 (and two other anti-CD44 mAbs) were shown to block binding of radiolabelled poliovirus to HeLa cells (which express PVR). Evidence for a direct interaction of poliovirus and CD44 has not been demonstrated and the mechanism by which anti-CD44 mAbs block infection remains to be determined.

β-2 microglobulin

The CELICS technique has been used to clone the ligand for a mAb that does not bind to DAF, but still blocks infection of RD cells by several echoviruses (T. Ward, D.J. Evans and J.W. Almond, unpublished observations). The mAb (originally raised to a HeLa cell membrane preparation) is specific for β-2 microglobulin, and the result has been confirmed by demonstration that other anti-β-2 microglobulin mAbs block infection by a wide range of echoviruses, including all those known to utilize DAF as the receptor, together with echovirus serotypes 1 and 8 (which use the $\alpha2$ subunit of VLA-2, see Table 1), and enterovirus 70. MAbs to β-2 microglobulin do not block infection by poliovirus, rhinovirus or coxsackie B viruses, although the DAF utilizing coxsackieviruses remain to be screened. At least for echovirus 7, the block is cell type specific, and not functional in HeLa cells which express high levels of DAF. However, β-2 microglobulin is not acting as a receptor; anti-β-2 microglobulin mAbs do not block virus binding, and soluble β-2 microglobulin does not inactivate virus or prevent virus binding to DAF-expressing cells. We have been unable to demonstrate a direct interaction between β-2 microglobulin and echovirus, and the role of β-2 microglobulin in echovirus infection therefore remains to be determined. It is interesting to note that antibodies to MHC I, with which β-2 microglobulin associates at the cell surface, are known to interfere with a number of receptor–ligand interactions, possibly by interrupting essential signalling events.

There is recent evidence to suggest that the $\alpha_V\beta_6$ fibronectin receptor may be a cofactor that enhances the replication of DAF-utilizing coxsackie B1 and B3 viruses in certain cell lines, including SW480 colon cancer cells (Agrez *et al.*, 1996). Deletion of the B subunit cytoplasmic tail, that has a critical role in signal transduction (Sheppard *et al.*, 1994; Agrez *et al.*, 1994), abrogated the enhancement effect in SW480 cells, although whether the effect is mediated at the level of virus entry or subsequent genome replication remains to be determined.

RECEPTOR EXPRESSION IN THE WHOLE ANIMAL

Molecular characterization of the cellular receptor provides the tools to investigate receptor expression at the organ, tissue or cellular level in the host. Amongst the picornaviruses, the most comprehensively studied is PVR. These studies represent the molecular extension of early work by Bodian and colleagues (Bodian, 1959), who investigated sites of poliovirus replication by autopsy of human poliomyelitis victims and experimental infection of non-human primates. Bodian demonstrated that poliovirus replicates exclusively in the tissue types of relevance to the disease, i.e. the motor cortex, cerebellum, brain stem and spinal cord. Parallel studies by Holland and colleagues demonstrated that only homogenates from susceptible tissues were capable of binding virus (McLaren, Holland & Syverton, 1959; Holland, 1961, 1962; Holland & McLaren, 1961), which contrasts with permissivity of the majority of cultured primate cell lines for poliovirus (Kaplan, Levy & Racaniello, 1989).

Molecular identification of the PVR allowed tissue distribution and expression of the receptor to be determined using both radiolabelled PVR cDNA probes in Northern blot analyses or mAbs specific for PVR in a Western blot. Although the poliovirus receptor is clearly the relevant mediator of poliovirus infection, studies of its distribution in human tissues are not consistent with the notion that it is the principal determinant of tissue tropism (Freistadt, Kaplan & Racaniello, 1990). Northern blot analysis revealed that PVR mRNA was expressed in a wide range of both susceptible and non-susceptible tissues including brain, leukocytes, kidney, liver and placenta (Koike *et al.*, 1990). Moreover, it is likely that both secreted forms and cell surface forms are expressed in these tissues. A similar picture has emerged from mice transgenic for the PVR which also express the receptor in many tissues (Koike *et al.*, 1991). Further detailed studies of the transgenic mice, by *in situ* hybridization (Ren & Racaniello, 1992), demonstrated a more limited tissue expression of PVR (in the CNS, thymus, lung, kidney and adrenal), which is limited to certain cell types (Bowman capsule in the kidney, neurones in the CNS). Detection of the PVR protein by Western blot analyses demonstrated both tissue-specific differences in immunoreactivity and heterogeneity of the reactive species (possibly due to differences in glycosylation or expression of variant spliced forms of the receptor), although the protein appears to be ubiquitously expressed in human tissues, and is clearly detectable in the kidney, liver, intestine, cerebellum and motor cortex.

It is therefore clear that the PVR is expressed on a number of tissues that do not normally support poliovirus replication *in vivo*, and further studies are required to determine whether the proteins expressed at these ectopic sites can function as virus receptors. Our understanding of poliovirus tropism, and therefore pathogenesis, is currently incomplete; is tropism determined solely by expression of a defined subset of poliovirus receptors or are there additional cellular determinants? It has been proposed that the expression of CD44 (see above) correlates well with known sites of poliovirus replication. However, there is little direct evidence to support CD44 being a molecular determinant of poliovirus tropism, and the marked serotype specificity with which anti-CD44 mAbs block poliovirus infection must be explained before this protein can be considered a credible component of the pathogenic determinants of poliovirus infection. Finally, it should be emphasized that additional post-entry blocks to poliovirus replication, and therefore tissue tropism, have also been extensively studied. The major attenuating mutation located within the 5′NCR (nt 472 in type 3 poliovirus) appears to operate at the level of translation through interaction with a cellular factor(s) which may be differentially expressed in the CNS and other sites of virus replication (Almond, 1987). Identification of these factors, and those that operate at the level of cell entry are required before we have a full understanding of the relationship between receptors, tropism and pathogenesis of poliovirus.

The other group of picornaviruses for which there is considerable information about the receptor are the major receptor group rhinoviruses. The pathogenesis of rhinovirus infection remains poorly understood even though this group of viruses account for at least 30% of upper respiratory tract infections. Since the host range of rhinovirus is restricted to higher primates, detailed studies of pathogenesis using an animal model are impractical. The major group receptor, ICAM-1, is expressed in a wide range of cell lineages and is unlikely to be a limiting factor that determines the virus tropism. Recent studies suggest that the pathogenesis of rhinovirus infections is a consequence of the host's immune response mediated by an as yet poorly characterized cytokine response to virus infection. It is known that a range of inflammatory mediators including kinins and interleukin 1 are enhanced in the nasal secretions of infected individuals and that the severity of nasopharyngeal symptoms correlates with the numbers of lymphocytes and neutrophils in nasal secretions (Levandowski, Weaver & Jackson, 1988; Proud *et al.*, 1990, 1994). A proper understanding of rhinovirus pathogenesis requires that the mediators of immune cell recruitment are properly identified and to this end Proud and colleagues have recently demonstrated the infection of a human respiratory epithelial cell line with rhinovirus type 14. These studies demonstrate that rhinovirus replication in the adeno/SV40 transformed BEAS-2B cell line was not cytopathic and that infection was enhanced by pre-treatment of the cells with TNF α, a known stimulator of ICAM-1 expression (Subauste, Jacoby & Proud, 1994). These studies were extended to demonstrate that rhinovirus infection increased the production of IL8, IL6 and GM-CSF from epithelial cells. A better understanding of the role of cytokines and immune mediators in rhinovirus pathogenesis is complicated by the fact that each may have both beneficial and detrimental effects upon virus replication. For example, Gern and colleagues (Gern *et al.*, 1996) have recently demonstrated that rhinovirus enters but cannot replicate within monocytes and airway macrophages, suggesting that certain immune cells lining the respiratory tract are not sites of virus replication. However, the same study demonstrated that interaction of rhinovirus with macrophages resulted in significant increase in the secretion of TNF α which, under certain conditions, is known to have a direct antiviral effect (Mestan *et al.*, 1986). However, TNF α may also have beneficial effects on virus replication, as it is known to increase the expression of ICAM-1 under experimental conditions (Mulligan *et al.*, 1993). Whether this is of physiological relevance remains to be determined.

Encephalomyocarditis virus (EMCV) is associated with a wide range of diseases including pancreatitis, insulin-dependent diabetes, paralysis and myocarditis, in which the virus is known to replicate in the targeted cells of the affected organ. Since VCAM-1 expression is restricted to endothelial cells, it is likely that the broad tissue tropism of EMCV is reflected in the ability to use receptors other than glycophorin A and VCAM-1. However, although the tissue tropism of this virus is likely to depend upon its ability to exploit more than one cellular receptor, the pathogenesis is probably a consequence of both the tropism and the host immune response. For example, the pathogenicity of experimental diabetes and pancreatitis in mice is dependent upon a functional T-lymphocyte response (Barger &

Craighead, 1991), which enhances virus infection of the target tissue resulting in more extensive damage. Furthermore, the immune response may indirectly influence virus tropism, in a manner similar to that suggested for rhinovirus. VCAM-1 expression is stimulated by a number of cytokines associated with an inflammatory immune response (IL-1 and 4, TNF-α, IF-γ), which may contribute to virus tropism by extending the range of permissive tissues surrounding a focal site of infection, and so further increase exposure of surrounding organs to virus (Huber, 1994).

CONCLUDING COMMENTS

Great advances have been made in the last three to five years on the molecular characterization of picornavirus receptors. These studies, coupled with parallel advances in our understanding of the structural transitions that occur within the virus capsid upon receptor binding (Fricks & Hogle, 1990; Wien, Chow & Hogle, 1996), are providing unique insights into the virus–receptor interactions required for cell infection. The dissection of the requirements for cell infection are also leading to the identification of accessory molecules which are a pre-requisite for delivery of the virus genome to the cell cytoplasm.

However, the molecular characterization of these events now generally exceeds our understanding of the factors that restrict the tissue tropism of viruses in the host, and in many cases are further advanced than our understanding of the molecular mechanisms of picornavirus pathogenesis. Observations of an increasing number of picornaviruses suggest that receptor expression *per se* is not the sole determinant of virus tropism. In many cases, receptor expression is more widespread in the host than our current understanding of virus tropism would suggest. This may reflect the means by which virus tropism or receptor expression is determined; PVR expression is relatively widespread, but it is not clear that all the detectable PVR can function as a virus receptor. The identification of accessory molecules required for cell infection *in vitro*, presumably reflecting a requirement *in vivo*, suggests instead that virus tropism is multi-factorial, the availability of both the attachment protein (receptor) and accessory proteins being required for virus entry. Although the tissue distribution of CD44 may reflect the tropism of poliovirus, the distribution of the other potential accessory molecules so far identified cannot explain the tropism of the viruses with which they have been associated. It should also be noted that the mechanistic role these accessory molecules have in cell infection remains unresolved and a physiological role must remain in doubt for those that are not obligately required. Other determinants, in particular the availability of factors within the cell that are subverted for virus replication, are likely to have a major influence on the tissues within a host that become productively infected, and consequently contribute to virus pathogenesis.

A better understanding of the molecular determinants of cell infection, the intracellular proteins exploited by the virus for replication, and the tissue and host distribution of functional versions of both classes of factors, are required before we can fully appreciate the role of receptors in virus tropism. The overall contribution of receptors to virus pathogenesis will also require a more complete understanding of the pathogenic mechanisms of this group of viruses.

REFERENCES

Agrez, M., Chen, A., Cone, R.I., Pytela, R. & Sheppard, D. (1994). The alpha-v-beta-6 integrin promotes proliferation of colon-carcinoma cells through a unique region of the beta-6 cytoplasmic domain. *Journal of Cell Biology*, **127**, 547–56.

Agrez, M., Cox, K., Shafren, D.R., Sheppard, D. & Barry, R. (1996). Integrin $\alpha v\beta 6$ (a fibronectin receptor) enhances replication of Coxsackieviruses B1 and B3. *Europic '96*, A7 (abstract).

Alkhatib, G., Combadiere, C., Broder, C.C., Feng, Y., Kennedy, P.E., Murphy, P.M. & Berger, E.A. (1996). CC CKR 5 – a RANTES, MIP-1-alpha, MIP-1-beta receptor as a fusion cofactor for macrophage-tropic HIV-1. *Science*, **272**, 1955–8.

Allaway, G.P. & Burness, A.T. (1986). Site of attachment of encephalomyocarditis virus on human erythrocytes. *Journal of Virology*, **59**, 768–70.

Allaway, G.P. & Burness, A.T. (1987). Analysis of the bond between encephalomyocarditis virus and its human erythrocyte receptor by affinity chromatography on virus-sepharose columns. *Journal of General Virology*, **68**, 1849–56.

Almond, J.W. (1987). The attenuation of poliovirus neurovirulence. *Annual Review of Microbiology*, **41**, 153–80.

Arthos, J., Deen, K.C., Chaikin, M.A., Fornwald, J.A., Sathe, G., Sattentau, Q.J., Clapham, P.R., Weiss, R.A., McDougal, J.S., Pietropaolo, C., Axel, R., Truneh, A., Maddon, P.J. & Sweet, R.W. (1989). Identification of the residues in human cd4 critical for the binding of HIV. *Cell*, **57**, 469–81.

Bai, M., Campisi, L. & Freimuth, P. (1994). Vitronectin receptor antibodies inhibit infection of HeLa and A549 cells by adenovirus type 12 but not by adenovirus type 2. *Journal of Virology*, **68**, 5925–32.

Barger, M.T. & Craighead, J.E. (1991). Immunomodulation of encephalomyocarditis virus-induced disease in A/J mice. *Journal of Virology*, **65**, 2676–81.

Bergelson, J.M., Shepley, M.P., Chan, B.M., Hemler, M.E. & Finberg, R.W. (1992). Identification of the integrin VLA-2 as a receptor for echovirus 1. *Science*, **255**, 1718–20.

Bergelson, J.M., St.John, N., Kawaguchi, S., Chan, M., Stubdal, H., Moldin, J. & Finberg, R.W. (1993*a*). Infection by Echovirus 1 and 8 depends on the alpha-2 subunit of human VLA-2. *Journal of Virology*, **67**, 6847–52.

Bergelson, J.M., Chan, B.M., Finberg, R.W. & Hemler, M.E. (1993*b*). The integrin VLA-2 binds echovirus 1 and extracellular matrix ligands by different mechanisms. *Journal of Clinical Investigation*, **92**, 232–9.

Bergelson, J.M., Chan, M., Solomon, K.R., St.John, N.F., Lin, H. & Finberg, R.W. (1994). Decay-accelerating factor (CD55), a glycosylphosphatidylinositol-anchored complement regulatory protein, is a receptor for several echoviruses. *Proceedings of the National Academy of Sciences, USA*, **91**, 6245–9.

Bernhardt, G., Bibb, J.A., Bradley, J. & Wimmer, E. (1994). Molecular characterization of the cellular receptor for poliovirus. *Virology*, **199**, 105–13.

Bodian, D. (1959). Poliomyelitis: pathogenesis and histopathology. In Rivers, T.M. & Horsfall, F.L. Jn. (eds.). *Viral and Rickettsial Infections of Man*. 3rd ed. pp. 497–518. J.B. Lippincott (pub).

Chang, K.H., Day, C., Walker, J., Hyypia, T. & Stanway, G. (1992). The nucleotide sequences of wild-type Coxsackievirus-A9 strains imply that an RGD motif in VP1 is functionally significant. *Journal of General Virology*, **73**, 621–6.

Clarkson, N.A., Kaufman, R., Lublin, D.M., Ward, T., Pipkin, P.A., Minor, P.D., Evans, D.J. & Almond, J.W. (1995). Characterization of the echovirus-7 receptor – domains of cd55 critical for virus binding. *Journal of Virology*, **69**, 5497–501.

Colonno, R.J., Callahan, P.L. & Long, W.J. (1986). Isolation of a monoclonal antibody that blocks attachment of the major group of human rhinoviruses. *Journal of Virology*, **57**, 7–12.

Colonno, R.J., Condra, J.H., Mizutani, S., Callahan, P.L., Davies, M.E. & Murcko, M.A. (1988). Evidence for the direct involvement of the rhinovirus canyon in receptor binding. *Proceedings of the National Academy of Sciences, USA*, **85**, 5449–53.

Colston, E. & Racaniello, V.R. (1994). Soluble receptor-resistant poliovirus mutants identify surface and internal capsid residues that control interaction with the cell-receptor. *EMBO Journal*, **13**, 5855–62.

Coyne, K.E., Hall, S.E., Thompson, E.S., Acre, M.A., Kinoshita, T., Fujita, T., Anstee, D.J., Rosse, W. & Lublin, D.M. (1992). Mapping of epitopes, glycosylation sites, and complement regulatory domains in human decay accelerating factor. *Journal of Immunology*, **149**, 2906–13.

Deng, H.K., Liu, R., Ellmeier, W., Choe, S., Unutmaz, D., Burkhart, M., Dimarzio, P., Marmon, S., Sutton, R.E., Hill, C.M., Davis, C.B., Peiper, S.C., Schall, T.J., Littman, D.R. & Landau, N.R. (1996). Identification of a major coreceptor for primary isolates of HIV-1. *Nature*, **381**, 661–6.

Feng, Y., Broder, C.C., Kennedy, P.E. & Berger, E.A. (1996). HIV-1 entry cofactor – functional cDNA cloning of a 7- transmembrane, G-protein-coupled receptor. *Science*, **272**, 872–7.

Fox, G., Parry, N.R., Barnett, P.V., McGinn, B., Rowlands, D.J. & Brown, F. (1989). The cell attachment site on foot-and-mouth disease virus includes the amino acid sequence RGD (arginine-glycine-aspartic acid). *Journal of General Virology*, **70**, 625–37.

Freistadt, M.S., Kaplan, G. & Racaniello, V.R. (1990). Heterogeneous expression of poliovirus receptor-related proteins in human cells and tissues. *Molecular and Cellular Biology*, **10**, 5700–6.

Fricks, C.E. & Hogle, J.M. (1990). Cell-induced conformational change in poliovirus: externalization of the amino terminus of VP1 is responsible for liposome binding. *Journal of Virology*, **64**, 1934–45.

Gern, J.E., Dick, E.C., Lee, W.M., Murray, S., Meyer, K., Handzel, Z.T. & Busse, W.W. (1996). Rhinovirus enters but does not replicate inside monocytes and airway macrophages. *Journal of Immunology*, **156**, 621–7.

Giachetti, C. & Semler, B.L. (1991). Role of a viral membrane polypeptide in strand-specific initiation of poliovirus RNA synthesis [published errata appear in *J Virol* 1991 Jul;65(7):3972 and 1991 Oct;65(10):5653]. *Journal of Virology*, **65**, 2647–54.

Giachetti, C., Hwang, S.S. & Semler, B.L. (1992). *cis*-acting lesions targeted to the hydrophobic domain of a poliovirus membrane protein involved in RNA replication. *Journal of Virology*, **66**, 6045–57.

Greve, J.M., Davis, G., Meyer, A.M., Forte, C.P., Yost, S.C., Marlor, C.W., Kamarck, M.E. & McClelland, A. (1989). The major human rhinovirus receptor is ICAM-1. *Cell*, **56**, 839–47.

Gruenberger, M., Wandl, R., Nimpf, J., Hiesberger, T., Schneider, W.J., Kuechler, E. & Blaas, D. (1995). Avian homologs of the mammalian low-density lipoprotein receptor family bind minor receptor group human rhinovirus. *Journal of Virology*, **69**, 7244–7.

Hodits, R.A., Nimpf, J., Pfistermueller, D.M., Hiesberger, T., Schneider, W.J., Vaughan, T.J., Johnson, K.S., Haumer, M., Kuechler, E., Winter, G. & Blaas, D. (1995). An antibody fragment from a phage display library competes for ligand binding to the low density lipoprotein receptor family and inhibits rhinovirus infection. *Journal of Biological Chemistry*, **270**, 24078–85.

Hofer, F., Berger, B., Gruenberger, M., Machat, H., Dernick, R., Tessmer, U., Kuechler, E. & Blaas, D. (1992). Shedding of a rhinovirus minor group binding protein: evidence for a $Ca^{(2+)}$-dependent process. *Journal of General Virology*, **73**, 627–32.

Hofer, F., Gruenberger, M., Kowalski, H., Machat, H., Huettinger, M., Kuechler, E. & Blass, D. (1994). Members of the low density lipoprotein receptor family mediate cell entry of a minor-group common cold virus. *Proceedings of the National Academy of Sciences, USA*, **91**, 1839–42.

Hogle, J.M., Chow, M. & Filman, D.J. (1985). Three-dimensional structure of poliovirus at 2.9 A resolution. *Science*, **229**, 1358–65.

Holland, J.J. (1961). Receptor affinities as major determinants of enterovirus tissue tropisms in humans. *Virology*, **15**, 312–26.

Holland, J.J. (1962). Irreversible eclipse of poliovirus by HeLa cells. *Virology*, **16**, 163–76.

Holland, J.J. & McLaren, L.C. (1961). The location and nature of enterovirus receptors in susceptible cells. *Journal of Experimental Medicine*, **114**, 161–71.

Holland, J.J., McLaren, L.C. & Syverton, J.T. (1959). Mammalian cell-virus relationhip. III. Poliovirus production by non-primate cells exposed to poliovirus ribonucleic acid. *Proceedings of the Society for Experimental Biology and Medicine*, **100**, 843–52.

Huber, S.A. (1994). VCAM-1 is a receptor for encephalomyocarditis virus on murine vascular endothelial cells. *Journal of Virology*, **68**, 3453–8.

Hyypia, T., Horsnell, C., Maaronen, M., Khan, M., Kalkkinen, N., Auvinen, P., Kinnunen, L. & Stanway, G. (1992). A distinct picornavirus group identified by sequence analysis. *Proceedings of the National Academy of Sciences, USA*, **89**, 8847–51.

Jackson, T., Ellard, F.M., Ghazaleh, R.A., Brookes, S.M., Blakemore, W.E., Corteyn, A.H., Stuart, D.I., Newman, J.W. & King, A.M.Q. (1996). Efficient infection of cells in culture by type-O foot-and-mouth- disease-virus requires binding to cell-surface heparan-sulfate. *Journal of Virology*, **70**, 5282–7.

Jones, E.Y., Harlos, K., Bottomley, M.J., Robinson, R.C., Driscoll, P.C., Edwards, R.M., Clements, J.M., Dudgeon, T.J. & Stuart, D.I. (1995). Crystal structure of an integrin-binding fragment of vascular cell adhesion molecule-1 at 1.8 A resolution. *Nature*, **373**, 539–44.

Kamata, T., Puzon, W. & Takada, Y. (1994). Identification of putative ligand binding sites within I domain of integrin alpha 2 beta 1 (VLA-2, CD49b/CD29). *Journal of Biological Chemistry*, **269**, 9659–63.

Kaplan, A.S. (1955). The susceptibility of monkey kidney cells to poliovirus *in vivo* and *in vitro*. *Virology*, **1**, 377–92.

Kaplan, G., Levy, A. & Racaniello, V.R. (1989). Isolation and characterization of HeLa cell lines blocked at different steps in the poliovirus life cycle. *Journal of Virology*, **63**, 43–51.

Karnauchow, T.M., Tolson, D.L., Harrison, B.A., Altman, E., Lublin, D.M. & Dimock, K. (1996). The HeLa cell receptor for enterovirus 70 is decay accelerating factor (CD55). *Journal of Virology*, **70**, 5143–52.

Kitamura, N., Semler, B.L., Rothberg, P.G., Larsen, G.R., Adler, C.J., Dorner, A.J., Emini, E.A., Hanecak, R., Lee, J.L., Van der Werf, S., Anderson, C.W. & Wimmer, E. (1981). Primary structure, gene organisation and polypeptide expression of poliovirus RNA. *Nature*, **291**, 547–53.

Koike, S., Horie, H., Ise, I., Okitsu, A., Yoshida, M., Iizuka, N., Takeuchi, K., Takegami, T. & Nomoto, A. (1990). The poliovirus receptor protein is produced both as membrane-bound and secreted forms. *EMBO Journal*, **9**, 3217–24.

Koike, S., Taya, C., Kurata, T., Abe, S., Ise, I., Yonekawa, H. & Nomoto, A. (1991). Transgenic mice susceptible to poliovirus. *Proceedings of the National Academy of Sciences, USA*, **88**, 951–5.

La Monica, N., Kupsky, W.J. & Racaniello, V.R. (1987). Reduced mouse neurovirulence of poliovirus type 2 Lansing antigenic variants selected with monoclonal antibodies. *Virology*, **161**, 429–37.

Levandowski, R.A., Weaver, C.W. & Jackson, G.G. (1988). Nasal-secretion leukocyte populations determined by flow cytometry during acute rhinovirus infection. *Journal of Medical Virology*, **25**, 423–32.

McClelland, A., Debear, J., Yost, S.C., Meyer, A.M., Marlor, C.W. & Greve, J.M. (1991). Identification of monoclonal antibody epitopes and critical residues for rhinovirus binding in domain-1 of intercellular adhesion molecule-1. *Proceedings of the National Academy of Sciences, USA*, **88**, 7993–7.

McLaren, L.C., Holland, J.J. & Syverton, J.T. (1959). The mammalian cell-virus relationship: I Attachment of poliovirus to cultivated cells of primate and non-primate origin. *Journal of Experimental Medicine*, **109**, 475–85.

Mendelsohn, C., Johnson, B., Lionetti, K.A., Nobis, P., Wimmer, E. & Racaniello, V.R. (1986). Transformation of a human poliovirus receptor gene into mouse cells. *Proceedings of the National Academy of Sciences, USA*, **83**, 7845–9.

Mendelsohn, C.L., Wimmer, E. & Racaniello, V.R. (1989). Cellular receptor for poliovirus: molecular cloning, nucleotide sequence, and expression of a new member of the immunoglobulin superfamily. *Cell*, **56**, 855–65.

Mestan, J., Digel, W., Mittnacht, S., Hillen, H., Blohm, D., Moller, A., Jacobsen, H. & Kirchner, H. (1986). Antiviral effects of recombinant tumour necrosis factor *in vitro*. *Nature*, **323**, 816–19.

Minor, P.D. (1990). Antigenic structure of picornaviruses. In V.R. Racaniello (ed.). *Current Topics in Microbiology and Immunology 161: Picornaviruses*. pp. 122–154. Berlin Heidelberg: Springer-Verlag.

Minor, P.D., Pipkin, P.A., Hockley, D., Schild, G.C. & Almond, J.W. (1984). Monoclonal antibodies which block cellular receptors of poliovirus. *Virus Research*, **1**, 203–12.

Morrison, M.E., He, Y.J., Wien, M.W., Hogle, J.M. & Racaniello, V.R. (1994). Homolog-scanning mutagenesis reveals poliovirus receptor residues important for virus binding and replication. *Journal of Virology*, **68**, 2578–88.

Mulligan, M.S., Vaporciyan, A.A., Miyasaka, M., Tamatani, T. & Ward, P.A. (1993). Tumor-necrosis-factor-alpha regulates *in vivo* intrapulmonary expression of icam-1. *American Journal of Pathology*, **142**, 1739–49.

Murray, M.G., Bradley, J., Yang, X.F., Wimmer, E., Moss, E.G. & Racaniello, V.R. (1988). Poliovirus host range is determined by a short amino acid sequence in neutralization antigenic site I. *Science*, **241**, 213–15.

Olson, N.H., Kolatkar, P.R., Oliveira, M.A., Cheng, R.H., Greve, J.M., McClelland,

A., Baker, T.S. & Rossmann, M.G. (1993). Structure of a human rhinovirus complexed with its receptor molecule. *Proceedings of the National Academy of Sciences, USA*, **90**, 507–11.

Pelletier, J. & Sonenberg, N. (1988). Internal initiation of translation of eukaryotic mRNA directed by a sequence derived from poliovirus RNA. *Nature*, **334**, 320–5.

Pelletier, J., Kaplan, G., Racaniello, V.R. & Sonenberg, N. (1988). Cap-independent translation of poliovirus mRNA is conferred by sequence elements within the 5′ noncoding region. *Molecular and Cellular Biology*, **8**, 1103–12.

Proud, D., Naclerio, R.M., Gwaltney, J.M. & Hendley, J.O. (1990). Kinins are generated in nasal secretions during natural rhinovirus colds. *Journal of Infectious Diseases*, **161**, 120–3.

Proud, D., Gwaltney, J.M., Hendley, J.O., Dinarello, C.A., Gillis, S. & Schleimer, R.P. (1994). Increased levels of interleukin-1 are detected in nasal secretions of volunteers during experimental rhinovirus colds. *Journal of Infectious Diseases*, **169**, 1007–13.

Racaniello, V.R. & Baltimore, D. (1981). Molecular cloning of poliovirus cDNA and determination of the complete nucleotide sequence of the viral genome. *Proceedings of the National Academy of Sciences, USA*, **78**, 4887–91.

Register, R.B., Uncapher, C.R., Naylor, A.M., Lineberger, D.W. & Colonno, R.J. (1991). Human-murine chimeras of ICAM-1 identify amino acid residues critical for rhinovirus and antibody binding. *Journal of Virology*, **65**, 6589–96.

Ren, R. & Racaniello, V.R. (1992). Human poliovirus receptor gene expression and poliovirus tissue tropism in transgenic mice. *Journal of Virology*, **66**, 296–304.

Ren, R., Costantini, F., Gorgacz, E.J., Lee, J.J. & Racaniello, V.R. (1990). Transgenic mice expressing a human poliovirus receptor: a new model for poliomyelitis. *Cell*, **63**, 353–62.

Rohll, J.B., Moon, D.H., Evans, D.J. & Almond, J.W. (1995). The 3′-untranslated region of picornavirus RNA – features required for efficient genome replication. *Journal of Virology*, **69**, 7835–44.

Roivainen, M., Hyypia, T., Piirainen, L., Kalkkinen, N., Stanway, G. & Hovi, T. (1991). RGD-dependent entry of coxsackievirus A9 into host cells and its bypass after cleavage of VP1 protein by intestinal proteases. *Journal of Virology*, **65**, 4735–40.

Roivainen, M., Piirainen, L., Hovi, T., Virtanen, I., Riikonen, T., Heino, J. & Hyypia, T. (1994). Entry of Coxsackievirus A9 into host cells: specific interactions with $\alpha v\beta 3$ integrin, the vitronectin receptor. *Virology*, **203**, 357–65.

Rossmann, M.G. (1989*a*). The canyon hypothesis. *Viral Immunology*, **2**, 143–61.

Rossmann, M.G. (1989*b*). The canyon hypothesis. Hiding the host cell receptor attachment site on a viral surface from immune surveillance. *Journal of Biological Chemistry*, **264**, 14587–90.

Shafren, D.R., Bates, R.C., Agrez, M.V., Herd, R.L., Burns, G.F. & Barry, R.D. (1995). Coxsackieviruses B1, B3, and B5 use decay accelerating factor as a receptor for cell attachment. *Journal of Virology*, **69**, 3873–7.

Sheppard, D., Chen, A., Weinacker, A. & Agrez, M. (1994). Distinct regions of the cytoplasmic domain of the integrin beta-6 subunit are required for localization to focal contacts and for stimulation of epithelial-cell proliferation. *Journal of Cellular Biochemistry*, **No. S18C SIC, p. 256**.

Smith, T.J., Chase, E.S., Schmidt, T.J., Olson, N.H. & Baker, T.S. (1996). Neutralizing antibody to human rhinovirus-14 penetrates the receptor- binding canyon. *Nature*, **383**, 350–4.

Stanway, G., Cann, A.J., Hauptmann, R., Hughes, P., Clarke, L.D., Mountford, R.C., Minor, P.D., Schild, G.C. & Almond, J.W. (1983). The nucleotide sequence of poliovirus type 3 leon 12 a1b: comparison with poliovirus type 1. *Nucleic Acids Research*, **11**, 5629–43.

Stanway, G., Kalkkinen, N., Roivainen, M., Ghazi, F., Khan, M., Smyth, M., Meurman, O. & Hyypia, T. (1994). Molecular and biological characteristics of echovirus 22, a representative of a new picornavirus group. *Journal of Virology*, **68**, 8232–8.

Staunton, D.E., Merluzzi, V.J., Rothlein, R., Barton, R., Marlin, S.D. & Springer, T.A. (1989). A cell adhesion molecule, ICAM-1, is the major surface receptor for rhinoviruses. *Cell*, **56**, 849–53.

Staunton, D.E., Dustin, M.L., Erickson, H.P. & Springer, T.A. (1990). The arrangement of the immunoglobulin-like domains of ICAM-1 and the binding sites for LFA-1 and rhinovirus [published erratum appears in *Cell* 1990 Jun 15;61(2):1157]. *Cell*, **61**, 243–54.

Subauste, M.C., Jacoby, D.B. & Proud, D. (1994). *In-vitro* infection of a human respiratory epithelial-cell line (beas-2b) with rhinovirus. *Journal of Allergy and Clinical Immunology*, **93**, 167.

Teterina, N.L., Kean, K.M., Gorbalenya, A.E., Agol, V.I. & Girard, M. (1992). Analysis of the functional significance of amino acid residues in the putative NTP-binding pattern of the poliovirus-2C protein. *Journal of General Virology*, **73**, 1977–86.

Tomassini, J.E., Graham, D., DeWitt, C.M., Lineberger, D.W., Rodkey, J.A. & Colonno, R.J. (1989). cDNA cloning reveals that the major group rhinovirus receptor on HeLa cells is intercellular adhesion molecule 1. *Proceedings of the National Academy of Sciences, USA*, **86**, 4907–11.

Uncapher, C.R., DeWitt, C.M. & Colonno, R.J. (1991). The major and minor group receptor families contain all but one human rhinovirus serotype. *Virology*, **180**, 814–17.

Ward, T., Pipkin, P.A., Clarkson, N.A., Stone, D.M., Minor, P.D. & Almond, J.W. (1994). Decay accelerating factor (CD55) identified as the receptor for echovirus 7 using CELICS, a rapid immuno-focal cloning method. *EMBO Journal*, **13**, 5070–4.

Wien, M.W., Chow, M. & Hogle, J.M. (1996). Poliovirus – new insights from an old paradigm. *Structure*, **4**, 763–7.

Wudunn, D. & Spear, P.G. (1989). Initial interaction of herpes-simplex virus with cells is binding to heparan-sulfate. *Journal of Virology*, **63**, 52–8.

Yoshii, T., Natori, K. & Kono, R. (1977). Replication of enterovirus 70 in nonprimate cell cultures. *Journal of General Virology*, **36**, 377–84.

Zajac, I. & Crowell, R.L. (1965). Effects of enzymes on the interaction of enteroviruses with living HeLa cells. *Journal of Bacteriology*, **89**, 1097–100.

Zajac, A.J., Amphlett, E.M., Rowlands, D.J. & Sangar, D.V. (1991). Parameters influencing the attachment of hepatitis A virus to a variety of continuous cell lines. *Journal of General Virology*, **72**, 1667–75.

CROSS-TALK BETWEEN *YERSINIA* AND EUKARYOTIC CELLS

G. R. CORNELIS

Microbial Pathogenesis Unit, International Institute of Cellular and Molecular Pathology and Faculté de Médecine, Université Catholique de Louvain, Brussels, Belgium

INTRODUCTION

Invasive pathogenic bacteria have in common the capacity to overcome the defence mechanisms of their animal host and to proliferate in its tissues. They all have their own lifestyle and target organs, leading to a variety of symptoms and diseases, which suggested the existence of great diversity among the bacterial virulence strategies. However, recent data from several laboratories contradict this view and reveal the existence of related major virulence systems in various pathogenic bacteria, including phytopathogens. One of these systems delivers bacterial proteins inside eukaryotic cells by extracellularly located bacteria that are in close contact with the target cell's surface. The Yop virulon of yersiniae, presented below, represents an archetype for this new family of systems. Literature describing homologues to the Yop virulon, in several other bacterial pathogens, is now abundant and complete citations of the similarities are outside the scope of this review. Only where knowledge of homologues contributes to a better understanding of the *Yersinia* virulon, will they be mentioned.

THE *YERSINIA* LIFESTYLE

The genus *Yersinia* includes three species that are pathogenic for rodents and humans; *Yersinia pestis* is the agent of black death, *Yersinia pseudotuberculosis* is an agent of mesenteric adenitis and septicaemia and *Yersinia enterocolitica*, the most prevalent in humans, causes a broad range of gastrointestinal syndromes, ranging from an acute enteritis to mesenteric lymphadenitis. *Y. pestis* is generally inoculated by a flea bite, while the two others are food-borne pathogens. In spite of these differences in the infection routes, all three share a common tropism for lymphoid tissues and a common capacity to resist the non-specific immune response. *Y. pestis*, *Y. pseudotuberculosis* and even *Y. enterocolitica* under appropriate experimental conditions cause fulminant infections of the mouse. Anatomo-pathological examinations show that yersiniae, when they have reached lymphoid tissues, are largely extracellular (Hanski *et al.*, 1989; Simonet, Richard & Berche,

1990). In accordance with these *in vivo* observations, *Yersinia* manifests some resistance to phagocytosis *in vitro*, by macrophages (Rosqvist, Bölin & Wolf-Watz, 1988; Fällman *et al.*, 1995) or by polymorphonuclear leukocytes (Burrows & Bacon, 1956; China *et al.*, 1994; Visser, Annema & van Furth, 1995; Ruckdeschel *et al.*, 1996). Once they are phagocytozed, *Y. pseudotuberculosis* and *Y. enterocolitica* generally do not survive. These observations led to the concept that *Y. pseudotuberculosis* and *Y. enterocolitica* are actually extracellular pathogens and that their survival strategy basically consists in escaping phagocytosis.

THE CLUE: CALCIUM DEPENDENCY

It has been known since the mid-1950s that *Y. pestis* is unable to grow at 37 °C in Ca^{2+}-deprived media (Higushi & Smith, 1961). It has also been known for decades that this unusual property can be lost and that its loss correlates with a loss of virulence. This Ca^{2+}-dependency phenotype offered an extraordinary clue to the pathogenicity apparatus because non-virulent mutants could be easily detected and even selected for. It rapidly appeared that virulence and Ca^{2+}-dependency are encoded by a 70-kb plasmid called pYV (Gemski *et al.*, 1980; Zink *et al.*, 1980). The genetic organization of the pYV plasmid of serotype O:9 *Y. enterocolitica* strains is shown in Fig. 1. Under conditions of growth restriction, this plasmid governs the synthesis (Portnoy, Moseley & Falkow, 1981; Straley & Brubaker, 1981) and release (Heesemann *et al.*, 1986; Michiels *et al.*, 1990) of a set of about 12 proteins called 'Yops' for '*Yersinia* outer proteins'. The LcrV protein, a *Y. pestis* antigen already discovered in the mid-1950s (Burrows & Bacon, 1956), turned out to be one of these Yops (Straley & Brubaker, 1981; Bölin *et al.*, 1988; Mulder *et al.*, 1989). Genetic analysis revealed that most of these proteins are essential for virulence. However, three observations were enigmatic: (i) Yops form large and insoluble aggregates in the culture medium, which is puzzling for virulence effectors; (ii) Yops have no toxic activity on their own; and (iii) Yops are not produced in the presence of the mM concentrations of Ca^{2+} that prevail in the extracellular fluid, where *Yersinia* are expected to be.

FROM THE YOP SOUP TO A MODEL

Light was shed on these enigmatic findings by the work of Rosqvist and colleagues in Umeå, Sweden using *Y. pseudotuberculosis*-infected cell cultures. They first showed that extracellular adherent yersiniae were cytotoxic for HeLa cells and that the YopE protein was involved in this action (Rosqvist *et al.*, 1990). However, crude Yop preparations containing the YopE protein had no cytotoxic effect, unless they were microinjected into HeLa cells, indicating that the target of the YopE protein was intracellular. The seminal discovery was that a *yopD* mutant was also unable to affect

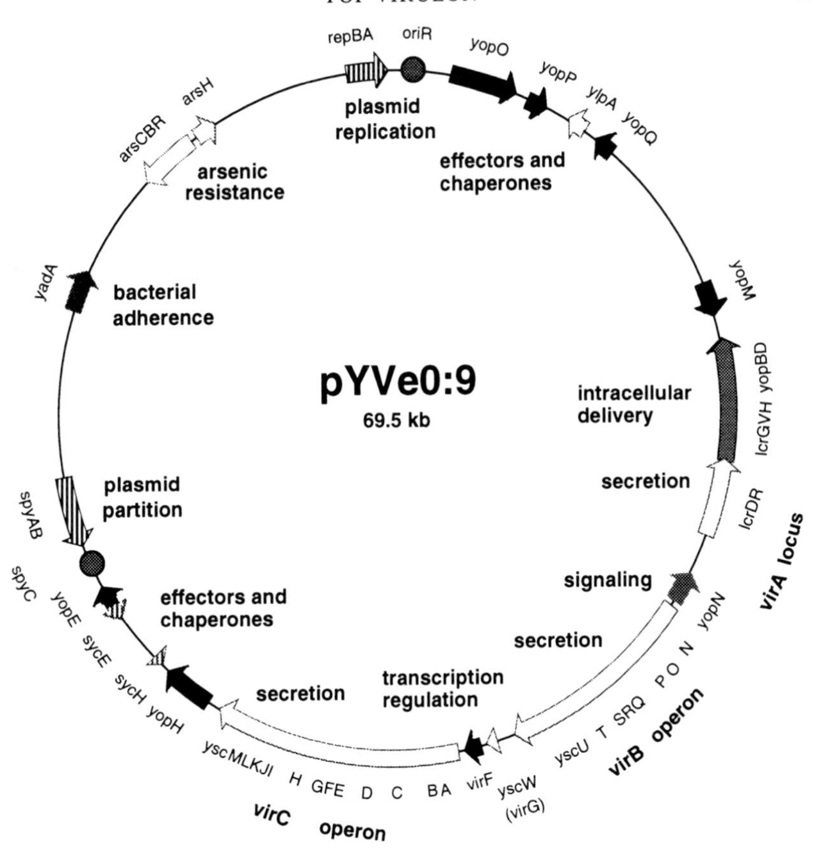

Fig. 1. Genetic map of the pYV plasmid from *Y. enterocolitica* serotype O:9. The complete plasmid has been sequenced.

HeLa cells but a preparation of Yops secreted by this mutant was cytotoxic when microinjected into the cytosol of HeLa cells. Rosqvist, Forsberg & Wolf-Watz (1991) logically concluded from this that the YopD protein plays a role in translocating the YopE protein across the plasma membrane of the eukaryotic target cell to reach the cytosolic compartment. Since intracellular bacteria are not cytotoxic, one can rule out that the YopE protein is translocated across the membrane of an endosome.

The evidence for YopD-mediated translocation of the YopE protein was essentially genetical. In 1994, two independent approaches allowed these enigmatic observations to be developed into a coherent model. One of the confirmations came from immunofluorescence and confocal laser scanning microscopy examinations performed by the Umeå group. The YopE protein appeared in the perinuclear region of HeLa cells infected with wild-type *Y. pseudotuberculosis*; in contrast, when infection was done with a mutant

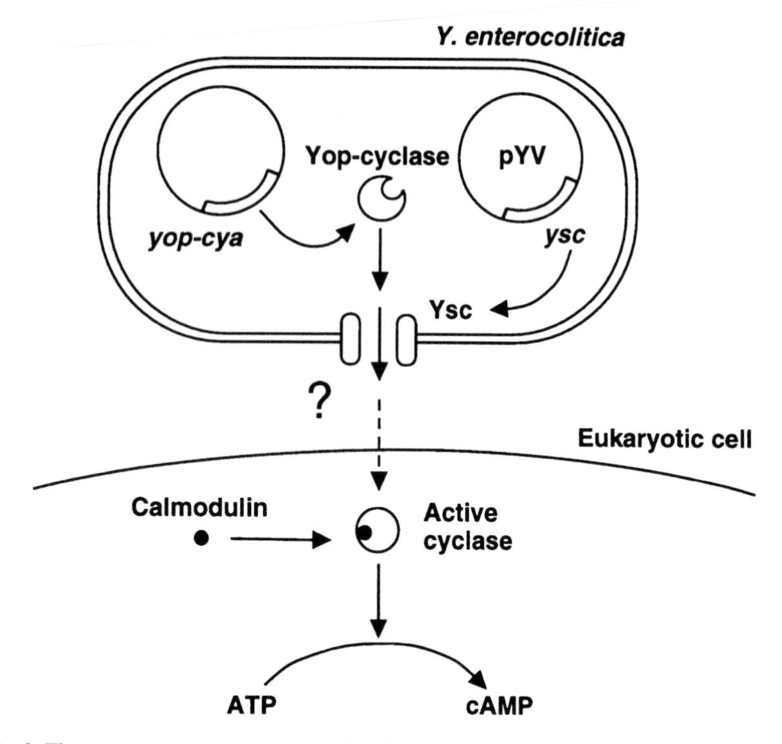

Fig. 2. The reporter-enzyme strategy used to demonstrate translocation of Yop proteins across the plasma membrane of eukaryotic cells. The pYV plasmid encodes the Yop secretion (Ysc) and translocation apparatus. The second plasmid encodes the hybrid reporter made of a truncated Yop and the calmodulin-dependent adenylate cyclase domain of cyclolysin (Cya) from *Bordetella pertussis*. (Reprinted from Sory & Cornelis, 1994; by permission.)

strain of *Y. pseudotuberculosis* unable to produce YopD, the YopE protein was only found as spots localized in the vicinity of bacteria adhering to the cell's surface, showing that the YopD protein was essential for the translocation of the YopE protein across the target cell membrane (Rosqvist, Magnusson & Wolf-Watz, 1994). The other confirmation was based on a reporter enzyme strategy introduced by the Louvain group (Sory & Cornelis, 1994) (Fig. 2). The reporter system consists of the calmodulin-activated adenylate cyclase domain (called Cya) of the *Bordetella pertussis* cyclolysin. The rationale is the following: the Yop-Cya hybrid enzyme introduced into the cytosol of eukaryotic cells produces cAMP while the intrabacterial Yop-Cya hybrid does not because of the absence of calmodulin in the bacterial cytoplasm. Since the catalytic domain of cyclolysin is unable to enter eukaryotic cells by itself, accumulation of cAMP essentially reflects Yop internalization. Infection of HeLa cells with recombinant *Y. enterocolitica* producing a hybrid YopE-Cya protein resulted in a marked increase in cAMP even when internalization of the bacteria themselves was prevented by cytochalasin D. Infection with a *Y. enterocolitica* mutant unable to secrete both the YopD and YopB proteins did not lead to cAMP accumulation, confirming the involvement of YopD or YopB or both proteins in translocation of the YopE protein across eukaryotic membranes (Sory & Cornelis, 1994).

A coherent picture emerged from these two approaches and a mechanistic insight now takes precedence over phenomenology. According to the model (Fig. 3), the Yops form two distinct groups of proteins. Some Yops are intracellular effectors delivered inside eukaryotic cells by extracellular yersiniae adhering at the cell's surface, while other Yops form a delivery apparatus. If this model is correct, effector Yops must end up in the cytosol of the target cell and not in endocytic vacuoles. This prediction is supported by immunological observations. While antigens processed in phagocytic vacuoles of phagocytes are cleaved and presented by MHC class II molecules, epitope 249–257 of the YopH protein is presented by MHC class I molecules, like cytosolic proteins (Starnbach & Bevan, 1994).

The pYV plasmid, thus, encodes an integrated anti-host system that could be referred to as the Yop virulon. The Yop virulon consists of a set of effector Yops delivered inside eukaryotic cells, a delivery apparatus and a specialized secretion system. These various components will be described hereafter. As will also be seen, the system is tightly regulated, both at the gene and protein levels.

INTRACELLULAR EFFECTORS

So far, four Yop effectors have been identified, namely YopE, YopH, YpkA (YopO) and YopM. The information on their intracellular and physiological roles is still scarce.

The activity of the 23-kDa YopE protein leads to disruption of the actin microfilament structure of cultured HeLa cells. However, it does not disrupt actin filaments polymerized *in vitro*, even in the presence of NAD^+, suggesting that its action is indirect (Rosqvist *et al.*, 1991).

The YopH effector is a 51-kDa protein tyrosine phosphatase (PTPase) related to eukaryotic PTPases (Guan & Dixon, 1990) that acts on tyrosine-phosphorylated proteins of macrophages (Bliska *et al.*, 1991); it contributes to the inhibition of bacterial uptake by cultured macrophages (Rosqvist *et al.*, 1988) and to inhibition of the oxidative burst of macrophages triggered by opsonized bacteria (Bliska & Black, 1995).

The YpkA (YopO) protein is an 81-kDa serine/threonine kinase which also shows extensive similarity to eukaryotic counterparts (Galyov *et al.*, 1993). This Yop is less abundant than the two preceding ones in *Yersinia* culture supernatant fluids. It was recently shown to be internalized (Håkansson *et al.*, 1996; M-P. Sory *et al.*, unpublished data) and targeted to the inner surface of the plasma membrane of eukaryotic cells (Håkansson *et al.*, 1996).

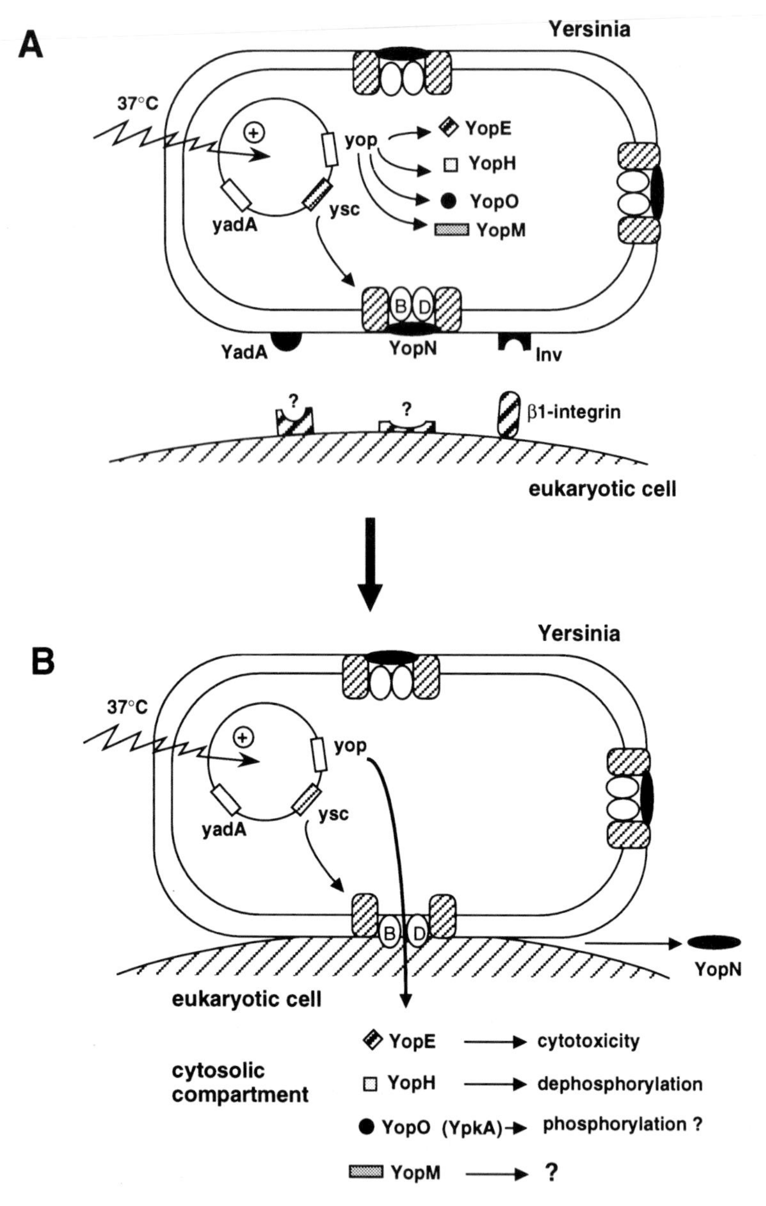

Fig. 3. Model showing *Y. enterocolitica* facing a eukaryotic cell. Before contact with the eukaryotic cell, the YopN protein acts as a stop-valve limiting the release of Yop effectors into the surrounding environment (A). Upon YadA- and Inv-mediated intimate contact with the eukaryotic cell, the YopN protein is removed and transfer of the YopE, YopH, YopM and YopO effectors across the plasma membrane of the eukaryotic cell occurs (B); the complete process of translocation requires the secretion apparatus (Ysc) and both the YopB and YopD proteins. (Reprinted from Boland *et al.*, 1996; by permission of Oxford Universty Press.)

The YopM effector is a 41-kDa protein that contains a succession of six repeated motifs made of 19 amino acids rich in leucine residues and related to the very common leucine-rich repeat motifs (Leung & Straley, 1989). As a consequence of the presence of these motifs, the YopM protein presents some similarity with a large number of proteins, among which is the α-chain of the platelet membrane glycoprotein Ib, a protein which binds thrombin and the von Willebrand factor (Leung & Straley, 1989). As anticipated from this similarity, the YopM protein was found to bind thrombin and, in doing so, to inhibit platelet aggregation (Reisner & Straley, 1992). Thus, the YopM effector would interfere with platelet-mediated events of the inflammatory response and play a role in the initial stages of an infection (Reisner & Straley, 1992). According to this hypothesis, the YopM protein would be released into the extracellular compartment, while other Yop effectors are specifically delivered into eukaryotic cells. It seemed possible that the YopM protein could also be either a translocator (like the YopD protein) or an intracellular effector (like the YopE, YopH and YpkA/YopO proteins). In *yopM* mutants, translocation of the YopE-Cya hybrid protein occurred equally as well as in wild-type bacteria, which ruled out the possibility that the YopM protein could be a translocator (Boland *et al.*, 1996). Constructed YopM-Cya derivatives were also observed to be internalized into macrophages by *Y. enterocolitica* infecting the monolayer (Boland *et al.*, 1996). Thus, the YopM protein is also an intracellular effector, although its intracellular target remains to be identified.

Thus, four Yops (YopE, YopH, YopM, and YpkA/YopO) have so far been shown to translocate into eukaryotic cells (Fig. 3).

THE DELIVERY OF EFFECTORS

Components of the translocation apparatus

As seen in the previous section, the early experiments on the YopE effector led to the hypothesis that the YopD protein is involved in translocation of the YopE protein (Rosqvist *et al.*, 1991;1994). Later, the YopD protein was also shown to be individually required for translocation of the YopH and YopM effectors (Boland *et al.*, 1996) as well as the YopO (YpkA) effector (M-P. Sory, unpublished data). In line with this, antiphagocytic effects (Fällman *et al.*, 1995) and macrophage protein dephosphorylation (Hartland *et al.*, 1994) were abolished in *yopD* mutants. The YopD protein is probably also involved in translocation of other Yop effectors. Indeed, it is required for phenomena which are independent of the YopE and YopH effectors, such as inhibition of the oxidative burst of zymosan-stimulated bone marrow macrophages (Hartland *et al.*, 1994) and inhibition of IL-8 secretion by epithelial cells (Schulte *et al.*, 1996).

The YopD translocator is encoded, together with two other Yops, LcrV and YopB, by the large *lcrGVHyopBD* operon. Since the first *yopD* mutants available were insertion mutants, it was only possible to draw conclusions about the YopD protein, the product of the last gene of the operon (Sory & Cornelis, 1994; Persson *et al.*, 1995). Recently, non-polar *yopB* and *yopD* mutants were constructed in *Y. pseudotuberculosis* and in *Y. enterocolitica.* Their analysis showed that the YopB protein is required, together with the YopD protein for translocation of the YopE, YopH and YopM effectors (Boland *et al.*, 1996) as well as the YpkA (YopO) protein (Håkansson *et al.*, 1996; M-P. Sory, unpublished data) (Fig. 3). In agreement with their involvement in translocation, the YopD and YopB proteins were both found to be individually required for the inhibition of phagocytosis by macrophages (Fällman *et al.*, 1995). In accordance with their putative role as translocators, the YopB and YopD proteins appear to be different from the effector Yop proteins. They contain hydrophobic domains and analysis of their amino acid sequences by the Eisenberg algorithm gives the pattern characteristic of transmembrane proteins (Håkansson *et al.*, 1993).

The *lcrGVHyopBD* operon thus encodes two proteins involved in translocation of the intracellular effectors. Since the LcrV protein, a Yop already discovered in the mid-1950s (Burrows & Bacon, 1956) is also encoded by this operon, one may wonder whether or not this protein is a third component of an organized delivery apparatus. So far, no model exists of the structure of this putative *Yersinia* translocation apparatus. However, the YopE protein can be internalized into HeLa cells by recombinant *Salmonella typhimurium* making use of the *Salmonella* invasion machinery (Rosqvist *et al.*, 1995). One could thus speculate that the Yop delivery apparatus resembles the *Salmonella* delivery apparatus, which could be in the 60-nm thick and 0.3–1 μm-long appendages described by Galan and coworkers (Ginocchio *et al.*, 1994). However, there is no proof yet that these appendages, called invasomes, are involved in the injection of bacterial proteins by *Salmonella* inside eukaryotic cells.

Intimate contact requirement for Yop translocation

One could argue that Yops formally resemble exotoxins, which are usually composed of two functionally distinct domains, namely a toxic moiety (domain A) and a vector (domain B) (for review, see Alouf & Freer, 1991). If Yops were simply a mixture of proteins of the A and B types, the only difference would be the requirement for more than one B protein for translocation of a single A protein. In what ways does the Yop delivery system differ from the internalization of an exotoxin? Two arguments support the concept that it is different from internalization of exotoxins. The first one is the requirement for adherence of bacteria to their target cell.

The early work of Rosqvist *et al.* (1990) already showed that the adhesins InvA or YadA are both required for YopE-induced cytotoxicity of *Y. pseudotuberculosis*. This was confirmed later by confocal microscopy studies (Rosqvist *et al.*, 1994), as well as by the YopE-Cya reporter strategy (Sory & Cornelis, 1994). The second argument comes from co-infection experiments. Monolayers of HeLa cells or macrophages were infected simultaneously with a *Yersinia* strain secreting all the Yops but not the YopE-Cya hybrid protein and with a *Yersinia* mutant producing the YopE-Cya hybrid but not the YopB and YopD proteins. No internalization of the effector Yop could be detected during these mixed infections showing that an effector Yop cannot be internalized by 'soluble' YopBD complexes (Sory *et al.*, 1995). In conclusion, translocation of effector Yops is achieved by extracellular bacteria adhering to the plasma membrane of the target cell and producing a translocation apparatus composed of the YopB and YopD proteins and possibly a few other Yops, like the LcrV protein (Fig. 3).

Signalling

As mentioned previously, extracellular yersiniae adhering to cultured HeLa cells or macrophages translocate Yops across the plasma membrane into the cytosol. Are Yops exclusively targeted to the eukaryotic cell or are they also released in the culture supernatant? The question was first addressed for the YopE protein. The Umeå group infected HeLa cells with *Y. pseudotuberculosis* and assayed the culture supernatant fluids as well as the cytosol of eukaryotic cells for the presence of the YopE protein by immunoprecipitation. The YopE protein was detected in the HeLa cell extracts but not in the L-15 culture medium (Rosqvist *et al.*, 1994). The same findings were obtained in the case of YopH, assayed by its PTPase activity; more than 99% of the total YopH protein secreted was found inside HeLa cells (Persson *et al.*, 1995), indicating that the effector Yops are injected into HeLa cells without any significant leakage into the culture medium. The same targeting phenomenon was also observed in Louvain with Yop-Cya hybrids and *Y. enterocolitica* but the results were less clear-cut; between only 20% (for YopM-Cya hybrid) to 70% (for YopO-Cya hybrid) of the reporters was targeted into the eukaryotic cells, while the remaining part was lost in the medium (Boland *et al.*, 1996; M-P. Sory *et al.*, unpublished data). The reason for the discrepancy between the findings with *Y. pseudotuberculosis* and *Y. enterocolitica* and for the variations between Yops are unknown. However, in the absence of eukaryotic cells, very little of the Yops was released from the bacteria into the medium, which leads to the conclusion that the actual signal which triggers Yop secretion *in vivo* would be contact with a eukaryotic cell.

Mutants that are unable to express the YopN protein have a different phenotypic behaviour. They release large amounts of Yops into the eukaryotic cell culture medium while they effect translocation of little of these Yops into eukaryotic cells (Rosqvist *et al.*, 1994; Persson *et al.*, 1995; Boland *et al.*, 1996; M-P. Sory *et al.*, unpublished data); for the Yop-Cya hybrids, the relative amount of reporters directed into the eukaryotic cells drops from 20–70% down to 1–7%. This confirms that there is a control mechanism for delivery and it shows that the YopN protein is involved in this process. The YopN protein blocks release of Yops unless there is close contact between the bacterium and eukaryotic cell. The YopN protein, thus, represents an element controlling the translocation apparatus and release of the effectors. It could be viewed as a stop-valve, anchored to the more external part of the secretion apparatus (Fig. 3). Not surprisingly, YopN-Cya hybrids are not internalized to the cytosol of eukaryotic cells, indicating that the YopN protein is not an effector (Boland *et al.*, 1996).

Since bacteria trigger translocation upon target cell contact, they must interact with a specific cell-surface receptor and respond accordingly. The YopN protein could be an interactive ligand of the bacterium in this process or it could be controlled by another molecule, which is the actual sensor.

Ca^{2+} chelation *in vitro* leads to massive release of Yops, independently of the presence of eukaryotic cells. Bearing in mind the proposed model, this raises a question related to the functioning of the YopN protein. Mutants that are unable to secrete the YopN protein are 'Ca^{2+}-blind', that is they secrete all Yops, save the YopN protein, in the presence as well as in the absence of Ca^{2+} (Forsberg *et al.*, 1991). Thus, control of Yop secretion by Ca^{2+} also involves the YopN protein. Possibly, Ca^{2+}-chelation specifically removes the stop-valve YopN protein and so leads to a totally deregulated release of Yops.

THE YSC SECRETION PATHWAY

Effector and translocator Yops are transported across the two bacterial membranes and the bacterial cell wall by a specialized secretion system that represents the first recognized member of the 'type III' family of secretion systems (Michiels *et al.*, 1990). The secretion apparatus is encoded by about 20 genes clustered in three operons of the pYV plasmid called *virA* (*lcrA*), *virB* (*lcrB*) and *virC* (*lcrC*) (Michiels *et al.*, 1991; Plano & Straley, 1993; Allaoui *et al.*, 1994; Bergman *et al.*, 1994; Fields, Plano & Straley, 1994; Allaoui, Schulte & Cornelis, 1995*b*).

Only a few of these gene products have been characterized so far. Locus *virA* encodes the LcrD protein, an inner membrane protein with eight membrane-spanning domains and a cytoplasmic C-terminal tail (Plano & Straley, 1993). Locus *virB*, comprising the eight *yscN-yscU* genes, encodes at least two other inner membrane proteins (YscR and YscU) and an ATP-binding protein called YscN (Allaoui *et al.*, 1994; Bergman *et al.*, 1994; Fields *et al.*, 1994; Woestyn *et al.*, 1994). Among the thirteen *virC* gene products, there is an outer membrane protein (YscC), a lipoprotein (YscJ), and a secreted Yop (YopR, the *yscH* gene product) (Michiels *et al.*, 1991; Rimpiläinen *et al.*, 1992; Allaoui *et al.*, 1995*b*). The YscC protein has homologues not only in other type III secretion systems but also in the main terminal branch of the general secretory pathway (GspD family) and in the extrusion system of the filamentous phages (protein PIV). By analogy with these homologues, the YscC protein presumably assembles into very stable multimers and forms a large gated channel in the outer membrane (Russel, 1994). It presumably requires a lipoprotein chaperone for its stability and insertion into the outer membrane, as was recently shown for the PulD protein (Hardie, Lory & Pugsley, 1996). The outer membrane lipoprotein VirG, which is required for secretion of the YopB and YopD translocators (Allaoui *et al.*, 1995*a*), could be this chaperone. Since VirG belongs to the secretion apparatus, a more appropriate designation might be the YscW protein. Gene *yscW (virG)* is localized between the *virB* and *virC* operons, close to the *virF* gene.

Some of the products encoded by the *virA* and *virC* operons behave as Yops, in the sense that they are released when Ca^{2+} ions are chelated. In particular, the *virA* operon encodes not only the LcrD protein (inner membrane) but also the YopN sensor (see above). Similarly, the *virC* operon encodes the YopR protein (Allaoui *et al.*, 1995*b*). This raises some questions about the independence of the secretion and delivery systems and suggests that both systems are intimately coupled at the bacterial surface.

Homologues of the Ysc proteins, as well as of the LcrD protein have been identified in the flagellum secretion and assembly apparatus (Albertini *et al.*, 1991), suggesting that the Yop secretion and delivery systems might have evolved from that involved in flagellum biogenesis. Thus, yersiniae, which are motile bacteria, possess two closely related secretion machineries (Kapatral & Minnich, 1995). Simultaneous expression of both probably needs to be prevented by appropriate switches. Not surprisingly, production of flagella only occurs below 30 °C and requires sigma factor 28 of the RNA polymerase, whereas Yop secretion only occurs at 37 °C, independently of sigma factor 28 (Iriarte *et al.*, 1995; Kapatral & Minnich, 1995).

MODULAR STRUCTURE OF THE YOP EFFECTORS

The signal required to secrete a Yop is localized in the N-terminal region of the Yop protein but it does not have the features of a classical signal peptide and is not cleaved off during secretion (Forsberg & Wolf-Watz, 1988; Michiels *et al.*, 1990; Michiels & Cornelis, 1991). The analysis of a panoply of *yopE'-cya A'* and *yopH'-cyaA'* deletion mutants by Sory *et al.* (1995) showed that this signal is contained within the first 15 and 17 residues of the

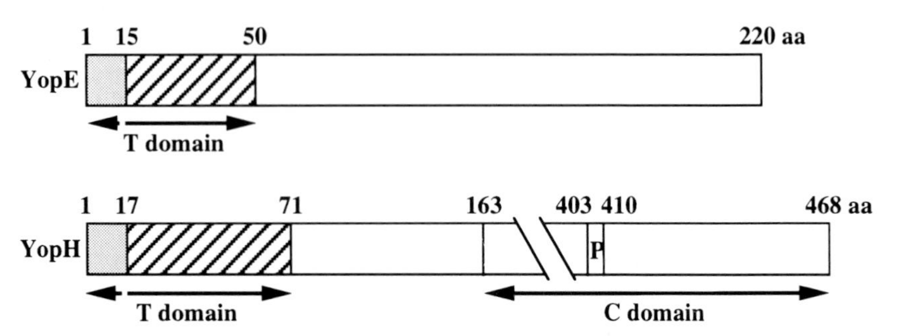

Fig. 4. Structure of the YopE and YopH effectors. S = secretion domain; T = translocation domain; C = catalytic domain of the YopH effector including the P-loop (P). (Reprinted from Sory *et al.*, 1995; by permission.)

YopE and YopH effectors, respectively. The secretion signal is also contained within the 40 N-terminal residues of the YopM effector, within the 47 N-terminal residues of the the YopN stop-valve protein (Boland *et al.*, 1996), within the 68 N-terminal residues of the LcrV protein (A. Boland & G.R. Cornelis, unpublished data) and within the 77 N-terminal residues of the large YopO effector (M-P. Sory *et al.*, unpublished data). Surprisingly, comparison of the N-terminal regions of all of these Yops did not allow definition of either a consensus sequence or a common structural motif.

The effector Yops that need to be internalized into a eukaryotic cell must also be recognized by the delivery apparatus. Does this recognition involve a discrete domain of the effector Yops? The Yop-Cya strategy has been applied to define this putative translocation signal on the YopE and YopH effectors. A panoply of *yop-cyaA′* hybrid genes was constructed by gradual deletions starting from the junction between the *yopE′* and *cyaA′* genes or the *yopH′* and *cyaA′* genes and secretion and internalization of the hybrid proteins was monitored (Sory *et al.*, 1995). It turned out that internalization into cultured macrophages, revealed by cAMP production, only required the 50 N-terminal amino acid residues of the YopE effector and the 71 N-terminal amino acid residues of the YopH effector. YopE and YopH effectors are thus modular proteins composed of a secretion domain, a translocation domain and an effector domain (Sory *et al.*, 1995) (Fig. 4). For the YopM effector, the translocation signal is not yet defined so accurately: it ends between amino acid residues 40 (partial translocation) and 100 (full translocation) (Boland *et al.*, 1996). The same general structure is also found for the YopO/YpkA effector; the translocation signal is included in the 77 N-terminal amino acids of the YopO protein (M.-P. Sory *et al.*, unpublished data).

The fact that the secretion signal is immediately followed by the translocation signal in the N-terminal domain of the effector Yops suggests that the two processes are coupled and that the Yops are probably delivered in a unfolded form and subsequently properly folded inside the eukaryotic cytosol. This, in my view, might well explain why these Yops are insoluble in the bacterial culture medium.

THE SYC CYTOSOLIC BODYGUARDS

Secretion of several Yops requires individual cytosolic chaperones called 'Syc' proteins for '*S*pecific *Y*op *c*haperone' (for review see Wattiau, Woestyn & Cornelis, 1996). Three Syc proteins have been described so far. The SycE chaperone specifically assists YopE secretion (Wattiau & Cornelis, 1993), the SycH protein is the chaperone of the YopH effector and the SycD chaperone is required for secretion of both the YopB and YopD translocators (Wattiau *et al.*, 1994). In the absence of the chaperone SycE, the YopE effector is no longer secreted and is degraded by proteases (Frithz-Lindsten *et al.*,1995). In contrast, disruption of the *sycH* gene causes a severe reduction in the amount of extracellular YopH protein and a substantial increase in YopH protein in the cytosol (Wattiau *et al.*, 1994; Persson *et al.*, 1995).

In *Y. enterocolitica*, the *syc* genes are localized immediately next to the gene encoding their cognate Yop but in *Y. pseudotuberculosis*, a genetic reshuffling has separated the *sycH* gene from the *yopH* gene (Persson *et al.*, 1995). The SycD chaperone is encoded by the *lcrH* gene, localized immediately upstream the *yopB* and *yopD* genes in the *lcrGVHyopBD* operon.

The Syc proteins specifically bind their cognate Yop partner (Wattiau & Cornelis, 1993, Wattiau *et al.*, 1994). Binding of the SycE chaperone to the YopE effector can be readily monitored by affinity chromatography using the YopE-Cya hybrid proteins (Sory & Cornelis, 1994; Sory *et al.*, 1995). By the analysis of a series of $YopE_n$-Cya hybrid proteins (where n corresponds to the number of YopE amino acid residues) as well as deletion derivatives, Woestyn *et al.* (1996) observed that there is only one Syc-binding domain on the YopE protein and that this domain is localized between amino acid residues 15 and 50. Similarly, the YopH effector contains a unique SycH-binding domain localized between amino acid residues 20 and 70 (Woestyn *et al.*, 1996). Since these binding sites are localized outside the secretion signal, the Syc chaperones cannot act as secretion pilots as hypothesized initially. This conclusion is reinforced by the observation that YopH and YopE derivatives deprived of the Syc-binding site maintain a normal secretion level (Woestyn *et al.*, 1996). The latter observation also indicates that it is the binding domain that makes the Syc chaperone necessary for Yop secretion, which suggests that the Syc-binding domain of Yops might interact with another protein in the absence of the Syc chaperone and so hinder Yop secretion. The SycH and SycE chaperones could, thus, act as anti-association factors, a role reminescent of that of the PapD chaperone which caps the pilin subunits in the periplasm and, in doing so, impedes non-productive interactions (Hultgren, Normark & Abraham, 1991). Since the SycE- and SycH-binding domains match the domains required for translocation of the YopE and YopH effectors across eukaryotic cell membranes as determined by Sory *et al.* (1995), the chaperones probably prevent premature interaction of the Yop translocation domains with the translocation apparatus. This hypothesis, however, awaits demonstration.

The SycD chaperone appears to be slightly different from the SycE and SycH chaperones in that it protects more than one Yop, namely, the SycD chaperone binds not only the YopD translocator (Wattiau *et al.*, 1994) but also the YopB translocator (C. Neyt & G.R. Cornelis, unpublished data). Since the YopB and YopD proteins are both required for translocation of all the effectors, they are likely to associate in order to build the delivery apparatus. One hypothesis might be that the SycD chaperone prevents the intracellular association of the YopB and YopD translocators. This role is very reminiscent of the IpgC protein from *Shigella flexneri* that prevents premature oligomerization of the IpaB and IpaC proteins (Menard *et al.*, 1994).

The three Syc proteins are dissimilar in terms of amino acid sequence but they share a few common properties, namely, an acidic pI (around 4.5), a molecular mass in the range of 15–20 kDa, a putative amphipathic α-helix in the C-terminal portion, and the capacity to bind their cognate Yop. The SycE chaperone forms a homodimer (Wattiau & Cornelis, 1993).

REGULATION OF THE YOP VIRULON SYSTEM

Most, if not all, of the genes involved in Yop synthesis and delivery are organized as a single regulon placed under a dual transcriptional control. The first regulatory level puts the genes on the alert when the environment is ideal and the temperature reaches 37 °C, while the second regulatory level prevents their full expression as long as the secretion apparatus is not operating.

The first control mechanism involves the temperature-influenced interplay between the VirF protein, a transcriptional activator of the AraC family (Cornelis *et al.*, 1989), and chromatin structure. DNaseI footprinting experiments carried out on four *yop* genes showed that the VirF protein binds to a 40 bp region localized immediately upstream from the RNA polymerase-binding site. These VirF-binding sequences are located in an AT-rich region and comprise two sites, each containing a 13 bp consensus sequence, either alone or invertedly repeated (Wattiau & Cornelis, 1994). The VirF protein activates transcription at 37 °C but not at low temperature, suggesting that chromatin structure influences the susceptibility of *yop* gene promoters to VirF activation. Rohde, Fox & Minnich (1994) confirmed that temperature alters DNA supercoiling and DNA bending in *Y. enterocolitica*, which leads to the hypothesis that temperature dislodges a repressor, perhaps the histone-like protein YmoA (Cornelis *et al.*, 1991), bound on promoter regions of VirF-sensitive genes and of some other thermoregulated genes.

The second control mechanism is feedback inhibition of Yop synthesis in the absence of Yop secretion. It has been known for a long time that, *in vitro*, transcription of the *yop* genes is down-regulated in the presence of Ca^{2+} ions (Straley & Bowmer, 1986; Cornelis, Vanooteghem & Sluiters, 1987; Forsberg & Wolf-Watz, 1988) or when the secretion apparatus is destroyed by mutation (Michiels *et al.*, 1991). These observations at first created persistent confusion between a regulatory network and the secretion system but ultimately led to the hypothesis of feed-back inhibition in the absence of Yop secretion (Cornelis *et al.*, 1987). Recently, the Umeå group elegantly demonstrated that contact with eukaryotic target cells, which allows release of Yop effectors from the bacterial cytosol into another compartment, also stimulates transcription of the *yopE* effector gene (Petterson *et al.*, 1996). Feedback inhibition thus presumably also exists *in vivo*. By analogy with the secreted anti-sigma factor involved in regulation of flagellum synthesis (Hughes *et al.*, 1993; Kutsukake, 1994), the most likely hypothesis is that feedback inhibition is mediated by an inhibitor that is normally expelled via the Yop secretion apparatus. The Umeå group suggested that the LcrQ protein, encoded downstream from the *virC* operon, could be this hypothetical negative regulator because overproduction of this 12 kDa protein abolished Yop production (Rimpiläinen *et al.*, 1992). Although this hypothesis is undoubtedly very appealing, it has not as yet been formally confirmed. One should also bear in mind that a few genes from the pYV plasmid have not yet been characterized.

THE pYV PLASMID

The system described above is highly conserved in *Y. enterocolitica, Y. pestis* and *Y. pseudotuberculosis*: most of the elements have been characterized in the three species and no significant differences have been identified. Although the genes are very conserved, the pYV plasmid in the three species suffered some reshuffling. On completion of the nucleotide sequence analysis of the *Y. enterocolitica* O:9 pYV plasmid, a potentially interesting evolutionary clue was revealed. Genes encoding Yop synthesis and secretion are tightly clustered in three quadrants of the pYV plasmid, while the fourth quadrant contains a class II transposon that confers arsenite and arsenate resistance (Neyt *et al.*, 1997) (Fig. 1). The *ars* operon is present in the pYV plasmids of all low virulence *Y. enterocolitica* strains tested. It was not detected in the pYV plasmid of the more virulent American strains of *Y. enterocolitica* and in the pYV plasmids from *Y. pseudotuberculosis* and *Y. pestis* (Neyt *et al.*, 1997). This suggests that the low virulence strains, which are distributed worldwide, constitute a single phylum that probably emerged more recently. One may wonder whether the acquisition of the *ars* genes coincided with the conquest of a new ecological habitat that ensured the recent worldwide spread of these low virulence strains. At the present time, pigs represent the major reservoir of pathogenic strains of *Y. enterocolitica* and contaminated pork meat is recognized as the major human source of infection (Tauxe *et al.*, 1987). One may, thus, wonder whether the *ars* transposon favoured the settling of a strain of *Y. enterocolitica* in pigs. In this regard, one should remember that arsenic compounds were largely used before World War II as therapeutic agents to treat pigs infected with *Serpulina hyodysenteriae*.

PROSPECTS

The Yop virulon constitutes a new and sophisticated type of bacterial weapon. Although the mechanism is now understood in its broad lines, many interesting questions remain regarding the Ysc secretion apparatus and the exact structure of the delivery apparatus. The role of the effectors is no less fascinating. The study of the exact role of this fifth column could indeed lead to a better understanding of the cellular processes that are sabotaged. Since the Yop virulon appears to be the archetype of a family of related systems encountered in other pathogenic Gram-negative bacteria, its understanding is of broad interest. The system has also interesting laboratory applications, namely *Yersinia* possessing the secretion/translocation apparatus but devoid of effectors could be engineered to introduce various proteins into eukaryotic cells. The efficacy of the system is illustrated by translocation of the adenylate cyclase. One could even think of using the system to deliver DNA into eukaryotic cells.

The pYV plasmid, which encodes the Yop virulon system, appears to be a particularly accomplished genetic element; the genes are organized according to their function (secretion, translocators, effectors) in very compact operons, and there is little non-coding sequence between these operons. This high degree of organization fits very well with the fact that yersiniae are probably the oldest pathogenic bacteria known in human history. The pYV plasmid is, nevertheless, quite striking if one considers that it results from the assembly of components presumably of eukaryotic origin (the effector-encoding genes) and of typical prokaryotic elements (type III secretion genes, translocator-encoding genes).

ACKNOWLEDGEMENTS

I thank Marie-Paule Sory for a critical reading of this manuscript and Maite Iriarte for the recalibrated, unpublished, genetic map. This research was supported by grants from the Belgian FRSM (Fonds National de la Recherche Scientifique Médicale, convention 3.4627.93) and the 'Actions de Recherche Concertées' 94/99–172 from the 'Direction générale de la Recherche Scientifique – Communauté Française de Belgique'.

REFERENCES

Albertini, A. M., Caramori, T., Crabb, W. D., Scoffone, F. & Galizzi, A. (1991). The *flaA* locus of *Bacillus subtilis* is part of a large operon coding for flagellar structures, motility functions, and an ATPase-like polypeptide. *Journal of Bacteriology*, **173**, 3573–9.

Allaoui, A., Scheen, R., Lambert de Rouvroit, C. & Cornelis, G. R. (1995a). VirG, a *Yersinia enterocolitica* lipoprotein involved in Ca^{2+} dependency, is related to ExsB of *Pseudomonas aeruginosa. Journal of Bacteriology*, **177**, 4230–7.

Allaoui, A., Schulte, R. & Cornelis, G. R. (1995b). Mutational analysis of the *Yersinia enterocoliticia virC* operon: characterization of *yscE, F, G, I, J, K* required for Yop secretion and *yscH* encoding YopR. *Molecular Microbiology*, **18**, 343–55.

Allaoui, A., Woestyn, S., Sluiters, C. & Cornelis, G. R. (1994). YscU, a *Yersinia enterocolitica* inner membrane protein involved in Yop secretion. *Journal of Bacteriology*, **176**, 4534–42.

Alouf, J. E. & Freer, J. H. (1991). *Sourcebook of Bacterial Protein Toxins*. Academic Press London.

Bergman, T., Erickson, K., Galyov, E., Persson, C. & Wolf-Watz, H. (1994). The *lcrB (yscN/U)* gene cluster of *Yersinia pseudotuberculosis* is involved in Yop secretion and shows high homology to the *spa* gene clusters of *Shigella flexneri* and *Salmonella typhimurium. Journal of Bacteriology*, **176**, 2619–26.

Bliska, J. B. & Black, D. S. (1995). Inhibition of the Fc receptor-mediated oxidative burst in macrophages by the *Yersinia pseudotuberculosis* tyrosine phosphatase. *Infection and Immunity*, **63**, 681–5.

Bliska, J. B., Guan, K. L., Dixon, J. E. & Falkow, S. (1991). Tyrosine phosphate hydrolysis of host proteins by an essential *Yersinia* virulence determinant. *Proceedings of the National Academy of Sciences, USA*, **88**, 1187–91.

Boland, A., Sory, M.-P., Iriarte, M., Kerbouch, C., Wattiau, P. & Cornelis, G. R. (1996). Status of YopM and YopN in the *Yersinia* Yop virulon: YopM of *Y. enterocolitica* is internalized inside the cytosol of PU5–1.8 macrophages by the YopB, D, N delivery apparatus. *EMBO Journal*, **15**, 5191–201.

Bölin, I., Forsberg, Å., Norlander, L., Skurnik, M. & Wolf-Watz, H. (1988). Identification and mapping of the temperature-inducible, plasmid-encoded proteins of *Yersinia* spp. *Infection and Immunity*, **56**, 343–8.

Burrows, T. W. & Bacon, G. A. (1956). The basis of virulence in *Pasteurella pestis*: an antigen determining virulence. *British Journal of Experimental Pathology*, **37**, 481–93.

China, B., N'Guyen, B. T., de Bruyere, M. & Cornelis, G. R. (1994). Role of YadA in resistance of *Yersinia enterocolitica* to phagocytosis by human polymorphonuclear leukocytes. *Infection and Immunity*, **62**, 1275–81.

Cornelis, G. R., Sluiters, C., Delor, I., Geib, D., Kaniga, K., Lambert de Rouvroit, C., Sory, M.-P., Vanooteghem, J.-C. & Michiels, T. (1991). *ymoA*, a *Yersinia enterocolitica* chromosomal gene modulating the expression of virulence functions. *Molecular Microbiology*, **5**, 1023–34.

Cornelis, G. R., Sluiters, C., Lambert de Rouvroit, C. & Michiels, T. (1989). Homology between *virF*, the transcriptional activator of the *Yersinia* virulence regulon, and AraC, the *Escherichia coli* arabinose operon regulator. *Journal of Bacteriology*, **171**, 254–62.

Cornelis, G. R., Vanootegem, J.-C. & Sluiters, C. (1987). Transcription of the *yop* regulon from *Y. enterocolitica* requires trans acting pYV and chromosomal genes. *Microbial Pathogenesis*, **2**, 367–79.

Fällman, M., Andersson, K., Håkansson, S., Magnusson, K.-E., Stendahl, O. &

Wolf-Watz, H. (1995). *Yersinia pseudotuberculosis* inhibits Fc receptor-mediated phagocytosis in J774 cells. *Infection and Immunity*, **63**, 3117–24.

Fields, K. A., Plano, G. V. & Straley, S. C. (1994). A low-Ca^{2+} response (LCR) secretion (*ysc*) locus lies within the *lcrB* region of the LCR plasmid in *Yersinia pestis*. *Journal of Bacteriology*, **176**, 569–79.

Forsberg, Å., Viitanen, A-M., Skurnik, M. & Wolf-Watz, H. (1991). The surface-located YopN protein is involved in calcium signal transduction in *Yersinia pseudotuberculosis*. *Molecular Microbiology*, **5**, 977–86.

Forsberg, Å. & Wolf-Watz, H. (1988). The virulence protein Yop5 of *Yersinia pseudotuberculosis* is regulated at transcriptional level by plasmid-pIBI-encoded *trans*-acting elements controlled by temperature and calcium. *Molecular Microbiology*, **2**, 121–33.

Frithz-Lindsten, E., Rosqvist, R., Johansson, L. & Forsberg, Å. (1995). The chaperone-like protein YerA of *Yersinia pseudotuberculosis* stabilizes YopE in the cytoplasm but is dispensible for targeting to the secretion loci. *Molecular Microbiology*, **16**, 635–47.

Galyov, E. E., Håkansson, S., Forsberg, Å. & Wolf-Watz, H. (1993). A secreted protein kinase of *Yersinia pseudotuberculosis* is an indispensable virulence determinant. *Nature*, **361**, 730–2.

Gemski, P., Lazere, J. R., Casey, T. & Wohlhieter, J. A. (1980). Presence of a virulence-associated plasmid in *Yersinia pseudotuberculosis*. *Infection and Immunity*, **28**, 1044–7.

Ginocchio, C. C., Olmsted, S. B., Wells, C. L. & Galan, J. E. (1994). Contact with epithelial cells induces the formation of surface appendages on *Salmonella typhimurium*. *Cell*, **76**, 717–24.

Guan, K. L. & Dixon, J. E. (1990). Protein tyrosine phosphatase activity of an essential virulence determinant in *Yersinia*. *Science*, **249**, 553–6.

Håkansson, S., Bergman, T., Vanooteghem, J-C., Cornelis, G. R. & Wolf-Watz, H. (1993). YopB and YopD constitute a novel class of *Yersinia* Yop proteins. *Infection and Immunity*, **61**, 71–80.

Håkansson, S., Galyov, E. E., Rosqvist, R. & Wolf-Watz, H. (1996). The *Yersinia* YpkA Ser/Thr kinase is translocated and subsequently targeted to the inner surface of the HeLa cell plasma membrane. *Molecular Microbiology*, **20**, 593–603.

Hanski, C., Kutschka, U., Schmoranzer, H. P., Naumann, M., Stallmach, A., Hahn, H., Menge, H. & Riecken, E. O. (1989). Immunohistochemical and electron microscopic study of interaction of *Yersinia enterocolitica* serotype O8 with intestinal mucosa during experimental enteritis. *Infection and Immunity*, **57**, 673–8.

Hardie, K. R., Lory, S. & Pugsley, A. P. (1996). Insertion of an outer membrane protein in *Escherichia coli* requires a chaperone-like protein. *EMBO Journal*, **15**, 978–88.

Hartland, E. L., Green, S. P., Phillips, W. A. & Robins-Browne, R. M. (1994). Essential role of YopD in inhibition of the respiratory burst of macrophages by *Yersinia enterocolitica*. *Infection and Immunity*, **62**, 4445–53.

Heesemann, J., Gross, U., Schmidt, N. & Laufs, R. (1986). Immunochemical analysis of plasmid-encoded proteins released by enteropathogenic *Yersinia* sp. grown in calcium-deficient media. *Infection and Immunity*, **54**, 561–7.

Higuchi, K. & Smith, J. L. (1961). Studies on the nutrition and physiology of *Pasteurella pestis*. VI. A differential plating medium for the estimation of the mutation rate to avirulence. *Journal of Bacteriology*, **81**, 605–8.

Hughes, K. T., Gillen, K. L., Semon, M. J. & Karlinsey, J. E. (1993). Sensing structural intermediates in bacterial flagellar assembly by export of a negative regulator. *Science*, **262**, 1277–80.

Hultgren, S. J., Normark, S. & Abraham, S. N. (1991). Chaperone-assisted assembly and molecular architecture of adhesive pili. *Annual Review of Microbiology*, **45**, 383–415.

Iriarte, M., Stainier, I., Mikulskis, A. V. & Cornelis, G. R. (1995). The *fliA* gene encoding sigma 28 in *Yersinia enterocolitica. Journal of Bacteriology*, **177**, 2299–304.

Kapatral, V. & Minnich, S. A. (1995). Co-ordinate, temperature-sensitive regulation of the three *Yersinia enterocolitica* flagellin genes. *Molecular Microbiology*, **17**, 49–56.

Kutsukake, K. (1994). Excretion of the anti-sigma factor through a flagellar substructure couples flagellar gene expression with flagellar assembly in *Salmonella typhimurium. Molecular and General Genetics*, **243**, 605–12.

Leung, K. Y. & Straley, S. C. (1989). The *yopM* gene of *Yersinia pestis* encodes a released protein having homology with the human platelet surface protein GPIb alpha. *Journal of Bacteriology*, **171**, 4623–32.

Menard, R., Sansonetti, P., Parsot, C. & Vasselon, T. (1994). Extracellular association and cytoplasmic partitioning of the IpaB and IpaC invasins of *S. flexneri. Cell*, **79**, 515–25.

Michiels, T. & Cornelis, G.R. (1991). Secretion of hybrid proteins by the *Yersinia* Yop export system. *Journal of Bacteriology*, **173**, 1677–85.

Michiels, T., Vanooteghem, J-C., Lambert de Rouvroit, C., China, B., Gustin, A., Boudry, P. & Cornelis, G. R. (1991). Analysis of *virC*, an operon involved in the secretion of Yop proteins by *Yersinia enterocolitica. Journal of Bacteriology*, **173**, 4994–5009.

Michiels, T., Wattiau, P., Brasseur, R., Ruysschaert, J-M. & Cornelis, G. R.(1990). Secretion of Yop proteins by yersiniae. *Infection and Immunity*, **58**, 2840–9.

Mulder, B., Michiels, T., Simonet, M., Sory, M.-P. & Cornelis, G. (1989). Identification of additional virulence determinants on the pYV plasmid of *Yersinia enterocolitica* W227. *Infection and Immunity*, **57**, 2534–41.

Neyt, C., Iriarte, M., Ha Thi, V. & Cornelis, G. R. (1997). Virulence and arsenic resistance in yersiniae. *Journal of Bacteriology*, **179**, in press.

Persson, C., Nordfelth, R., Holmström, A., Håkansson, S., Rosqvist, R. & Wolf-Watz, H. (1995). Cell-surface-bound *Yersinia* translocate the protein tyrosine phosphatase YopH by a polarized mechanism into the target cell. *Molecular Microbiology*, **18**, 135–50.

Pettersson, J., Nordfelth, R., Dubinina, E., Bergman, T., Gustafsson, M., Magnusson, K.E. & Wolf-Watz, H. (1996). Modulation of virulence factor expression by pathogen target cell contact. *Science*, **273**, 1231–3.

Plano, G. V. & Straley, S. C. (1993). Multiple effects of *lcrD* mutations in *Yersinia pestis. Journal of Bacteriology*, **175**, 3536–45.

Portnoy, D. A., Moseley, S. L. & Falkow, S. (1981). Characterization of plasmids and plasmid-associated determinants of *Yersinia enterocolitica* pathogenesis. *Infection and Immunity*, **31**, 775–82.

Reisner, B. S. & Straley, S. C. (1992). *Yersinia pestis* YopM: thrombin binding and overexpression. *Infection and Immunity*, **60**, 5242–52.

Rimpiläinen, M., Forsberg, Å. & Wolf-Watz, H. (1992). A novel protein, LcrQ, involved in the low-calcium response of *Yersinia pseudotuberculosis* shows extensive homology to YopH. *Journal of Bacteriology*, **174**, 3355–63.

Rohde, J. R., Fox, J. M. & Minnich, S. A. (1994). Thermoregulation in *Yersinia enterocolitica* is coincident with changes in DNA supercoiling. *Molecular Microbiology*, **12**, 187–99.

Rosqvist, R., Bölin, I. & Wolf-Watz, H. (1988). Inhibition of phagocytosis in *Yersinia pseudotuberculosis*: a virulence plasmid-encoded ability involving the Yop2b protein. *Infection and Immunity*, **56**, 2139–43.

Rosqvist, R., Forsberg, Å., Rimpiläinen, M., Bergman, T. & Wolf-Watz, H. (1990). The cytotoxic protein YopE of *Yersinia* obstructs the primary host defence. *Molecular Microbiology*, **4**, 657–67.

Rosqvist, R., Forsberg, Å. & Wolf-Watz, H. (1991). Intracellular targeting of the *Yersinia* YopE cytotoxin in mammalian cells induces actin microfilament disruption. *Infection and Immunity*, **59**, 4562–9.

Rosqvist, R., Håkansson, S., Forsberg, Å. & Wolf-Watz, H. (1995). Functional conservation of the secretion and translocation machinery for virulence proteins of yersiniae, salmonellae and shigellae. *EMBO Journal*, **14**, 4187–95.

Rosqvist, R., Magnusson, K.-E. & Wolf-Watz, H. (1994). Target cell contact triggers expression and polarized transfer of *Yersinia* YopE cytotoxin into mammalian cells. *EMBO Journal*, **13**, 964–72.

Ruckdeschel, K., Roggenkamp, A., Schubert, S. & Heesemann, J. (1996). Differential contribution of *Yersinia enterocolitica* virulence factors to evasion of microbicidal action of neutrophils. *Infection and Immunity*, **64**, 724–33.

Russel, M. (1994). Phage assembly: a paradigm for bacterial virulence factor export? *Science*, **265**, 612–4.

Schulte, R., Wattiau, P., Hartland, E. L., Robins-Browne, R. M. & Cornelis, G. R. (1996). Differential secretion of interleukin-8 by human epithelial cell lines upon entry of virulent or nonvirulent *Yersinia enterocolitica*. *Infection and Immunity*, **64**, 2106–13.

Simonet, M., Richard, S. & Berche, P. (1990). Electron microscopic evidence for *in vivo* extracellular localization of *Yersinia pseudotuberculosis* harboring the pYV plasmid. *Infection and Immunity*, **58**, 841–5.

Sory, M-P., Boland, A., Lambermont, I. & Cornelis, G. R. (1995). Identification of the YopE and YopH domains required for secretion and internalization into the cytosol of macrophages, using the *cyaA* gene fusion approach. *Proceedings of the National Academy of Sciences, USA*, **92**, 11998–2002.

Sory, M-P. & Cornelis, G. R. (1994). Translocation of a hybrid YopE-adenylate cyclase from *Yersinia enterocolitica* into HeLa cells. *Molecular Microbiology*, **14**, 583–94.

Starnbach, M. N. & Bevan, M. J. (1994). Cells infected with *Yersinia* present an epitope to class I MHC-restricted CTL. *Journal of Immunology*, **153**, 1603–12.

Straley, S. C. & Bowmer, W. S. (1986). Virulence genes regulated at the transcriptional level by Ca^{2+} in *Yersinia pestis* include structural genes for outer membrane proteins. *Infection and Immunity*, **51**, 445–54.

Straley, S. C. & Brubaker, R. R. (1981). Cytoplasmic and membrane proteins of yersiniae cultivated under conditions simulating mammalian intracellular environment. *Proceedings of the National Academy of Sciences, USA*, **78**, 1224–8.

Tauxe, R. V., Vandepitte, J., Wauters, G., Martin, S. M., Goossens, V., de Mol, P., van Noyen, R. & Thiers, G. (1987). *Yersinia enterocolitica* infections and pork: the missing link. *Lancet*, **ii**, 1129–32.

Visser, L. G., Annema, A. & van Furth, R. (1995). Role of Yops in inhibition of phagocytosis and killing of opsonized *Yersinia enterocolitica* by human granulocytes. *Infection and Immunity*, **63**, 2570–5.

Wattiau, P., Bernier, B., Deslee, P., Michiels, T. & Cornelis, G. R. (1994). Individual chaperones required for Yop secretion by *Yersinia*. *Proceedings of the National Academy of Sciences, USA*, **91**, 10493–7.

Wattiau, P. & Cornelis, G. R. (1993). SycE, a chaperone-like protein of *Yersinia enterocolitica* involved in the secretion of YopE. *Molecular Microbiology*, **8**, 123–31.

Wattiau, P. & Cornelis, G. R. (1994). Identification of DNA sequences recognized by VirF, the transcriptional activator of the *Yersinia yop* regulon. *Journal of Bacteriology*, **176**, 3878–84.

Wattiau, P., Woestyn, S. & Cornelis, G. R. (1996). Customized secretion chaperones in pathogenic bacteria. *Molecular Microbiology*, **20**, 255–62.

Woestyn, S., Allaoui, A., Wattiau, P. & Cornelis, G. R. (1994). YscN, the putative energizer of the *Yersinia* Yop secretion machinery. *Journal of Bacteriology*, **176**, 1561–9.

Woestyn, S., Sory, M-P., Boland, A., Lequenne, O. & Cornelis, G. R. (1996). The cytosolic SycE and SycH chaperones of *Yersinia* protect the region of YopE and YopH involved in translocation across eukaryotic cell membranes. *Molecular Microbiology*, **20**, 1261–71.

Zink, D. L., Feeley, J. C., Wells, J. G., Vanderzant, C., Vickery, J. C., Roof, W. D. & O'Donovan, G. A. (1980). Plasmid-mediated tissue invasiveness in *Yersinia enterocolitica*. *Nature*, **283**, 224–6.

HOST–PATHOGEN PROTEIN–PROTEIN INTERACTIONS IN *STAPHYLOCOCCUS*

T. J. FOSTER, O. HARTFORD AND D. O'CONNELL

Department of Microbiology, Moyne Institute of Preventive Medicine, University of Dublin, Trinity College, Dublin 2, Republic of Ireland

INTRODUCTION

Staphylococcus aureus can cause various infections both in healthy individuals and in hospital patients ranging from superficial skin infections to life-threatening diseases. *S. epidermidis*, once regarded as a non-pathogenic commensal of human skin, is now the major cause of infections associated with indwelling medical devices. *S. aureus* also causes a significant number of device-related infections, often with more severe consequences than *S. epidermidis*. Haematogenous spread of *S. aureus* from a superficial infection can lead to infections at sites such as the heart (endocarditis), bone (osteomyelitis) and joints (septic arthritis) or may develop into septicaemia.

S. aureus expresses several proteins that interact with host proteins. These can be divided into two broad categories: (i) the extracellular proteins coagulase and staphylokinase which are stoichiometric activators that bind prothrombin and plasminogen, respectively, and activate these host proteins' protease activities; (ii) bacterial cell surface proteins which can bind to proteins present in plasma or in the extracellular matrix, or which are deposited onto indwelling medical devices. The latter category includes the collagen-, fibronectin- and fibrinogen-binding proteins which have been studied in some detail and which will be the main focus of this article. *S. aureus* can bind to a wide range of other host proteins but the molecular bases of these interactions are less well understood. In some cases a bacterial protein has been identified as the ligand-binding determinant.

ATTACHMENT OF PROTEINS TO THE BACTERIAL CELL SURFACE

It is now apparent that there is a class of surface-associated proteins in Gram-positive cocci that attach to the cell wall by a common mechanism (Fig. 1). They can be recognized by amino acid sequences in their C-termini which comprise a motif LPXTG followed by a hydrophobic membrane-spanning region and a positively charged terminus which is thought to stop protein secretion (Schneewind, Model & Fischetti, 1992; Schneewind, Mihaylova-Petkov & Model, 1993). The LPXTG sequence is cleaved between the threonine and glycine residues by a protease called sortase

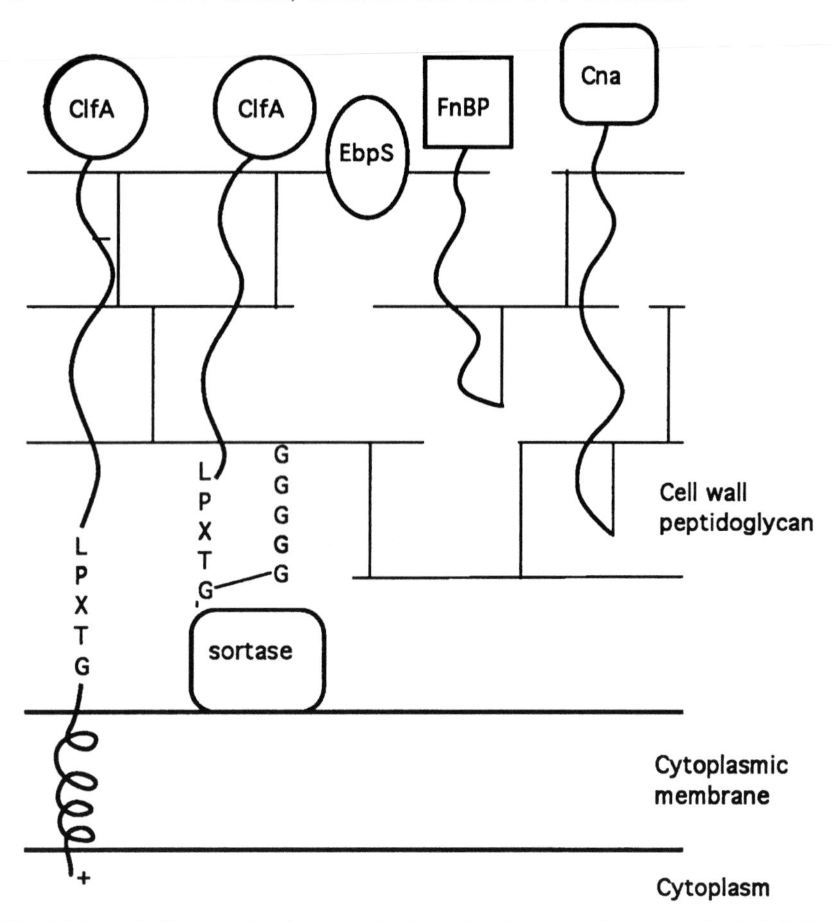

Fig. 1. Schematic diagram of surface-associated proteins. Some proteins are covalently linked to peptidoglycan by sortase: fibrinogen-binding protein, ClfA; fibronectin-binding protein, FnBP; collagen-binding protein, Cna. Others are non-covalently linked to peptidoglycan, for example, elastin-binding protein, EbpS.

(Navarre & Schneewind, 1994). The threonine is then covalently linked to the pentaglycine bridge of uncross-linked peptidoglycan, thereby anchoring the protein to the cell wall. The biologically active N-terminal domain of the peptidoglycan-associated protein is exposed on the bacterial cell surface.

Wall-associated proteins also have a region N-terminal to the LPXTG motif which is often rich in proline and glycine residues and is thought to be a cell wall-spanning domain (region W) (Fig. 2). The length of region W varies from one protein to another. The minimum number of residues necessary for proper exposure of the respective biologically active domains is probably between 70 and 110 depending on the bacterial host and the protein concerned. Studies with deletion mutants of the collagen-binding protein Cna indicate that region W (78 residues) is sufficient to allow normal

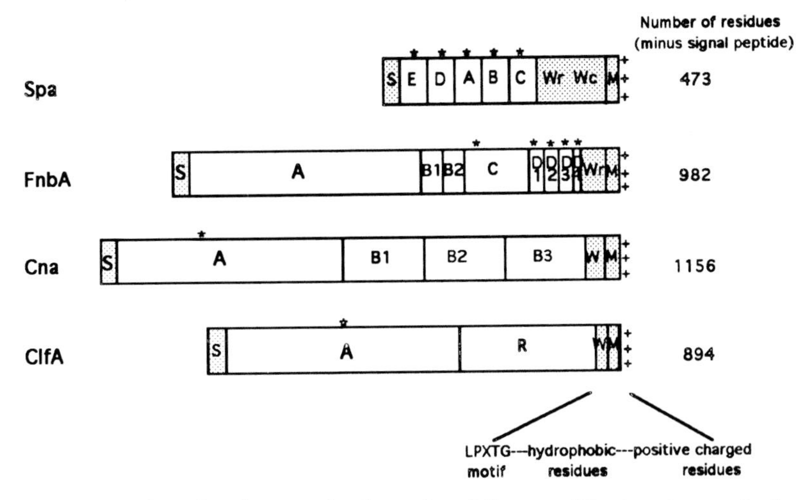

Fig. 2. Organization of surface-associated proteins of *S. aureus*. The domain organization of protein A (Spa), the fibronectin-binding protein A (FnbA), the collagen-binding protein (Cna) and the fibrinogen-binding protein (clumping factor, ClfA). The signal sequences (S) are removed during secretion across the cytoplasmic membrane. Each protein has common features at the C-terminus: a LPXTG motif, membrane anchor (M) and positively charged residues. Region W is the putative peptidoglycan spanning region. * indicates the position of the ligand-binding domains. Reproduced from Foster & McDevitt (1994), with permission of the publisher.

expression of the ligand-binding activity of region A (O. Hartford & T.J. Foster, unpublished data). In contrast, with the fibrinogen-binding protein ClfA at least 110 residues are required between the ligand binding region A and the LPXTG motif to permit bacterial clumping in fibrinogen whereas shorter sequences (50–70 residues) permit binding of soluble fibrinogen without permitting clumping (O. Hartford & T.J. Foster, unpublished data). Surface expression of an enzymatically active lipase from a hybrid lipase-fibronectin-binding protein gene on the surface of *S. carnosus* requires at least 92 residues (Strauss & Gotz, 1996).

Not all surface-exposed proteins in Gram-positive bacteria are linked to peptidoglycan or have a membrane anchor. Neither the 25 kDa elastin-binding protein (Park *et al.*, 1996) nor the 77 kDa MHC class II-adherent protein Map (Jönsson *et al.*, 1995) have sequences equivalent to regions W and M or the LPXTG motif. Map can be released from the cell surface by washing with LiCl. Members of the anchored family of surface proteins can only be released by enzymic degradation of peptidoglycan.

PROTEIN A, AN IMMUNOGLOBULIN-BINDING PROTEIN

Structural analysis

The archetype of the family of cell wall-associated proteins which has a membrane anchor and a LPXTG transpeptidase recognition site is protein A

(Spa). It comprises five (in some isolates four) repeats of 58 residues (Fig. 2), each of which can bind the Fc region of IgG. Structural analysis of repeat unit B (Fig. 2) identified three α-helices in an anti-parallel bundle when complexed with the Fc fragment and demonstrated that two of the α-helices make several contacts with the C_H2 and C_H3 domains of the Fc fragment (Deisenhofer, 1981; Gouda *et al.*, 1992; Jendeberg *et al.*, 1996).

Comparison of the amino acid sequences of the repeats suggests that protein A evolved by tandem duplication followed by genetic drift resulting in a gradient of amino acid similarity between repeats with the outermost repeats being most distantly related and neighbouring repeats being most closely related (Uhlén *et al.*, 1984). Furthermore, α-helices 1 and 2 are highly conserved and the residues in direct contact with the Fc region of IgG are almost completely identical. This suggests strong selective pressure for retention of α-helices and Fc binding.

Role of protein A in virulence

The ability of bacteria to bind immunoglobulins onto their surface by a non-immune mechanism might be expected to help thwart the immune system during infection. The IgG binds to protein A in such a way that Fc receptors on phagocytic cells cannot bind their target in the Fc region of the bound IgG. Also, a coating of IgG might mask binding sites for opsonins present in normal sera. This notion is supported by *in vitro* opsonophagocytosis experiments where a protein A-deficient (Spa^-), site-specific mutant isolated by allele replacement was taken up more efficiently by polymorphonuclear leucocytes than the wild-type strain (Gemmell *et al.*, 1990). The Spa^- mutant was also less virulent that the wild-type strain in infection models in mice (Patel *et al.*, 1987; Foster *et al.*, 1990).

FIBRONECTIN-BINDING PROTEIN

Structural and functional analysis of FnBP–fibronectin interaction

Fibronectin is a dimeric glycoprotein that occurs in soluble form in body fluids and in fibrillar form in the extracellular matrix (Fig. 3) (Proctor, 1987; Hynes, 1993). Its primary biological function is to serve as a substratum for the adhesion of animal cells (Yamada, 1989). This adhesion is mediated by integrin receptors that bind to a specific site in the central part of fibronectin. In contrast, the primary binding site for fibronectin-binding proteins of *S. aureus* is located in the 29 kDa N-terminal domain termed N29 (Sottile *et al.*, 1991). It is composed of five type I modules each of about 45 amino acid residues. The N29 domain is a rigid structure comprising a series of anti-parallel β-sheets stabilized by disulphide bonds (Potts & Campbell, 1994). A

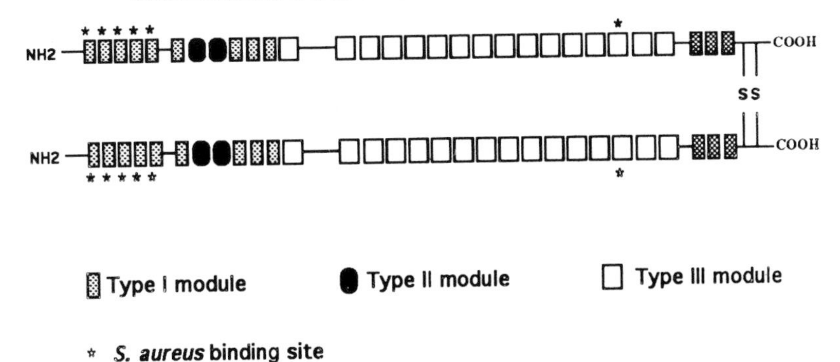

Fig. 3. Structure of fibronectin. The binding sites for *S. aureus* are indicated. (Adapted from Proctor, 1987.)

second weaker binding site for *S. aureus* is located in a C-terminal repeat type III (Bozzini *et al.*, 1992).

Binding of *S. aureus* to fibronectin is mediated by two fibronectin-binding proteins FnBPA and FnBPB (Signas *et al.*, 1989; Jönsson *et al.*, 1991). *S. aureus* strains 8325–4 and Newman have two closely linked genes that express FnBPA and FnBPB. Many strains of human and bovine origin also have two *fnb* genes but some strains have a single *fnb* gene that is closely related to *fnbA* (Greene *et al.*, 1996; T.J. Foster, unpublished data). The N-terminal regions A of the FnBPA and FnBPB proteins share only 45% identical residues but the C-terminal regions C, D, W and M are 95% similar in amino acid sequence (Fig. 2). The functions of regions A of FnBPA and FnBPB are unknown but it is conceivable they could have biological activities. The equivalent region of a FnBP analogue of *Streptococcus pyogenes* has enzymatic activity which gives rise to the opacity phenotype (Kreikemeyer, Talay & Chhatwal, 1995; Rakonjac, Robbins & Fischetti, 1995). Regions W and M have properties consistent with the protein being cell wall-associated (Schneewind *et al.*, 1992, 1993).

The main fibronectin-binding domain in FnBPs is in region D (Fig. 2). It comprises three complete repeats of 38 residues (D1, D2, D3) and a partial repeat, D4. Synthetic peptides corresponding to each individual repeat can inhibit fibronectin–FnBP interactions suggesting that each repeat binds the host protein (Raja, Raucci & Höök, 1990). Repeat D3 has the strongest interaction. A recombinant protein carrying the complete D region is a much more potent inhibitor than any single D peptide (Raja *et al.*, 1990). This suggests that multiple contacts occur between region D ligand-binding domains and fibronectin, and is consistent with the finding that FnBPs recognize type I modules in the N-terminal region of fibronectin. Mutations in any single type I module of a recombinant fibronectin reduced the interaction with FnBPA indicating that each is important in the reaction (Sottile *et al.*, 1991).

Several species of *Streptococcus* express FnBPs which have a similar domain organization to FnBPs of *S. aureus* (Joh *et al.*, 1994; Patti *et al.*, 1994*a*). Recombinant proteins comprising the putative fibronectin-binding repeats of each species blocked interactions between bacteria and fibronectin suggesting that the FnBPs react with identical regions. By comparing the sequences of fibronectin-binding repeats it is possible to identify amino acid residues that are conserved and to suggest a consensus sequence: EDT/S – (X9, 10) – GG – (X3,4) – I/VDF (where X is any residue). Inhibition studies with synthetic peptides have identified the minimum fibronectin-binding region of repeat D3 of *S. aureus* FnBPs between residues 15–36. Chemical modification of the D3 peptide suggested that the EDT motif was important but this was not confirmed using peptide analogues lacking one or more of these residues (McGavin *et al.*, 1991). However, the conserved GG and I/VDF residues were shown to be important by this approach (McGavin *et al.*, 1993).

The N29 domain of fibronectin is a rigid rod-like structure. A recombinant protein comprising the *S. aureus* fibronectin-binding regions D1-D4 or equivalents from other bacterial species lacked discernible secondary structure (House-Pompeo *et al.*, 1996). However, in the presence of fibronectin the D1-D4 protein assumed an ordered structure comprising β-sheets. The flexible unordered binding regions were induced to form a specific conformation by fibronectin binding. This notion is supported by the properties of a monoclonal antibody which only bound to *Streptococcus dysgalactiae* FnBP in the presence of fibronectin. Its epitope on FnBP is presumably conformational and is only formed in the presence of the ligand (Speziale *et al.*, 1996). Ligand-induced conformational changes may be the mechanism whereby the fibronectin-binding site is not exposed to antibodies which could block ligand-receptor interactions.

Role of FnBP in virulence

Binding of *S. aureus* to fibronectin may allow bacteria to colonize various sites in the body, for example, blood clots and damaged tissue, vegetations on damaged heart valves and implanted medical devices that have been conditioned by a coating of host plasma proteins. There is considerable evidence that implanted biomaterial is rapidly coated with host proteins which can promote bacterial attachment and biofilm formation (Vaudaux *et al.*, 1989, 1993). Fibronectin is a significant component of the conditioning layer. *S. aureus* and *S. epidermidis* bound to fibronectin-coated plastic *in vitro*, to *ex vivo* biomaterial and to plasma clots. The latter reaction was, at least in part, dependent on the presence of fibronectin in the clots (Raja *et al.*, 1990; Valentin-Weigand *et al.*, 1993).

Allele replacement mutants of *S. aureus* strain 8325–4 lacking both FnBPA and FnBPB were defective in binding to fibronectin-coated plastic surfaces *in vitro* (Greene *et al.*, 1995). The *fnbA fnbB* double mutant was also defective in adherence to *ex vivo* biomaterial removed from subcutaneous chambers implanted in guinea-pigs, a foreign body model involving long-term host contact with biomaterial (Greene *et al.*, 1995). This indicates that FnBPs contribute to adherence to biomaterial which has become coated with host proteins *in vivo*. In contrast, adhesion to biomaterial which had been in contact with blood for only a short time period (<60 min) was not promoted by FnBPs (Vaudaux *et al.* 1995). Instead the fibrinogen-binding protein was shown to be the major adhesin (see below).

S. aureus can adhere to primary cultures of endothelial cells and can be rapidly internalized. It is possible that this event is important *in vivo* in the initiation of invasive endocarditis of native valves where bacteria could attach to the undamaged endothelium of a heart valve (Tompkins *et al.*, 1990). It has recently been shown that fibrillar fibronectin present on the surface of cultured human umbilical vein endothelial cells promotes bacterial attachment. The *fnbA fnbB* mutant of *S. aureus* strain 8325–4 was defective in attachment and was complemented to a higher level of adherence than the parental strain by a multicopy plasmid expressing elevated levels of either FnBPA or FnBPB (S. Peacock, T.J. Foster & A. Berendt, unpublished data).

In vivo experiments with mutants defective in fibronectin binding have produced conflicting results. Kuypers and Proctor (1989) showed that a mutant of the phage group I strain 879RF defective in fibronectin binding had a markedly reduced ability to adhere to traumatized rat heart valves *in vivo*, indicating that binding to fibronectin is important in initiation of endocarditis. However, no difference in virulence in this model was seen between the *fnbA fnbB* allele replacement mutant of strain 8325–4 and the wild-type strain (Flock *et al.*, 1996). These conflicting data could be ascribed to differences between the parental strains or differences in the animal model. Immunizing rats with FnBP gave significant protection against valve colonization (Schennings *et al.*, 1993), but this could have been due to opsonization rather than blocking adherence.

COLLAGEN-BINDING PROTEIN

Domain structure of Cna

Collagen is the major structural protein of mammals (Olsen & Ninomiya, 1993). The architecture of various tissues is dependent on collagen and is regulated by intricate interactions between different collagen molecules and non-collagenous components (Heinegard & Oldberg, 1989). There are many distinct collagen types. Some have a tissue specific distribution. Thus, type II collagen is preferentially found in cartilage while type IV collagen is found almost exclusively in basement membranes (Miller & Gay, 1983; Olsen & Ninomiya, 1993). Type II collagen fibres are composed of a triple helix with each chain comprising repeating triplets of the amino acids Gly-X-Y where X is often proline and Y is often hydroxyproline.

The ability of *S. aureus* to bind collagen is conferred by a surface-located collagen-binding protein, Cna, which recognizes several types of collagens including types I, II, III and IV (Switalski, Speziale & Höök, 1989). The Cna protein has C-terminal sequences typical of surface-located proteins including an LPXTG motif, a hydrophobic membrane domain and a positively charged C-terminus (Fig. 2; Patti *et al.*, 1992). The collagen-binding domain is located in the N-terminal 504 residue region A (Patti, Bols & Höök, 1993). In the Cna protein isolated from strains Cowan and FDA574 there are three repeats of the 187 residue-long region B, while in other strains one copy or two copies of region B are present (Switalski *et al.*, 1993; Patti *et al.*, 1995). The function of region B is unknown. It is not required for spanning the cell wall of *S. aureus* strain 8325–4 because a variant Cna lacking region B still bound collagen (O. Hartford, M. Höök & T.J. Foster, unpublished data).

The collagen-binding domain was localized between residues 151 and 318 of region A by testing a series of truncated derivatives for collagen-binding activity (Patti *et al.*, 1995). Further reduction in size resulted in loss of binding. The non-binding derivatives lacked the secondary structure detectable in polypeptide 151–318 by CD spectroscopy, suggesting that the collagen-binding domain is conformational. Both the intact region A and the polypeptide 151–318 recognized several sites on type II collagen but the smaller molecules bound less avidly than the full-length protein (M. Höök, personal communication).

Residues of importance in collagen binding were localized by synthesizing a series of overlapping peptides spanning amino acid residues 151–318 and testing each for its ability to block collagen binding to *S. aureus* cells (Patti *et al.*, 1995). One peptide (amino acid residues 209–233) inhibited collagen binding. Residues Asn232 and Try233 were shown to be important because a shorter peptide lacking these residues no longer blocked collagen binding. Conversion of these residues to Ala in the collagen-binding region A truncate 151–318 by site-directed mutagenesis reduced the affinity of the protein for collagen. The three-dimensional structure of the Cna polypeptide 151–318 is currently being investigated. This will allow details of the Cna–collagen interaction to be determined.

Role of Cna in virulence

S. aureus is the most frequent cause of bacterial arthritis and osteomyelitis, infections which are initiated by haematogenous spread or by direct inoculation after trauma or surgery. Scanning electron microscopy of infected tissue showed bacteria directly attached to collagen fibres (Gristina *et al.*, 1985; Voytek *et al.*, 1988). In one survey over 90% of strains isolated from bone or joint infections expressed Cna, while only 30% of strains from soft tissue infections had the *cna* gene and expressed the Cna protein (Switalski *et al.*, 1993). This is consistent with Cna determining bacterial tropism for collagen-containing tissues, a notion further supported by the reduced virulence of a *cna* mutant, isolated by allelic replacement, in a septic arthritis infection model (Patti *et al.*, 1994).

The collagen-binding protein might contribute to the pathogenesis of infections at other sites, for example, where basement membranes are exposed following tissue damage. This is consistent with the behaviour of a *cna* mutant in the rat endocarditis model (Hienz *et al.*, 1996). The mutant attached to vegetations as efficiently as the wild-type strain (presumably mediated by the fibronectin- or fibrinogen-binding proteins, or both), but subsequent proliferation on infected heart valves, where adhesion to exposed basement membrane might be an advantage, was reduced.

FIBRINOGEN-BINDING PROTEIN

Structural and functional analysis of ClfA – fibrinogen interaction

Fibrinogen is a large protein of M_r 340 000. It is composed of pairs of three polypeptide chains (α,β and γ) that are extensively linked by disulphide bonds to form an elongated dimeric structure (Fig. 4; Doolittle, 1984; Ruggeri, 1993). It is the most abundant ligand for the integrin αIIb/β3 (glycoproteins gpIIb/IIIa) receptor on the surface of platelets. The binding of fibrinogen to the integrin receptor on activated platelets results in platelet aggregation *in vitro* and the formation of platelet-fibrin thrombi *in vivo* (Hawiger, 1995). Fibrinogen is converted to insoluble fibrin by proteolytic cleavage by thrombin.

The fibrinogen-binding protein (clumping factor, ClfA) of *S. aureus* was initially identified by analysis of transposon Tn*917* mutants that were defective in clumping in fibrinogen (McDevitt *et al.*, 1994). Sequencing of the region carrying the insertions revealed an open reading frame with the potential to encode a wall-associated protein. This protein had a typical wall attachment region comprising an LPDTG motif, membrane anchor and positively charged C-terminus (Fig. 2). However, it had a very small region W compared to other surface proteins (Fig. 2). A remarkable repeat sequence (region R) comprising 154 repeats of mainly Ser and Asp residues occurs immediately upstream of region W.

The repeated dipeptide Asp-Ser is encoded by an 18 bp DNA repeat GAY TCN GAY TCN GAY AGY where N is any base and Y is a pyrimidine. Codon AGY encodes every third Ser while the others are specified by TCN. Such a repeated sequence might be expected to be unstable with unequal

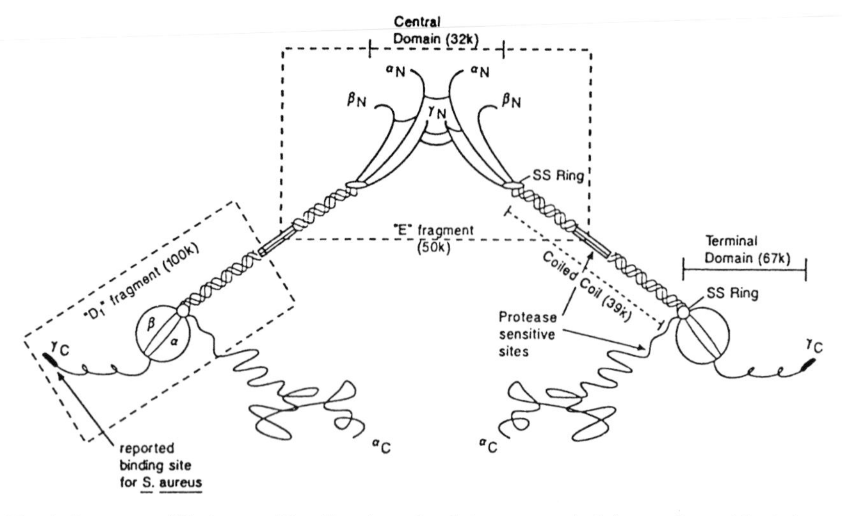

Fig. 4. Structure of fibrinogen. The dimeric molecule is composed of three polypeptide chains, α (66 kDa), β (54 kDa) and γ (48 kDa). The N-terminal ends are held together by disulphide bridges. Protease treatment generates a number of fragments including D and E, which are indicated. (Reproduced from Lantz *et al.* (1990), with permission of the publisher.)

crossing over or replication slippage resulting in variants with shorter or longer region R segments. Indeed, the length of the region R-encoding DNA domain varies from 580–1320 bp in different clinical isolates. No evidence for rapid changes in repeat length in strain Newman was found (McDevitt & Foster, 1995). Examination of codon usage within region R of strain Newman showed that the variable third position of codons is saturated with silent mutations giving maximum sequence divergence consistent with retention of the ability to encode Asp–Ser residues (Shields, McDevitt & Foster, 1995). This may be a mechanism to reduce the frequency of unequal crossing over and to stabilize region R whilst retaining the Ser–Asp dipeptide. Replication slippage may also be suppressed by the sequence divergence in the third position and by the 18 bp (rather than 6 bp) repeated sequence encoding $[\text{Ser–Asp}]_3$. Slippage has been observed to occur at high frequencies between identical repeats of 4 and 5 bp in Gram-negative pathogens (High, Jennings & Moxon, 1996; Stern *et al.*, 1986).

The length of region R varies from strain to strain. The shortest region R in a ClfA^+ natural isolate was ca. 580 bp (193 amino acid residues), approximately half the length of region R in strain Newman (McDevitt & Foster, 1995). A set of mutants of the Newman *clfA* gene with deletions in region R was constructed. Strains expressing ClfA lacking region R altogether or with a region R of 40 amino acid residues or less failed to form clumps in fibrinogen and had reduced fibrinogen-binding ability compared to the wild-type strain (O. Hartford & T.J. Foster, unpublished data). Thus, a very short region R is still capable of supporting surface expression of the fibrinogen-binding region A. The reason for the very long repeats in the wild-type proteins is not clear.

The fibrinogen-binding domain is located in the 520 residue region A (McDevitt *et al.*, 1995). Recombinant region A polypeptide bound fibrinogen and strongly inhibited clumping and binding to fibrinogen, as did anti-region A serum. The ligand-binding domain was shown to be located between amino acid residues 220–550 by analysis of truncated recombinant derivatives. However, removal of amino acid residues from the C-terminus or the N-terminus of the amino acid 220–550 truncated form caused loss of fibrinogen-binding activity, suggesting that, like Cna, conformation is crucial for the structure of the ligand-binding domain.

Does ClfA bind to fibrinogen by the same mechanisms as integrins?

Calcium-binding EF-hand in platelet integrin

Fibrinogen binds to an integrin receptor ($\alpha IIb/\beta 3$) on the surface of activated platelets resulting in aggregation *in vitro* and the formation of platelet-fibrin thrombi *in vivo*. Twelve residues at the C-terminus of the γ-chain of fibrinogen mediate the initial contact with non-stimulated platelets and on activation are sufficient to promote stable adhesion to fibrinogen (Savage, Bottini & Ruggeri, 1995).

Pioneering work by Hawiger and colleagues showed that the same residues are involved in the binding of fibrinogen to the fibrinogen receptor, clumping factor, on the surface of *S. aureus* (Hawiger *et al.*, 1982; Strong *et al.*, 1982). Recently, it was shown that the purified ClfA protein also reacted with the C-terminus of the γ-chain of fibrinogen. Recombinant mutant γ-chains with alterations at the C-terminus failed to bind ClfA and ClfA protein also inhibited ADP-induced platelet aggregation (D. McDevitt & M. Höök, personal communication).

The site within the α-subunit of platelet integrin that reacts with the γ-chain peptide contains a putative divalent cation-binding sequence that resembles an EF-hand, Ca^{2+}-binding motif (D'Souza *et al.* 1990). An EF-hand consists of 13 amino acids, with five appropriately spaced oxygenated residues at positions 1, 3, 5, 7 and 12 and a solvent molecule bonded at position 9. These act as Ca^{2+}–coordination sites and are flanked by α-helices. EF-hands are found in many eukaryotic Ca^{2+}-binding proteins such as calmodulin and troponin C (Tuckwell, Brass & Humphries, 1992). Integrins do not have a coordinating residue at position 12 but instead have a hydrophobic residue.

Analysis of the primary structure of region A of the ClfA protein identified four potential EF-hand divalent-cation-binding motifs, one of which differs from the EF-hand consensus at only one residue, a non-cation coordination site (Fig. 5; D. O'Connell & T.J. Foster, unpublished data). Secondary structure analysis predicts that this is flanked by α-helices. Substitution of

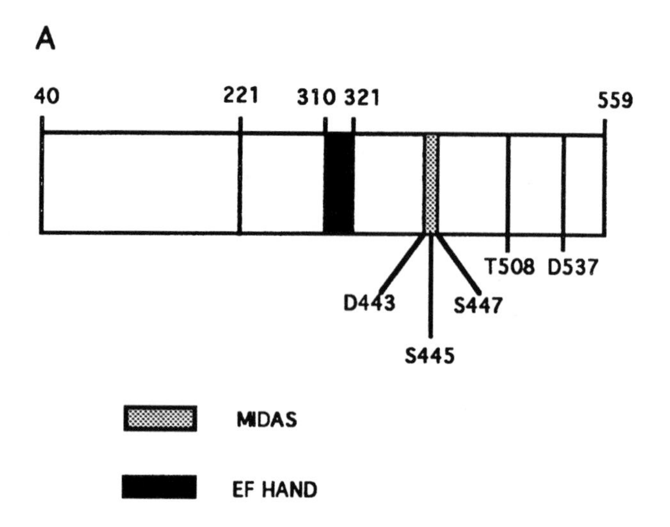

B. Consensus EF-hand motif

1 2 **3** 4 **5** 6 **7**
D - x - (**DNS**) - {ILVFYW} - (**DENSTG**) - (DNQGHKR) - {**GP**} -

8 **9** 10 11 **12** 13
-(ILVMC) - (**DENQTSGCA**) - x - x - (**DE**) - (ILVMFYW)

C. Calcium-binding motif (EF-hand) in some integrins and ClfA

	1	2	**3**	4	**5**	6	**7**	8	**9**	10	11	**12**	13
α_MI449-558	**D**	V	**D**	S	**N**	G	**S**	T	**D**	L	V	L	I
α_MII513-522	**D**	V	**N**	G	**D**	K	**L**	T	**D**	-	V	**A**	I
α_MIII576-595	**D**	L	**T**	M	**D**	G	**L**	V	**D**	L	T	**V**	G
αIIb 297-308	**D**	V	**N**	G	**D**	G	**R**	H	**D**	L	-	**L**	
ClfA 310-321	**D**	S	**D**	G	**N**	V	**I**	Y	**T**	F	T	**D**	Y

D. Divalent cation-binding (MIDAS) motif

D X **S** X **S** - (~64) - **T** - (~33) - **D**

Fig. 5.

cation-coordinating residues in EF-hand I of the ClfA protein with alanine residues caused a significant reduction in fibrinogen-binding activity (D. O'Connell & T.J. Foster, unpublished data).

The displacement model proposed for the interaction between integrin $\alpha IIb/\beta 3$ with fibrinogen suggests that an intermediate complex is formed in which the fibrinogen, integrin and cation interact. This complex is unstable and the cation is displaced (D'Souza *et al.*, 1994). An important prediction of the cation displacement hypothesis is that excess cation should drive ligand binding in reverse and prevent formation of complexes. A high concentration of Ca^{2+} reversed bacterial clumps formed with fibrinogen, prevented bacterial clumping in a solution of fibrinogen and prevented ClfA protein binding to fibrinogen in an *in vitro* assay (D. O'Connell, M. Höök & T.J. Foster, unpublished data). This suggests that the ClfA protein interacts with fibrinogen by a similar mechanism to platelet integrin.

Metal ion-dependent adhesion site in leucocyte integrin

Many mammalian proteins involved in cell–cell, cell–matrix and matrix–matrix interactions carry a ca. 200 residue-long I (or A) domain. The I (A) domain of the major leucocyte integrin complement receptor type 3, CR3 (also called CD16/CD18, Mac-1 or $\alpha M\beta 2$), has been studied by X-ray crystallographic analysis (Lee *et al.*, 1995). It was shown to contain an unusual Mg^{2+} coordination site composed of non-contiguous residues. Comparison with other I (A) domains showed that amino acid residues are highly conserved and a consensus metal ion-dependent adhesion site (MIDAS motif) was proposed. Bound Mg^{2+} seems to be important in determining the correct structure of the domain, facilitating interactions of CR3 with ligands such as fibrinogen (Altieri *et al.*, 1988), iC3b (Arnaout *et al.*, 1983), intracellular adhesion molecule-1 (Diamond *et al.*, 1991) and factor X (Altieri & Edgington, 1988).

A MIDAS motif was identified in the primary sequence of region A of the ClfA protein (Fig. 5). Potential cation-coordinating residues have been changed for Ala by site-directed mutagenesis. Preliminary analysis of mutant D443A suggests that the protein has reduced affinity for fibrinogen,

Fig. 5. Metal ion-binding motifs in integrins and *S. aureus* clumping factor. A, Schematic drawing of the fibrinogen-binding region A of the ClfA protein. The smallest fibrinogen-binding truncated polypeptide is from residues 221–550 (McDevitt *et al.*, 1995). The positions of one putative EF-hand motif and a MIDAS motif are shown. B, The consensus EF-hand showing preferred residues at each position (Bairoch, 1989; Tuckwell *et al.*, 1992). (...) acceptable residues; {...} unacceptable residues; x, any residue. Integrin EF-hands do not have a coordinating residue at position 12, which is often a small hydrophobic amino acid. The bold letters indicate cation-coordinating residues. C, Calcium-binding motifs in the α_M subunit of leucocyte integrins (Corbi *et al.*, 1988), the αIIb-subunit of platelet integrin (D'Souza *et al.*, 1990, 1994) and the ClfA protein (McDevitt *et al.* 1994). The bold letters indicate cation-coordinating residues. D, The consensus metal ion-dependent adhesion site found in the I domain of integrins (Lee *et al.*, 1995). The bold letters indicate cation-coordinating residues.

indicating a role for the MIDAS motif as well as the EF-hand in ClfA binding to fibrinogen (D. O'Connell, M. Höök & T.J. Foster, unpublished data).

Recently an integrin-like protein (αInt1p) from *Candida albicans* which binds fibrinogen was characterized (Gale *et al.*, 1996; Hostetter, 1996). The protein has some amino acid sequence similarity with the I domain of integrin CR3 and contains three putative MIDAS and one EF-hand consensus sequences. The protein also has 25% amino acid identity with region A of the ClfA protein of *S. aureus*. This strengthens the notion of common mechanisms involved in ligand-binding interactions in mammalian cells, in the lower eukaryote *C. albicans* and in the prokaryote *S. aureus*.

Role of ClfA in virulence

A *clfA* mutant of strain Newman has been tested in several *in vitro* and *in vivo* models. The mutant showed reduced adherence to blood-conditioned catheters that had been in contact with the host for periods of less than 1 h (Vaudaux *et al.*, 1995). This is consistent with observations that fibrinogen is the major blood protein deposited on biomaterial (Vaudaux *et al.*, 1993). However, fibrinogen is eventually degraded and loses its ability to promote bacterial attachment. Fibronectin, which is present in much lower amounts, then becomes the major ligand for promoting bacterial adherence. Also, the *clfA* mutant had reduced binding to blood clots formed *in vitro* (Moreillon *et al.*, 1995).

The rat endocarditis model has been used to demonstrate a role for the ClfA protein in pathogenesis (Moreillon *et al.*, 1995). The mutant had a significantly reduced ability to bind to damaged heart tissue after intravenous administration at the ID_{80}. The infection rate was restored to that of the wild-type when the *clfA* mutant was complemented by a copy of the wild-type gene. However, the reduced virulence of the *clfA* mutant was not detectable at higher infection doses. This, and the relatively small reduction in infection at the ID_{80}, suggest that other adhesion mechanisms are important in the pathogenesis of endocarditis. This could very well be the fibronectin-binding proteins (Kuypers & Proctor, 1989) or other fibrinogen-binding proteins (see below).

FIBRINOGEN-BINDING PROTEINS OTHER THAN CLUMPING FACTOR

The ClfA protein is both necessary and sufficient for the formation of clumps when bacterial cells are mixed with soluble fibrinogen (McDevitt *et al.*, 1994). The ClfA-deficient insertion mutant was completely defective in clumping and clumping was restored by introduction of the cloned *clfA* gene. Also, polystyrene beads coated with recombinant ClfA region A protein formed clumps in a fibrinogen solution.

Interactions of *S. aureus* with fibrinogen are complicated by the presence of several different cell wall-associated and extracellular fibrinogen-binding proteins and by apparent differences in the role of these factors in binding to fibrinogen deposited on a solid surface, in bacterial clumping in a solution of fibrinogen and in binding of ^{125}I-labelled fibrinogen to the bacterial cell.

Coagulase is a predominantly extracellular protein that binds prothrombin stoichiometrically forming a complex called staphylothrombin which activates the conversion of fibrinogen to fibrin clots (Hemker, Bas & Muller, 1975; Kawabata *et al.*, 1985). The prothrombin-binding domain is in the N-terminus of the protein while the C-terminal repeated region binds to fibrinogen (McDevitt, Vaudaux & Foster, 1992). In addition, a small fraction of coagulase remains firmly attached to the bacterial cell surface where it can still bind to and activate prothrombin (Bodén & Flock, 1989). However, this bound form of coagulase is not the clumping factor as a coagulase-deficient deletion mutant had the same clumping phenotype as the wild-type strain (McDevitt *et al.*, 1992). In addition, the mutant still adhered normally to fibrinogen-coated surfaces in a dose-dependent fashion. This adherence was reduced to background levels in the *clfA* mutant (McDevitt *et al.*, 1994). Thus, it was suggested that the fibrinogen-binding domain of coagulase is not exposed on the cell surface. However, site-specific mutants of *S. aureus* strain Newman defective in clumping factor still bound significant amounts of soluble fibrinogen, this being attributable in part to the presence of coagulase (Wolz *et al.*, 1996). The effect was more pronouced in an *agr* mutant.

Coagulase (Coa) also seems to have a role in promoting bacterial attachment to a fibrinogen-coated surface in a radial flow chamber under conditions of shear (Dickinson *et al.*, 1995). A ClfA$^-$ Coa$^-$ double mutant adhered weakly and was resistant to detachment from the purified fibrinogen-coated surface, suggesting the presence of an additional adherence factor. One of the other known fibrinogen-binding proteins, namely Fib, FbpA or Map (Table 1 and see below) could perhaps be responsible. However, a role for coagulase in promoting attachment of bacteria to damaged rat heart valves in the endocarditis model was not apparent by comparing a Coa$^-$ mutant with the wild-type strain (Baddour *et al.*, 1994; Moreillon *et al.*, 1995).

THE MHC CLASS II-ANALOGOUS PROTEIN

A remarkable protein called Map can be released from the cell wall of *S. aureus* by treatment with LiCl (Homonylo McGavin *et al.*, 1993; Jönsson *et al.*, 1995). It does not have a wall-spanning or membrane anchor region or an LPXTG motif. The Map protein comprises six repeated domains each of 110 residues. Thirty-one amino acid residues in each repeat have sequence identity with the mammalian MHCII protein which is involved in antigen presentation to B cells. Conserved residues include those involved in MHCII peptide binding.

Purified Map protein can bind many different host proteins in ligand-binding assays and may be responsible for previously observed bacterial binding to vitronectin, thrombospondin and bone sialoprotein (Table 1), although this needs to be verified by isolating a site-specific Map$^-$ mutant or by testing ligand binding in the presence of anti-Map serum. The Map protein may also contribute to the interactions of bacteria with soluble and deposited fibrinogen.

THE ELASTIN-BINDING PROTEIN

Elastin fibres are components of the mammalian extracellular matrix and are present in abundance in tissues that require elasticity (Mecham, 1993). Mature elastin is a polymer of tropoelastin, the monomeric form secreted from the cell (Rosenbloom, 1987). *S. aureus* binds strongly to tropoelastin and to elastin peptides (Park *et al.*, 1991). Elastin binding may contribute to the infection by *S. aureus* of elastin-containing organs, such as the lung, skin and blood vessels.

The elastin-binding protein EbpS is a 25 kDa bacterial surface protein (Park *et al.*, 1996). Interestingly, this protein has neither a secretory signal sequence nor the C-terminal wall – membrane anchoring structure seen in wall-associated proteins such as protein A (Fig. 2). It appears to be processed post-translationally from a 44 kDa precursor. The role of this process in surface localization and the mechanism of secretion is unknown. It would be intriguing to think that *S. aureus* possesses a Sec-independent secretion mechanism analogous to those of Gram-negative bacteria.

INTERACTIONS OF *S. AUREUS* WITH PLASMINOGEN AND PLASMIN

Plasminogen is a 90 kDa plasma protein which is a key component of the mammalian fibrinolytic system (Stephens & Vaheri, 1993). It is the precursor of plasmin, a broad spectrum serine protease. Plasminogen is normally activated by host activators urokinase and tissue plasminogen activator (tPA), both of which are tightly controlled by specific regulators. In addition, in the absence of fibrin, plasmin is rapidly inactivated by α_2-antiplasmin (Dano *et al.*, 1985).

S. aureus expresses an extracellular protein staphylokinase (Sak) which undergoes a complex series of reactions with plasminogen and plasmin (Sako *et al.*, 1983; Behnke & Gerlach, 1987). It binds plasminogen/plasmin with 1:1 stoichiometry (Collen *et al.*, 1993). In the current model for staphylokinase action, Sak initially binds to trace amounts of plasmin in plasma. In the absence of fibrin this complex is rapidly inactivated by α_2-antiplasmin resulting in dissociation of Sak which is recycled (Lijnen *et al.*, 1991).

However, in the presence of fibrin the α_2-antiplasmin inhibition of Sak-plasmin is reduced 100-fold (Lijnen *et al.*, 1992). The now active Sak-plasmin complex rapidly converts both free plasminogen to plasmin and Sak-plasminogen to Sak-plasmin and fibrin clot lysis ensues (Silence *et al.*, 1995).

In common with some other invasive pathogens, *S. aureus* cells can bind plasminogen on their surface and subsequently convert it to plasmin (Kuusela & Saksela, 1990; Christner & Boyle, 1995). This may be a common mechanism for pathogens to acquire invasive potential (Lottenberg, Minning-Wenz & Boyle, 1994). Studies with purified components showed that *S. aureus* cells bound plasminogen which could subsequently be activated by tPA (Kuusela & Saksela, 1990). Binding and activation also occurred with Sak-producing bacteria growing in human serum or plasma (Christner & Boyle, 1995). Either Sak secreted by bacteria can bind plasminogen in the fluid phase and the complex is then captured by the bacterial cell or some of the normally extracellular Sak is actually surface-bound and this directly captures plasminogen and binds it to the bacterial cell surface. Cell-associated plasmin is not regulated by α_2-antiplasmin and is very effective at solubilizing fibrin clots.

OTHER SURFACE-LOCATED, LIGAND-BINDING PROTEINS OF *S. AUREUS*

S. aureus can bind to several mammalian proteins (Table 1) but in many cases the bacterial ligand-binding proteins have not yet been characterized. Even if a protein with binding activity can be purified from bacterial cells, it may not necessarily be responsible for the bacterial ligand-binding activity. The ligand-binding domain must be shown to be exposed on the bacterial cell surface. Site-specific mutants defective in the putative ligand-binding protein should lack binding activity. In addition, the purified ligand-binding domain must be able to block bacterial interactions with the host protein.

Another approach to identifying potential surface proteins is to clone loci with DNA sequence similarity with genes for known wall-associated proteins. Several loci with DNA sequences homologous to the repeated region R of clumping factor occur in the genome of *S. aureus*. One locus has been cloned and sequenced and was shown to comprise an open-reading frame with potential to express a typical wall-associated protein with a domain organization similar to the ClfA protein (D. Ní Edhin & T.J. Foster, unpublished data). However, region A of the novel protein bears very little similarity to region A of the ClfA protein. Antibodies raised against region A of the novel protein reacted with a high molecular weight surface protein expressed by strain Newman. Recombinant protein will be tested for ligand-binding activity with a variety of mammalian proteins.

Once the genome sequence of *S. aureus* (McCarthy, 1996) is in the public domain the identification of other region R-like, repeat-bearing loci as well as

Table 1. S. aureus *proteins that bind to host proteins*

Host protein	Cell-associated binding protein[a]	Extracellular binding protein	References
Fibronectin	Fibronectin-binding proteins A and B[b] FnBPA 108 kDa (> 200 kDa) FnBPB 98 kDa (> 200 kDa)		Signas *et al.*, 1989 Jönsson *et al.*, 1991 Greene *et al.*, 1995
Fibrinogen	Fibrinogen-binding protein (clumping factor) ClfA 92 kDa (190 kDa)[b]		McDevitt *et al.*, 1992
	Coagulase 69 kDa (87 kDa)[b]	Coagulase 69 kDa (87 kDa)[b]	Bodén & Flock, 1989 McDevitt *et al.*, 1992
	Fibrinogen-binding protein Fib 19 kDa	Fib 19 kDa	Bodén & Flock, 1989, 1994
	Fibrinogen-binding protein A FbpA 70 kDa		Cheung *et al.*, 1995
Vitronectin,[c] heparan sulphate	(60 kDa)		Chhatwal *et al.*, 1987 Liang *et al.*, 1992 Liang, Flock & Wadström, 1995
Bone sialoprotein	(97 kDa)		Rydén *et al.*, 1989 Yacoub *et al.*, 1994
Thrombospondin			Hermann *et al.*, 1991
Collagen	Collagen-binding protein Cna 135 kDa, 110 kDa		Switalski *et al.*, 1989, 1993

IgG	Protein A Spa 58 kDa (48 kDa)		Löfdahl *et al.*, 1983 Uhlén *et al.*, 1984
Elastin	Elastin-binding protein EbpS 44 kDa (25 kDa)		Park *et al.*, 1991, 1996
Prothrombin	Coagulase Coa 69 kDa (87 kDa)[d]	Coagulase Coa 69 kDa (87 kDa)[d]	Bodén & Flock 1989 Phonimdaeng *et al.*, 1990 McDevitt *et al.*, 1992
Plasminogen	Plasminogen-binding factor	Staphylokinase Sak 15.5kDa	Sako *et al.*, 1983 Behnke & Gerlach, 1987
Laminin	(52 kDa)		Lopes, dos Reis & Brentani, 1985 Mota *et al.*, 1988
Mucin	(138 kDa, 127 kDa)		Shuter, Hatcher & Lowy, 1996
Fibrinogen, vitronectin, thrombospondin bone sialoprotein	MHC class II analogous protein Map 72 kDa[d]		Homonylo McGavin *et al.*, 1993 Jönsson *et al.*, 1995

[a] The numbers in parentheses are the observed molecular weights. The adjacent numbers not in parentheses are the predicted molecular weights. Where a single number occurs without parenthesis, there is little or no difference between predicted and observed molecular weights.
[b] There is a large discrepancy between the predicted and observed molecular weights of the FnBPs and ClfA, and a smaller discrepancy for coagulase. Western blotting reveals several polypeptides. The size of the largest is cited. The molecular mass of coagulase varies with the serotype and the number of C-terminal repeats. The size of the strain Newman and strain 8325–4 coagulases are given.
[c] These two ligand-binding activities are specified by the same protein.
[d] It has been suggested that Map is responsible for vitronectin-, thrombospondin- and bone sialoprotein-binding by *S. aureus* (Jönsson *et al.*, 1995). The discrepancies between the molecular mass of Map and the binding proteins could be due to differences in the number of repeats in Map.

genes encoding proteins with C-terminal wall attachment motifs (LPXTG, hydrophobic anchor, positively charged C-terminus) will be facilitated.

PROTEINS FROM OTHER STAPHYLOCOCCI

Many *S. epidermidis* isolates can bind to fibronectin, but generally less avidly than *S. aureus* (Valentin-Weigand *et al.*, 1993; Vaudaux *et al.*, 1989). *S. epidermidis* is the major cause of infections associated with indwelling devices and binding to fibronectin is probably an important factor in promoting colonization. A fibronectin-binding protein has not yet been identified in *S. epidermidis* but PCR primers based on the sequences encoding the D repeats of FnBP of *S. aureus* amplified a short DNA sequence which could be part of an *fnb* gene (Minhas *et al.*, 1995). Similar results were obtained with several *S. epidermidis* isolates and with representatives of *S. haemolyticus, S. simulans, S. hominis, S. warneri, S. cohnii* and *S. lugdunensis*. These findings need to be verified by cloning *fnb* genes and showing directly that the FnBPs are responsible for bacterial interactions with fibronectin. The 160 kDa haemagglutinin of *S. saprophyticus* is also a fibronectin-binding protein (Gatermann & Meyer, 1994).

CONCLUSIONS AND FUTURE PROSPECTS

One of the hopes of current research investigating the molecular basis of staphylococcal adherence to host tissues and cells is the possibility of devising new approaches to combatting staphylococcal infections. This is particularly relevant with the increase in multiple resistance to antibiotics and the distinct possibility that resistance will develop to vancomycin, the last remaining effective anti-staphylococcal drug (Dixon, 1995).

It is becoming apparent that staphylococcal adhesion to host tissues is multifactorial with several different mechanisms contributing to adherence to a specific site. In the case of fibrinogen, several different bacterial proteins can bind to the host protein. This may explain the disparity between the high reductions in adherence to a purified host protein *in vitro* seen with mutants or by addition of polyclonal anti-ligand-binding protein serum and the small difference in infectivity between mutants and wild-type strains in the endocarditis infection model.

There is renewed interest in the possibility of vaccination to combat staphylococcal infection, particularly among hospital patients. This could involve passive immunization with hyperimmune serum from human donors for people at immediate risk, or active immunization preceding elective surgery. Capsular polysaccharide serotypes 5 and 8 conjugated to *Pseudomonas aeruginosa* exotoxin A toxoid as a carrier stimulated an antibody response in human volunteers (Fattom *et al.*, 1993, 1996; Welch *et al.*, 1996). Antibodies were opsonic in *in vitro* phagocytosis tests and conferred passive immunity in experimental infections in mice and rats. Active immunity was also generated in experimental animals (Fattom *et al.*, 1996; Lee, 1996).

Another approach is to vaccinate with antigens derived from surface proteins. Indeed, a fusion protein comprising part of the fibronectin-binding protein stimulated protective immunity against endocarditis in rats and mastitis in mice and cows (Nelson *et al.* 1992; Rozalska & Wadström, 1993; Schennings *et al.*, 1993). Because of the multifactorial nature of *S. aureus* adhesion to host tissues, several surface protein antigens may be required for protection.

It may be possible to substitute *P. aeruginosa* exotoxin A toxoid in the capsular polysaccharide conjugate vaccine with a staphylococcal protein, perhaps one of the surface antigens or genetically engineered α-toxoid (Menzies & Kernodle, 1996). To be effective, the vaccine may need to block colonization and promote opsonization (Lee, 1996).

It may also be possible to block infection of wounds with small molecule inhibitors of bacteria–host interactions. Synthetic peptides corresponding to the ligand-binding D repeats of FnBP and a peptide corresponding to part of the Cna protein inhibit bacterial interactions with the appropriate host protein *in vitro* (Raja *et al.*, 1990; Patti *et al.*, 1995). This raises the possibility of mimetic peptides, engineered to remove labile bonds to increase potency and retention time in the body, being used to treat staphylococcal infections.

ACKNOWLEDGEMENTS

The Wellcome Trust is thanked for financial support and Magnus Höök for providing unpublished data and for many stimulating discussions.

REFERENCES

Altieri, D.C., Bader, R., Mannucci, P.M. & Edgington, T.S. (1988). Oligospecificity of the cellular adhesion receptor Mac-1 encompasses an inducible recognition specificity for fibrinogen. *Journal of Cell Biology*, **107**, 1893–900.

Altieri, D.C. & Edgington, T.S. (1988). The saturable high affinity association of factor X to ADP-stimulated monocytes defines a novel function of the Mac-1 receptor. *Journal of Biological Chemistry*, **263**, 7007–15.

Arnaout, M.A., Todd, R.F., Dana, N., Melamed, J., Schlossman, S.F. & Colten, H.R. (1983). Inhibition of phagocytosis of complement C3- or immunoglobulin G-coated particles and of C3bi binding by monoclonal antibodies to a monocyte-granulocyte membrane glycoprotein (Mo1). *Journal of Clinical Investigation*, **72**, 171–9.

Baddour, L.M., Tayidi, M.M., Walker, E., McDevitt, D. & Foster, T.J. (1994). Virulence of coagulase-deficient mutants of *Staphylococcus aureus* in experimental endocarditis. *Journal of Medical Microbiology*, **41**, 259–63.

Bairoch, A. (1989). *PROSITE: a dictionary of protein sites and patterns.* 4th edn. Geneva: University of Geneva.

Behnke, D. & Gerlach, D. (1987). Cloning and expression in *Escherichia coli, Bacillus subtilis* and *Streptococcus sanguis* of a gene for staphylokinase – a bacterial plasminogen activator. *Molecular and General Genetics*, **210**, 528–34.

Bodén, M.K. & Flock, J-I. (1989). Fibrinogen-binding protein/clumping factor from *Staphylococcus aureus*. *Infection and Immunity*, **57**, 2358–63.

Bodén, M.K. & Flock, J.-I. (1994). Cloning and characterization of a gene for a 19 kDa fibrinogen-binding protein from *Staphylococcus aureus*. *Molecular Microbiology*, **12**, 599–606.

Bozzini, S., Visai, L., Pignatti, P., Petersen, T.E. & Speziale, P. (1992). Multiple binding sites in fibronectin and the staphylococcal fibronectin receptor. *European Journal of Biochemstry*, **207**, 327–33.

Cheung, A.I., Projan, S.J., Edelstein, R.E. & Fischetti, V. (1995). Cloning, expression, and nucleotide sequence of a *Staphylococus aureus* gene (*fbpA*) encoding a fibrinogen binding protein. *Infection and Immunity*, **63**, 1914–20.

Chhatwal, G.S., Preissner, K.T., Müller-Berghaus, G. & Blobel, H. (1987). Specific binding of the human S protein (vitronectin) to streptococci, *Staphylococcus aureus* and *Escherichia coli*. *Infection and Immunity*, **55**, 1878–83.

Christner, R.B. & Boyle, M.D.P. (1995). Role of staphylokinase in the acquisition of plasmin(ogen)-dependent enzymatic activity by staphylococci. *Journal of Infectious Diseases*, **173**, 104–12.

Collen, D., Schlott, B., Engelborghs, Y., van Hoef, B., Hartmann, M., Lijnen, H.R. & Behnke, D. (1993). On the mechanism of activation of human plasminogen by recombinant staphylokinase. *Journal of Biological Chemistry*, **268**, 8284–9.

Corbi, A.l., Kishimoto, L., Miller, L.J. & Springer, T.A. (1988). The human leukocyte adhesion glycoprotein Mac-1 (complement receptor type 3, CD11b) alpha subunit. Cloning, primary structure, and relation to the integrins, von Willebrand factor and factor B. *Journal of Biological Chemistry*, **263**, 12403–11.

Dano, K., Andreasen, P.A., Grondahl-Hansen, J., Kristensen, P., Nielsen, L.S. & Striver, L. (1985). Plasminogen activators, tissue degradation, and cancer. *Advances in Cancer Research*, **44**, 139–266.

Deisenhofer, J. (1981). Crystallographic refinement and atomic models of a human Fc fragment and its complex with fragment B of protein A from *Staphylococcus aureus* at 2.9 and 2.8 Å resolution. *Biochemistry*, **20**, 2361–70.

Diamond, M.S., Staunton, D.E., Marlin, S.D. & Springer, T.A. (1991). Binding of the integrin Mac-1 (CD11b/CD18) to the third immunoglobulin-like domain of ICAM-1 (CD54) and its regulation by glycosylation. *Cell*, **65**, 961–71.

Dickinson, R.B., Nagel, J.A., McDevitt, D., Foster, T.J., Proctor, R.A. & Cooper, S.L. (1995). Quantitative comparison of clumping factor- and coagulase-mediated *Staphylococcus aureus* adhesion to surface-bound fibrinogen under flow. *Infection and Immunity*, **63**, 3143–50.

Dixon, B. (1995). Horror bacteria. *Molecular Medicine Today*, **1**, 105.

Doolittle, R.F. (1984). Fibrinogen and fibrin. *Annual Review of Biochemistry*, **53**, 195–229.

D'Souza, S.E., Ginsberg, M.H., Burke, T.A. & Plow, E.F. (1990). The ligand binding site of the platelet integrin receptor GPIIb-IIIa is proximal to the second calcium binding domain of its α subunit. *Journal of Biological Chemistry*, **265**, 3440–6.

D'Souza, S.E., Haas, T.A., Piotrowicz, R.S., Byers-Ward, V., McGrath, D.E., Soule, H.R., Cierniewski, C., Plow, E.F. & Smith, J.W. (1994). Ligand and cation binding are dual functions of a discrete segment of the integrin β_3 subunit: cation displacement is involved in ligand binding. *Cell*, **79**, 659–67.

Fattom, A., Schneerson, R., Watson, D.C., Karakawa, W.W., Fitzgerald, D., Pastan, I., Li, X., Shiloach, J., Bryla, D.A. & Robbins, J.B. (1993). Laboratory and clinical evaluation of conjugate vaccines composed of *Staphylococcus aureus* type 5 and type 8 capsular polysaccharides bound to *Pseudomonas aeruginosa* recombinant exotoxin A. *Infection and Immunity*, **61**, 1023–32.

Fattom, A.I., Sarwar, J., Ortiz, A. & Naso, R. (1996). A *Staphylococcus aureus* capsular polysaccharide (CP) vaccine and CP-specific antibodies protect mice against bacterial challenge. *Infection and Immunity*, **64**, 1659–65.

Flock, J.-I., Hienz, S.A., Heimdahl, A. & Schennings, T. (1996). Reconsideration of the role of fibronectin binding in endocarditis caused by *Staphylococcus aureus*. *Infection and Immunity*, **64**, 1876–8.

Foster, T.J. & McDevitt, D. (1994). Surface-associated proteins of *Staphylococcus aureus*: their possible roles in virulence. *FEMS Microbiology Letters*, **118**, 199–206.

Foster, T.J., O'Reilly, M., Phonimdaeng, P., Cooney, J., Patel, A.H. & Bramley, A.J. (1990). Genetic studies of virulence factors of *Staphylococcus aureus*. Properties of coagulase and gamma-toxin and the role of alpha-toxin, beta-toxin and protein A in the pathogenesis of *S. aureus* infections. In Novick, R.P., ed. *Molecular Biology of the Staphylococci*, pp. 403–17. Cambridge, New York: VCH.

Gale, C., Finkel, D., Tao, N., Meinke, M., McClellan, M., Olson, J., Kendrick, K. & Hostetter, M. (1996). Cloning and expression of a gene encoding an integrin-like protein in *Candida albicans*. *Proceedings of the National Academy of Sciences, USA*, **93**, 357–61.

Gatermann, S. & Meyer, H-G. W. (1994). *Staphylococcus saprophyticus* haemagglutinin binds fibronectin. *Infection and Immunity*, **62**, 4556–63.

Gemmell, C.G., Tree, R., Patel, A., O'Reilly, M. & Foster T.J. (1990). Susceptibility to opsonophagocytosis of protein A, alpha-haemolysin and beta-toxin deficient mutants of *Staphylococcus aureus* isolated by allele-replacement. *Zentralblatt für Bakteriologie*, Suppl. **21**, 273–7.

Gouda, H., Torigoe, H., Saito, A., Sato, M., Arata, Y. & Shimada, I. (1992). Three-dimensional solution structure of the B domain of staphylococcal protein A: comparisons of the solution and crystal structures. *Biochemistry*, **31**, 9665–72.

Greene, C., McDevitt, D., Francois, P., Vaudaux, P.E., Lew, D.P. & Foster, T.J. (1995). Adhesion properties of mutants of *Staphylococcus aureus* defective in fibronectin-binding proteins and studies on the expression of the *fnp* genes. *Molecular Microbiology*, **17**, 1143–52.

Greene, C., Vaudaux, P.E., Francois, P., Proctor, R.A., McDevitt, D. & Foster, T.J. (1996). A low-fibronectin-binding mutant of *Staphylococcus aureus* 879R4S has Tn*918* inserted in its single *fnb* gene. *Microbiology*, **142**, 2153–60.

Gristina, A.G., Oga, M., Webb, L.X. & Hobgood, C.D. (1985). Adherent bacterial colonization in the pathogenesis of osteomyelitis. *Science*, **228**, 990–3.

Hawiger, J. (1995). Adhesive ends of fibrinogen and its anti-adhesive peptides: the end of a saga? *Seminars in Haematology*, **32**, 99–109.

Hawiger, J., Timmons, S., Strong, D.D., Cottrell, B.A., Riley, M. & Doolittle, R.F. (1982). Identification of a region of human fibrinogen interacting with staphylococcal clumping factor. *Biochemistry*, **21**, 1407–13.

Heinegard, D. & Oldberg, A. (1989). Structure and biology of cartilage and bone matrix non-collagenous macromolecules. *Federation of American Societies for Experimental Biology Journal*, **3**, 2042–51.

Hemker, H.C., Bas, B.M. & Müller, A.D. (1975). Activation of a pro-enzyme by stoichiometric reaction with another protein. The reaction between prothrombin and staphylocoagulase. *Biochimica et Biophysica Acta*, **379**, 180–8.

Hermann, M., Suchard, S.J., Boxer, L.A., Waldvogel, F.A. & Lew, D.P. (1991). Thrombospondin binds to *Staphylococcus aureus* and promotes staphylococcal adherence to surfaces. *Infection and Immunity*, **59**, 279–88.

Hienz, S., Schennings, T., Heimdahl, A. & Flock, J-I. (1996). Collagen binding of

Staphylococcus aureus is a virulence factor in experimental endocarditis. *Journal of Infectious Diseases*, **174**, 83–8.

High, N.J., Jennings, M.P. & Moxon, E.R. (1996). Tandem repeats of the tetramer 5′-CAAT-3′ present in *lic2A* are required for phase variation but not lipopolysaccharide biosynthesis in *Haemophilus influenzae. Molecular Microbiology*, **20**, 165–74.

Homonylo McGavin, M., Krajewska-Pietrasik, D., Rydén, C. & Höök, M. (1993). Identification of a *Staphylococcus aureus* extracellular matrix-binding protein with broad specificity. *Infection and Immunity*, **61**, 2479–85.

Hostetter, M.K. (1996). An integrin-like protein in *Candida albicans*: implications for pathogenesis. *Trends in Microbiology*, **4**, 242–6.

House-Pompeo, K., Xu, Y., Joh, D., Speziale, P. & Höök, M. (1996). Conformational changes in the fibronectin binding MSCRAMMs are induced by ligand binding. *Journal of Biological Chemistry*, **271**, 1379–84.

Hynes, R. (1993). Fibronectins. In Kreis, T. & Vale, R. eds. *Guidebook to the Extracellular Matrix and Adhesion Proteins*, pp. 56–8. Oxford: Oxford University Press.

Jendeberg, L., Tashiro, M.,Tejero, R., Lyons, B.A., Uhlén, M., Montelione, G.T. & Nilsson, B. (1996). The mechanism of binding staphylococcal protein A to immunoglobulin G does not involve helix unwinding. *Biochemistry*, **35**, 22–31.

Joh, H.J., House-Pompeo, K., Patti, J., Gurusiddappa, S. & Höök, M. (1994). Fibronectin receptors from Gram-positive bacteria: comparison of sites. *Biochemistry*, **33**, 6086–92.

Jönsson, K., Signas, C., Müller, H.P. & Lindberg, M (1991). Two different genes encode fibronectin binding proteins in *Staphylococcus aureus*. The complete nucleotide sequence and characterization of the second gene. *European Journal of Biochemstry*, **202**, 1041–8.

Jönsson, K., McDevitt, D., Homonylo McGavin, M., Patti, J.M. & Höök, M. (1995). *Staphylococcus aureus* expresses a major histocompatibility complex class II analog. *Journal of Biological Chemistry*, **270**, 21457–60.

Kawabata, S., Morita, T., Iwanaga, S. & Igarashi, H. (1985). Enzymatic properties of staphylothrombin, an active molecular complex formed between staphylocoagulase and human prothrombin. *Journal of Biochemistry*, **98**, 1603–14.

Kreikemeyer, B., Talay, S.R. & Chhatwal, G.S. (1995). Characterization of a novel fibronectin-binding surface protein in group A streptococci. *Molecular Microbiology*, **17**, 137–45.

Kuusela, P. & Saksela, O. (1990). Binding and activation of plasminogen at the surface of *Staphylococcus aureus*. Increase in affinity after conversion to the Lys form of the ligand. *European Journal of Biochemistry*, **193**, 759–65.

Kuypers, J.M. & Proctor, R.A. (1989). Reduced adherence to traumatized rat heart valves by a low-fibronectin-binding mutant of *Staphylococcus aureus*. *Infection and Immunity*, **57**, 2306–12.

Lantz, M.S., Allen, R.D., Bounelis, P., Switalski, L. & Höök, M. (1990). *Bacteroides gingivalis* and *Bacteroides intermedius* recognize different sites on human fibrinogen. *Journal of Bacteriology*, **172**, 716–26.

Lee, J.C. (1996). The prospects for developing a vaccine against *Staphylococcus aureus*. *Trends in Microbiology*, **4**, 162–6.

Lee, J.O., Rieu, P., Arnaout, M.A. & Liddington, R. (1995). Crystal structure of the A-domain from the α-subunit of integrin CR3 (CDIIb/CD18). *Cell*, **80**, 631–8.

Liang, O.D., Ascencio, F., Fransson, L.A. & Wadström, T. (1992). Binding of heparan sulfate to *Staphylococcus aureus*. *Infection and Immunity*, **60**, 899–906.

Liang, O.D., Flock, J.-I. & Wadström, T. (1995). Isolation and characterization of a vitronectin-binding surface protein from *Staphylococcus aureus*. *Biochimica et Biophysica Acta*, **1250**, 110–16.

Lijnen, H.R., van Hoef, B., de Cock, F., Okada, K., Ueshima, S., Matsuo, O. & Collen, D. (1991). On the mechanism of fibrin-specific plasminogen activation by staphylokinase. *Journal of Biological Chemistry*, **266**, 11826–32.

Lijnen, H.R., van Hoef, B., Matsuo, O. & Collen, D. (1992). On the molecular interactions between plasminogen-staphylokinase, α_2-antiplasmin and fibrin. *Biochimica et Biophysica Acta*, **1118**, 144–8.

Löfdahl, S., Guss, B., Uhlén, M., Philpson, L. & Lindberg, M. (1983). Gene for staphylococcal protein A. *Proceedings of the National Academy of Sciences, USA*, **80**, 697–701.

Lopes, J.D., dos Reis, M. & Brentani, R.R. (1985). Presence of laminin receptors in *Staphylococcus aureus*. *Science*, **229**, 275–7.

Lottenberg, R., Minning-Wenz, D. & Boyle, M.D.P. (1994). Capturing host plasmin(ogen): a common mechanism for invasive pathogenesis. *Trends in Microbiology*, **2**, 20–4.

McCarthy, M. (1996). *Staphylococcus aureus* genome sequence. *Lancet*, **347**, 251.

McDevitt, D. & Foster, T.J. (1995). Variation in the size of the repeat region of the fibrinogen receptor (clumping factor) of *Staphylococcus aureus* strains. *Microbiology*, **141**, 937–43.

McDevitt, D., Vaudaux, P. & Foster, T.J. (1992). Genetic evidence that bound coagulase of *Staphylococcus aureus* is not clumping factor. *Infection and Immunity*, **60**, 1514–23.

McDevitt, D., Francois, P, Vaudaux, P. & Foster, T.J. (1994). Molecular characterization of the fibrinogen receptor (clumping factor) of *Staphylococcus aureus*. *Molecular Microbiology*, **11**, 237–48.

McDevitt, D., Francois, P., Vaudaux, P. & Foster, T.J. (1995). Identification of the ligand-binding domain of the surface-located fibrinogen receptor (clumping factor) of *Staphylococcus aureus*. *Molecular Microbiology*, **16**, 895–907.

McGavin, M.J., Raucci, G., Gurusiddappa, S. & Höök, M. (1991). Fibronectin binding determinants of the *Staphylococcus aureus* fibronectin receptor. *Journal of Biological Chemistry*, **266**, 8343–7.

McGavin, M.J., Gurusiddappa, S., Lindgren, P.E., Lindberg, M., Raucci, G. & Höök, M. (1993). Fibronectin receptors from *Streptococcus dysgalactiae* and *Staphylococcus aureus*. Involvement of conserved residues in ligand binding. *Journal of Biological Chemistry*, **268**, 23946–53.

Mecham, R.P. (1993). Elastin. In Kreis, T. & Vale, R., eds. *Guidebook to the Extracellular Matrix and Adhesion Proteins*, pp. 50–2. Oxford: Oxford University Press.

Menzies, B.E. & Kernodle, D.S. (1996). Passive immunization with antiserum to a nontoxic alpha-toxin mutant from *Staphylococcus aureus* is protective in a murine model. *Infection and Immunity*, **64**, 1839–41.

Miller, E.J. & Gay, S. (1983). The collagens: an overview and update. *Methods in Enzymology*, **144**, 3–41.

Minhas, T., Ludlam, H.A., Wilks, M. & Tabaqchali, S. (1995). Detection by PCR and analysis of the distribution of a fibronectin-binding protein gene (*fbn*) among staphylococcal isolates. *Journal of Medical Microbiology*, **42**, 96–101.

Moreillon, P., Entenza, J.M., Francioli, P., McDevitt, D., Foster, T.J., Francois, P. & Vaudaux, P. (1995). Role of *Staphylococcus aureus* coagulase and clumping factor in the pathogenesis of experimental endocarditis. *Infection and Immunity*, **63**, 4738–43.

Mota, G.F.A., Carniero, C.R.W., Gomes, L. & Lopes, J.D. (1988). Monoclonal antibodies to *Staphylococcus aureus* laminin-binding protein cross-react with mammalian cells. *Infection and Immunity*, **56**, 1580–4.

Navarre, W.W. & Schneewind, O. (1994). Proteolytic cleavage and cell wall anchoring at the LPXTG motif of surface proteins in Gram-positive bacteria. *Molecular Microbiology*, **14**, 115–21.

Nelson, L., Flock, J.-I., Höök, M., Lindberg, M., Muller, H.P. & Wadström, T. (1992). Adhesins in staphylococcal mastitis as vaccine components. *Flemish Veterinary Journal*, **62**, Suppl. 1, 111–25.

Olsen, B.R. & Ninomiya (1993). Collagens. In Kreis, T. & Vale, R., eds. *Guidebook to the Extracellular Matrix and Adhesion Proteins*, pp. 32–48. Oxford: Oxford University Press.

Park, P.W., Roberts, D.D., Grosso, L.E., Parks, W.C., Rosenbloom, J., Abrams, W.R. & Mecham, R.P. (1991). Binding of elastin to *Staphylococcus aureus*. *Journal of Biological Chemistry*, **266**, 23399–406.

Park, P.W., Rosenbloom, J., Abrams, W.R., Rosenbloom, J. & Mecham, R.P. (1996). Molecular cloning and expression of the gene for elastin binding protein (EbpS) in *Staphylococcus aureus*. *Journal of Biological Chemistry*, **271**, 15803–9.

Patel, A.H., Nowlan, P., Weavers, E.D. & Foster, T.J. (1987). Virulence of protein A-deficient and alpha-toxin-deficient mutants of *Staphylococcus aureus* isolated by allele replacement. *Infection and Immunity*, **55**, 3103–10.

Patti, J.M., Jönsson, H., Guss, B., Switalski, L.M., Wiberg, K., Lindberg, M. & Höök, M. (1992). Molecular characterization and expression of a gene encoding *Staphylococcus aureus* collagen adhesin. *Journal of Biological Chemistry*, **267**, 4766–72.

Patti, J.M., Boles, J.O. & Höök, M. (1993). Identification and biochemical characterization of the ligand binding domain of the collagen adhesin from *Staphylococcus aureus*. *Biochemistry*, **32**, 11428–35.

Patti, J.M., Allen, B.A., McGavin, M.J. & Höök, M. (1994*a*). MSCRAMM-mediated adherence of microorganisms to host tissues. *Annual Review of Microbiology*, **45**, 585–617.

Patti, J.M., Bremell, T., Krajewska-Pietrasik, D., Abdelnour, A., Tarkowski, A., Rydén, C. & Höök, M. (1994*b*). The *Staphylococcus aureus* collagen adhesin is a virulence determinant in experimental septic arthritis. *Infection and Immunity*, **62**, 152–61.

Patti, J.M., House-Pompeo, K., Boles, J.O., Garza, N., Gurusiddappa, S. & Höök, M. (1995). Critical residues in the ligand-binding site of the *Staphylococcus aureus* collagen-binding adhesin (MSCRAMM). *Journal of Biological Chemistry*, **270**, 12005–11.

Phonimdaeng, P., O'Reilly, M., Nowlan, P., Bramley, A.J. & Foster, T.J. (1990). The coagulase of *Staphylococcus aureus* 8325–4. Sequence analysis and virulence of site-specific coagulase-deficient mutants. *Molecular Microbiology*, **4**, 393–404.

Potts, J. R. & Campbell, I.D. (1994). Fibronectin structure and assembly. *Current Opinions in Cell Biology*, **6**, 648–55.

Proctor, R.A. (1987). Fibronectin: a brief overview of its structure, function and physiology. *Reviews of Infectious Diseases*, **9**, Suppl. 4, S317–21.

Raja, R.H., Raucci, G. & Höök, M. (1990). Peptide analogs to a fibronectin receptor inhibit attachment of *Staphylococcus aureus* to fibronectin-coated substrates. *Infection and Immunity*, **58**, 2593–8.

Rakonjac, J.V., Robbins, J.C. & Fischetti, V.A. (1995). DNA sequence of the serum opacity factor of group A streptococci: identification of a fibronectin-binding repeat domain. *Infection and Immunity*, **63**, 622–31.

Rozalska, B. & Wadström, T. (1993). Protective opsonic activity of antibodies against fibronectin-binding proteins (FnBPs) of *Staphylococcus aureus*. *Scandinavian Journal of Immunology*, **37**, 575–80.

Rosenbloom, J. (1987). Elastin: an overview. *Methods in Enzymology*, **144**, 172–96.

Ruggeri, Z.M. (1993). Fibrinogen/fibrin. In Kreis, T. & Vale, R., eds. *Guidebook to the Extracellular Matrix and Adhesion Proteins*, pp. 52–3. Oxford: Oxford University Press.

Rydén, C., Yacoub, A.I., Maxe, I., Heinegard, D., Oldberg, A., Franzén, A., Ljungh, Å. & Rubin, K. (1989). Specific binding of bone sialoprotein to *Staphylococcus aureus* isolated from patients with osteomyelitis. *European Journal of Biochemistry*, **184**, 331–6.

Sako, T., Sawaki, S., Sakurai, T., Ito, S., Yoshizawa, Y. & Kondo, I. (1983). Cloning and expression of the staphylokinase gene of *Staphylococcus aureus* in *Escherichia coli*. *Molecular and General Genetics*, **190**, 271–7.

Savage, B., Bottini, E. & Ruggeri, Z.M. (1995). Interaction of integrin $\alpha_{IIb}\beta_3$ with multiple fibrinogen domains during platelet activation. *Journal of Biological Chemistry*, **270**, 28812–17.

Schennings, T., Heimdahl, A., Coster, K. & Flock, J-I. (1993). Immunization with fibronectin binding protein from *Staphylococcus aureus* protects against experimental endocarditis in rats. *Microbial Pathogenesis*, **15**, 227–36.

Schneewind, O., Model, P. & Fischetti, V. A. (1992). Sorting of protein A to the staphylococcal cell wall. *Cell*, **70**, 267–81.

Schneewind, O., Mihaylova-Petkov, D. & Model, P. (1993). Cell wall sorting signals in surface proteins of Gram-positive bacteria. *EMBO Journal*, **12**, 4803–11.

Shields, D.C., McDevitt, D. & Foster, T.J. (1995). Evidence against concerted evolution in a tandem array in the clumping factor gene of *Staphylococcus aureus*. *Molecular Biology and Evolution*, **12**, 963–5.

Shuter, J., Hatcher, V.B. & Lowy, F.D. (1996). *Staphylococus aureus* binding to human nasal mucin. *Infection and Immunity*, **64**, 310–18.

Signas, C., Raucci, G., Jönsson, K., Lindgren, P.E., Anantharamaiah, G.M., Höök, M. & Lindberg, M. (1989). Nucleotide sequence of the gene for a fibronectin-binding protein from *Staphylococcus aureus*: use of this peptide sequence in the synthesis of biologically active peptides. *Proceedings of the National Academy of Sciences, USA*, **86**, 699–703.

Silence, K., Hartmann, M., Guhrs, K-H., Gase, A., Schlott, B., Collen, D. & Lijnen, H.R. (1995). Structure-function relationships in staphylokinase as revealed by 'clustered charge to alanine' mutagenesis. *Journal of Biological Chemistry*, **270**, 27192–8.

Sottile, J., Schwarzbauer, J., Selegue, J. & Mosher, D.F. (1991). Five type I modules of fibronectin form a functional unit that binds to fibroblasts and to *Staphylococcus aureus*. *Journal of Biological Chemistry*, **266**, 12840–3.

Speziale, P., Joh, D., Visai, L., Bozzini, S., House-Pompeo, K., Lindberg, M. & Höök, M. (1996). A monoclonal antibody enhances ligand binding of fibronectin MSCRAMM (adhesin) from *Streptococcus dysgalactiae*. *Journal of Biological Chemistry*, **271**, 1371–8.

Stephens, R.W. & Vaheri, A. (1993). Plasminogen. In Kreis, T. & Vale, R., eds. *Guidebook to the Extracellular Matrix and Adhesion Proteins*, pp. 81–2. Oxford: Oxford University Press.

Stern, A., Brown M., Nickel, P. & Meyer, T.F. (1986). Opacity genes in *Neisseria gonorrhoeae:* control of phase and antigenic variation. *Cell*, **47**, 61–71.

Strauss, A. & Gotz, F. (1996). *In vivo* immobilization of enzymatically active polypeptides on the cell surface of *Staphylococcus carnosus*. *Molecular Microbiology*, **21**, 491–500.

Strong, D.D., Laudano, A.P., Hawiger, J. & Doolittle, R.F. (1982). Isolation, characterization and synthesis of peptides from human fibrinogen that block the staphylococcal clumping reaction and construction of a synthetic clumping particle. *Biochemistry*, **21**, 1414–20.

Switalski, L.M., Speziale, P. & Höök, M. (1989). Isolation and characterization of a putative collagen receptor from *Staphylococcus aureus* strain Cowan 1. *Journal of Biological Chemistry*, **264**, 21080–6.

Switalski, L.M., Patti, J.M., Butcher, W., Gristina, A.G., Speziale, P. & Höök, M. (1993). A collagen receptor on *Staphylococcus aureus* strains isolated from patients with septic arthritis mediates adhesion to cartilage. *Molecular Microbiology*, **7**, 99–107.

Tompkins, D.C., Hatcher, V.B., Patel, D., Orr, G.A., Higgins, L.L. & Lowy, F.D. (1990). A human endothelial cell membrane protein that binds *Staphylococcus aureus in vitro*. *Journal of Clinical Investigation*, **85**, 1248–54.

Tuckwell, D.S., Brass, A. & Humphries, M.J. (1992). Homology modelling of integrin EF-hands. *Biochemical Journal*, **285**, 325–31.

Uhlén, M., Guss, B., Nilsson, B., Gatenbeck, S., Philipson, L. & Lindberg, M. (1984). Complete sequence of the staphylococcal gene encoding protein A. A gene evolved through multiple duplications. *Journal of Biological Chemistry*, **259**, 1695–702.

Valentin-Weigand, P., Timmis, K.N. & Chhatwal, G.S. (1993). Role of fibronectin in staphylococcal colonization of fibrin thrombi and plastic surfaces. *Journal of Medical Microbiology*, **38**, 90–5.

Vaudaux, P., Pittet, D., Haeberli, A. Huggler, E., Nydegger, U.E., Lew, D.P. & Waldvogel, F.A. (1989). Host factors selectively increase staphylococcal adherence on inserted catheters: a role for fibronectin and fibrinogen or fibrin. *Journal of Infectious Diseases*, **160**, 865–75.

Vaudaux, P., Pittet, D., Haeberli, A., Lerch, P.G., Morgenthaler, J.J., Proctor, R.A., Waldvogel, F.A. & Lew, D.P. (1993). Fibronectin is more active than fibrin or fibrinogen in promoting *Staphylococcus aureus* adherence to inserted intravascular devices. *Journal of Infectious Diseases*, **167**, 633–41.

Vaudaux, P.E., Francois, P., Proctor, R.A., McDevitt, D., Foster, T.J., Albrecht, R.M., Lew, D.P., Wabers, H. & Cooper, S.L. (1995). Use of adhesion-defective mutants of *Staphylococcus aureus* to define the role of specific plasma proteins in promoting bacterial adhesion to canine arterio-venous shunts. *Infection and Immunity*, **63**, 585–90.

Voytek, A., Gristina, A.G., Barth, E., Myrvik, Q., Switalski, L.M., Höök, M. & Speziale, P. (1988). Staphylococcal adhesion to collagen in intraarticular sepsis. *Biomaterials*, **9**, 107–10.

Welch, P.G., Fattom, A., Moore, J., Schneerson, R., Shiloach, J., Bryla, D.A., Li, X. & Robbins, J.B. (1996). Safety and immunogenicity of *Staphylococcus aureus* type 5 capsular polysaccharide-*Pseudomonas aeruginosa* recombinant exoprotein A conjugate vaccine in patients on hemodialysis. *Journal of the American Society of Nephrology*, **7**, 247–53.

Wolz, C., McDevitt, D., Foster, T.J. & Cheung, A. (1996). Influence of *agr* on fibrinogen binding in *Staphylococcus aureus* Newman. *Infection and Immunity*, **64**, 3142–7.

Yacoub, A., Lindahl, P., Rubin, K., Wendel, M., Heinegard, D. & Rydén, C. (1994). Purification of a bone sialoprotein-binding protein from *Staphylococcus aureus*. *European Journal of Biochemistry*, **222**, 919–25.

Yamada, K.M. (1989). Fibronectins: structure, function and receptors. *Current Opinion in Cell Biology*, **1**, 956–63.

MECHANISMS OF MICROBIAL ADHESION; THE PARADIGM OF *NEISSERIAE*

M. VIRJI

Department of Paediatrics, University of Oxford, John Radcliffe Hospital, Oxford OX3 9DU, UK[1]

Microbial adhesion to host tissue is a primary event in colonization and an important stage in microbial pathogenesis. Several common mechanisms of microbial adhesion have been reported. In this review, an overview of some of these mechanisms involving interactions of microbial surface located adhesins with their host target receptors are discussed. A large part of the review presents studies on adhesins and invasins of *Neisseria meningitidis*, the interplay between surface-located virulence factors and proposed mechanisms of host colonization and invasion.

Adhesive ligands on bacteria range from hair or rod-like polymeric structures (pili or fimbriae) that extend long distances from bacterial surface to integral outer-membrane proteins and polysaccharides. Individual bacteria may possess multiple adhesins that target distinct host cell molecules and deliver diverse signals resulting in extracellular location or internalization. To this end, not only the nature but also the density of the target receptor on the host cell may be a determining factor in the outcome of bacteria/host interaction.

AN OUTLINE OF SOME COMMON MECHANISMS INVOLVED IN MICROBE–HOST INTERACTIONS[2]

Available targets on host cell surfaces for microbial adhesins include the cell–cell or cell–matrix interaction molecules (adhesion receptors belonging to the Integrin and Immunoglobulin superfamilies, the Selectins and the Cadher-Cadherins), which are subverted by many pathogenic organisms whose ligands often mimic the natural ligands. Besides true ligand mimicry, other mechanisms, including other forms of mimicry, that are favoured in microbial interactions with the host, are described below.

[1] *Present address:* School of Animal and Microbial Sciences, The University of Reading, Reading RG6 6AJ, UK; [2] For a detailed review, see Virji, 1996.

True ligand mimicry

This involves direct interactions with receptors via mechanisms that mimic those of natural ligands. The interactions with the receptor in these cases occur at the normal host ligand recognition site and may utilise the ligand recognition motif, for example, the sequence RGD (Arg-Gly-Asp). This motif is present on a wide variety of microbial structures exemplified by *Bordetella pertussis* filamentous haemagglutinin FHA and pertactin (Leininger *et al.*, 1992; Sandros & Tuomanen, 1993). Since RGD is a recognition sequence for several receptors ($\beta1$, $\beta2$, $\beta3/\beta5$ integrin groups), the possession of this sequence has the potential to mediate interactions with multiple receptors.

Pseudo-ligand mimicry

This describes microbial adhesion to host ligands and to their natural receptors by a bridging or sandwich mechanism. This mode of interaction has several advantages in that adhesion to host ligands (which in many cases are extracellular matrix (ECM) proteins) allows an organism not only to adhere to ECM, which may become exposed on damage resulting from mechanical injury or by other infectious agents, but also to the host cell via these proteins. As in the case of the leukocyte complement receptor CR3 (integrin $\alpha m\beta2$), the coating of microbes by the ligand complement component C3bi, renders the microbe resistant to intracellular killing. Several microbes interact with CR3 by prior engagement with the ligand C3bi. In addition to the natural complement deposition that may opsonise many microbes via CR3 (in addition to the non-integrin receptor CR1), some microbial proteins specifically engage with C3 components. *Legionella pneumophila*, which invades and grows within alveolar macrophages and causes pneumonia, appears to localize complement components on its major outer membrane proteins (Bellinger-Kawahara & Horwitz, 1990). Not all sandwich adhesion results in ligand/microbe internalization. It is reported that covering of bacterial surface with fibronectin results in some cases in extracellular location only (Isberg, 1991).

Receptor mimicry

Receptor mimicry or microbial mimicry of integrin structure has been described in *Mycobacteria avium/intracellulare* and in *Candida albicans* and *C. tropicalis*. Both *Mycobacteria* and *Candida* express $\beta1$-like integrins (Rao, Gehlsen & Catanzaro, 1992; Hostetter, 1994). In addition, *Candida albicans* expresses antigenically, structurally and functionally similar protein to the α subunit of the integrins CR3 and CR4. These microbial integrins have been implicated in host cell and extracellular matrix targeting (Hostetter, 1994).

Multiple/complex mimicry

The most striking example of the mimicry of host cellular recognition molecules is seen in *Bordetella pertussis* FHA (Sandros & Tuomanen, 1993). This protein contains several distinct sites involved in interactions with glycoconjugates, heparin and CR3 (an RGD motif). It also contains sequence similarities to endothelial ligands, to factor X and to C3bi and represents mimicry of multiple ligands recognized by the leukocyte CR3 (Sandros & Tuomanen, 1993). *B. pertussis* also expresses other virulence related proteins including pertussis toxin (PT). PT, a hexameric protein, contains subunits S2 and S3 that share similarities with selectins. E-, P-, and L-selectins are involved in inflammatory responses and exhibit lectin-like activity. The PT subunits S2 and S3 mimic this carbohydrate binding property. They recognize lactosylceramide and gangliosides from epithelial cells and macrophages. They also inhibit neutrophil adhesion to purified selectins and leukocyte adhesion to endothelial cells in vitro. Other studies suggest that the subunits S2 and S3 also upregulate CR3 in macrophages, in a manner similar to that observed with selectins. CR3 integrin is also the receptor for *B.pertussis* FHA, thus PT and FHA function as co-operative adhesins (Sandros & Tuomanen, 1993).

Receptor modulation

Microbial modulation of host cell receptors has been observed frequently and has important implications in pathogenesis. For example, increased receptor density may conceivably render host cells more susceptible to invasion by organisms with pathogenic potential able to use these receptors (for example, meningococci and pneumococci that colonise many healthy individuals); since receptor density may play a role in microbial location in or out of the host cell (discussed below). Certain microbes induce their own receptors by supplementary mechanisms. *Bordetella pertussis* appears to upregulate CR3 via FHA interactions involving leukocyte signal transduction complex (comprising a β3 integrin and CD47, Ishibashi, Claus & Relman, 1994) as well as via selectin-like function of PT (Sandros & Tuomanen, 1993). Upregulation of receptors has also been observed during malaria infection. Brain endothelium from patients dying of malaria was shown to express increased levels of several of the implicated malarial receptors (CD36, ICAM-1, ELAM-1, VCAM-1). The upregulation of some of these may result from increased levels of TNFα observed in malaria patients (Pasloske & Howard, 1994).

FACTORS THAT MAY DETERMINE BETWEEN ADHESION AND INVASION

Yersinia with its multiple virulence mechanism provides an interesting example of complex adhesion/invasion events manipulated via encounter with cellular integrins. Chromosomally coded invasin interacts with several $\beta 1$ integrins resulting in bacterial internalization (Isberg, 1991). The factors that determine uptake of yersiniae via $\beta 1$ integrins (which primarily mediate adherence to their ECM ligands) were investigated by Tran van Nhieu and Isberg (1993). One of the mechanisms leading to invasion may be multiple receptor occupancy (as described for CR3 and *Leishmania* interactions which result in internalization, Talamas-Rohana *et al.*, 1990). This does not occur in the case of yersinia invasin/$\beta 1$ interactions since a small region of invasin mediates both binding and invasion (Isberg, 1991). However, unlike the natural ligand, invasin engages with $\beta 1$ integrins via high affinity interactions. This may allow efficient competition with other ligands and result in 'zippering': a process by which the host cell forms a series of contacts over the surface of the microorganism leading to uptake (Isberg, 1991; Tran Van Nhieu & Isberg, 1993). The studies indicated that the site of attachment to integrins was unimportant; high affinity interactions to any site on $\alpha 5\beta 1$ integrins resulted in bacterial internalization. Additionally, receptor clustering achieved by cross-linking of ligands was also a requirement for signal leading to uptake (Tran Van Nhieu & Isberg, 1993). It is also suggested that high receptor density may achieve the same final end. Thus, upregulation of integrin receptors during many viral and other (e.g. malarial) diseases may help augment widespread invasion. In the absence of the invasin protein, YadA, a plasmid coded protein of Yersiniae, can mediate interactions and invasion also via $\beta 1$ integrins and this may occur via a bridging molecule such as fibronectin which binds to YadA. Interestingly, internalization by invasion proteins of Yersiniae may be inhibited by other plasmid coded proteins, Yops (*Yersinia* outer membrane proteins with homologies to eukaryotic signal transduction proteins) that appear to act on host cytoskeletal protein phosphorylation. Thus Yersiniae can bind integrins for cellular invasion, but also have the capacity to modulate subsequent events by interfering with signal transduction via accessory proteins (Bliska, Galan & Falkow, 1993).

MECHANISMS OF INTER-CELLULAR SPREAD

Shigella flexneri enters epithelial cells via basolateral membrane, having first traversed the intestinal epithelial barrier via M cells. The organism lyses of the phagocytic vacuole, grows within the cytoplasm and recruits cellular actin to attain movement within the cytoplasm and for inter-cellular spread. Sansonetti *et al.* (1994) have shown that intercellular spread of *S. flexneri* requires the expression of Cadherins that maintain the intercellular junctional integrity within intact epithelium. Cadherin expression appears to be required for efficient bacterial anchorage to the internal face of the cytoplasmic membrane of infected cells as well as in homotypic interactions between the surface of the protrusions and that of adjacent cell to facilitate internalization by the adjacent cell.

NEISSERIA MENINGITIDIS INTERACTIONS WITH HUMAN CELLS: COMPLEX INTERPLAY BETWEEN SURFACE-LOCATED STRUCTURES, MULTIPLE ADHESIVE LIGANDS AND MOLECULAR MIMICRY IN PATHOGENESIS

N. meningitidis resides in the nasopharynx of its human host and further sequestration occurs from this primary and specific site of colonization. Factors necessary for colonization and for epithelial invasion that lead to serious pathogenic conditions remain to be described fully. Clinical observations suggest that dissemination to the central nervous system occurs via the haematogenous route. Therefore, bacteria must traverse the epithelial and endothelial barriers. The possible routes include direct intra- or inter-cellular translocation in addition to carriage via phagocytic cells. In considering the molecular mechanisms of meningococcal interactions with human target cells, investigations on three adhesins/invasins pili (Opc and Opa) will be described. In addition, modulation of their function by surface polysaccharides will be addressed.

Meningococcal surface polysaccharides: capsule and LPS

Capsule and sialylated lipopolysaccharides are expressed in disseminated isolates and are believed to protect the organism against antibody/complement and phagocytosis. They are also expressed by a number of carrier isolates and may have functions that allow the organism to exist in the nasopharynx (allow avoidance of mucosal immunity) or are physically protective against extra-host environment (anti-desiccation property of capsular polysaccharide). However, acapsulate meningococci are isolated frequently from the nasopharynx (Cartwright, 1995). *In vitro* studies show that adhesion and particularly invasion of epithelial cells is enhanced (aided by some opacity proteins) in the absence of capsule (Virji *et al.*, 1992*b*, 1993*a*). This invites the hypothesis that loss of capsulation may help establish long-term nasopharyngeal carriage where intracellular state would potentially provide protection from host defences. Whether factors in the nasopharynx trigger down-modulation of capsulation is not known, but one study suggests that environmental factors may regulate capsule expression (Masson & Holbein, 1985). In such cases, dissemination from the site of colonization would require upregulation of capsulation since acapsulate bacteria are unlikely to survive in the blood. Alternatively, since blood provides an environment in which meningococci can grow rapidly, it is possible that a small number of capsulate organisms arising as a result of natural phase variation, will be selected for in the blood. In the case of *Haemophilus influenzae*, studies on the infant rat model of *Haemophilus* bacteraemia and meningitis have shown that bacteraemia may arise as a result of survival of a single organism in the blood stream (Moxon & Murphy, 1978).

Meningococci from the nasopharynx often express the L8 LPS immunotype that resists sialylation due to the absence of lacto-*N*-neotetraose structure, a receptor for sialic acid (Cartwright, 1995). Sialylation of LPS has functional consequences similar to capsule and it imparts resistance to immune mechanisms of the host (Hammerschmidt *et al.*, 1993) and in doing so, masks the functions of many outer membrane proteins (Virji *et al.*, 1991, 1993*a*, 1995*a*). The interplay between surface polysaccharides and various adhesins and invasins is a complex area of investigation with antigenically and phase varying components adding to the complexity. Some aspects of this interplay are addressed below.

Pili and their importance in multiple cellular targeting and in potentiation of cellular damage

Both carrier and disease isolates are usually piliated. However, pili are lost on non-selective subculture (Virji *et al.*, 1995*b*), which suggests that pili are selected for *in vivo*. Pili have been implicated in mediating epithelial interactions (Stephens & Farley, 1991) and were shown to mediate haemagglutination (Trust *et al.*, 1983). Pili also mediate adhesion both to human umbilical vein and microvascular endothelial cells (Virji *et al.*, 1991, 1993*b*, 1995*a*). One consequence of pilus-mediated adhesion to endothelial cells is increased cellular damage which is primarily mediated by LPS and is dependent on the presence of serum CD14 (Dunn, Virji & Moxon, 1995; Dunn and Virji, 1996). These *in vitro* toxic effects reflect the acute toxicity of meningococci for endothelial cells observed during vascular dissemination.

Structure/function relationships of meningococcal pili, pilus-associated adhesins

Although two structurally distinct classes of pili occur in meningococci, no discernible functional difference has been assigned to either class. Both undergo antigenic variation which alters their tissue tropism. Studies using adhesion variants (derived by single colony isolation, with or without prior selection on host cells) implied that structural variations in pilin affect epithelial interactions significantly, but have lesser effect on endothelial interactions (Virji *et al.*, 1992*b*). Thus the pilin subunit may contain a

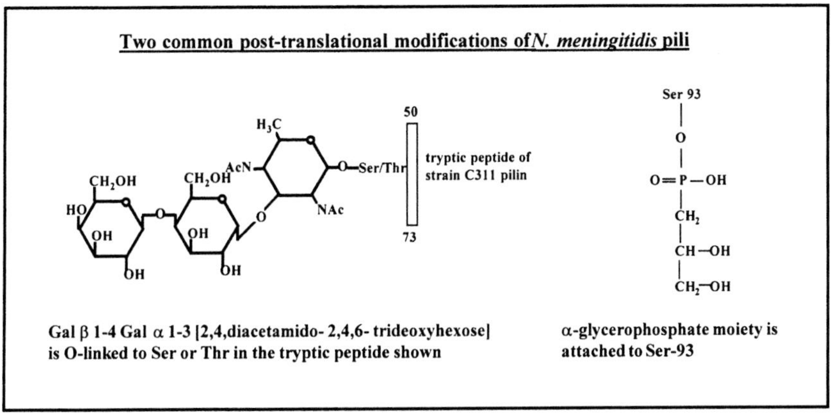

Two common post-translational modifications of *N. meningitidis* pili

Gal β 1-4 Gal α 1-3 [2,4,diacetamido- 2,4,6- trideoxyhexose] is O-linked to Ser or Thr in the tryptic peptide shown

α-glycerophosphate moiety is attached to Ser-93

Fig. 1.

human cellular binding domain, or at least has influence on adhesion if mediated by an accessory protein such as PilC, which has been implicated in cellular adhesion and in biogenesis in both meningococci and gonococci (Jonsson, Nyberg & Normark, 1991; Rudel *et al.*, 1992; Nassif *et al.*, 1994; Virji *et al.*, 1995*b*). At present, how pilin structural variations modulate PilC or other pilus-associated adhesion functions is not clear.

Studies on adhesion variants of meningococci have revealed that meningococcal pili are subject to post-translational modifications (Virji *et al.*, 1993*b*). Also, they contain unusual substitutions. A trisaccharide structure (Galβ1-4,Galα1-3-2,4diacetamido-2,4,6-trideoxyhexose) is present on all variant pili of strain C311 (Stimson *et al.*, 1995). Further recent studies have shown that, at a distinct site, meningococcal pili contain what is, perhaps, a unique substitution, α-glycerophosphate (Stimson *et al.*, 1996). In addition to these, meningococcal pili may contain distinct variant-specific substitutions (Stimson *et al.*, 1995). Such additional modifications have been demonstrated by fast atom bombardment- and electro spray-mass spectrometry, but the structures concerned are not yet elucidated. The functional consequences of pilin modifications are not understood at present (Fig. 1).

Phenotypic requirements for interactions mediated via outer-membrane proteins

In fully capsulate bacteria, only pili appear to be effective in mediating cellular adhesion to human epithelial and endothelial cells. Using acapsulate derivatives of a serogroup A strain (C751), the invasive potential of the meningococcal proteins Opc and Opa was demonstrated (Virji *et al.*, 1992*b*, 1993*a*). More recently, a library of variants and mutants (varying in expression of capsule, LPS, pili, Opa and Opc) was created in strain MC58 (serogroup B). The use of these derivatives has established that Opc can act as an invasin in distinct serogroups, and that surface sialic acids inhibit Opc-mediated invasion. Also, pili may potentiate Opc-mediated invasion of some cells (Virji *et al.*, 1995*a*). As asialylated phenotypes occur in the nasopharynx, and Opc and Opa are expressed in many nasopharyngeal isolates, these proteins may be important in interactions with nasopharyngeal epithelial cells.

Interactions via the outer-membrane protein Opc: molecular mechanisms of interactions with host cell receptors

Opc is a basic protein which is expressed by many meningococcal strains but has not been described in gonococci. It appears to have the capacity to bind to multiple extracellular matrix (ECM) components and serum proteins (Virji *et al.*, 1994*a*) giving the organism the potential to interact with several different integrins by bridging via their respective ECM or soluble ligands. This could be an effective strategy for the organism for cellular invasion and to adhere to substrata of damaged mucosa as well as to penetrate deeper tissues after cellular invasion. Indeed, Opc mediates invasion of cultured epithelial and endothelial cells (Virji *et al.*, 1992*b*). These interactions can be inhibited by monoclonal antibodies against Opc which appears to be the major requirement on bacteria for this interaction. However, the cloned Opc protein does not confer invasive property to the host *E. coli* even though the protein is surface expressed and is immunologically similar to that of *N. meningitidis*. This suggests that additional bacterial factors may be required in host cell interactions mediated by Opc. It is also possible that the level of Opc expressed by *E. coli* is not optimum, since efficient interactions of *N. meningitidis* via Opc require the protein to be expressed at a high density on bacterial surface (Virji *et al.*, 1995*a*) (Fig. 2).

RGD-dependent cellular invasion mediated by Opc

Studies using cultured human endothelial cells have shown that interactions of Opc-expressing *N. meningitidis* with the apical surface of polarised host cells require serum-derived factors (Virji *et al.*, 1994*b*). These factors appear to be RGD-containing proteins and RGDS but not RGES peptides inhibit

---

Fig. 2. Serum-dependent cellular invasion mediated by *N. meningitidis* Opc protein. A, Scanning electron micrograph (SEM) showing acapsulate, Opc-expressing bacteria adherent to human endothelial cell surface and membrane protrusions forming around adherent bacteria (b). B, SEM of an infected endothelial cell in which the cellular plasmalemma has fractured during the preparative procedure revealing intracellular bacteria. C, Transmission electron micrograph showing large numbers of internalized bacteria, typically seen when Opc-expressingg meningococci are incubated with human endothelial cells in the presence of serum. Bars represent 1 μm.

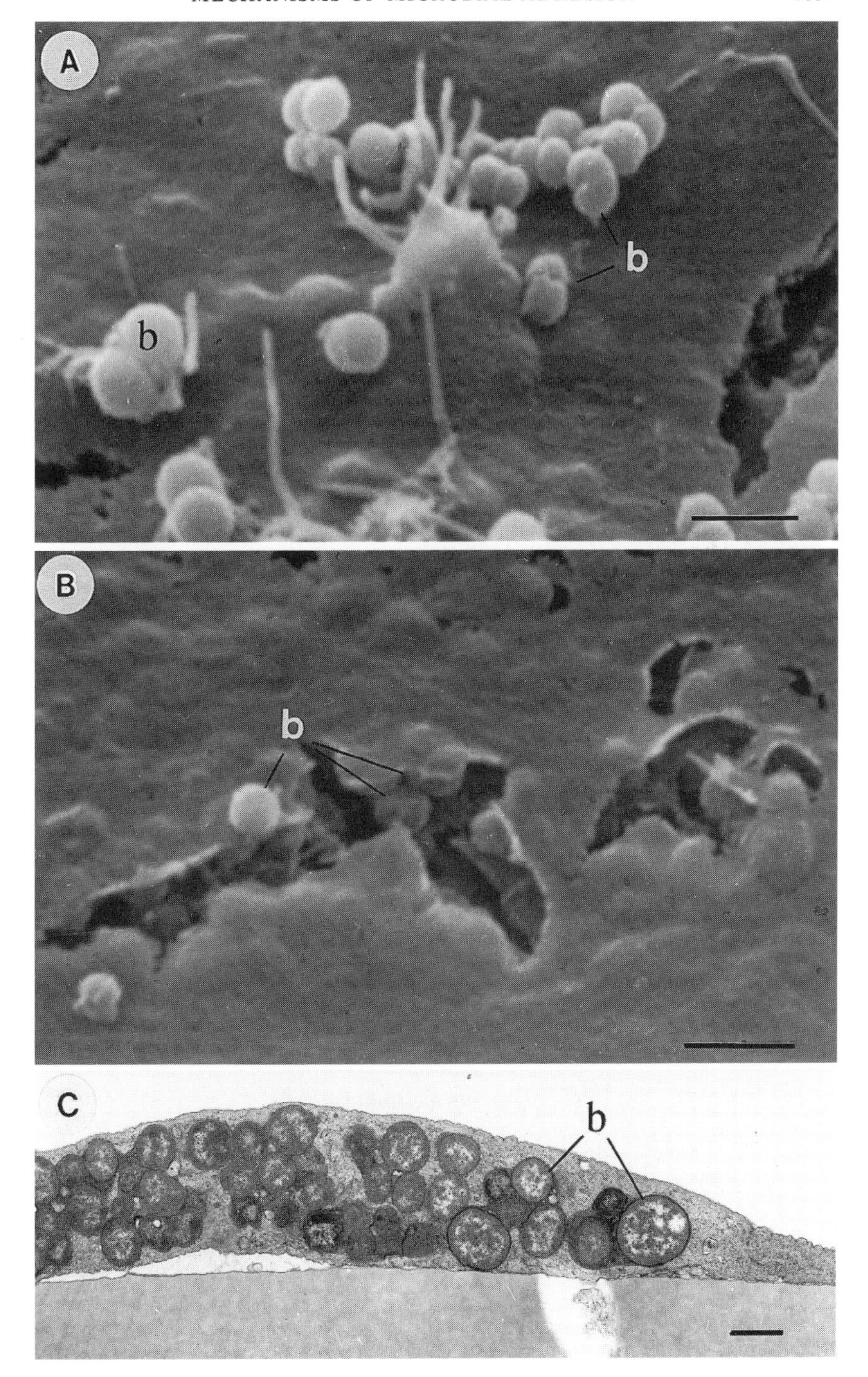

Fig. 2.

bacterial invasion of human endothelial cells. Moreover, antibodies against vitronectin receptor ($\alpha v \beta 3$, VNR) and fibronectin receptor ($\alpha 5 \beta 1$, FNR) inhibit adherence and invasion. VNR appears to be the major receptor involved in serum-dependent apical interactions of *N. meningitidis* (Virji *et al.*, 1994*b*). Some preparations of vitronectin and fibronectin can replace serum factors for cellular invasion.

As discussed above, although pseudo-ligand mimicry appears to be utilized by Opc expressing *N. meningitidis*, further factors may be involved in the interactions via the VNR. In analogy with the complement receptor CR3, which has been shown to interact simultaneously with C3bi-coated particles and with microbial glycolipids at distinct sites (Wright *et al.*, 1989), VNR may require multiple ligand engagement. Indeed the vitronectin receptor also exhibits binding sites for ganglioside GD2 (Cheresh *et al.*, 1987). Gangliosides and LPS share structural similarities in that both are amphipathic with strongly anionic hydrophilic groups and it is tempting to speculate that some manner of LPS interaction with the vitronectin receptor may be an additional factor required.

An interesting feature of Opc interaction was the requirement for host cytoskeletal function. Attempts to inhibit host cell invasion by the use of cytochalasin D resulted in inhibition not only of invasion but also of total cell association. This observation is in contrast to cell adhesion mediated by *N. meningitidis* Opa proteins, which apparently increases in the presence of cytochalasin D (Virji *et al.*, 1994*b*). It has been suggested that efficient bacterial internalization requires direct adherence of bacteria to host receptors (Isberg, 1991) and fibronectin-dependent sandwich interactions do not mediate uptake unless receptors are disengaged from matrix binding. Affinity of ligand/receptor interactions may also determine attachment/uptake via the same receptor (Isberg, 1991). In the case of *N. meningitidis*, serum protein dependence for interactions which is an invasion event, may suggest high affinity interactions with the receptors; the details of which remain to be elucidated.

The role of Opa proteins

Opa proteins are a family of related proteins expressed both by meningococci and gonococci. In contrast to gonococci which may encode up to 12 distinct Opa proteins, a single meningococcal strain encodes fewer (< 4) Opa proteins. Studies on meningococcal serogroup A strain C751, have shown that, of the three Opa proteins expressed by this strain, OpaB and OpaD are effective in epithelial adhesion and invasion whereas OpaA is ineffective (Virji *et al.*, 1993*a*). A receptor for a gonococcal Opa protein on some epithelial cells has been described (van Putten & Paul, 1995). Whether OpaB and OpaD of C751 also engage this receptor is not known at present.

However, on certain epithelial cells and PMN, another receptor may be targeted by meningococcal Opa proteins (see below).

Interactions of meningococci with monocytes and polymorphs: the role of surface virulence factors (Capsule, LPS, pili, Opa and Opc)

Bacterial components that mediate cellular interactions in the absence of added opsonins are of importance from the point of view of phagocytes acting as potential 'Trojan horse' carriers of bacteria. Moreover, up to 16% of the cells present in the nasal mucosa are monocytes, and inflammation increases PMN infiltration. In studies to identify bacterial factors that increase phagocyte interactions, it was shown that capsulate bacteria resist phagocytosis in the absence of opsonins, but acapsulate bacteria are internalized and opacity proteins Opa and Opc mediate bacterial uptake. Pili of distinct structural make-up, or the pilus-associated protein PilC, were ineffective in mediating interactions with phagocytic cells (McNeil, Virji & Moxon, 1994; McNeil & Virji, 1996).

Neutrophil and some epithelial cell receptors for Opa proteins

COS cells transfected with cloned cDNA encoding several distinct surface molecules which include constitutively expressed or inducible human cellular adhesion molecules were used to identify receptors for meningococcal adhesins. These studies showed that some meningococci adhered only to transfected cells that expressed the CD66a molecule at high levels. CD66a (BGP, biliary glycoprotein) is a member of the Immunoglobulin superfamily. It belongs to carcinoembryonic antigen (CEA or CD66) family, which comprises numerous structurally related, secreted, GPI-anchored or transmembrane proteins (Watt *et al.*, 1994, Stanners *et al.*, 1995) (Fig. 3).

In studies using transformed chinese hamster ovary cells expressing CD66a or CD33, specific interaction of CD66a with Opa proteins of a number of distinct strains was observed. Its adherence to distinct Opa proteins of a single strain was demonstrated in an ELISA using defined derivatives of strain C751. These results showed that all three Opa proteins of C751 interacted with CD66a. A monoclonal antibody against the N-terminal domain of CD66a and soluble chimeric receptors inhibited interactions of several Opa-expressing meningococci with PMN. PMN and some epithelial cells express several members of the CD66 family including BGP (Stanners *et al.*, 1995; Teixeira *et al.*, 1994). These have highly homologous N-terminal domains. Therefore, meningococci may target multiple molecules by interacting via common determinants on distinct members of CD66. Indeed, Opa-mediated meningococcal interactions with a colonic carcinoma cell line HT29 as well as a lung carcinoma cell line A549 were demonstrated by the

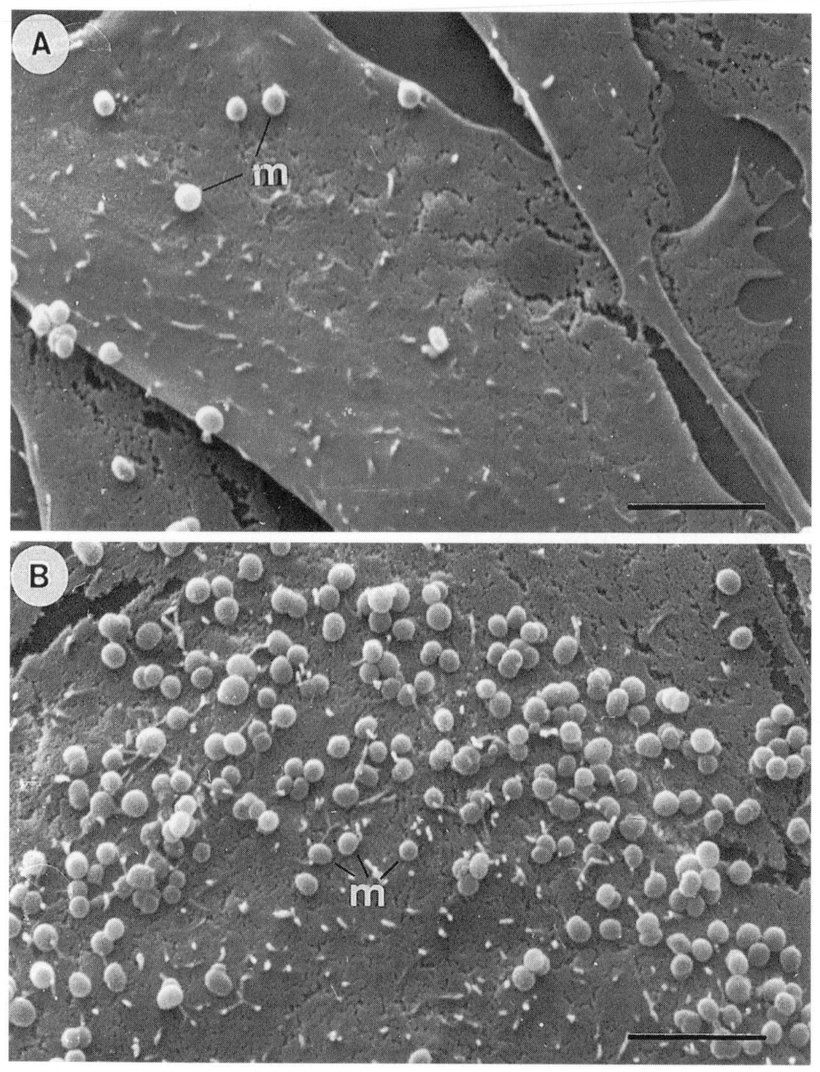

Fig. 3. Opa targeting of human CD66. Scanning electron micrographs showing the adherence of an acapsulate, Opa-expressing variant of meningococci (m) to transformed Chinese hamster ovary cells expressing human CD33 (A) or human CD66 molecules (B). Bars represent 5 μm.

use of antibodies against CD66 and soluble CD66a-Fc chimeric constructs (Virji *et al.*, 1996).

Numerous human mucosal and disease isolates (including 50 strains each of gonococci and meningococci), were examined for their interactions with soluble CD66a-Fc. Specific adherence of the N-terminal domain of CD66a to $>$ 95% of Opa-expressing pathogenic *Neisseriae* was observed. No commensal *Neisseriae* (14 strains) or other human commensals or pathogens (including *E. coli*, *Pseudomonas* and *Haemophilus*: a total of 16 strains tested) adhered significantly to CD66a-Fc. These studies show that CD66a is a target for a conserved epitope present on the majority of Opa proteins of gonococci and meningococci.

An important observation in these studies was that some capsulate bacteria expressing certain Opa proteins with high affinity for CD66 interacted with cells expressing large numbers of these receptors. Therefore, inhibitory effects of surface sialic acids may be overcome to some extent when appropriate ligand/receptor pairs are present at the required density. Thus, targeting of the CEA members of cell adhesion molecules that are upregulated during inflammation (Dansky-Ullmann *et al.*, 1995) may be critical to pathogenesis of meningococcal infection and may shift the balance from carrier state to dissemination. A low level constitutive expression of the receptor, for example, on epithelial cells, may favour attachment to epithelial cells without invasion. Viral infections or other conditions during which cytokines may be upregulated, could result in increased expression of CD66 and related adhesion molecules increasing the potential of meningococci to enter both phagocytic cells as well as mucosal epithelial cells. Massive epithelial cellular invasion could be injurious to the host, that of phagocytic cells could result in incomplete elimination of bacteria and possibility of transmission within phagocytic cells. Indeed, epidemiological investigations have implicated increased susceptibility to meningococcal infection following viral and other infections (Cartwright, 1995; Achtman, 1995). These may contribute to meningococcal dissemination by increasing inflammation. Whether this represents a possible mechanism which determines host susceptibility to meningococcal invasion, and perhaps also contributes to the development of serious complications during some gonococcal infections, remains to be shown.

ACKNOWLEDGEMENTS

I would like to acknowledge my colleagues and collaborators, particularly Professors E. Richard Moxon, Jon Saunders and Anne Dell. I would also like to thank Dr David J. P. Ferguson for his help with electron microscopy.

REFERENCES

Achtman, M. (1995). Epidemic spread and antigenic variability of *Neisseria meningitidis*. *Trends in Microbiology*, **3**, 186–92.

Bellinger-Kawahara, C. & Horwitz, M.A. (1990). Complement component C3 fixes selectively to the major outer membrane protein (MOMP) of *Legionella pneumophila* and mediates phagocytosis of liposome-MOMP complexes by human monocytes. *Journal of Experimental Medicine*, **172**, 1201–10.

Bliska, J.B., Galan, J.E. & Falkow, S. (1993). Signal transduction in the mammalian cell during bacterial attachment and entry. *Cell*, **73**, 903–20.

Cartwright, K. ed. (1995). *Meningococcal Disease*. Chichester, UK: John Wiley and Sons.

Cheresh, D.A., Pytela, R., Pierschbacher, M.D., Klier, F.G., Ruoslahti, E. & Reisfeld, R.A. (1987). An Arg-Gly-Asp-directed receptor on the surface of human melanoma cells exists in an divalent cation-dependent functional complex with the disialoganglioside GD2. *Journal of Cell Biology*, **105**, 1163–73.

Dansky-Ullmann, C., Salgaller, M., Adams, S., Schlom, J. & Greiner, J. W. (1995). Synergistic effects of IL-6 and IFN-γ on carcinoembryonic antigen (CEA) and HLA expression by human colorectal carcinoma cells: role for endogenous IFN-β. *Cytokine*, **7**, 118–29.

Dunn, K.L.R & Virji, M. (1996). *Neisseria meningitidis* toxicity for cultured human endothelial cells requires soluble CD14. In Zollinger, W.D., Frasch, C.E. & Deal, C.D., eds. *Pathogenic Neisseria*. pp. 275–276. Bethesda, ND, USA: NIH.

Dunn, K.L.R, Virji, M. & Moxon, E.R. (1995). Investigations into the molecular basis of meningococcal toxicity for human endothelial and epithelial cells: the synergistic effect of LPS and pili. *Microbial Pathogenesis*, **18**, 81–96.

Hammerschmidt, S., Birkholz, C., Zahringer, U., Robertson, B.D., van Putten, J., Ebeling, O. & Frosch, M. (1993). Contribution of genes from the capsule gene complex *(cps) to* lipooligosaccharide biosynthesis and serum resistance in *Neisseria meningitidis*. *Molecular Microbiology*, **11**, 885–96.

Hostetter, M.K. (1994). Adhesins and ligands involved in the interaction of *Candida* spp. with epithelial and endothelial surfaces. *Clinical Microbiology Reviews*, **7**, 29–42.

Isberg, R.R. (1991). Discrimination between intracellular uptake and surface adhesion of bacterial pathogens. *Science*, **252**, 934–8.

Ishibashi, Y., Claus, S. & Relman, D.A. (1994). *Bordetella pertussis* filamentous hemagglutinin interacts with a leukocyte signal transduction complex and stimulates bacterial adherence to monocyte CR3 (CD11b/CD18). *Journal of Experimental Medicine*, **180**, 1225–33.

Jonsson, A-B., Nyberg, G. & Normark, S. (1991). Phase variation of gonococcal pili by frameshift mutation in *pilC*, a novel gene for pilus assembly. *EMBO Journal*, **10**, 477–88.

Leininger, E., Ewanowich, C.A., Bhargava, A., Peppler, M.S., Kenimer, J.G. & Brennan, M.J. (1992). Comparative roles of the Arg-Gly-Asp sequence present in the *Bordetella pertussis* adhesins pertactin and filamentous hemagglutinin. *Infection and Immunity*, **60**, 2380–5.

Masson, L. & Holbein, B.E. (1985). Influence of environmental conditions on serogroup B *Neisseria meningitidis* capsular polysaccharide levels. In Schoolnik, G.K. *et. al.*, eds. *The Pathogenic Neisseriae*. pp. 571–8. Washington DC: American Society for Microbiology.

McNeil, G. & Virji, M. (1996). Meningococcal interactions with human phagocytic cells: a study on defined phenotypic variants. In Zollinger, W.D., Frasch, C.E. & Deal, C.D., eds. *Pathogenic Neisseria*. pp. 320–2. Bethesda, ND, USA: NIH.

McNeil, G., Virji, M. & Moxon, E.R. (1994). Interactions of *Neisseria meningitidis* with human monocytes. *Microbial Pathogenesis*, **16**, 153–63.

Moxon, E.R. & Murphy, P.A. (1978). *Haemophilus influenzae* bacteraemia and meningitis resulting from survival of a single organism. *Proceedings of the National Academy of Sciences*, USA, **75**, 1534–6.

Nassif, X., Beretti, J-L., Lowy, J., Stenberg, P., O'Gaora, P., Pfeifer, J., Normark, S. & So, M. (1994). Roles of pilin and PiIC in adhesion of *Neisseria meningitidis* to human epithelial and endothelial cells. *Proceedings of the National Academy of Sciences*, USA, **91**, 3769–73.

Pasloske, B.L. & Howard, R.J. (1994). Malaria, the red cell, and the endothelium. *Annual Review in Medicine*, **45**, 283–95.

Rao, S.P., Gehlsen, K.R. & Catanzaro, A. (1992). Identification of a b1 integrin on *Mycobacterium avium–mycobacterium intracellulare. Infection and Immunity*, **60**, 3652–7.

Rudel, T., van Putten, J.P.M., Gibbs, C.P., Haas, R. & Meyer, T.F. (1992). Interaction of two variable proteins (pilE and pilC) required for pilus-mediated adherence of *Neisseria gonorrhoeae* to human epithelial cells. *Molecular Microbiology*, **6**, 3439–50.

Sandros, J. & Tuomanen, E. (1993). Attachment factors of *Bordetella pertussis*: mimicry of eukaryotic cell recognition molecules. *Trends in Microbiology*, **1**, 192–6.

Sansonetti, P.J., Mounier, J., Prevost, M.C. & Mege, R.M. (1994). Cadherin expression is required for the spread of *Shigella flexneri* between epithelial cells. *Cell*, **76**, 829–39.

Stanners, C.P., DeMarte, L., Rojas, M., Gold, P. & Fuchs A. (1995). Opposite functions for two classes of the human carcinoembryonic antigen family. *Tumor Biology*, **16**, 23–31.

Stephens, D.S. & Farley, M.M. (1991). Pathogenic events during infection of the human nasopharynx with *Neisseria meningitidis* and *Haemophilus influenzae*. *Reviews in Infectious Diseases*, **13**, 22–33.

Stimson, E., Virji, M., Barker, S., Panico, M., Blench, I., Saunders, J., Payne, G., Moxon, E.R., Dell, A. & Morris, H.R. (1996). Discovery of a novel protein modification: a-glycerophosphate is a substituent of meningococcal pilin. In Zollinger, W.D., Frasch, C.E. & Deal, C.D., eds. *Pathogenic Neisseria*. pp. 275–6. Bethesda, ND, USA: NIH.

Stimson, E., Virji, M., Makepeace, K., Dell, A., Morris, H., Payne, G., Saunders, J., Jennings, M., Barker, S., Panico, M., Blench, I. & Moxon, E.R. (1995). Meningococcal pilin: a glycoprotein substituted with digalactosyl 2,4-diacetamido-2,4,6-trideoxyhexose. *Molecular Microbiology*, **17**, 1201–14.

Talamas-Rohana, P., Wright, S.D., Lennartz, M.R. & Russell, D.G. (1990). Lipophosphoglycan from *Leishmania mexicana* promastigotes binds to members of the CR3, p150,95 and LFA-1 family of leukocyte integrins. *Journal of Immunology*, **144**, 4817–24.

Teixeira, A.M., Fawcett, J., Simmons, D.L. & Watt, S.M. (1994). The N-domain of the biliary glycoprotein (BGP) adhesion molecule mediates homotypic binding: domain interactions and epitope analysis of CD66a. *Blood*, **84**, 211–19.

Tran Van Nhieu, G. & Isberg, R.R. (1993). Bacterial internalization mediated by β1 chain integrins is determined by ligand affinity and receptor density. *EMBO Journal*, **12**, 1887–95.

Trust, T.J., Gillespie, R.M., Bhatti, A.R. & White, L.A. (1983). Differences in the adhesive properties of *Neissseria meningitidis* for human buccal epithilial cells and erythrocytes. *Infection and Immunity*, **41**, 106–13.

van Putten, J.P.M. & Paul, S.M. (1995). Binding of syndecan-like cell surface proteoglycan receptors is required for *Neisseria gonorrhoeae* entry into human mucosal cells. *EMBO Journal*, **14**, 2144–54.

Virji, M., Kayhty, H., Ferguson, D.J.P., Alexandrescu, C., Heckels, J.E. & Moxon, E.R. (1991). The role of pili in the interactions of pathogenic *Neisseria* with cultured human endothelial cells. *Molecular Microbiology*, **5**, 1831–41.

Virji, M., Alexandrescu, C., Ferguson, D.J.P., Saunders, J.R. & Moxon, E.R. (1992*a*). Variations in the expression of pili: the effect on adherence of *Neisseria* *meningitidis* to human epithelial and endothelial cells. *Molecular Microbiology*, **6**, 1271–9.

Virji, M., Makepeace, K., Ferguson, D.J.P, Achtman, M., Sarkari, J. & Moxon, E.R. (1992*b*). Expression of the Opc protein correlates with invasion of epithelial and endothelial cells by *Neisseria meningitidis*. *Molecular Microbiology*, **6**, 2785–95.

Virji, M., Makepeace, K., Ferguson, D.J.P., Achtman, M. & Moxon, E.R. (1993*a*). Meningococcal Opa and Opc proteins: role in colonisation and invasion of human epithelial and endothelial cells. *Molecular Microbiology*, **10**, 499–510.

Virji, M., Saunders, J.R., Sims, G., Makepeace, K., Maskell, D. & Ferguson, D.J.P. (1993*b*). Pilus-facilitated adherence of *Neisseria meningitidis* to human epithelial and endothelial cells: modulation of adherence phenotype occurs concurrently with changes in amino acid sequence and the glycosylation status of pilin. *Molecular Microbiology*, **10**, 1013–28.

Virji, M., Makepeace, K. & Moxon, E.R. (1994*a*). Distinct mechanisms of interaction of Opc-expressing meningococci at apical and basolateral surfaces of human endothelial cells; the role of integrins in apical interactions. *Molecular Microbiology*, **14**, 173–84.

Virji, M., Makepeace, K. & Moxon, E.R. (1994*b*). Meningococcal outer membrane protein Opc mediates interactions with multiple extracellular matrix components. In Evans, J.S., Yost, S.E., Maiden, M.C.J. & Feavers, I.M., eds. *Neisseria 94*. pp. 263–4. England: NIBSC.

Virji, M., Makepeace, K., Peak, I.R.P., Payne, G., Saunders, J.R., Ferguson, D.J.P. & Moxon, E.R. (1995*a*). Functional implications of the expression of PilC proteins in meningococci. *Molecular Microbiology*, **16**, 1087–97.

Virji, M., Makepeace, K., Peak, I.R.A., Ferguson, D.JP., Jennings, M.P. & Moxon, E.R. (1995*b*). Opc- and pilus-dependent interactions of meningococci with human endothelial cells: molecular mechanisms and modulation by surface polysaccharides. *Molecular Microbiology*, **18**, 741–54.

Virji, M. (1996). Adhesion receptors in microbial pathogenesis. In Horton, M.A., ed. *Molecular Biology of Cell Adhesion Molecules*. Chichester, UK: John Wiley and Sons (in press).

Virji, M., Watt, S. M., Barker, S. & Makepeace, K. (1996). The N-domain of the human CD66a adhesion molecule is a target for Opa proteins of *Neisseria meningitidis*. In Zollinger, W.D., Frasch, C.E. & Deal, C.D., eds. *Pathogenic Neisseria*. pp. 316–19. Bethesda, ND, USA: NIH.

Watt, S.M., Fawcett, J., Murdoch, S.J., Teixeira, A.M., Gschmeissner, S.E., Hajibagheri N.M. & Simmons, D.L. (1994). CD66 identifies the biliary glycoprotein (BGP) adhesion molecule: cloning, expression, and adhesion functions of the CD66a splice variant. *Blood*, **84**, 200–10.

Wright, S.D., Levin, S.M., Jong, M.T., Chad, Z. & Kabbash, L.G. (1989). CR3 (CD11b/CD18) expresses one binding site for Arg-Gly-Asp-containing peptides and a second site for bacterial lipopolysaccharide. *Journal of Experimental Medicine*, **169**, 175–83.

VIRUS–HOST INTERACTIONS IN THE CONTROL OF THE GENE EXPRESSION OF NUCLEAR REPLICATING DNA VIRUSES

R. D. EVERETT

MRC Virology Unit, Church Street, Glasgow G11 5JR, UK

INTRODUCTION

This chapter aims to give an overview of the interactions between viral and cellular proteins that modulate the control of gene expression of nuclear replicating DNA viruses. This is a huge field, and space limitations preclude the presentation of a comprehensive review. Instead, the broad principles involved will be discussed, mainly concentrating on herpes simplex virus. However, a number of examples from other herpesviruses, adenoviruses and papovaviruses will also be presented where appropriate. Comprehensive recent reviews of most of the topics covered in this chapter can be found in Fields (1996).

Interactions between viral and cellular proteins play key roles in the control of viral gene expression. Indeed, it is a general principle that nuclear replicating DNA viruses modify or re-direct the host's transcriptional apparatus in order to express their genes, and without such interactions the virus would simply be unable to replicate. Unlike the cytoplasmically replicating DNA or RNA viruses, the nuclear DNA viruses do not encode their own RNA polymerase and they generally utilize host transcription factors which bind to sequences in the viral promoters. Another general principle to be emphasized is that there are a number of different stages at which a virus can control its gene expression, and a number of different strategies have been adopted. The control of viral gene expression, or even simply of transcription, is not just a case of interactions within the transcriptional initiation complex itself.

THE STAGES OF HERPESVIRUS GENE EXPRESSION

The herpesviruses comprise a large family with members that infect vertebrates from fish and amphibians to mammals. On the basis of genome arrangement and biological properties, the herpesviruses can be sub-divided into three sub-families, namely the alpha-, beta- and gammaherpesviruses. All have large double-stranded DNA genomes ranging from about 100 kb to over 200 kb. The complete genomic sequences of several members of the family have been determined, revealing some genes which are conserved in all three sub-families and others which are sub-family specific. In the case of herpes simplex virus type 1 (HSV-1) at least 76 distinct genes have been identified. In this chapter, as well as a detailed considereration of HSV-1, pseudorabies virus (PRV) will also be discussed which, like HSV-1 is an alphaherpesvirus, while from the beta- and gammaherpesvirus sub-families human cytomegalovirus (HCMV), and Epstein-Barr virus (EBV) will be referred to, respectively.

Viral gene expression can generally be divided into three phases; immediate–early (IE), early and late. The immediate–early phase is characterized by the production of viral regulatory proteins whose roles concern the efficient and controlled transcription of the other viral genes. Therefore, IE proteins are often transactivators of early and late promoters, but the mechanisms that have been adopted to achieve this vary significantly between different transactivators and viruses. Early gene products include those that bring about viral DNA replication, and the late class of promoters are most efficiently utilized only after replication of the parental viral genomes. While it is possible to separate these three phases using viral mutants or metabolic inhibitors, it is important to realize that, in practice, at least in an HSV-1 infection in cultured cells, the onset of viral DNA replication is rapid and all three classes of genes can be expressed almost simultaneously. Other herpesviruses may replicate more slowly, but even so the expression of their IE, early and late gene products are not rigidly separable during the course of a normal infection.

Amongst the questions that immediately arise from these observations are (i) what allows the IE promoters to be efficiently recognized and utilized at the onset of virus infection, and (ii) how do the viral IE regulators recognise and activate viral early and late promoters? The answers to these questions are now reasonably well understood, but in the examples that follow it will be seen that the various viral modulators of gene expression utilize a variety of different mechanisms to achieve similar goals. However, the control of gene expression does not simply concern the mechanism of the onset of specific transcription because the primary transcripts must be polyadenylated, spliced and exported from the nucleus. All of these stages are subject to control, and interactions between viral and host factors frequently determine the efficiency of these processes. Finally, there are perhaps other, more global, levels of control which could involve the structural organization of the nucleus. While still rather speculative and poorly understood, control at this level involves intimate interactions between the virus and the cell and is a very exciting research area of the future.

THE CONTROL OF IE TRANSCRIPTION

Enhancers

When the incoming viral genome has finally arrived in the nucleus, after adsorption, penetration and uncoating of the virus particle, how does the virus ensure that its genome is transcribed to initiate the replicative cycle? This is obviously a crucial stage, even more so when one considers that the viral genome comprises a minute proportion of all potentially active genes in the host cell. The answer lies in enhancer elements, sequences present in the IE promoters of all nuclear replicating DNA viruses which direct the host's transcriptional apparatus onto the viral genome. Enhancers were first discovered in the papovaviruses SV40 and polyoma, but one of the strongest is that of HCMV (Boshart *et al.*, 1985; for review see Nelson, Gnann & Ghazal, 1990). An underlying principle of enhancers is that they contain closely packed and often multiple binding sites for a variety of host transcription factors; binding of these factors stimulates the formation of active preinitiation complexes at the nearby IE promoter proximal elements through interactions between the transcription factors and components of the basal transcription machinery.

Regulated IE promoters

While many enhancers use the unmodified host transcription machinery which simply binds to the viral DNA, some viruses have evolved a system in which a viral protein becomes part of the host factor complex and boosts even further IE gene transcription. The best understood example is that of the regulation of HSV-1 IE transcription, which is now a paradigm for virus–host interactions in transcriptional control. All HSV-1 IE promoters contain one or more copies of an IE-specific regulatory sequence of consensus TAATGARAT (where R is a purine). These motifs are normally located within a few hundred base pairs of the proximal IE promoter sequences, but in conjuction with their flanking sequences they are discrete functional entities which can confer IE-specific regulation to other proximal promoter elements of different temporal class (for review see O'Hare, 1993).

The viral protein which is crucial for HSV-1 IE-specific regulation is VP16 (also known as Vmw65 or αTIF) which is present in the viral tegument (a complex structure located between the viral capsid and the outer envelope). VP16 is therefore imported into the nucleus with the viral genome before viral gene expression has begun. VP16 has little DNA binding activity by itself, but overlapping the 5′ side of the TAATGARAT sequence is a binding site for the cellular transcription factor Oct-1. VP16 binds to Oct-1 in a tripartite complex with another cellular factor, HCF (Wilson *et al.*, 1993), and the formation of this complex results in efficient stimulation of IE promoters (for review see O'Hare, 1993). Therefore, the host factor Oct-1 provides the specific DNA binding function while transcriptional stimulation is brought about by recruitment of VP16 which has a very strong transactivation domain in its acidic C-terminal region. The role of HCF is not well understood, but it is required for the binding of VP16 to Oct-1 and it may play a structural role to ensure correct presentation of the acidic tail of VP16 to the transcriptional initiation complex. The subtleties of these interactions are further underlined by the observation that the GARAT part of the IE consensus sequence, while not an obligate part of the Oct-1 binding site, appears to alter the conformation of Oct-1 such that it can bind VP16 (Walker, Hayes & O'Hare, 1994). Thus, not only does the virus recruit a cellular transcription factor, the viral DNA also modifies it so that the functional activation complex with the viral transactivator will be formed only on the viral IE-specific response sequences, and not on the related cellular Oct-1 recognition sites.

There is another possible consequence of the use of cellular transcription factors in the regulation of IE gene expression. Oct-1 is one of a family of related transcription factors and while many of its relatives have been shown to bind to TAATGARAT sequences, not all can also bind VP16. Obviously, binding of an octamer factor which cannot recruit VP16 could repress IE promoter activity. Therefore, in principle, the balance of particular octamer factors within a cell could influence the efficiency of IE gene expression. This idea forms the basis of an attractive but controversial model to explain the establishment of herpesvirus latency in neuronal cells (see, for example, Lillycrop & Latchman, 1992, but compare with Hagmann *et al.*, 1995).

EARLY PROMOTERS, HOST TRANSCRIPTION FACTORS AND VIRAL TRANSACTIVATORS

The protein products of many viral IE genes are transactivators of viral gene expression which are required for transcription from early and late promoters. However, examination of the majority of early and late promoter sequences of herpes and adenoviruses reveals the presence of binding sites for host transcription factors but little evidence of virus specific regulatory sequences exemplified by the HSV IE response element. A number of examples of promoter stimulation by viral transactivators are now quite well understood. All require interactions between viral and host proteins but a number of different strategies have been adopted to achieve similar goals.

Viral transactivator binding to host transcription factors

This mechanism is closely related to the example of IE promoter activation by VP16 and Oct-1 discussed above. An example where it is used for the activation of early viral promoters is that of EBNA2 responsive genes in EBV. Like VP16, the EBV transactivator EBNA2 does not bind to DNA directly but its responsive promoters contain binding sites for the cellular factor RBP-Jκ. EBNA2 forms a complex with RBP-Jκ, so binding of RBP-Jκ in the promoter region results in the recruitment of EBNA2 and presentation of the latter's transactivation domain to the transcriptional initiation complex (Henkel *et al.*, 1994; Waltzer *et al.*, 1994). A further subtlety of this system is that RBP-Jκ contains a domain which normally represses transcription, and binding of EBNA2 masks the repression domain (Hsieh & Hayward, 1995). Again, the concept of regulated viral promoter activity resulting from the balance of viral and cellular factors is an obvious consequence of these findings.

Viral transactivator binding to host transcription factor repressors

The activity of many host transcription factors can be modulated by their sequestration in inactive forms. For example, NFκB is retained in the cytoplasm in a complex with its inhibitor IκB and can only migrate in an active form into the nucleus after IκB has been specifically phosphorylated, ubiquitinated and degraded (Scherer *et al.*, 1995). The possibility exists that a virus could utilize such a transcription factor by specifically inactivating the inhibitor. This basic principle is utilized by the adenovirus E1A protein in order to activate transcription via transcription factor E2F. The role of E2F in the cell is in the regulated expression of many cell cycle-related genes and its activity is therefore also regulated in a cell cycle-dependent manner. This is achieved by the binding of E2F to the Rb family of proteins, such that the E2F-Rb complex is unable to activate transcription (Dalton, 1992). Adenoviruses have developed an extremely elegant mechanism to harness the properties of E2F to activate both its own promoters and also to stimulate the cell cycle in order to induce synthesis of host proteins which are required for viral replication. This is achieved by the simple process of E1A binding to the same site on Rb to which E2F binds (Bagchi, Raychaudhuri & Nevins, 1990; Bandara & La Thangue, 1991). The competition between the two releases free E2F which is then able to activate both viral and cellular promoters which contain its cognate recognition sites. This mechanism may not be limited to the adenoviruses since the HCMV ie2 transactivator also binds to Rb to release free E2F (Hagemeier *et al.*, 1994).

Viral transactivator stimulation of the basal transcription initiation complex

The initiation of transcription requires the binding to the TATA box of the TATA binding protein (TBP) in association with its various associated factors (TAFs; see Zhou, Boeyer & Berk, 1993) in a complex termed TFIID. Subsequently the other components of the active complex are assembled in a stepwise fashion: firstly TFIIB, then TFIIF in complex with RNA polymerase II (which itself comprises a large number of distinct polypeptides), then finally TFIIE, TFIIH and TFIIJ (for review see Conaway & Conaway, 1993). The mechanisms involved are probably similar on both viral and cellular promoters and an obvious target for a viral transactivator is to improve the rate of assembly of the complex on the viral genome. This subject has been a very active area of research.

Since the initial and rate-limiting step of assembly of the transcription initiation complex is the binding of TFIID to the TATA box, this is an attractive target for viral regulation. Early experiments using *in vitro* transcription methods established that the pseudorabies virus major transactivator increased the efficiency of this process, especially when it was made intrinsically inefficient by low template concentrations, competition by non-specific DNA binding proteins or reconstitution of nucleosomes (Abmayr, Workman & Roeder, 1988; Workman *et al.*, 1988). Later experiments using the corresponding major HSV-1 transactivator, Vmw175, extended these findings by showing that a tri-partite complex of Vmw175, TFIID and TFIIB formed on promoter DNA and that the stimulation of transcription *in vitro* by Vmw175 was dependent on the formation of this complex (Smith *et al.*, 1993; Gu & DeLuca, 1994). More detailed studies have shown that Vmw175 interacts directly with the TAF250 component of TFIID, and that this interaction correlates with transcriptional activation (Carrozza & DeLuca, 1996). These experiments suggest a clear and logical mechanism underlying the activation of transcription by the Vmw175 family of transactivators that are expressed by all alphaherpesviruses. However, there is a complication which has yet to be fully resolved. The Vmw175 family of proteins bind to specific DNA sequences using a conserved DNA binding domain which has been shown to be important for the transactivation of transcription. Paradoxically, Vmw175 and its related proteins activate a variety of promoters from diverse origins, whether or not they contain binding sites with high homology to the consensus in their vicinity (see Smiley *et al.*, 1992 for a detailed discussion of this point). Therefore the question arises of the role of the DNA binding properties of Vmw175 in its mechanism of action, since high affinity binding sites are not generally present in its target promoters. A possible reconciliation of these facts comes from observations that, within an order of magnitude or so of DNA binding affinity, Vmw175 is able to bind to a range of sequences which can be quite divergent from the original consensus (Everett, Orr & Elliott, 1991 and references therein). Perhaps it is this relatively non-specific DNA binding activity which is involved in the formation of the tripartite complex of Vmw175, TFIID and TFIIB on promoter DNA.

In contrast, high affinity Vmw175 binding sites have been found at the cap sites of promoters that are down-regulated by Vmw175, including that of its own transcription unit (Muller, 1987). The presence of this site and the ability of Vmw175 to bind to it are both essential for the observed autoregulation of Vmw175 expression (Roberts *et al.*, 1988). In this instance, locking Vmw175 onto a strong binding site at a specific location may either impede the passage of the transcription complex or result in the formation of a tripartite complex which is inappropriately positioned to allow assembly of the rest of the transcription machinery (Gu, Kuddus & DeLuca, 1995; Kuddus, Gu & DeLuca, 1995).

Viral transactivator interactions with components of the basal transcription machinery

The above example of the Vmw175 family of proteins provides a relatively well-understood example of a viral transactivator influencing the activity of the basal transcription machinery, and the conclusions have been supported by a combination of binding and functional assays. However, the study of the interactions between viral and cellular regulators and individual components of the transcription initiation complex has exploded into an extensive industry, fuelled by the ease of simple protein–protein interaction assays. The genes for most of the components have been cloned and the proteins can be expressed in high amounts, allowing interactions to be studied by a battery of *in vitro* methods. Centre stage has been TBP itself, and especially its conserved region which is thought to provide a protein interaction surface. In addition, interactions with individual TAFs, TFIIB and other components of the basal transcription machinery have been extensively explored, and the result is a bewildering tangle of putative interactions. For this reason, a description of all the published interactions between viral transactivators and TBP and other basal transcription factors is beyond the scope of this review.

However, the data from simple protein–protein binding assays must be considered carefully before it can be concluded the observed interaction is biologically meaningful. Some assays, such as GST pull-down experiments, are subject to non-specific binding effects (especially when the two target proteins are the only components of the assay present at high concentration), while others, such as far-Western blotting, depend on the assumption of correct re-folding of denatured protein. So, while these approaches are certainly very useful and allow rapid experimentation, it is always reassuring if the interaction can be confirmed by independent methods and its consequences can be determined in a functional assay. An example of the complex problems which can arise concerns the activation domain of VP16 which has been shown to bind *in vitro* to TBP, TFIIB and TFIIH. Recent work has questioned which of these interactions is important for transactivation by VP16 by comparing the effects of mutations in VP16 on its ability to activate transcription and to bind to the various factors. Since the VP16 mutations that were investigated in this study reduced transactivation and also reduced binding to TBP and TFIIH, but not to TFIIB, it seems unlikely that TFIIB is a functional target of the VP16 transactivation domain (Gupta *et al.*, 1996). Similarly, another recent study has called into question the *in vivo* significance of the *in vitro* interaction of VP16 with TBP (Tansey & Herr, 1995).

However, doubts cast on the relevance of some interactions between viral transactivators and TBP or other basal transcription factors should by no means be generally extended. There are good reasons for believing that interactions between viral regulatory proteins and TBP, TAFs in the TFIID complex or the other basal factors play important roles in the control of viral transcription, but a future emphasis must be to provide detailed evidence that these interactions are required for functional activity.

Modification of RNA polymerase II

RNA polymerase II is a complex assembly of many subunits, the largest of which includes the catalytic active sites. When the polymerase is recruited into the preinitiation complex in association with TFIIF, a heptad repeat sequence in its carboxy-terminal domain (CTD) is unphosphorylated. After promoter binding has occurred, the CTD becomes highly phosphorylated in a process which requires TFIIH and associated kinase activities. The role of this phosphorylation is not clear since in some experimental systems kinase-deficient derivatives of TFIIH can still support activated transcription. However, it has been suggested that phosphorylation might allow the polymerase to disengage from the initiation complex to commence transcription. Such a process seems a likely target for transcriptional control, and therefore it is interesting that the HSV-1 immediate–early protein Vmw68 has been found to induce an altered, intermediate phosphorylation status of the CTD (Rice *et al.*, 1995). Vmw68 is required for fully efficient viral gene transcription in some cell lines, and while no clear correlation can be drawn between this phenotype and its ability to induce abnormal phosphorylation of the CTD, these observations at least open the possibility that viral transcription could be modulated by modification of RNA polymerase II.

DNA BINDING VIRAL TRANSACTIVATORS

The examples of transcriptional control mechanisms discussed above have concerned sequences within viral promoters that bind host transcription factors and how viral transactivators modify these interactions to stimulate transcription. However, it may be thought simpler for the virus to encode its own DNA binding transcription factors that would bind to virus-specific sequences in viral promoters. Perhaps surprisingly, this has not been a commonly adopted strategy. While the alphaherpesviruses encode a member of the DNA-binding Vmw175 family of transactivators, consensus DNA binding sites are not generally present in responsive promoters, nor are these sites required for transcriptional activation (Smiley *et al.*, 1992). Similarly the HCMV ie2 transactivator protein binds to a specific DNA sequence which is not obviously present in all HCMV early and late promoters, but in a way analogous to the situation with Vmw175 discussed above, the presence of an ie2 recognition site at the cap site of its own promoter is required for autoregulation of ie2 expression (Lang & Stamminger, 1993; Macias & Stinski, 1993). There has been no convincing demonstration that the adenovirus E1A transactivator binds efficiently or specifically to DNA, and the DNA binding properties of the large T-antigens of SV40 and polyoma virus are required for their DNA replication properties rather than any direct effect on transcriptional control through promoter-specific sequences.

Examples of more specific transactivators have however been found in the papillomaviruses and in the gamma herpesvirus EBV. The papillomavirus E2 protein binds strongly and specifically to a number of sites within the promoter/regulatory region of the viral genome, and binding to these sites can modulate transcription either positively or negatively depending on the position of the site or the particular papillomavirus being studied (McBride, Romanczuk & Howley, 1991). EBV encodes a number of transactivators amongst which are Zta and Rta, both of which bind to specific sequences within responsive promoters (Farrell *et al.*, 1989; Gruffat & Sergeant, 1994). In these cases, the mechanism of transactivation probably involves binding of these factors to regulatory sites upstream of the TATA box region, thereby presenting activation domains which interact with components of the basal transcription machinery.

POST-TRANSCRIPTIONAL REGULATION

Viral gene expression is also subject to various post-transcriptional controls. In general, these are less well understood than regulation at the level of transcription, but none the less they are important and an active area of current research. Due to limitations of space, I can only briefly discuss these topics.

Splicing

The degree of splicing of the primary transcripts varies between the different DNA viruses from a small minority of messages (for example, the alphaherpesviruses) to extensive splicing of most or all mRNAs (for example, the adenoviruses). Alternative splicing can produce related proteins of differing functions (for example the adenovirus E1 region gene products and the HCMV ie1 and ie2 proteins) and enables the use of different reading frames in the same stretch of DNA. There are examples in which splice site selection varies as viral infection proceeds, which strongly suggests that splicing can be a regulated process, yet the mechanisms by which this occurs remain very poorly understood.

An example of a viral gene product which interacts with host factors to influence splicing efficiency is provided by Vmw63 of HSV-1. This IE protein is essential for virus replication and is required for fully efficient late gene expression (McCarthy, McMahan & Schaffer, 1989). It has become clear that Vmw63 most likely acts at the post-transcriptional level and it has been found to inhibit splicing (Sandri-Goldin & Mendoza, 1992). Since most HSV-1 mRNAs are unspliced, this activity will inhibit cellular but not viral gene expression. Interestingly, HSV-1 infection results in the aggregation of splicing factors into discrete foci in the nucleus and Vmw63 is both necessary and sufficient for this effect (Phelan *et al.*, 1993). Recently it has been found that Vmw63 is an RNA binding protein which also co-immune precipitates with snRNPs (ribonucleoprotein complexes involved in mRNA processing) (Sandri-Goldin & Hibbard, 1996), thus providing evidence for direct physical associations which undoubtedly will be found to play a crucial role in the functions of Vmw63.

Polyadenylation

Before a primary transcript can be exported from the nucleus it must be cleaved and polyadenylated, so gene expression levels also depend on how efficiently polyA sites are selected and used. HSV-1 infection has been shown to increase the efficiency of usage of certain weak viral polyA sites through induction of an as yet poorly defined activity called LPF ('late polyadenylation factor'). The induction of LPF also depends on the synthesis of Vmw63 (see above) although in this case the role of Vmw63 is thought to be indirect, perhaps as a consequence of de-coupling splicing from polyadenylation which might release additional polyadenylation factors for processing of unspliced messages (McGregor *et al.*, 1996).

RNA transport

The final step before a mature mRNA can be translated is its export from the nucleus into the cytoplasm. The mechanisms by which this occurs are again poorly understood, but viral gene products can again influence this process. Perhaps the best documented example in a DNA virus is that of the adenovirus E1B 55 kD protein, which is necessary for efficient export of late adenovirus messages (Pilder *et al.*, 1986).

HSV-1 VMW110: AN ENIGMATIC TRANSACTIVATOR

The examples presented above illustrate how interactions between viral transactivators and host factors or processes can modulate viral gene expression at each stage from the recruitment of host transcription factors to the DNA through to export of the final mRNA. However, there are other possible levels of control which might play crucial roles in the efficiency of viral infection. As an illustration, transactivator Vmw110 of HSV-1 will be discussed, which has turned up a steady stream of surprises over the past 10 years.

Vmw110 is a non-specific activator of gene expression

Vmw110 is the product of gene IE-1 of HSV-1. Early transfection assays showed that it was an activator of gene expression, and that in some circumstances it could work synergistically with the major viral transactivator, Vmw175. It was rapidly discovered that Vmw110 could activate expression from all promoters that have been tested, and despite extensive efforts, no indication for the requirement for any particular promoter sequence for its activity could be found (for review see Everett, Preston & Stow, 1991). Consistent with these results, purified Vmw110 failed to exhibit any significant or specific DNA binding activity in solution (Everett, Orr & Elliott, 1991). Taken with the finding that viruses from which the gene encoding Vmw175 have been deleted fail to activate early and late promoters despite expressing abundant amounts of Vmw110 (DeLuca, McCarthy & Schaffer, 1985), these results seem to exclude the possibility that Vmw110 acts as a classical transcription factor.

Another pertinent result is that viruses which fail to express Vmw110 synthesize essentially normal amounts of viral gene products during a high multiplicity infection, but they have a multiplicity-dependent defect in the onset of virus infection (Stow & Stow, 1986). In other words, a virus particle which does not express Vmw110 has a low probability of entering the lytic cycle, but once viral gene expression has been successfully initiated the replication cycle proceeds essentially normally. That normal amounts of most or all viral transcripts and proteins can be produced in the absence of Vmw110 suggests that it does not have an obligate role in post-transcriptional processing or translation of viral mRNAs.

An intriguing observation is that the defect of Vmw110 deficient viruses can be relieved in a cell cycle- or cell type-dependent manner, thus suggesting that cellular factors play a significant role in the mechanism of activation of gene expression of Vmw110 (Cai & Schaffer, 1991; Yao & Schaffer, 1995). The reconciliation of all of the above observations is not easy, and also rather controversial, and it must be admitted that it is not universally accepted that Vmw110 does not function directly at the level of transcription. Nevertheless I consider that the following data strongly point towards a mechanism of action for Vmw110 which does not involve direct or specific control of the initiation of transcription.

Vmw110 localizes to discrete domains within the nucleus

Vmw110 is a phosphoprotein which at early times of infection localizes to discrete punctate domains within the nucleus. A significant advance occured when it was found that these local accumulations were not random, but corresponded (at least in part) with defined nuclear structures (herein called ND10) which contain at least five cellular proteins (Maul, Guldner & Spivak, 1993). Although the functions of most of these cellular proteins are unknown, one of them has generated much interest since it has been implicated in the onset of promyelocytic leukaemia (and is therefore known as PML). The affected leukaemic blasts have suffered a chromosomal translocation, as a result of which the N-terminal portion of PML becomes linked to the retinoic acid receptor. The consequence is uncontrolled proliferation of the cells, perhaps caused by disturbance of the retinoic acid control pathways, but in addition the PML-containing ND10 structures are modified from their normal appearance (on average about ten discrete foci per nucleus) so that a multitude of dispersed micropunctate structures are formed (see, for example, Dyck *et al.*, 1994). Hence it is generally held that the normal function of the PML domains is involved in some way with the growth status of the cell. However, ND10 do not correspond to abundant sites of transcription, DNA replication, splicing or any other previously described sub-nuclear structure (Weis *et al.*, 1994).

Interestingly, the consequence of the initial localization of Vmw110 to ND10 is the dispersal of the constituent proteins (Maul *et al.*, 1993). That this property of Vmw110 is functionally significant is indicated by the observation that mutations which eliminate the ability of Vmw110 to localize to these sites, or to disrupt them, invariably compromise the ability of Vmw110 to activate gene expression (Everett & Maul, 1994). While these results do not further the understanding of the mechanism of activation of gene expression by Vmw110 at the molecular level, they suggest the possibility that the organization of specific nuclear structures is an important factor in the control of viral gene expression. This concept is further underlined by the observation that incoming viral genomes preferentially localize to ND10, and that the viral 'replication factories' (which contain accumulations of replicated viral DNA and replication proteins) also preferentially form adjacent to ND10 (Maul, Ishov & Everett, 1996).

The modification of ND10 by Vmw110 could result in the activation of gene expression by a variety of mechanisms. For example, if the functions of ND10 would normally inhibit virus infection (and it may be relevant that PML has been suggested to be a growth suppressor; Mu *et al.*, 1994), their destruction by Vmw110 would relieve this repression. Alternatively, Vmw110 may release factors which enhance gene expression and which are normally sequestered in ND10. An even more speculative explanation is that some more general modification of nuclear architecture or the nuclear matrix may be brought about by Vmw110.

Vmw110 binds to a novel member of the ubiquitin specific protease family

The results discussed above clearly show that Vmw110 interacts with cellular structures, and therefore it is likely that Vmw110 binds to specific cellular proteins. The identification of these proteins would clearly enhance our understanding of the function of Vmw110 at the molecular level, and could also further the analysis of the role of ND10 during normal cell growth. Based on the observation that sequences within the C-terminal 180 residues of Vmw110 are required for migration of Vmw110 to ND10, we set out to identify cellular proteins that interact with this region of Vmw110. One such protein was detected which bound to Vmw110 very strongly and specifically, not only in *in vitro* assays but also in immune precipitates from virus-infected cells (Meredith, Orr & Everett, 1994). Furthermore, Vmw110 deletion mutants unable to establish this interaction had significantly reduced abilities to activate gene expression (Meredith *et al.*, 1995). We were able to purify the bound cellular protein and obtain peptide sequence information which was used to clone the encoding cDNA. Rather surprisingly, this identified the protein as a member of the ubiquitin specific protease (USP) family (Everett *et al.*, 1997). Even more interestingly, we found that this protein is a component of at least some of the ND10 structures.

Ubiquitin-dependent pathways have been found to play essential roles in an increasing number of cellular control mechanisms, amongst which are cell cycle progression and transcription factor activation (for review see Hochstrasser, 1995). A substrate protein which becomes ubiquitinated (often in a specific, regulated manner) is targeted for destruction by the proteasome, but ubiquitinated substrates may be protected from this fate by USP enzymes which remove the ubiquitin adducts. Cells express many USPs, some of which are involved in the production of free ubiquitin from its precursors while others regenerate ubiquitin from degraded substrates (for review see Wilkinson, 1995). However, there is increasing evidence that certain USPs have specific regulatory functions, for example, in the cell cycle or development, and two have been implicated in human oncogenesis (for references, see Everett *et al.*, 1997). It is possible that binding of Vmw110 to this particular USP may inhibit, activate or re-direct its activity, thereby either stabilizing or de-stabilizing its normal substrate and, directly or indirectly, allowing increased viral gene expression. The details of this putative mechanism await identification of the normal substrate for the USP which binds to Vmw110, but it is intriguing that a ubiquitin homology protein has very recently been identified as a component of ND10 (Boddy *et al.*, 1996). Since both Vmw110 and its associated USP can localize to ND10, it seems likely that ubiquitin-dependent control mechanisms are occuring at these nuclear sites. Whatever the details of the processes involved, these studies suggest that herpes simplex virus gene expression may be modulated by Vmw110 by mechanisms which, at present, appear to be quite distinct from the actual process of transcription.

CONCLUSIONS

Gene expression during nuclear DNA virus infection can be regulated at a great many different stages, from structural organization of nuclear components, through modulation of host transcription factor and basal transcriptional machinery activity, to processing of the pre-mRNAs and their transport to the cytoplasm. Of course, a variety of translational controls may also be applied in certain instances. While it is tempting to believe that control at the level of transcription itself is the dominant factor, it should be remembered that mutations in viral proteins which are involved in the other steps of the overall pathway of gene expression can be very deleterious and even fatal for the virus. However, the guiding principle at every stage of gene expresssion is that the virus does not function independently of the cell, but must adopt and adapt the machinery of the cell for its own ends. At the centre of these processes are interactions between viral and cellular proteins at the molecular level.

ACKNOWLEDGEMENTS

While attempts have been made to describe the wide range of mechanisms underlying the control of DNA virus gene expression, this presentation cannot be considered an exhaustive review. Therefore, the selection of references is far from complete, and apologies are offered to those whose work has not been cited. The helpful comments of Nigel Stow, Michayla Meredith and Jane Parkinson are much appreciated. The recent work from the authors own laboratory that has been summarized here was done in collaboration with Michayla Meredith, Anne Orr and Anne Cross and was supported by the Medical Research Council.

REFERENCES

Abmayr, S.M., Workman, J.L. & Roder, R.G. (1988). The pseudorabies immediate early protein stimulates *in vitro* transcription by facillitating TFIID: promoter interactions. *Genes and Development*, **2**, 542–53.

Bagchi, S., Raychaudhuri, P. & Nevins, J.R. (1990). Adenovirus E1A protein can dissociate heteromeric complexes involving the E2F transcription factor: a novel mechanism for E1A transactivation. *Cell*, **62**, 659–69.

Bandara, L.R. & La Thangue, N.B. (1991). Adenovirus E1A prevents the retinoblas-

toma gene product from complexing with a cellular transcription factor. *Nature*, **351**, 494–7.
Boddy, M.N., Howe, K., Etkin, L.D., Solomon, E. & Freemont, P.S. (1996). PIC1, a novel ubiquitin-like protein which interacts with the PML component of a multiprotein complex that is disrupted in acute promyelocytic leukaemia. *Oncogene,* **13**, 971–82.
Boshart, M., Weber, F., Jahn, G., Dorch-Hasler, K., Fleckinstein, B. & Schaffner, W. (1985). A very strong enhancer is located upstream of an immediate early gene of cytomegalovirus. *Cell*, **41**, 521–30.
Cai, W. & Schaffer, P.A. (1991). A cellular function can enhance gene expression and plating efficiency of a mutant defective in the gene for ICP0, a transactivating protein of herpes simplex virus type 1. *Journal of Virology*, **65**, 4078–90.
Carrozza, M.J. & DeLuca, N.A. (1996). Interaction of the viral activator protein ICP4 with TFIID through TAF250. *Molecular and Cellular Biology*, **16**, 3085–93.
Conaway, R.C. & Conaway, J.W. (1993). General initiation factors for RNA polymerase II. *Annual Reviews of Biochemistry*, **62**, 161–90.
Dalton, S. (1992). Cell cycle regulation of the human cdc2 gene. *The EMBO Journal*, **11**, 1797–804.
DeLuca, N.A., McCarthy, A.M. & Schaffer, P.A. (1985). Isolation and characterization of deletion mutants of herpes simplex virus type 1 in the gene encoding immediate–early regulatory protein ICP4. *Journal of Virology*, **56**, 558–70.
Dyck, J.A., Maul, G.G., Miller, W.H.Jr., Chen, J.D., Kakizuka, A. & Evans, R.M. (1994). A novel macromolecular structure is a target of the promyelocytic-retinoic acid receptor oncoprotein. *Cell*, **76**, 333–43.
Everett, R.D., Elliott, M., Hope, G. & Orr, A. (1991). Purification of the DNA binding domain of herpes simplex virus type 1 immediate early protein Vmw175 as a homodimer and extensive mutagenesis of its DNA recognition site. *Nucleic Acids Research*, **19**, 4901–8.
Everett, R.D., Orr, A. & Elliott, M. (1991). High level expression and purification of herpes simplex virus type 1 immediate–early polypeptide Vmw110. *Nucleic Acids Research*, **19**, 6155–61.
Everett, R.D., Preston, C.M. & Stow, N.D. (1991). Functional and genetic analysis of the role of Vmw110 in herpes simplex virus replication. In Wagner, E.K., ed. *The Control of Herpes Simplex Virus Gene Expression*. pp. 50–76. Boca Raton: CRC Press Inc.
Everett, R.D. & Maul, G.G. (1994). HSV-1 IE protein Vmw110 causes redistribution of PML. *The EMBO Journal*, **13**, 5062–9.
Everett, R.D., Meredith, M., Orr, A., Cross, A., Kathoria, M. & Parkinson, J. (1997). A novel ubiquitin-specific protease is dynamically associated with the PML nuclear domain and binds to a herpesvirus regulatory protein. *The EMBO Journal*, in press.
Farrell, P.J., Rowe, D.T., Rooney, C.M. & Kouzarides, T. (1989). Epstein–Barr virus BZLF1 transactivator specifically binds to a consensus AP1 site and is related to c-fos. *The EMBO Journal*, **8**, 127–32.
Fields, B.N., ed. (1996). *Virology*, 3rd edition. Philadelphia: Lippincott-Raven Publishers.
Gruffat, H. & Sergeant, A. (1994). Characterisation of the DNA binding site repertoire for the Epstein–Barr virus transcription factor R. *Nucleic Acids Research*, **22**, 1172–8.
Gu, B. & DeLuca, N.A. (1994). Requirements for activation of the herpes simplex virus glycoprotein C promoter *in vitro* by the viral regulatory protein ICP4. *Journal of Virology*, **68**, 7953–65.
Gu, B., Kuddus, R. & DeLuca, N.A. (1995). Repression of activator mediated

transcription by herpes simplex virus ICP4 via a mechanism involving interactions with the basal transcription factors TATA-binding protein and TFIIB. *Molecular and Cellular Biology*, **15**, 3618–26.
Gupta, R. Emili, A., Pan, G., Xioa, H., Shales, M., Greenblat, J. & Ingles, C.J. (1996). Characterisation of the interaction between the acidic activation domain of VP16 and the RNA polymeerase II initiation factor TFIIB. *Nucleic Acids Research*, **24**, 2324–30.
Hagemeier, C., Caswell, R., Hayhurst, G., Sinclair, J. & Kouzarides, T. (1994). Functional interaction between the HCMV IE2 transactivator and the retinoblastoma protein. *The EMBO Journal*, **13**, 2897–903.
Hagmann, M., Georgiev, O., Schaffner, W. & Douville, P. (1995). Transcription factors interacting with herpes simplex virus α gene promoters in sensory neurones. *Nucleic Acids Research*, **23**, 4978–85.
Henkel, T., Ling, P.D., Hayward, S.D. & Peterson, M.G. (1994). Mediation of Epstein–Barr virus EBNA2 transactivation by recombination signal binding protein Jκ. *Science*, **265**, 92–5.
Hochstrasser, M. (1995). Ubiquitin, proteasomes, and the regulation of intracellular protein degradation. *Current Opinion in Cell Biology*, **7**, 215–23.
Hsieh, J.J.D. & Hayward, S.D. (1995). Masking of the CBF1/RBP-Jκ transcriptional repression domain by Epstein–Barr virus EBNA2. *Science*, **268**, 560–3.
Kuddus, R., Gu, B. & DeLuca, N.A. (1995). Relationship between TATA-binding protein and herpes simplex virus type 1 ICP4 DNA binding sites in complex formation and repression of transcription. *Journal of Virology*, **69**, 5568–75.
Lang, D. & Stamminger, T. (1993). The 86-kilodalton IE-2 protein of human cytomegalovirus is a sequence specific DNA binding protein that interacts directly with the negative autoregulatory response element located near the cap site of the IE1/2 enhancer–promoter. *Journal of Virology*, **67**, 323–31.
Lillycrop, K.A. & Latchman, D.S. (1992). Alternative splicing of the Oct-2 transcription factor RNA is differentially regulated in neuronal cells and B cells and results in protein isoforms with opposite effects on the activity of octamer/TAATGARAT-containing promoters. *Journal of Biological Chemistry*, **267**, 24960–5.
McBride, A.A., Romanczuk, H & Howley, P.M. (1991). The papillomavirus E2 regulatory proteins. *Journal of Biological Chemistry*, **266**, 18411–14.
McCarthy, A.M., McMahan, L. & Schaffer, P.A. (1989). Herpes simplex virus type 1 ICP27 deletion mutants exhibit altered patterns of transcription and are DNA deficient. *Journal of Virology*, **63**, 18–27.
McGregor, F., Phelan, A., Dunlop, J. & Clements, J.B. (1996). Regulation of herpes simplex virus poly(A) site usage and the action of immediate–early protein IE63 in the early–late switch. *Journal of Virology*, **70**, 1931–40.
Macias, M.P. & Stinski, M.F. (1993). An *in vitro* system for human cytomegalovirus immediate early 2 protein (IE-2). mediated site dependent repression of transcription and biding of IE2 to the major immediate early promoter. *Proceedings of the National Academy of Sciences, USA*, **90**, 707–11.
Maul, G.G., Guldner, H.H. & Spivack, J.G. (1993). Modification of discrete nuclear domains induced by herpes simplex virus type 1 immediate–early gene 1 product ICP0. *Journal of General Virology*, **74**, 2679–90.
Maul, G.G., Ishov, A. & Everett, R.D. (1996). Nuclear domain 10 as preexisting potential replication start sites of herpes simplex virus type 1. *Virology*, **217**, 67–75.
Meredith, M., Orr, A. & Everett, R.D. (1994). Herpes simplex virus type 1 immediate–early protein Vmw110 binds strongly and specifically to a 135 kD cellular protein. *Virology*, **200**, 457–69.
Meredith, M.R., Orr, A., Elliott, M. & Everett, R.D. (1995). Separation of the

sequence requirements for HSV-1 Vmw110 multimerisation and interaction with a 135 kD cellular protein. *Virology*, **209**, 174–87.

Mu, Z., Chin, K., Liu, J., Lozano, G. & Chang, K. (1994). PML, a growth suppressor disrupted in acute promyelocytic leukaemia. *Molecular and Cellular Biology*, **14**, 6858–67.

Muller, M. (1987). Binding of herpes simplex type 1 gene product ICP4 to its own transcription start site. *Journal of Virology*, **61**, 858–65.

Nelson, J.A., Gnann, J.J. & Ghazal, P. (1990). Regulation and tissue-specific expression of human cytomegalovirus. *Current Topics in Microbiology and Immunology*, **154**, 75–100.

O'Hare, P. (1993). The virion transactivator of herpes simplex virus. *Seminars in Virology*, **4**, 145–55.

Phelan, A., Carmo-Fonseca, M., McLaughlan, J., Lamond, A.I. & Clements, J.B. (1993). A herpes simplex virus type 1 immediate–early gene product, IE63, regulates small nuclear ribonucleoprotein distribution. *Proceedings of the National Academy of Sciences, USA*, **90**, 9056–60.

Pilder, S., Moore, M., Logan, J. & Shenk, T. (1986). The adenovirus E1B-55K transforming polypeptide modulates transport or cytoplasmic stabilization of viral and host mRNAs. *Molecular and Cellular Biology*, **6**, 470–6.

Rice, S.A., Long, M.C., Lam, V., Schaffer, P.A. & Spencer, C.A. (1995). Herpes simplex virus immediate–early protein ICP22 is required for viral modification of host RNA polymerase II and establishment of the normal viral transcriptional program. *Journal of Virology*, **69**, 5550–9.

Roberts, M.S., Boundy, A., O'Hare, P., Pizzorno, M.C., Ciufo, D.M. & Hayward, G.S. (1988). Direct correlation between a negative autoregulatory responce element at the cap site of the herpes simplex virus type 1 IE175 (α4). promoter and a specific binding site for the IE175 (ICP4). protein. *Journal of Virology*, **62**, 4307–20.

Sandri-Goldin, R.M. & Mendoza, G.E. (1992). A hepesvirus regulatory protein appears to act post-transcriptionally by affecting RNA processing. *Genes and Development*, **6**, 848–63.

Sandri-Goldin, R.M. & Hibbard, M. (1996). The herpes simplex virus type 1 regulatory protein ICP27 coimmuneprecipitates with anti-Sm antiserum, and the C terminus appears to be required for this interaction. *Journal of Virology*, **70**, 108–18.

Scherer, D.C., Brockman, J.A., Chen, Z., Maniatis, T. & Ballard, D.W. (1995). Signal-induced degradation of IκBα requires site-specific ubiquitination. *Proceedings of the National Academy of Sciences, USA*, **92**, 11259–63.

Smiley, J.R., Johnson, D.C., Pizer, L. & Everett, R.D. (1992). The ICP4 binding sites in the herpes simplex virus type 1 glycoprotein D (gD) promoter are not essential for efficient gD transcription during virus infection. *Journal of Virology*, **66**, 623–31.

Smith, C.A., Bates, P., Rivera-Gonzales, R., Gu, B. & DeLuca, N.A. (1993). ICP4, the major transcriptional regulatory protein of herpes simplex virus type 1, forms a tri-partite complex with TATA-binding protein and TFIIB. *Journal of Virology*, **67**, 4676–87.

Stow, N.D. & Stow, E.C. (1986). Isolation and characterisation of a herpes simplex virus type 1 mutant containing a deletion within the gene encoding the immediate–early polypeptide Vmw110. *Journal of General Virology*, **67**, 2571–85.

Tansey, W.P. & Herr, W. (1995). The ability to associate with activation domains *in vitro* is not required for the TATA box binding protein to support activated transcription *in vivo*. *Proceedings of the National Academy of Sciences, USA*, **92**, 10550–4.

Walker, S., Hayes, S. & O'Hare, P. (1994). Site-specific conformational alteration of the Oct-1 POU domain-DNA complex as the basis for differential recognition by Vmw65 (VP16). *Cell*, **79**, 841–52.

Waltzer, L., Logeat, F., Brou, C., Israel, A., Sergeant, A. & Manet, E. (1994). The human Jκ recombination signal sequence binding protein (RBP-Jκ) targets the Epstein–Barr virus EBNA2 protein to its responsive elements. *The EMBO Journal*, **13**, 5633–8.

Weis, K., Rambaud, S., Lavau, C., Jansen, J., Carvalho, T., Carmo-Fonseca, M., Lamond, A. & DeJean, A. (1994). Retinoic acid regulates aberrant nuclear localization of PML-RARα in acute promyelocytic leukaemia cells. *Cell*, **76**, 345–56.

Wilkinson, K.D. (1995). Roles of ubiquitination in proteolysis and cellular regulation. *Annual Reviews of Nutrition*, **15**, 161–89.

Wilson, A.C., LaMarco, K., Peterson, M.G. & Herr, W. (1993). The VP16 accessory protein HCF is a family of polypeptides processed from a large precursor protein. *Cell*, **74**, 115–25.

Workman, J.L., Abmayr, S.M., Cromlish, W.A. & Roeder, R.G. (1988). Transcriptional regulation by the immediate–early protein of pseudorabies virus during *in vitro* nucleosome assembly. *Cell*, **55**, 211–19.

Yao, F. & Schaffer, P.A. (1995). An activity specified by the osteosarcoma line U2OS can substitute functionally for ICP0, a major regulatory protein of herpes simplex virus type 1. *Journal of Virology*, **69**, 6249–58.

Zhou, Q., Boeyer, T.G. & Berk, A.J. (1993). Factors (TAFs). required for activated transcription interact with TATA box binding protein conserved core domain. *Genes and Development*, **7**, 180–7.

REGULATION OF HUMAN IMMUNODEFICIENCY VIRUS GENE EXPRESSION BY TAT

J. KARN, N. J. KEEN, M. J. CHURCHER, F. ABOUL-ELA, G. VARANI AND M. J. GAIT

MRC Laboratory of Molecular Biology, Hills Road, Cambridge CB2 2QH, UK

INTRODUCTION

One of the most surprising results to emerge from the intensive research into the molecular biology of human immunodeficiency virus (HIV), which was prompted by the AIDS crisis, was the discovery that the regulation of HIV gene expression is mediated by the binding of viral regulatory proteins to specific RNA target sequences, rather than to DNA targets. In this chapter, we describe how one of these RNA-binding proteins, the viral *trans*-activator protein, Tat, regulates HIV transcription through a novel elongation mechanism.

The first evidence that gene expression in HIV requires virally encoded *trans*-acting factors came from experiments in 1985 by Haseltine, Sodroski and colleagues in which it was noted that transcription of reporter genes placed under the control of the viral long terminal repeat (LTR) was stimulated 200 to 300-fold in cells which had been previously infected by HIV (Sodroski *et al.*, 1985*a*,*b*). The *trans*-acting factor, now known to be encoded by the HIV *tat* gene, is a small, basic nuclear protein. As shown in Fig. 1, the *tat* gene contains two exons which partially overlap the *rev* and *env* genes.

THE TAT–TAR INTERACTION

Deletion analysis of the HIV LTR demonstrated that Tat activity requires the *trans*-activation-responsive region (TAR), a regulatory element of 59 nucleotides (nt) located immediately downstream of the initiation site for transcription (Fig. 1) (Cullen, 1986; Feng & Holland, 1988; Muesing, Smith & Capon, 1987; Rosen, Sodroski & Haseltine, 1985). TAR functions primarily as an RNA element rather than a DNA element. In contrast to enhancer elements, the TAR element is only active when it is placed downstream of the start of HIV transcription and in the correct orientation and position (Berkhout, Silverman & Jeang, 1989). As is also shown in Fig. 1, TAR RNA forms a highly stable, nuclease-resistant, stem–loop structure (Muesing *et al.*, 1987). Point mutations that disrupt base-pairing in the upper

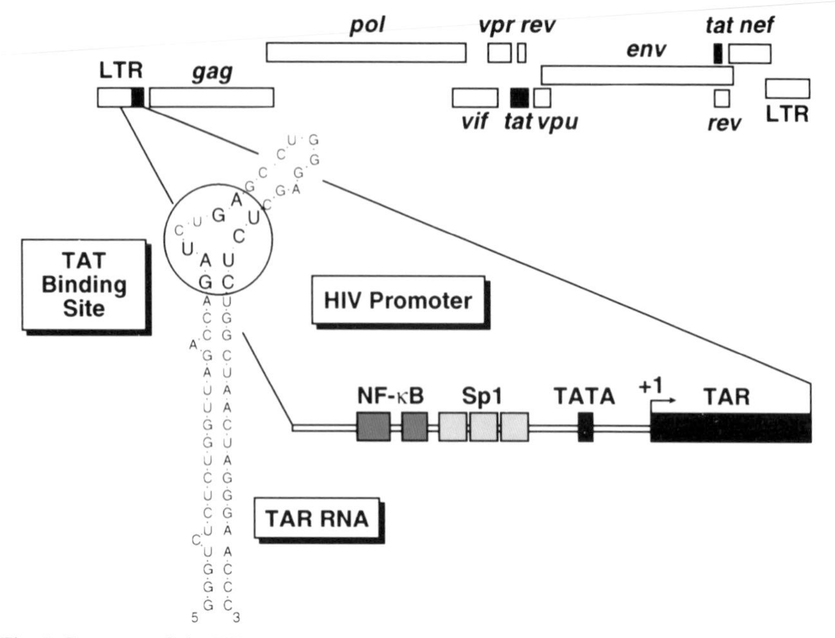

Fig. 1. Structure of the HIV genome (top), the viral promoter (bottom right) and TAR RNA (bottom left). The Tat gene is encoded by two exons which are highlighted on the HIV genome map. The viral promoter has a structure typical of promoters activated by RNA polymerase II. Immediately upstream of the TATA box are two tandem NF-κB binding sites and three tandem SP-1 binding sites. Immediately downstream of the start of transcription is the *trans*-activation response region (TAR). TAR encodes an RNA which can fold into the stem–loop structure shown at right. Tat is able to bind directly to the highlighted region near to the apex of the stem containing a U-rich bulge. Critical residues for Tat binding are shown in bold. A cellular co-factor is believed to bind to the apical loop structure. (Figure after Karn *et al.*, 1996.)

TAR RNA stem invariably abolish Tat-stimulated transcription (Feng & Holland, 1988; Garcia *et al.*, 1989; Selby *et al.*, 1989).

RNA binding by the Tat protein

The genetic evidence strongly suggested that Tat interacts directly with TAR RNA. In 1989 we provided the first demonstration that recombinant Tat expressed in *Escherichia coli* could bind specifically to a trinucleotide bulge located near the apex of the TAR RNA stem–loop structure (Dingwall *et al.*, 1989). Figure 2 shows an example of a competition binding experiment similar to the experiments we used to define the Tat binding site on TAR RNA (Churcher *et al.*, 1993). In the absence of competitor RNA, Tat formed a one-to-one complex with TAR RNA with a binding constant of approximately 3 nM. Unlabelled TAR RNA acts as an effective competitor, but TAR RNAs carrying mutations in the binding site, such as the U23 → C

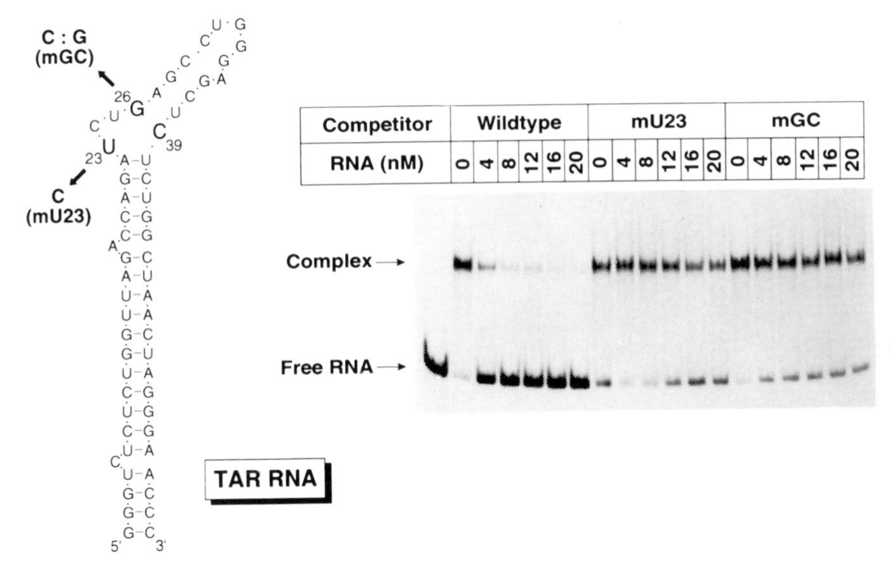

Fig. 2. Tat binds specifically to TAR RNA *in vitro*. (*Left*) Sequence of TAR RNA and mutations that affect Tat binding. Bases in the stem are numbered from the start of the TAR RNA sequence, at the initiation site for transcription. (*Right*) Gel mobility shift assays for Tat binding to TAR RNA. Free RNA was fractionated from the Tat–TAR RNA complex by electrophoresis on non-denaturing polyacrylamide gels. Complexes were formed between 2 nM $[^{35}S]$-labelled TAR RNA and 5 nM Tat protein in the presence of 0 to 20 nM unlabelled competitor TAR RNA. TAR RNAs carrying the mU23 ($U_{23} \rightarrow C$) and mGC ($G_{26}•C_{39} \rightarrow C•G$) mutations competed with more than 10-fold lower affinity than the wild-type sequence. (Figure from Karn *et al.*, 1996 after Churcher *et al.*, 1993.)

(mU23) and G26•C39 → C•G (mGC) mutations were poor competitors for protein binding.

Extensive mutagenesis and chemical probing studies have now defined the key elements required for high affinity binding of Tat to TAR RNA. Tat recognition requires both the presence of the bulged nucleotides and base pairs in the stem (Calnan *et al.*, 1991; Churcher *et al.*, 1993; Delling *et al.*, 1992; Dingwall *et al.*, 1989, 1990; Hamy *et al.*, 1993; Roy *et al.*, 1990*a*; Sumner-Smith *et al.*, 1991; Weeks *et al.*, 1990; Weeks & Crothers, 1991, 1992). The bulge can be varied in size between two and four residues, but the 5′ residue in the bulge must always be a uridine (U23) (Delling *et al.*, 1992; Dingwall *et al.*, 1990; Roy *et al.*, 1990*a*). The other residues in the bulge, C24 and U25, appear to act predominantly as spacers and may be replaced by other nucleotides, or even by non-nucleotide linkers (Churcher *et al.*, 1993; Sumner-Smith *et al.*, 1991). Tat recognition also requires two base pairs in the stem above the U-rich bulge, G26•C39 and A27•U38 (Churcher *et al.*, 1993; Delling *et al.*, 1992). Critical phosphate contacts involve phosphates P21, P22, and P40, which are located below the bulge on both strands

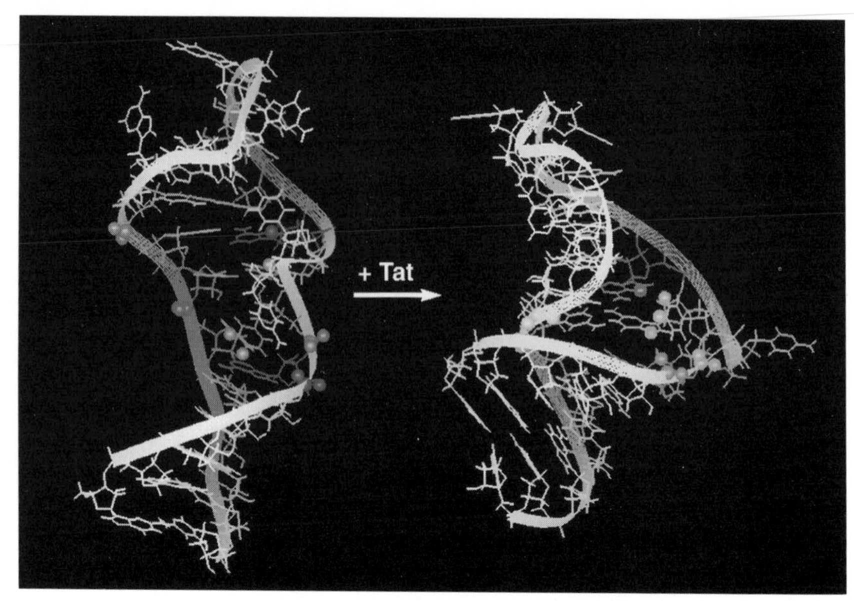

Fig. 3. Major groove view of the free TAR RNA (*left*) and bound TAR RNA (*right*) structures. Functional groups which have been identified as critical for Tat binding, including the backbone phosphates, are highlighted by van der Waals spheres. (Figure adapted from Aboul-ela *et al.*, 1996.)

(Calnan *et al.*, 1991; Churcher *et al.*, 1993; Hamy *et al.*, 1993; Pritchard *et al.*, 1994).

Structure of TAR RNA

Detailed structural information about the Tat binding site on TAR is now available (Fig. 3). Recent NMR studies of TAR RNA demonstrate that the accessibility of the critical functional groups recognised by Tat is enhanced by a local conformational rearrangement (Aboul-ela, Karn & Varani, 1995, 1996; Puglisi *et al.*, 1992). This refolding process involves one of the arginine side chains present in the basic binding domain of the Tat protein (Aboul-ela *et al.*, 1995; Puglisi *et al.*, 1992). In the presence of the arginine, the stacking of the bulged residues U23 on A22 and C24 on U23 is disrupted and A22 becomes juxtaposed to G26. This creates a binding pocket where the guanidinium and εNH groups of the arginine are placed within hydrogen-bonding distance of G26–N^7 and U23–O^4, respectively. The conformational change in TAR RNA also repositions the P22, P23 and P40 phosphates, which provide energetically important contacts with Tat (Pritchard *et al.*, 1994). Model building has shown that these phosphates can be easily contacted by other basic residues found in the TAR RNA binding region.

Contacts between these phosphates and amino acid side chains from Tat contribute not only to the affinity of the interaction, but also to its specificity, by providing discrimination with respect to other bulged RNA structures. The Tat–TAR interaction therefore provides a clear example of the 'indirect readout' of nucleic acid sequences through recognition of backbone phosphates. The importance of the conformational change in TAR for Tat binding is confirmed by the observation that the mutations that produce the most severe reductions in TAR activity involve G26 and U23 and disrupt the intermolecular interactions that are responsible for the folding transition (Churcher *et al.*, 1993; Weeks & Crothers, 1991).

RNA binding is required for Tat activity

Normally, Tat is brought in close proximity to the transcribing polymerase because of its ability to bind to TAR RNA (Keen, Gait & Karn, 1996). There is a direct correlation between Tat binding to TAR RNA and *trans*-activation. Mutations in TAR that show reduced binding to Tat protein also produce a corresponding reduction in the levels of Tat-stimulated transcription *in vivo* (Churcher *et al.*, 1993; Delling *et al.*, 1992; Dingwall *et al.*, 1990; Roy *et al.*, 1990*a*) and *in vitro* (Graeble *et al.*, 1993; Kato *et al.*, 1992; Laspia, Wendel & Mathews, 1993; Marciniak *et al.*, 1990).

Experiments using hybrid proteins have also provided evidence that TAR RNA functions primarily as a 'loading site' for Tat (Madore & Cullen, 1993; Selby & Peterlin, 1990; Southgate, Zapp & Green, 1990). Fusion proteins containing sequences from Tat and bacteriophage R17 coat protein can stimulate transcription from HIV LTRs when the TAR RNA sequence is replaced entirely by an RNA stem–loop structure carrying R17 operator sequence (Selby & Peterlin, 1990). Similar results are obtained when the HIV-Rev protein is used as a fusion partner and an RNA binding site derived from the Rev-response element (RRE) is used to replace TAR (Madore & Cullen, 1993; Southgate *et al.*, 1990).

Drug design

Important themes for anti-viral drug development are also emerging from studies of the interactions of HIV Tat and TAR RNA. For example, it is possible to inhibit virus replication by over-expression of TAR RNA 'decoy' sequences (Lisziewicz, Rappaport & Dhar, 1991; Sullenger *et al.*, 1990). This suggests that it may eventually be possible to inhibit virus replication by a 'gene therapy' strategy aimed at interfering with regulatory protein function through overexpression of competitor RNA molecules. A more immediate prospect is to look for small molecules that inhibit Tat activity, since the biochemical assays for binding can be easily scaled up and automated to permit screening of large libraries of compounds.

Two strategies for drug design are suggested by the structural studies on TAR RNA. First, small molecules that induce TAR RNA refolding and target the bound conformation of TAR RNA are expected to behave as competitive inhibitors of Tat binding. A complementary strategy for drug discovery is to target the free TAR RNA structure. A small molecule that is able to bind TAR RNA with high affinity and block the conformational change could also be an effective inhibitor of the Tat–TAR interaction, and consequently, of HIV growth.

In order to discover effective inhibitors of the Tat–TAR RNA interaction, we adopted a combinatorial chemistry approach based on the synthesis of oligomeric peptoids in collaboration with a group at CIBA-Geigy, Basel (Hamy, F., Felder, E., Heizmann, G., Lazdins, J., Aboul-ela, F., Varani, G., Karn, J. & Klimkait, T., unpublished observations). Peptoids are isomers of peptides which carry N-substituted glycines in their backbones (Simon *et al.*, 1992). Peptoids tend to be more flexible than peptides since intramolecular CO-HN hydrogen bonds are removed and there are differences in the steric interactions that induce secondary structure in peptides. For pharmacological applications, peptoids have the additional advantage of being stabilized against enzymatic degradation.

Starting from a pool of 3.2×10^6 individual chemical entities, we were able to rapidly screen and select a nonapeptoid compound, CGP-64222, that efficiently blocked the formation of the Tat/TAR RNA complex *in vitro*. Unique residues were selected at each position of the peptoid; simple insertion of a positive charge at each position does not produce a tight binding ligand. Perhaps surprisingly, most of these residues were analogues of the natural amino acid side chains. N-Arg residues were optimal at the second, third and fifth positions, while an analogue of N-Lys with an extended aliphatic chain was optimal at the first position, and a N-Phe residue was the optimal residue at the fourth position.

NMR studies demonstrate that the peptoid binds similarly to the Tat protein and induces the conformational change in TAR RNA. Micromolar concentrations of CGP-64222 specifically inhibited Tat activity in a Tat-dependent cellular *trans*-activation assay and also blocked HIV-1 replication in primary human lymphocytes.

CONTROL OF TRANSCRIPTIONAL ELONGATION BY TAT

Tat activation of transcription

How does Tat stimulate transcription? In the absence of Tat, the majority of RNA polymerases initiating transcription stall near the promoter (Feinberg, Baltimore & Frankel, 1991; Kao *et al.*, 1987; Kato *et al.*, 1992; Laspia, Rice &

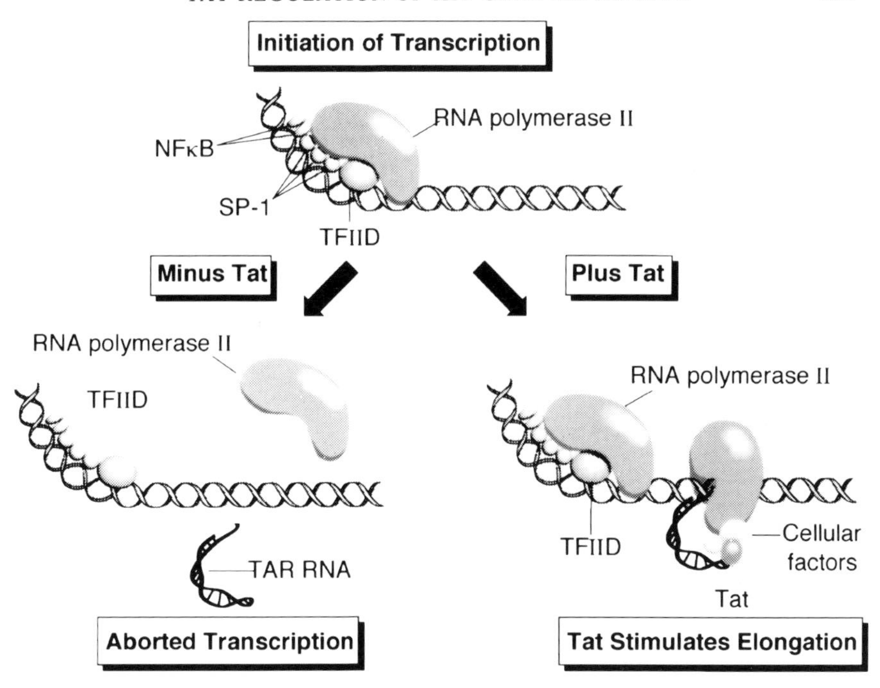

Fig. 4. Model for the control of transcription elongation by Tat. In a newly infected cell, the binding of cellular transcription factors to the viral LTR DNA sequences stimulates a basal level of transcription. In the absence of Tat the polymerase is unstable and disengages from the template at random sites downstream of TAR. As a result, the majority of the transcripts are short and non-polyadenylated. When Tat is present in the cell, it can bind to nascent TAR RNA sequences and form a complex with RNA polymerase, and cellular co-factors. This new complex is capable of efficient elongation and leads to the production of high levels of the HIV mRNAs. (Figure from Karn *et al.*, 1996.)

Mathews, 1989, 1990; Marciniak & Sharp, 1991). Following addition of Tat, there is a dramatic increase in the density of RNA polymerases found downstream of the promoter (Feinberg *et al.*, 1991; Kao *et al.*, 1987; Kato *et al.*, 1992; Laspia *et al.*, 1989, 1990; Marciniak & Sharp, 1991; Rittner *et al.*, 1995).

The observations described above have led us to the working hypothesis for the mechanism of action of Tat illustrated in Fig. 4. According to this proposal, the RNA polymerase engaged by the HIV core promoter is intrinsically unstable. Soon after transcription through TAR, the polymerase either pauses or falls off the template. If Tat is present in the cell, it can associate with TAR RNA and the transcribing RNA polymerase. It seems likely that this reaction involves not only the binding of Tat to TAR RNA but also the recruitment of cellular cofactors, most likely including the TAR RNA loop-binding proteins and cellular elongation factors. The modified transcription complex is then able to transcribe the remainder of the HIV genome efficiently.

The anti-termination mechanism envisaged in Fig. 4 is closely analogous to the mechanism used by the bacteriophage λN protein (Dingwall *et al.*, 1990; Graeble *et al.*, 1993; Kao *et al.*, 1987). However, in the λN system anti-termination takes place at a specific ρ-dependent termination site, whereas in the HIV system, Tat stimulates a general activation of RNA polymerase II and stimulates elongation through a variety of distal sites, including TAR (Dingwall *et al.*, 1990; Feinberg *et al.*, 1991; Graeble *et al.*, 1993; Laspia *et al.*, 1993; Marciniak & Sharp, 1991; Ratnasabapathy *et al.*, 1990; Rittner *et al.*, 1995).

Trans-*activation* in vitro

Detailed analysis of the *trans*-activation mechanism is now possible because of the availability of efficient cell-free transcription systems that respond to Tat (Graeble *et al.*, 1993; Kato *et al.*, 1992; Laspia *et al.*, 1993; Marciniak *et al.*, 1990; Marciniak & Sharp, 1991; Rittner *et al.*, 1995). The experiment shown in Fig. 5 demonstrates that the Tat-stimulated RNA polymerase has an intrinsic anti-termination activity (Rittner *et al.*, 1995). To measure the elongation capacity of RNA polymerase at a downstream site, a synthetic terminator sequence (τ) consisting of a stable RNA stem–loop structure followed by a tract of 9 uridine residues was placed approximately 200 nts downstream of the start of transcription (Fig. 5(a)). The presence of τ caused approximately 30% of the transcribing polymerases to disengage prematurely from the template. Addition of recombinant Tat protein (Fig. 5(b)) to the reaction stimulated the production of run-off product (ρ) by more than 25-fold.

Stimulation of transcription by Tat in the cell-free system requires a functional TAR element. Templates carrying the ΔU mutation in TAR (a mutation that abolishes specific Tat binding *in vitro*) were stimulated less than 3-fold by the addition of Tat (Fig. 5(b)). Studies using an extensive series of templates carrying mutations in TAR have shown that Tat-dependent

Fig. 5. Tat stimulates transcriptional elongation but not initiation *in vitro*. (a) The structure of template DNAs. Each plasmid carried synthetic terminators (τ) inserted at the *Nar*I site, 183 nt downstream of the start of transcription. Templates were linearized by cleavage at the *Xba*I site at 668 nt downstream of the start of transcription. (b) Transcription assay. Transcription reactions were performed in the presence of 0 (−), or 200 ng (+) recombinant Tat protein. Templates carried either an intact TAR element or the ΔU (ΔU_{23-25}) mutations in the TAR element. Addition of Tat protein increased the synthesis of the run-off product (ρ) approximately 25-fold from the wild-type template, but the level of transcription products ending at the terminator (τ) remained relatively constant in the presence or absence of Tat. (c) RNase protection assay. Templates carrying either a wild-type (WT) TAR sequence or TAR carrying the ΔU (ΔU_{23-25}) mutation were transcribed *in vitro* and the products were hybridized to [^{32}P]-labelled antisense transcripts. The probes used were complementary to HIV-1 sequences −10 to +59 (Probe I), +100 to +175 (Probe II) and +470 to +570 (Probe III). After digestion with T_1 ribonuclease, the protected fragments were fractionated by polyacrylamide gel electrophoresis. (Figure from Rittner *et al.*, 1995.)

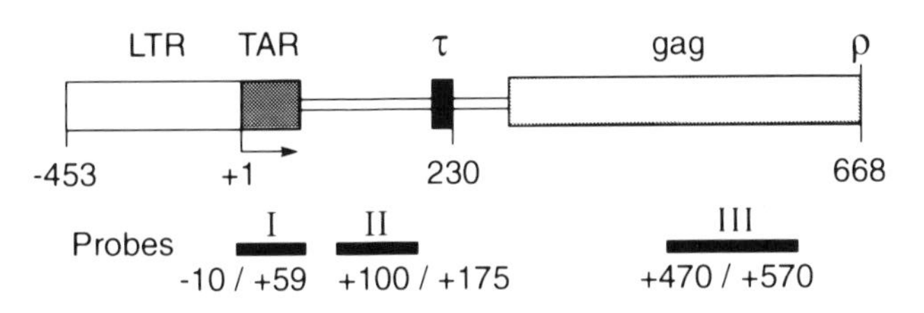

(b) Transcription

	Template			
TAR	WT		ΔU	
Tat	–	+	–	+

ρ —
τ —

(c) RNase protection

	Probes							
	I + II				I + III			
TAR	WT		ΔU		WT		ΔU	
Tat	–	+	–	+	–	+	–	+

Fragment length (nt)
100 —
75 —
59 —
36 —
23 —

Fig. 5.

trans-activation *in vitro* has the same sequence requirements seen *in vivo*. Templates carrying either mutations that reduce Tat affinity for TAR or mutations in the apical loop of TAR RNA show poor responses to Tat in the cell-free system (Churcher *et al.*, 1995; Garcia-Martinez *et al.*, 1995; Graeble *et al.*, 1993; Rittner *et al.*, 1995).

Several recent experiments using the cell-free systems suggest that Tat acts exclusively at the level of elongation (Garcia-Martinez *et al.*, 1995; Graeble *et al.*, 1993; Kato *et al.*, 1992; Laspia *et al.*, 1993; Marciniak *et al.*, 1990; Marciniak & Sharp, 1991; Rittner *et al.*, 1995). The relative effects of Tat on initiation and elongation are demonstrated by RNase protection experiments shown in Fig. 5(c). In these assays, the distribution of transcripts which extend to various points downstream of the start of the transcription are measured by hybridising the transcribed RNAs to a series of ^{32}P-labelled complementary probes. Probe I overlaps the start of transcription and TAR and provides a measure of the total initiation events. In the presence or absence of Tat the levels of transcripts which can hybridise to Probe I are constant, suggesting that initiation rates are constant. By contrast, in the presence of Tat, there is a significant increase in the levels of transcripts which extend to the more distal Probes II and III. Addition of Tat stimulates an approximately five-fold increase in the number of transcripts which extend through the Probe II region. Tat simultaneously stimulates a 20-fold increase in the number transcripts hybridising to Probe III, which is located downstream of the terminator sequence. The results of these experiments are consistent with the observation of Marciniak & Sharp (1991) that relatively few transcripts from the LTR are able to extend more than 200 nts in the absence of Tat, whereas in the presence of Tat, the majority of the transcripts initiated from the LTR are able to extend until the end of the template.

TAR is functional at a distance

If Tat binds to the TAR RNA which is produced by an elongating RNA polymerase, as postulated by the mechanism depicted in Fig. 4, then it should also be possible to introduce Tat when TAR is placed considerably downstream of the promoter. However, in an early experiment, Selby *et al.* (1989) found that maximal promoter activity required TAR to be located immediately downstream of the start of transcription. This experiment has been interpreted to mean that TAR needs to be near the start of transcription in order to allow Tat to interact with the upstream elements in the promoter (Berkhout *et al.*, 1990). Unfortunately, the experiment of Selby *et al.* (1989) is ambiguous because during the construction of the templates with displaced TAR elements critical promoter elements within the HIV LTR structure were also disrupted.

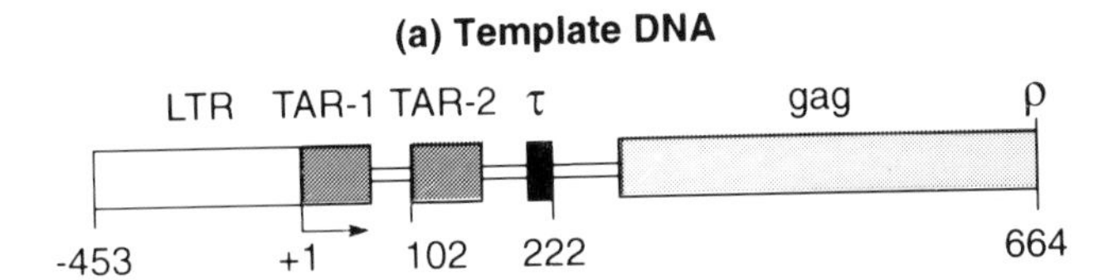

(b) Transcription

	Template							
	MTX 53		MTX 60		MTX 61		MTX 62	
TAR-1	WT		WT		mGC		mGC	
TAR-2	WT		mGC		WT		mGC	
Tat	–	+	–	+	–	+	–	+

ρ —

τ —

Fig. 6. TAR is functional at a distance. (a): Structure of template DNAs carrying a duplicated TAR (TAR-2) element inserted 100 nts downstream of TAR (TAR-1). (b): The effects of mutations in both elements were analysed in cell-free transcription reactions. A template carrying the mGC (G_{26}•C_{39} → C•G) mutation in TAR-1 and a wild-type sequence in TAR-2 produced approximately 10-fold more run-off product (ρ) after addition of Tat. When both TAR-1 and TAR-2 were inactivated by the mGC mutation, there was no significant production of the run-off product. (Figure after Churcher *et al.*, 1995.)

In order to re-examine whether Tat could be introduced at a distance, under conditions where the sequences near the start of HIV transcription were unaltered, we prepared templates carrying duplicated TAR elements (Churcher *et al.*, 1995; Fig. 6). The first TAR carried the mGC mutation and is therefore unable to bind Tat efficiently. The second TAR carried either an intact Tat binding site, or as a control, the mGC mutation in the Tat binding site. As shown in Fig. 6(b), when the first TAR is inactivated by the mGC mutation, but a wild-type sequence is present in the second TAR, Tat is able to efficiently stimulate transcription. The double mutation has negligible activity. Because the ability of RNA polymerases initiating transcription from the HIV-1 LTR to elongate is reduced until the polymerase reaches TAR, the position of the active TAR RNA element relative to the start site of transcription has an effect on the level of transcription observed in this experiment. The amount of run-off transcripts produced from templates where only TAR-2 is active is approximately 50% that obtained from templates where TAR-1 is active. When TAR-2 is displaced further downstream, there is an even greater redution in the amount of run-off transcripts produced.

Tat is an integral component of the activated transcription–elongation complex

In addition to pausing at RNA stem–loop structures, elongation by RNA polymerase can be blocked by DNA-binding proteins, such as the *lac* repressor (Reines & Mote, 1993). The block imposed by the *lac* repressor protein can be used to 'trap' actively elongating transcription complexes because, in contrast to polymerases arrested by terminator sequences, the majority of the transcription complexes pausing at the *lac* repressor do not disengage from the template (Reines & Mote, 1993). As outlined in Fig. 7, we have taken advantage of this property to determine whether Tat becomes attached to transcription complexes that are formed on templates carrying a functional TAR element (Keen *et al.*, 1996).

Transcription reactions were performed using templates carrying a *lac* operator site inserted 221 nts downstream of the transcription start site and *lac* repressor protein (LacR). The arrested transcription complexes were then purified by binding biotinylated template DNAs to streptavidin-coated magnetic beads. After washing the beads extensively, the transcription complexes were released by cutting the template DNA immediately after TAR with *Hin*dIII (Fig. 8). Finally, the levels of RNA polymerase II, *lac* repressor and Tat in the released fractions were measured by immunoblotting.

Immunoblots of the proteins present in the released transcription complexes are shown in Fig. 8(b). Immunoblots for LacR provide an internal control for the release of proteins bound to the templates following the purification scheme. RNA polymerase is also readily detected by immunoblotting of the released fractions. In the HeLa cell extract, the majority of the RNA polymerase II is non-phosphorylated (IIa) (Fig. 8(b)). By contrast, all of the RNA polymerase II detected in the released fractions is phosphorylated (IIo) and migrates more slowly in the gels. Since the phosphorylation of the carboxy-terminal domain (CTD) of RNA polymerase is a post-initiation event, this result demonstrates that only transcription complexes that are actively engaged in transcription are purified by this procedure.

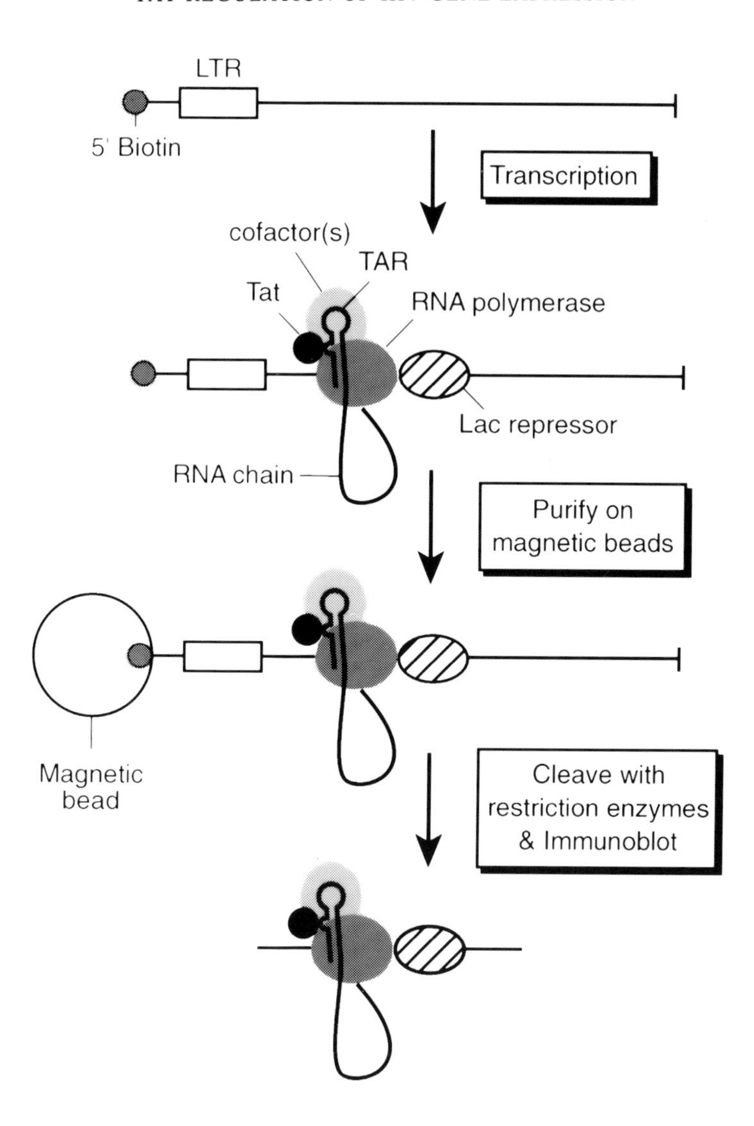

Fig. 7. Experimental strategy for analysing active transcription complexes. Step 1: Cell-free transcription reactions are performed using biotinylated templates in the presence of *lac* repressor protein (LacR). Step 2: Transcription complexes arrested by LacR are purified by binding to streptavidin-coated magnetic beads. The diagram depicts a complex between polymerase, Tat protein, cellular co-factors, and TAR RNA. The elongating RNA chain is shown forming a loop structure, but it is equally possible that Tat and its cellular co-factors are transferred from TAR to the transcription complex during elongation. Step 3: The transcription arrested complexes are released from the magnetic beads by cleavage of the templates with restriction enzymes. The protein composition of the released complexes can then be analysed by immunoblotting. (Figure from Keen *et al.*, 1996.)

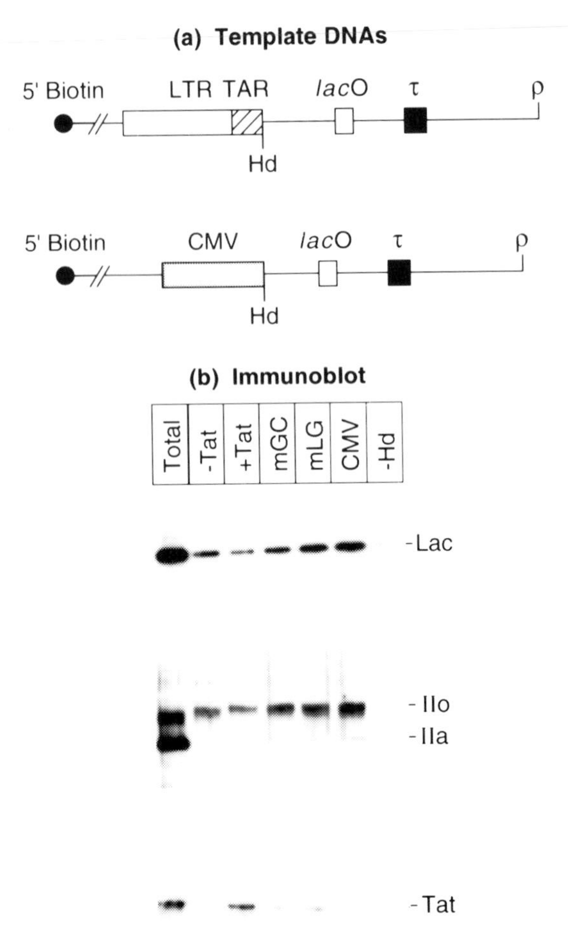

Fig. 8. Tat co-purifies with the arrested transcription complex. (a) Structures of the p10SLT template and the pCMT template which carries the CMV promoter in place of the HIV LTR. Transcription reactions were performed in the presence (+Tat) or absence (−Tat) of 200 ng recombinant Tat protein using the p10SLT template carrying a wild-type TAR RNA element. In parallel, transcription reactions were performed in the presence of 200 ng Tat using templates carrying the mGC and mLG mutations in TAR (Churcher *et al.*, 1995). After incubation for 30 min in the presence of 5 μM [α-^{32}P]UTP and 1 μg LacR, the templates were bound to streptavidin-coated magnetic beads and washed, and transcription complexes were released from the beads by cleavage with *Hin*dIII. (b) Immunoblots of the proteins released from the magnetic beads by *Hin*dIII cleavage. A marker lane containing 5% of the nuclear extract present in the transcription reaction was included as a control (Total). (Figure after Keen *et al.*, 1996.)

A strong Tat signal is found in association with transcription complexes purified on templates carrying a wild-type TAR RNA element (Fig. 8(b)). Templates carrying mutations in TAR RNA were used in control experiments designed to demonstrate that Tat association with transcription complexes is specific and dependent upon a functional TAR RNA element. TAR was inactivated by either the mGC mutation in the Tat binding site or by the $G_{32\text{-}34} \rightarrow$ UUU (mLG) mutation in the apical loop sequence. An additional control experiment was performed using the CMV promoter to direct transcription instead of the HIV-1 LTR.

Templates carrying either the mGC or the mLG mutations accumulate significantly less Tat than templates carrying a functional TAR element (Fig. 8(b)). Densitometry of the gels shows that the Tat levels on the wild-type template are 6.6-fold greater than for the mGC template, 4.3-fold greater than the mLG template and 10.7-fold greater than the template carrying the CMV promoter. The small amount of Tat that is detected in association with the templates that carry mutant TAR RNA elements probably represents a background of protein that is bound non-specifically to the transcription complexes. The background of Tat protein that is released from the beads during the restriction enzyme digestion is shown by the control experiment where *Hin*dIII was omitted (Fig. 8(b)).

Transfer of Tat from TAR RNA to the elongating polymerase

What happens to TAR RNA during transcription? Does TAR form a ternary complex with the elongating polymerase and Tat or does it dissociate? Figure 9 shows the outline of an experiment designed to answer this question (Keen, N. J., Churcher, M. J., Lowe, A. D., & Karn, J., unpublished observations). Transcription complexes carrying ^{32}P-labelled nascent chains were prepared, paused at LacR, and purified as described above. The nascent chains were then cleaved at distinct sites by hybridizing to DNA-oligonucleotides and digesting the hybridized segment with RNaseH. If TAR remains associated with the transcription complex because of its interactions with various proteins, then the 5′ end of the nascent chain should remain bound to the transcription complex following cleavage by RNase H. On the other hand, if Tat is transferred from TAR during the transcription elongation, then the 5′ end of the nascent RNA chain should be released. As shown in Fig. 10, cleavage of the nascent RNA with oligonucleotides and RNaseH produces distinct fragments corresponding to the 5′ and 3′ ends of the nascent RNA chain. After washing of the complexes the 3′ end of the nascent chain remains bound to the template, but the 5′ dissociates. It therefore seems likely that Tat is transferred from TAR to the transcription complex early during the elongation reaction.

FUTURE CHALLENGES

The studies of the HIV Tat protein described above have revealed a unique example of the regulation of transcription through control of elongation

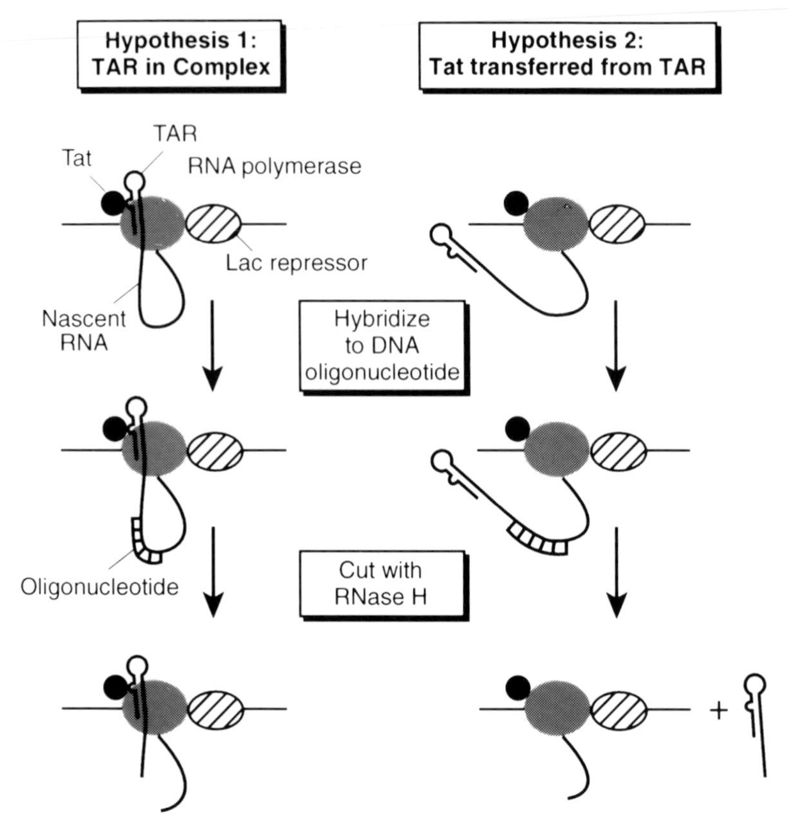

Fig. 9. Experimental strategy for testing whether TAR remains associated with Tat-activated transcription complexes. Transcription complexes carrying ^{32}P-labelled nascent chains were paused at the LacR site and purified as described in Fig. 7. The RNA chains were then cleaved at specific sites by hybridizing the nascent chain to complementary DNA oligonucleotides and treating the hybrids with RNaseH. Following the cleavage reaction, the complexes are washed extensively. If TAR remains associated with the transcription complex (*left panel*), both the 5′ and 3′ ends of the nascent chains will remain bound. However, if Tat is transferred from TAR to the transcription complex (*right panel*) then the 5′ end of the nascent chain, carrying TAR, will be released by RNaseH cleavage.

rates. Although there has been considerable progress in defining the detailed mechanism of action of the HIV *trans*-activator protein many key aspects of the mechanism remain poorly understood.

Control of basal transcription

A central feature of the *trans*-activation mechanism envisaged in Fig. 4 is that transcription from the HIV LTR is initiated by a polymerase with poor elongation capacity. What makes the transcription complex responsive to Tat?

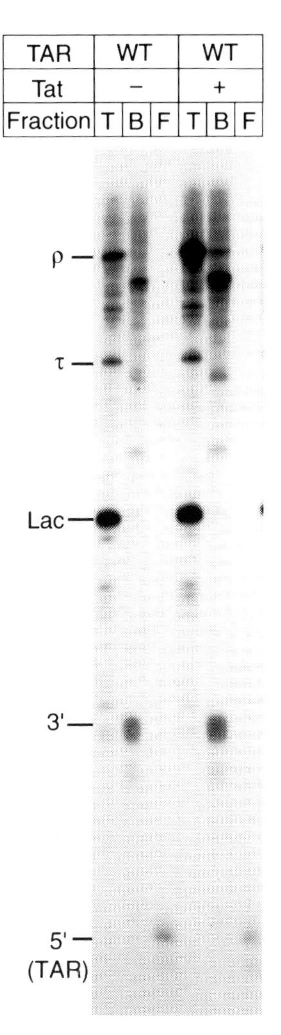

Fig. 10. TAR is released from the Tat-activated transcription complex. Transcription reactions were performed in the presence (+Tat) or absence (−Tat) of 200 ng recombinant Tat protein using the p10SLT template carrying a wild-type TAR RNA element. After incubation for 30 min in the presence of 5 μM [α-^{32}P]UTP and 1 μg LacR, the complexes were purified on magnetic beads. The transcripts prepared in this way are shown in lane 1 (T). The nascent RNA chains from the purified complexes were hybridized to RH7, an 11mer DNA oligonucleotide complementary to residues +75 to +86. The chain was then cut with RNaseH generating a 75 nt fragment corresponding to the 5′ end of the transcript and a 98 nt fragment corresponding to the 3′ end. Note that after washing the cleaved complexes, the 5′ end fragment appears in the free fraction (F), whereas the 3′ end fragment appears in the bound fraction (B).

Although the complete HIV LTR is approximately 600 nt in length, the core promoter which includes all the elements that are necessary for transcription initiation *in vitro* extends from only −78 to approximately +26 (Fig. 1) (Rittner *et al.*, 1995). The core promoter includes only three major elements: a trio of tandem SP1 binding sites (Jones *et al.*, 1986), a TATA element (Berkhout & Jeang, 1992; Olsen & Rosen, 1992) and a novel initiator sequence (Rittner *et al.*, 1995; Zenzie-Gregory *et al.*, 1993). Each of these elements is required for efficient initiation of transcription, but mutations in these elements do not affect elongation and a strong Tat response is retained even when initiation rates are significantly reduced by mutations in the core promoter elements.

Although it is tempting to speculate that the HIV core promoter elements specify that only Tat-responsive transcription complexes are able to initiate on the viral promoter, mutagenesis experiments have failed to define any unique features in the HIV core promoter elements that control the nature of the transcription complex. Experiments using hybrid promoters have shown that SP1 can be replaced by other types of activator elements, including sequences derived from the adenovirus major late promoter, the U2 snRNA promoter, as well as the α and β globin promoters from chicken, rabbit and man (Berkhout *et al.*, 1990; Kato *et al.*, 1992; Laspia *et al.*, 1993; Muesing *et al.*, 1987; Ratnasabapathy *et al.*, 1990). Similarly, artificial promoters which carry GAL-4 sites in place of the SP1 sites of HIV can be activated by a GAL-4-VP16 hybrid protein and produce a strong Tat response (Ghosh, Selby & Peterlin, 1993; Southgate & Green, 1991). Thus, there have been no reports of any mutation, or combination of mutations, in the HIV-1 LTR which permits initiation by a Tat-independent transcription complex.

Perhaps the observation that Tat can activate transcription from virtually any promoter placed upstream of TAR is simply an artefact arising because most of the promoters that have been analysed are relatively simple and do contain highly specialised enhancer elements. Most of the promoters that have been tested for Tat activation are derived from viruses and may already be designed to support rapid initiation. We would like to suggest that these promoters permit incompletely formed transcription complexes to initiate transcription. It is tempting to speculate that more elaborate promoters, such as those used by cellular genes that require stable production of full-length transcripts, may be designed to permit initiation of transcription using polymerases that carry stabilizing elongation factor(s). It will be of great interest to learn whether certain cellular promoters can specify initiation by a transcription complex that carries specialized elongation factor(s) that resemble the factor(s) recruited by Tat.

Cellular co-factor(s) for Tat

What are the identities of the cellular cofactors that interact with Tat, TAR and the elongating transcription complex? Sequences in the apical loop of TAR RNA are not required for Tat binding (Dingwall *et al.*, 1989, 1990; Roy *et al.*, 1990*a*). However, mutations in the TAR RNA loop strongly inhibit *trans*-activation (Berkhout & Jeang, 1989; Dingwall *et al.*, 1990; Feng & Holland, 1988; Garcia *et al.*, 1989; Roy *et al.*, 1990*b*; Selby *et al.*, 1989). Experiments using duplicated TAR sequences show that active TAR elements must carry both an intact loop sequence and an intact Tat binding site (Churcher *et al.*, 1995). Similarly, the insertion of extra Watson-Crick base pairs into the upper stem of TAR RNA, alters the spacing between the Tat binding site and the loop and inhibits Tat-stimulated transcription (Berkhout & Jeang, 1991; Churcher *et al.*, 1993). The simplest explanation for the behaviour of these mutations is that cellular RNA-binding proteins participate in *trans*-activation by binding to the loop sequences (Berkhout & Jeang, 1991; Cullen, 1990; Dingwall *et al.*, 1990).

Additional evidence supporting the view that cellular co-factors which recognise the TAR RNA loop sequences are also essential for Tat activity comes from experiments using TAR RNA as a competitive inhibitor. An excess of TAR RNA can effectively inhibit transcription from the HIV LTR (Lisziewicz *et al.*, 1991; Sullenger *et al.*, 1990). This activity requires an intact loop sequence, implying that TAR acts as an inhibitor by sequestering the cellular co-factor(s).

There has been a considerable effort to identify these possible co-factors for Tat which bind to the TAR loop sequence. The best candidate for a cellular co-factor that recognizes the TAR RNA loop is TRP-185, a protein that can be UV-crosslinked to TAR RNA (Sheline, Milocco & Jones, 1991; Wu *et al.*, 1991; Wu-Baer, Lane & Gaynor, 1995*a*; Wu-Baer, Sigman & Gaynor, 1995*b*). Efficient complex formation between TRP-185 and TAR has strict requirements for sequences of the TAR loop that closely match the requirements for efficient *trans*-activation *in vivo* (Wu-Baer *et al.*, 1995*a*,*b*).

One surprising feature of TRP-185 activity is that Tat is able to compete with TRP-185 for TAR binding. However, although it has often been assumed that a cellular co-factor for Tat should be able to form a stable ternary complex together with Tat and TAR, the genetic data is also consistent with sequential interactions taking place between Tat, TAR and TAR-binding co-factor(s). In fact, our recent experiments suggesting that Tat is displaced from TAR support the hypothesis that factors like TRP-185 play a role *trans*-activation by assisting the transfer of Tat from TAR to the elongating RNA polymerase. Consistent with this view, Wu-Baer *et al.* (1995*b*) have reported recently that TAR RNA can bind specifically to RNA polymerase II, but this complex is disrupted both by the addition of Tat and by the phosphorylation of the CTD region.

In addition to the factor(s) recognizing the TAR loop, there are also likely to be co-factors that interact directly with Tat. These include, TAK, a protein kinase that can bind to Tat (Herrmann & Rice, 1995), and a factor with Tat-complementing activity recently identified in partially reconstituted transcription systems (Zhou & Sharp, 1995). It is also likely that elongation factors normally associated with RNA polymerase, such as TFIIF, participate in the formation of the Tat-modified transcription complex (Kato *et al.*, 1992). It remains a significant challenge to devise biochemical experiments that will allow the identification of these novel components of Tat-modified transcription complexes. The identities of several of the key factors may only be discovered once the system is entirely reconstituted using purified components.

CONCLUSIONS

Historically, studies of viral control mechanisms have highlighted fundamental aspects of molecular biology. The HIV *trans*-activator protein, Tat, provides the first example of a protein that regulates transcriptional elongation in eukaryotic cells. Until Tat was discovered, elongation control was a relatively neglected area of transcription regulation. There is still a great deal of ignorance about the mode of action of the key elongation factors, but undoubtedly, identification of the co-factor(s) used by Tat will reveal a great deal about the normal cellular mechanisms of elongation control.

Studies of the interactions of Tat with TAR RNA are also providing new insights into the chemistry of nucleic acid recognition. There are many examples of RNA binding proteins that recognize bases displayed in apical loop and distorted bulge structures. Studies of the Tat–TAR interaction have revealed that the binding reaction involves two distinct steps. First, Tat induces a conformational changes in RNA structure at the bulge. Secondly, the structural rearrangement in TAR permits Tat to interact with specific functional groups displayed on base pairs in the major groove as well as on adjacent phosphate residues. This new principle of RNA recognition is likely to extend to many other protein–RNA interactions.

Finally, in addition to its academic interest, studies of the Tat–TAR interaction is providing a basis for drug discovery. In the 12 years since the discovery of the virus seven antiviral drugs have been licenced (three within the last year) and scores more have entered clinical trials. Yet in spite of this enormous effort, effective regimens for the prevention of progression to clinical symptoms of AIDS and the treatment of AIDS patients have not yet been achieved. Small molecules that selectively inhibit Tat activity have already been discovered. We are optimistic that these early leads can be developed into drugs that will make a useful contribution to the management of AIDS.

ACKNOWLEDGEMENTS

We thank the present and former members of the HIV group at LMB for their collaboration.

REFERENCES

Aboul-ela, F., Karn, J. & Varani, G. (1995). The structure of the human immunodeficiency virus type 1 TAR RNA reveals principles of RNA recognition by Tat protein. *Journal of Molecular Biology*, **253**, 313–32.

Aboul-ela, F., Karn, J. & Varani, G. (1996). Structure of HIV-1 TAR RNA in the absence of ligands reveals a novel configuration of the trinucleotide bulge. *Nucleic Acids Research*, **24**, 3974–81.

Berkhout, B. & Jeang, K.-T. (1989). *Trans*-activation of human immunodeficiency virus type 1 is sequence specific for both the single-stranded bulge and loop of the *trans*-acting responsive hairpin: a quantitative analysis. *Journal of Virology*, **63**, 5501–4.

Berkhout, B. & Jeang, K.-T. (1991). Detailed mutational analysis of TAR RNA: critical spacing between the bulge and loop recognition domains. *Nucleic Acids Research*, **19**, 6169–76.

Berkhout, B. & Jeang, K.-T. (1992). Functional roles for the TATA promoter and enhancers in basal and Tat-induced expression of the human immunodeficiency virus type 1 long terminal repeat. *Journal of Virology*, **66**, 139–49.

Berkhout, B., Silverman, R. H. & Jeang, K.-T. (1989). Tat *trans*-activates the human immunodeficiency virus through a nascent RNA target. *Cell*, **59**, 273–82.

Berkhout, B., Gatignol, A., Rabson, A. B. & Jeang, K.-T. (1990). TAR-independent activation of the HIV-1 LTR: evidence that Tat requires specific regions of the promoter. *Cell*, **62**, 757–67.

Calnan, B. J., Tidor, B., Biancalana, S., Hudson, D. & Frankel, A. D. (1991). Arginine-mediated RNA recognition: the arginine fork. *Science*, **252**, 1167–71.

Churcher, M., Lamont, C., Hamy, F., Dingwall, C., Green, S. M., Lowe, A. D., Butler, P. J. G., Gait, M. J. & Karn, J. (1993). High affinity binding of TAR RNA by the human immunodeficiency virus Tat protein requires amino acid residues flanking the basic domain and base pairs in the RNA stem. *Journal of Molecular Biology*, **230**, 90–110.

Churcher, M. J., Lowe, A. D., Gait, M. J. & Karn, J. (1995). The RNA element encoded by the *trans*-activation-responsive region of human immunodeficiency virus type 1 is functional when displaced downstream of the start of transcription. *Proceedings of the National Academy of Sciences, USA*, **92**, 2408–12.

Cullen, B. R. (1986). *Trans*-activation of human immunodeficiency virus occurs via a bimodal mechanism. *Cell*, **46**, 973–82.

Cullen, B. R. (1990). The HIV Tat protein: an RNA sequence-specific processivity factor? *Cell*, **63**, 655–7.

Delling, U., Reid, L. S., Barnett, R. W., Ma, M. Y.-X., Climie, S., Sumner-Smith, M. & Sonenberg, N. (1992). Conserved nucleotides in the TAR RNA stem of human immunodeficiency virus type 1 are critical for Tat binding and *trans*-activation: model for TAR RNA tertiary structure. *Journal of Virology*, **66**, 3018–25.

Dingwall, C., Ernberg, I., Gait, M. J., Green, S. M., Heaphy, S., Karn, J., Lowe, A. D., Singh, M., Skinner, M. A. & Valerio, R. (1989). Human immunodeficiency virus 1 Tat protein binds *trans*-activation-responsive region (TAR) RNA *in vitro*. *Proceedings of the National Academy of Sciences, USA*, **86**, 6925–9.

Dingwall, C., Ernberg, I., Gait, M. J., Green, S. M., Heaphy, S., Karn, J., Lowe, A. D., Singh, M. & Skinner, M. A. (1990). HIV-1 Tat protein stimulates transcription by binding to a U-rich bulge in the stem of the TAR RNA structure. *European Molecular Biology Organisation Journal*, **9**, 4145–53.

Feinberg, M. B., Baltimore, D. & Frankel, A. D. (1991). The role of Tat in the human immunodeficiency virus life cycle indicates a primary effect on transcriptional elongation. *Proceedings of the National Academy of Sciences, USA*, **88**, 4045–9.

Feng, S. & Holland, E. C. (1988). HIV-1 Tat *trans*-activation requires the loop sequence within TAR. *Nature (London)*, **334**, 165–8.

Garcia, J. A., Harrich, D., Soultanakis, E., Wu, F., Mitsuyasu, R. & Gaynor, R. B. (1989). Human immunodeficiency virus type 1 LTR TATA and TAR region sequences required for transcriptional regulation. *European Molecular Biology Organisation Journal*, **8**, 765–78.

Garcia-Martinez, L., Mavankal, G., Peters, P., Wu-Baer, F. & Gaynor, R. B. (1995). Tat functions to stimulate the elongation properties of transcription complexes paused by the duplicated TAR RNA element of human immunodeficiency virus 2. *Journal of Molecular Biology*, **254**, 350–63.

Ghosh, S., Selby, M. J. & Peterlin, B. M. (1993). Synergism between Tat and vp16 in *trans*-activation of HIV-1 LTR. *Journal of Molecular Biology*, **234**, 610–19.

Graeble, M. A., Churcher, M. J., Lowe, A. D., Gait, M. J. & Karn, J. (1993). Human immunodeficiency virus type 1 *trans*-activator protein Tat, stimulates transcriptional read-through of distal terminator sequences *in vitro*. *Proceedings of the National Academy of Sciences, USA*, **90**, 6184–8.

Hamy, F., Asseline, U., Grasby, J., Iwai, S., Pritchard, C., Slim, G., Butler, P. J. G., Karn, J. & Gait, M. J. (1993). Hydrogen-bonding contacts in the major groove are required for human immunodeficiency virus type-1 Tat protein recognition of TAR RNA. *Journal of Molecular Biology*, **230**, 111–23.

Herrmann, C. H. & Rice, A. P. (1995). Lentivirus Tat proteins specifically associate with a cellular protein kinase, TAK, that hyperphosphorylates the carboxyl-terminal domain of the large subunit of RNA polymerase II: candidate for a Tat cofactor. *Journal of Virology*, **69**, 1612–20.

Jones, K., Kadonaga, J., Luciw, P. & Tjian, R. (1986). Activation of the AIDS retrovirus promoter by the cellular transcription factor, Sp1. *Science*, **232**, 755–9.

Kao, S.-Y., Calman, A. F., Luciw, P. A. & Peterlin, B. M. (1987). Anti-termination of transcription within the long terminal repeat of HIV-1 by Tat gene product. *Nature (London)*, **330**, 489–93.

Karn, J., Churcher, M. J., Rittner, K., Keen, N. J. & Gait, M. J. (1996). Control of transcriptional elongation by the human immunodeficiency virus Tat protein. In Goodbourn, S., ed. *Eurkaryotic Gene Transcription*, pp. 256–88. Oxford: Oxford University Press.

Kato, H., Sumimoto, H., Pognonec, P., Chen, C.-H., Rosen, C. A. & Roeder, R. G. (1992). HIV-1 Tat acts as a processivity factor *in vitro* in conjunction with cellular elongation factors. *Genes & Development*, **6**, 655–66.

Keen, N. J., Gait, M. J. & Karn, J. (1996). Human immunodeficiency virus type-1 Tat is an integral component of the activated transcription-elongation complex. *Proceedings of the National Academy of Sciences, USA*, **93**, 2505–10.

Laspia, M. F., Rice, A. P. & Mathews, M. B. (1989). HIV-1 Tat protein increases transcriptional initiation and stabilizes elongation. *Cell*, **59**, 283–92.

Laspia, M. F., Rice, A. P. & Mathews, M. B. (1990). Synergy between HIV-1 Tat and adenovirus E1a is principally due to stabilization of transcriptional elongation. *Genes and Development*, **4**, 2397–408.

Laspia, M. F., Wendel, P. & Mathews, M. B. (1993). HIV-1 Tat overcomes inefficient transcriptional elongation *in vitro*. *Journal of Molecular Biology*, **232**, 732–46.

Lisziewicz, J., Rappaport, J. & Dhar, R. (1991). Tat-regulated production of multimerized TAR RNA inhibits HIV-1 gene expression. *New Biologist*, **3**, 82–9.

Madore, S. J. & Cullen, B. R. (1993). Genetic analysis of the cofactor requirement for human immunodeficiency virus type 1 Tat function. *Journal of Virology*, **67**, 3703–11.

Marciniak, R. A. & Sharp, P. A. (1991). HIV-1 Tat protein promotes formation of more-processive elongation complexes. *European Molecular Biology Organisation Journal*, **10**, 4189–96.

Marciniak, R. A., Calnan, B. J., Frankel, A. D. & Sharp, P. A. (1990). HIV-1 Tat protein *trans*-activates transcription *in vitro*. *Cell*, **63**, 791–802.

Muesing, M. A., Smith, D. H. & Capon, D. J. (1987). Regulation of mRNA accumulation by a human immunodeficiency virus *trans*-activator protein. *Cell*, **48**, 691–701.

Olsen, H. S. & Rosen, C. A. (1992). Contribution of the TATA motif to Tat-mediated transcriptional activation of the human immunodeficiency virus gene expression. *Journal of Virology*, **66**, 5594–7.

Pritchard, C. E., Grasby, J. A., Hamy, F., Zachareck, A. M., Singh, M., Karn, J. & Gait, M. J. (1994). Methylphosphonate mapping of phosphate contacts critical for RNA recognition by the human immunodeficiency virus Tat and Rev proteins. *Nucleic Acids Research*, **22**, 2592–600.

Puglisi, J. D., Tan, R., Calnan, B. J., Frankel, A. D. & Williamson, J. R. (1992). Conformation of the TAR RNA-arginine complex by NMR spectroscopy. *Science*, **257**, 76–80.

Ratnasabapathy, R., Sheldon, M., Johal, L. & Hernandez, N. (1990). The HIV-1 long terminal repeat contains an unusual element that induces the synthesis of short RNAs from various mRNA and snRNA promoters. *Genes and Development*, **4**, 2061–74.

Reines, D. & Mote, J. J. (1993). Elongation factor SII-dependent transcription by RNA polymerase II through a sequence-specific DNA-binding protein. *Proceedings of the National Academy of Sciences, USA*, **90**, 1917–21.

Rittner, K., Churcher, M. J., Gait, M. J. & Karn, J. (1995). The human immunodeficiency virus long terminal repeat includes a specialised initiator element which is required for Tat-responsive transcription. *Journal of Molecular Biology*, **248**, 562–80.

Rosen, C. A., Sodroski, J. G. & Haseltine, W. A. (1985). The location of *cis*-acting regulatory sequences in the human T cell lymphotropic virus type III (HTLV-III/LAV) long terminal repeat. *Cell*, **41**, 813–23.

Roy, S., Delling, U., Chen, C.-H., Rosen, C. A. & Sonenberg, N. (1990*a*). A bulge structure in HIV-1 TAR RNA is required for Tat binding and Tat-mediated *trans*-activation. *Genes and Development*, **4**, 1365–73.

Roy, S., Parkin, N. T., Rosen, C. A., Itovitch, J. & Sonenberg, N. (1990*b*). Structural requirements for *trans*-activation of human immunodeficiency virus type 1 long terminal repeat-directed gene expression by Tat: Importance of base-pairing, loop sequence and bulges in the Tat-responsive sequence. *Journal of Virology*, **64**, 1402–6.

Selby, M. J. & Peterlin, B. M. (1990). *Trans*-activation by HIV-1 Tat via a heterologous RNA binding protein. *Cell*, **62**, 769–76.

Selby, M. J., Bain, E. S., Luciw, P. & Peterlin, B. M. (1989). Structure, sequence and position of the stem-loop in TAR determine transcriptional elongation by Tat through the HIV-1 long terminal repeat. *Genes and Development*, **3**, 547–58.

Sheline, C. T., Milocco, L. H. & Jones, K. A. (1991). Two distinct nuclear transcription factors recognize loop and bulge residues of the HIV-1 TAR RNA hairpin. *Genes and Development*, **5**, 2508–20.

Simon, R. J., Kania, R. S., Zuckerman, R. N., Huebner, V. D., Jewell, D. A., Banville, S., Ng, S., Wang, L., Rosenberg, S., Marlowe, C. K., Spellmeyer, D. C., Tan, R., Frankel, A. D., Danti, D. V., Cohen, F. E. & Bartlett, P. A. (1992). Peptoids: a modular approach to drug discovery. *Proceedings of the National Academy of Sciences, USA*, **89**, 9367–71.

Sodroski, J., Patarca, R., Rosen, C., Wong-Staal, F. & Haseltine, W. A. (1985*a*). Location of the *trans*-acting region on the genome of human T-cell lymphotropic virus type III. *Science*, **229**, 74–7.

Sodroski, J. G., Rosen, C. A., Wong-Staal, F., Salahuddin, S. Z., Popovic, M., Arya, S., Gallo, R. C. & Haseltine, W. A. (1985*b*). *Trans*-acting transcriptional regulation of human T-cell leukemia virus type III long terminal repeat. *Science*, **227**, 171–3.

Southgate, C. D. & Green, M. R. (1991). The HIV-1 *tat* protein activates transcription from an upstream DNA-binding site: implications for *tat* function. *Genes and Development*, **5**, 2496–507.

Southgate, C., Zapp, M. L. & Green, M. R. (1990). Activation of transcription by HIV-1 Tat protein tethered to nascent RNA through another protein. *Nature, (London)*, **345**, 640–2.

Sullenger, B. A., Gallardo, H. F., Ungers, G. E. & Gilboa, E. (1990). Overexpression of TAR sequences renders cells resistant to human immunodeficiency virus replication. *Cell*, **63**, 601–8.

Sumner-Smith, M., Roy, S., Barnett, R., Reid, L. S., Kuperman, R., Delling, U. & Sonenberg, N. (1991). Critical chemical features in *trans*-acting-responsive RNA are required for interaction of human immunodeficiency virus type 1 Tat protein. *Journal of Virology*, **65**, 5196–202.

Weeks, K. M. & Crothers, D. M. (1991). RNA recognition by Tat-derived peptides: interaction in the major groove? *Cell*, **66**, 577–88.

Weeks, K. M. & Crothers, D. M. (1992). RNA binding assays for Tat-derived peptides: implications for specificity. *Biochemistry*, **31**, 10281–7.

Weeks, K. M., Ampe, C., Schultz, S. C., Steitz, T. A. & Crothers, D. M. (1990). Fragments of the HIV-1 Tat protein specifically bind TAR RNA. *Science*, **249**, 1281–5.

Wu, F., Garcia, J., Sigman, D. & Gaynor, R. (1991). Tat regulates binding of the human immunodeficiency virus *trans*-activating region RNA loop-binding protein TRP-185. *Genes and Development*, **5**, 2128–40.

Wu-Baer, F., Lane, W. S. & Gaynor, R. B. (1995*a*). The cellular factor TRP-185 regulates RNA polymerase II binding to HIV-1 TAR RNA. *European Molecular Biology Organisation Journal*, **14**, 5995–6009.

Wu-Baer, F., Sigman, D. & Gaynor, R. B. (1995*b*). Specific binding of RNA polymerase II to the human immunodeficiency virus trans-activating region RNA is regulated by cellular co-factors and Tat. *Proceedings of the National Academy of Sciences, USA*, **92**, 7153–7.

Zenzie-Gregory, B., Sheridan, P., Jones, K. A. & Smale, S. T. (1993). HIV-1 core promoter lacks a simple initiator element but contains bipartite activator at the transcription start site. *Journal of Biological Chemistry*, **268**, 15823–32.

Zhou, Q. & Sharp, P. A. (1995). Novel mechanism and factor for regulation by HIV-1 Tat. *European Molecular Biology Organisation Journal*, **14**, 321–8.

CHLAMYDIA HOST AND HOST CELL INTERACTIONS

M. E. WARD

Mailpoint 814, Molecular Microbiology Group, Southampton University Medical School, Southampton SO16 6YD, UK

INTRODUCTION

Chlamydia cause a wide range of important human and animal infections. In humans, trachoma, due to ocular infection with *Chlamydia trachomatis* serovars A, B, Ba or C, is currently one of the world's major causes of preventable blindness, with approximately 300 million cases in some of the world's poorest communities. Of these, up to 9 million people, mainly elderly, are blind or severely visually impaired. Although the incidence of acute trachoma is generally decreasing, paradoxically the overall prevalence of blindness is expected to increase as a result of demographic change in developing countries leading to increased survival into old age. Below the waist, *C. trachomatis* serovars D to K are one of the major causes of genital tract infection world-wide. These infections are far from insignificant, with women bearing a disproportionate share of the consequences of sexually transmitted *C. trachomatis* infection through well-known complications such as pelvic inflammatory disease (PID) and its sequelae, ectopic pregnancy or tubal infertility. In one large UK study, women with PID were six times more likely to suffer abdominal pain and ten times more likely to contract intra-tubal pregnancy or to subsequently require hysterectomy than control women (Buchan *et al.*, 1993). *C. trachomatis* is probably also a significant cause of male infertility. *C. trachomatis* infection of farm animals has been recognized relatively recently, with the organism able to colonize the large intestine of piglets (Zahn *et al.*, 1995). In contrast, *Chlamydia psittaci* and *Chlamydia pecorum* (Fukushi & Hirai, 1992) are primarily agents of veterinary disease. It has been considered that no other group of pathogenic micro-organisms is more widespread in nature than *C. psittaci*, with apparent chlamydial infections being reported amongst amoebae, coelenterates, arthropods, molluscs, some 130 species of bird and a wide range of mammals. The former TWAR agents, recently reclassified as *Chlamydia pneumoniae* (Grayston *et al.*, 1989) are now recognized as major causes of acute respiratory tract infections in humans (Grayston *et al.*, 1993) and are implicated in the pathogenesis of heart disease and sarcoidosis.

Despite the importance of chlamydial infections, little is known about the molecular basis of chlamydial interaction with host cells. This does not reflect indolence, but the difficulties of working with an intracellular pathogen for which techniques of genomic manipulation have yet to be devised, although transient expression of recombinant genes in chlamydiae has been reported (Tam, Davis & Wyrick, 1994). Given this lack of molecular detail, a necessarily broad view of the main symposium topic will be taken, covering chlamydial interactions with the host immune system as well as with host epithelia. This review extends a recent review by the author on the immunobiology of chlamydial infections (Ward, 1995).

CHLAMYDIAL TAXONOMY

The genus *Chlamydia* consists of highly specialized Gram-negative bacteria which are intracellular pathogens of eukaryotic cells. Chlamydiae are distinguished from bacteria such as *Listeria monocytogenes* or *Mycobacterium tuberculosis* by the fact that they are true, obligate intracellular pathogens that cannot replicate, and have never been cultured, outside the host cell. The chlamydiae are distinguished from rickettsiae by their characteristic dimorphic growth cycle and by their inability to synthesize high energy compounds such as ATP and GTP. Transit and survival through hostile extracellular space is ensured by the tough, 'spore-like' elementary bodies (EBs) while the fragile, reticulate bodies (RBs), ideal for interactions with the host cell, are responsible for intracellular replication. Properties of the four species which currently belong to the genus *Chlamydia* are summarized in Table 1.

EVOLUTIONARY ADAPTATION OF *CHLAMYDIA* TO THE HOST

Chlamydial evolution must inevitably be linked to the evolution of the cells and organisms which they infect; the importance of these interactions should be reflected in any satisfactory chlamydial taxa. Presently, the order Chlamydiales consists of one family, the Chlamydiaceae with one genus, *Chlamydia* within which are four species: C. *trachomatis, C. psittaci,* C. *pneumoniae* (Grayston *et al.*, 1989) and *C. pecorum* (Fukushi & Hirai, 1992) (see Table 1). Unfortunately this classification, with the exception of *C. pecorum*, is largely based on phenotypic characteristics, making any relationship between the evolution of chlamydiae and their hosts uncertain. Moreover, the original view that *C. trachomatis* and *C. pneumoniae* are primarily human pathogens is clearly less valid now that *C. trachomatis* has been recognized as an important pathogen of pigs, while *C. pneumoniae* also infects horses and causes ocular disease in koalas. At the genomic level, DNA hybridization clearly indicates that *C. psittaci* is much more heterogeneous than the other species. Inter-species DNA homology is generally less than 10%, leading to the view that chlamydiae converged in evolution from divergent, free-living ancestors. This makes little common sense, given that the *Chlamydia* share a

Table 1. *Characteristics of the four chlamydial species*[a]

	Chlamydial species			
Characteristics	*C. trachomatis*	*C. pneumoniae*	*C. psittaci*	*C. pecorum*
Natural hosts	Humans, mice, pigs	Humans, horses	Birds, mammals, occasionally humans	Cattle and sheep
EB morphology	Round	Round or pear shaped	Round	Round
Inclusion	Oval, vacuolar	Oval, dense	Variable, dense	Oval, dense
Iodine staining	Yes	No	No	No
Sulphonamide sensitive	Yes	No	No	No
Number of serovars	At least 15	1	Not defined	3
Characteristic infections	Genital and ocular mucosae. Often inapparent. Intermittent shedding. Rarely systemic	Wide range of respiratory tract infections. Often inapparent. Possible association with heart disease and sarcoidosis	Often systemic: pneumonia, abortion, arthritis	CNS, respiratory and gut. Often inapparent. Prolonged carriage
DNA: Mol % G+C	39.8	40.3	39.6	39.3
[b]Homology % relative to				
C. trachomatis	92			
C. pneumoniae	1–7	94–96		
C. psittaci	1–33	1–8	14–95	
C. pecorum	1–10	10	1–20	88–100

[a] Adapted from Fukushi & Hirai (1993).
[b] DNA hybridization studies on whole genomic DNA.

common characteristic developmental cycle, a genus-specific epitope on their lipopolysaccharide and closely related gene sequences for outer envelope proteins not found in other bacteria and have similar, unique, biological and metabolic activities.

A more satisfactory indication of how chlamydiae may have evolved in relationship with their hosts is provided by computer taxonomic studies of single chlamydial genes. The *omp1* gene which encodes chlamydial MOMP tends to be highly variable and is particularly useful for looking at strain or intra-species differences. The gene encoding 16S rRNA is used as the basis for the classification of other organisms, but it is relatively highly conserved within chlamydiae. This gene can be used at family level to differentiate the genus *Chlamydia*, within which sequence differences are generally less than 7%, from other 'chlamydia-like' organisms whose 16S sequences are over

15% different. The domain I segment of the gene encoding 23S rRNA is believed to be particularly appropriate for the speciation of organisms within the present genus *Chlamydia* (A. Andersen & K. D. E. Everett, personal communication). Computer taxonomy gives a broadly consistent picture of chlamydiae diverging in evolution from a common ancestor, with a single, rooted stem and two branches, one for *C. trachomatis* and a second for *C. psittaci*, the latter bearing sub-branches for the recently described species *C. pecorum* and *C. pneumoniae*. The most ancient *C. trachomatis* strains are the mouse pneumonitis agent and an isolate from pigs, suggesting perhaps that this predominantly human pathogen may have evolved from strains infecting non-human mammals. It has been suggested that *Chlamydia* might have speciated as their host evolved or as individual cell types differentiated. However, a detailed study of 24 amino acid sequences inferred from the *omp1* gene concluded that there has been only limited co-evolution of the parasite and of the host cell, organ or species that it infects. Moreover, there was little divergence of *omp1* sequence over the time span of isolation of these genes (Fitch, Peterson & de la Maza, 1993). Nevertheless, recent extensive computer taxonomic studies of domain I sequences of the gene encoding 23S rRNA indicate that the present genus *Chlamydia* can be divided into nine major groups (see Fig. 1) which frequently correspond to the host infected. Thus, within the present *C. trachomatis* there are three groups corresponding to rodent, porcine and human branches, suggesting that host selection has been a significant determinant of chlamydial evolution. Similarly, within the present *C. psittaci* (Fig. 1) there are clear groupings corresponding to avian, ovine abortion and feline hosts. Altogether, nine groups are suggested by the 23S sequence data (Fig. 1) although more data are needed for the *C. psittaci* guinea-pig isolates and *C. pneumoniae* horse isolates to determine if these warrant separate groupings. Clearly, the current species *C. psittaci* is too heterogeneous to warrant single species status. The current debate is whether the nine groups observed in sequencing the gene encoding 23S rRNA warrant the creation of nine chlamydial species constituting one, or possibly two, genera.

The key evolutionary step which the free-living ancestor of the *Chlamydia* must have taken was entry into, and survival within, eukaryotic cells. This

Fig. 1. Host-driven evolution of *Chlamydia*. The figure shows a Phylogenetic tree of chlamydiae based on numerical taxonomy of domain I of the gene encoding 23S rRNA. The closest relative to *Chlamydia* is the chlamydia-like agent Z1 which has subsequently been termed *Simkania* in Genbase. Another, less close relative, is *Pirellula marina*. There are two main branches corresponding to *C. psittaci* and *C. trachomatis*. *C. psittaci* (top of Figure), which can be divided into a number of separate groups related to the host species. Similarly, *C. trachomatis*, bottom of Figure, can be divided into human, mouse and porcine groups. Such observations support the notion that adaptation to the host has been a key feature of chlamydial evolution. *C. pneumoniae* and *C. pecorum* subsequently split from *C. psittaci*. The molecular bases for host tropisms among *Chlamydia* are not known. Data and analysis kindly provided by Dr Karin D. E. Everett and Dr Arthur A. Andersen, USDA, Agriculture Research Service, National Animal Disease Centre, Ames, Iowa USA.

Computer Taxonomy of Chlamydia Species
based on PAUP analysis of the
23S Gene, Domain I (620 ± 2 bp)

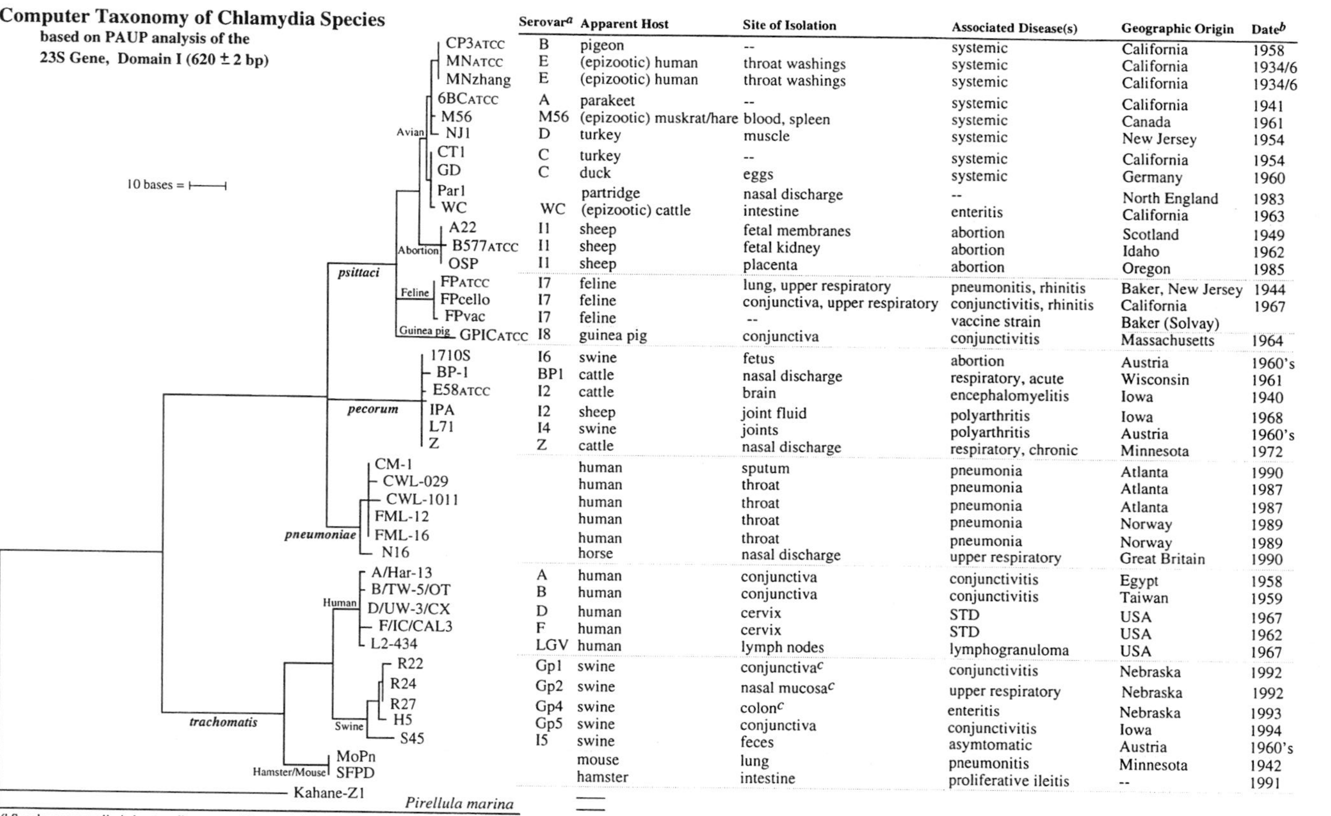

Strain	Serovar[a]	Apparent Host	Site of Isolation	Associated Disease(s)	Geographic Origin	Date[b]
CP3ATCC	B	pigeon	--	systemic	California	1958
MNATCC	E	(epizootic) human	throat washings	systemic	California	1934/6
MNzhang	E	(epizootic) human	throat washings	systemic	California	1934/6
6BCATCC	A	parakeet	--	systemic	California	1941
M56	M56	(epizootic) muskrat/hare	blood, spleen	systemic	Canada	1961
NJ1	D	turkey	muscle	systemic	New Jersey	1954
CT1	C	turkey	--	systemic	California	1954
GD	C	duck	eggs	systemic	Germany	1960
Par1		partridge	nasal discharge	--	North England	1983
WC	WC	(epizootic) cattle	intestine	enteritis	California	1963
A22	I1	sheep	fetal membranes	abortion	Scotland	1949
B577ATCC	I1	sheep	fetal kidney	abortion	Idaho	1962
OSP	I1	sheep	placenta	abortion	Oregon	1985
FPATCC	I7	feline	lung, upper respiratory	pneumonitis, rhinitis	Baker, New Jersey	1944
FPcello	I7	feline	conjunctiva, upper respiratory	conjunctivitis, rhinitis	California	1967
FPvac	I7	feline	--	vaccine strain	Baker (Solvay)	
GPICATCC	I8	guinea pig	conjunctiva	conjunctivitis	Massachusetts	1964
1710S	I6	swine	fetus	abortion	Austria	1960's
BP-1	BP1	cattle	nasal discharge	respiratory, acute	Wisconsin	1961
E58ATCC	I2	cattle	brain	encephalomyelitis	Iowa	1940
IPA	I2	sheep	joint fluid	polyarthritis	Iowa	1968
L71	I4	swine	joints	polyarthritis	Austria	1960's
Z	Z	cattle	nasal discharge	respiratory, chronic	Minnesota	1972
CM-1		human	sputum	pneumonia	Atlanta	1990
CWL-029		human	throat	pneumonia	Atlanta	1987
CWL-1011		human	throat	pneumonia	Atlanta	1987
FML-12		human	throat	pneumonia	Norway	1989
FML-16		human	throat	pneumonia	Norway	1989
N16		horse	nasal discharge	upper respiratory	Great Britain	1990
A/Har-13	A	human	conjunctiva	conjunctivitis	Egypt	1958
B/TW-5/OT	B	human	conjunctiva	conjunctivitis	Taiwan	1959
D/UW-3/CX	D	human	cervix	STD	USA	1967
F/IC/CAL3	F	human	cervix	STD	USA	1962
L2-434	LGV	human	lymph nodes	lymphogranuloma	USA	1967
R22	Gp1	swine	conjunctiva[c]	conjunctivitis	Nebraska	1992
R24	Gp2	swine	nasal mucosa[c]	upper respiratory	Nebraska	1992
R27	Gp4	swine	colon[c]	enteritis	Nebraska	1993
H5	Gp5	swine	conjunctiva	conjunctivitis	Iowa	1994
S45	I5	swine	feces	asymtomatic	Austria	1960's
MoPn		mouse	lung	pneumonitis	Minnesota	1942
SFPD		hamster	intestine	proliferative ileitis	--	1991
Pirellula marina	—	—				

[a] Serology generally is host or disease specific; established serovars are indicated with letters, as immunotypes (I), or as groups (Gp).

[b] Date of isolation or of earliest publication. [c] These strains were isolated from multiple sites, and represent a respiratory, conjunctivitis, enteritis syndrome.

provided pre-formed nutrients and bio-energetic molecules, a lack of competition from other micro-organisms and sequestration from the host immune response, but at the eventual cost of necessary structural specialization to ensure survival in the hostile extracellular milieu and loss of metabolic independence from the host cell. The identity of the free-living ancestor of the *Chlamydia* must inevitably be speculation but there are some intriguing clues. Sequencing studies of 16S rRNA of *C. psittaci* strain 6BC suggest that the ancestral chlamydial branch probably split off the main eubacterial trunk at a relatively early age unlike the rickettsiae which have rRNA sequences similar to the plant-associated genera *Rhizobium* and *Agrobacterium* (Weisburg *et al.*, 1991). On the basis of 16S rRNA sequences, *Planctomyces staleyi* is a close relative of *C. psittaci* strain 6BC. The Planctomycetales are free-living aquatic bacteria with motile buds, numerous fimbriae and a hold fast (Moulder, 1988). This relationship might reasonably be dismissed as coincidental were it not that, unusually, *Planctomyces*, like *Chlamydia*, has no peptidoglycan.

There are also accounts of 'chlamydia-like' organisms in invertebrates ranging from amoebae and *Hydra* through molluscs to the arthropods (Moulder, 1988; Kahane, Metzer & Friedman, 1995). Most of these agents, although repeatedly observed, have not been grown and their taxonomic relatedness to the *Chlamydia* is entirely unknown. The closest known relatives to the genus *Chlamydia* are the *Simkania*. Originally isolated as tissue culture contaminant Kahane-Z1, these organisms have a typical chlamydial growth cycle and the gene encoding 16S rRNA has 83% nucleotide sequence identity with the corresponding gene in *Chlamydia* (Kahane *et al.*, 1993, 1995), suggesting that this organism should eventually be placed within the Order Chlamydiales. This is supported by the sequencing data from domain I of the gene encoding 23S rRNA (Fig. 1). Another chlamydia-like organism, identified in clams, expresses the *Chlamydia*-specific lipopolysaccharide group antigen, produces glycogen in its inclusions and has been provisionally classified as *C. trachomatis*. Furthermore, there are several preliminary reports of 16S rRNA gene sequences from chlamydia-like organisms that have been isolated from amoebae. One speculation is that *Chlamydia* arose from some primitive, free-living aquatic bacterium, which adapted first to single cell or invertebrate hosts. It is likely peptidoglycan was already absent, as in the archaeobacteria and planctomyces, but it may have become lost as a consequence of intracellular residence, being no longer needed for osmotic protection. Instead, compensatory, reversible, disulphide bond cross-linking of the outer membrane proteins became an essential adaptation, permitting the peptidoglycan-free EBs to retain structural integrity in the extracellular environment. Divergence of *C. psittaci* and *C. trachomatis* occurred long ago, probably in invertebrate hosts, as suggested by the clam agent. Subsequently, *C. pneumoniae* and *C. pecorum* diverged within the *C. psittaci* group.

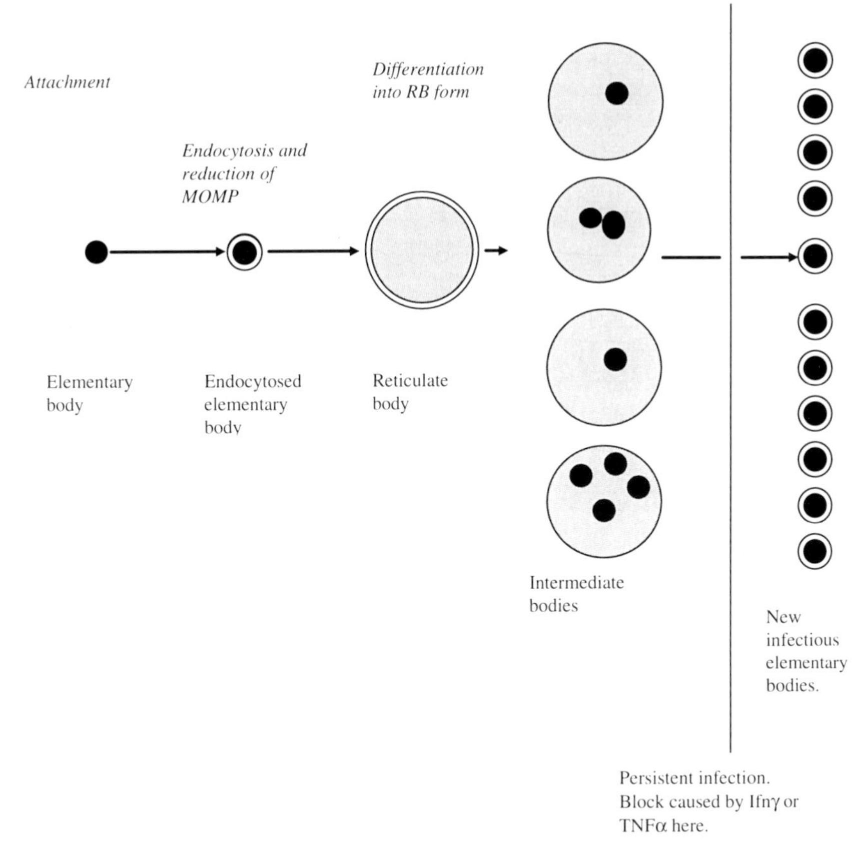

Fig. 2. Diagram showing the chlamydial replication cycle. Chlamydial infection begins with attachment of the elementary body (0.3 μm diam.) to uncharacterized host cell receptors. Entry into the host cell in an endosome is accompanied by tyrosine phosphorylation of host cell protein and unknown signal transduction events. The chlamydial endosome is not acidified and does not fuse with cell lysosomes. Following endocytosis, the elementary body is exposed to a reducing environment which is thought to trigger differentiation into the 1 μm diameter reticulate body. Differentiation is complete around 9 h after infection. The reticulate body divides by binary fission within the endosome, producing an immature reticulate body inclusion some 15 h after infection. Condensation of DNA on histone leads to the formation of intermediate bodies about 18 h after infection from which one or more elementary bodies are produced. In persistent infection, Ifnγ or other products of the cell-mediated immune system block the conversion of reticulate bodies into elementary bodies. These reticulate bodies may be passed on as the host cell divides or they may be lost as a result of epithelial turnover. In productive infection the cycle is completed in 40–72 h with the release of elementary bodies to initiate a new cycle.

THE CHLAMYDIAL GROWTH CYCLE

The elementary body (EB) and the reticulate body (RB)

A diagram of the chlamydial growth cycle is given in Fig. 2. The growth cycle commences with the infectious elementary body (EB), a rigid structure some 0.3 μm in diameter whose DNA nucleoid is highly condensed onto histone-like proteins called Hc1 and Hc2 (depending on the species) (Pedersen *et al.*, 1996*b*). The EB outer membrane is rigid and may be recovered in an insoluble outer membrane complex following Sarkosyl extraction. The Sarkosyl-insoluble proteins detected for *C. trachomatis* and *C. psittaci* in this complex include the 40 kDa major outer membrane protein (MOMP) and two cysteine-rich proteins, the 60 kDa outer envelope protein (EnvB) encoded by the *omp2* gene and the 12 kDa EnvA lipoprotein coded by the *omp3* gene (Raulston, 1995). *C. pneumoniae* EBs contain an additional 98 kDa protein. These outer membrane proteins are extensively cross-linked by disulphide bonds, conferring rigidity in the absence of peptidoglycan.

The reticulate body (RB) is some 1 μm in diameter, resembling a typical Gram-negative coccus in the electron microscope. The RB outer envelope is relatively pleomorphic; MOMP is present in RBs but in a reduced, non-disulphide-linked form, while the EnvA and EnvB outer membrane proteins are absent. The RB is non-infectious, being concerned solely with intracellular replication (for a detailed review of the chlamydial growth cycle see Ward, 1988).

Attachment to host cells

Attachment of an EB to the host cell is critical for initiation of the growth cycle, but no high avidity chlamydial adhesins or their corresponding host cell receptors have yet been identified. It seems likely that chlamydiae are not critically dependent on a single high avidity interaction, as chlamydial adhesion to relevant host cells derived from the genital tract does not show high binding affinity (Moorman, Sixbey & Wyrick, 1986; Wyrick *et al.*, 1989, 1993). Chlamydial adhesion may be the aggregate effect of multiple, weak affinity interactions involving several chlamydial surface components. Such a mechanism would explain the ability of chlamydiae to attach to a wide variety of host cells. In general, binding of labelled EBs to host cells is saturable with a large excess of unlabelled EBs and it is sensitive to trypsin treatment of the host cell. It has been suggested that the chlamydial MOMP, which represents up to 60% of the protein at the surface of the chlamydial EB, is one possible adhesin, although the evidence for this is largely circumstantial. Thus, trypsin treatment of viable *C. trachomatis* serovar B organisms leads to a loss of their ability to attach to and to infect host cells. This could be associated with *in situ* cleavage of MOMP in variable segments II and IV. Serovar L2 organisms were not so affected and their MOMP was cleaved in variable segment IV only (Su *et al.*, 1988). Cleavage of variable segment IV leads to exposure of a hitherto-buried species-specific epitope, presumably that identified by peptide mapping as LNPTIA (Conlan, Clarke & Ward, 1988). It was suggested that variable segments II and IV formed a combined binding site, with variable segment II involved in the actual binding but variable segment IV necessary to maintain conformational integrity. However, limited trypsin digestion of infectious EBs also removes a small terminal peptide of the *omp2*-encoded 60 kDa EnvB outer envelope protein. This protein was the dominant solubilized EB outer envelope protein binding to glutaraldehyde-fixed HeLa cells in a ligand-blot assay (Ting *et al.*, 1995). Thus, the role played by MOMP or the 60 kDa protein in chlamydial adhesion to host cells is uncertain.

Interestingly, MOMP is also a glycoprotein. The glycan moiety can be cleaved with *N*-glycanase and labelled with borotritide. The labelled glycan competes for binding sites on HeLa cells with native or UV-inactivated, but not heat-treated EBs and its binding is inhibited by high concentrations of galactose, mannose or *N*-acetyl glucosamine (Swanson & Kuo, 1994). The possibility that this glycan might interact with host-cell-surface carbohydrate receptors warrants serious investigation. However, to date, there is no compelling evidence that MOMP plays a key role as a chlamydial adhesin.

It has been known for a long time that the infectivity of oculogenital isolates of *C. trachomatis* may be increased by treatment of HeLa cells with cationic polymers such as DEAE-dextran, heparin or poly-L-lysine (Kuo, Wang & Grayston, 1973). These compounds have little effect on the strains of *C. trachomatis* which cause lymphogranuloma venereum (LGV). The most likely explanation was originally thought to be that these charged compounds abolished the net electrostatic repulsive barrier for chlamydial attachment between the negatively charged chlamydial surface and the negatively charged host cell surface (Ward, 1986). However, it has recently been suggested that a chlamydial heparan sulphate-like molecule mediates attachment, and perhaps subsequent infectivity, by bridging the EB outer membrane to receptors on the mammalian host cell (Zhang & Stephens, 1992). Thus, infectivity could be inhibited by heparin or heparan sulphate but not by other glycosaminoglycans. This suggests that charge alone was not the key factor. Fibronectin and other compounds capable of binding to heparan sulphate, or treatment with heparinitase, neutralized chlamydial infectivity if the chlamydiae, but not the host cell, were treated. Exogenous heparin bound strongly to chlamydiae with an apparent K_d of 10^{-9} to 10^{-10} M and restored their infectivity (Zhang & Stephens, 1992). The sulphated polysaccharides carrageenan, pentosan, fucoidan and dextran sulphate and the glycosaminoglycans heparin, heparan sulphate and dermatan sulphate were all capable of blocking *C. trachomatis* adhesion to cervix-derived human epithelia, whereas other, less negatively charged glycosaminoglycans such as chondroitin sulphates A or C, keratan sulphate and hyaluronic acid, characteristic of cartilage and joints, were not. Similar sulphated polysaccharides inhibit transmission of HIV and are being tested as vaginal formulations for prevention of HIV infection (Zaretzky, Pearce-Pratt & Phillips, 1995). Heparan sulphate-like molecules are similarly implicated in the binding or infectivity of a large number of pathogens to host cells, including *Bordetella pertussis*, *Streptococcus mutans*, HIV, herpes simplex virus, cytomegalovirus, pseudorabies virus, *Leishmania*, *Trypanosoma* and *Plasmodium* (see Zhang & Stephens, 1992). However, the unique feature of chlamydiae is that the organism itself is reported to synthesize a heparan sulphate-like molecule. This needs to be confirmed by identification of the chlamydial genes encoding the enzymes responsible for heparan biosynthesis and characterization of the host cell receptor involved. Perhaps the closest parallel is with *Neisseria gonorrhoeae*. One of the gonococcal opacity proteins associated with epithelial cell invasion rather than attachment, binds to the sulphated glycosaminoglycan syndecan-1, which occurs on both genital and ocular epithelia (van Putten & Paul, 1995).

It is likely that chlamydial adhesion to eukaryotic cells may involve several different chlamydial surface ligands, with the heparan-sulphate-based mechanism being only one factor and MOMP and the 60 kDa protein possibly also playing a role. Interestingly, an 82 kDa chlamydial heat-shock protein (Hsp70), reported in association with the outer membrane of *C. trachomatis*, has a contiguous 73 amino acid region with 51% amino acid sequence identity with the sperm receptor binding domain of the sea urchin, *Strongylocentrosus purpuratus* (Raulston *et al.*, 1993).

Endocytosis of chlamydial elementary bodies

Heparin inhibits the infectivity of trachoma biovar organisms to a greater extent than it inhibits attachment, suggesting that the heparan sulphate-like molecule might also be involved in entry. Entry into host cells involves both microfilament- and clathrin-dependent mechanisms (Reynolds & Pearce, 1993) and is accompanied by tyrosine kinase activation (Birkelund, Johnsen & Christiansen, 1994), although the signal transduction events involved have yet to be characterized. The tyrosine-phosphorylated proteins are associated spatially with the chlamydiae during the first eight hours post-infection and, by 16 hours post-infection, can be observed at the periphery of the inclusion. Tyrosine phosphorylation is inhibited by exogenous heparin, suggesting that glycosaminoglycan-mediated attachment of EBs to the host cell is required for its initiation. The particular organism, its presentation to the host cell, differences between host cells themselves, their receptors, their degree of morphological polarity and their hormone responsiveness are all considered important factors for chlamydial entry (Reynolds & Pearce, 1993; Stephens, 1994), influencing endosome selection, processing and other events that might modulate the intracellular fate of the internalized organism and subsequent antigen processing by the host cell.

Recently, Hsia *et al.* (1996) reported that the genome of a *C. psittaci* guinea-pig inclusion conjunctivitis strain encodes proteins analogous to enterobacterial type III secretion proteins. Unlike proteins secreted via the general secretion (type I) pathway, the type III proteins lack a leader sequence and are genetically linked to the *tts* locus. The interest in type III proteins is that they frequently include proteins involved in motility, invasion of eukaryotic cells, or subversion of host signal transduction mechanisms. Thus, in *Yersinia pestis* or *Y. enterocolitica*, YscU, LcrD and LcrE are all type III proteins. YscU and LcrD are cytoplasmic membrane proteins with a large cytoplasmic domain presumably involved in protein export, while LcrE (InvE in *Salmonella* species) is an outer membrane protein which is itself secreted by the type III pathway and which functions as negative regulator of secretion. The presence of the *lcrE* gene nucleotide sequence homologue in *C. psittaci* supports the concept that the type III pathway is active in chlamydiae. Furthermore, some type III-secreted products such as *Shigella flexneri* Lpa or *Yersinia* Yop proteins induce protective immunity and have been associated with bacterial invasion and virulence (Hsia *et al.*, 1996). Thus, characterization of the extended chlamydial *tts* locus and of genes utilizing the type III pathway may delineate new chlamydial genes involved in pathogenesis or immunity.

Intracellular differentiation and replication of reticulate bodies

The normal course of the chlamydial developmental cycle is shown in Fig. 2. Within 6–9 h after ingestion, a single EB within a single endosome has differentiated into the intracellular replicating reticulate body (RB) form. The trigger for initial differentiation is not known. EBs exposed to a reducing environment commence biochemical events such as glutamate oxidation that are thought to possibly be critical for differentiation (Bavoil, Ohlin & Schachter, 1984). However, low Eh does not trigger morphological differentiation. Limited evidence based on ultrastructural (Chang, Leonard & Arad, 1982) and biophysical (Bavoil *et al.*, 1984) observations suggest that MOMP may be an outer envelope porin, in which the pore structure is formed by an S-S linked trimer. This structure would be reduced when EBs are exposed to a reducing environment. This is postulated to cause relaxation of MOMP pore size and ingress of host-cell-derived ATP, GTP or other compounds necessary to initiate development (Bavoil *et al.*, 1984). Definitive evidence is lacking that MOMP is a porin. The protein lacks the characteristic structural motifs of porins like *E. coli* OmpF porin or the meningococcal class 1 protein, suggesting that MOMP may belong to a unique class of chlamydial porins.

Endosomes containing chlamydial EBs, but not those containing the rickettsia *Coxiella burnetii*, are not acidified by activation of the vacuolar ATPase-driven proton pump (Heinzen *et al.*, 1996). Thus, chlamydial growth is not inhibited by the vacuolar (H^+) ATPase enzyme inhibitor bafilomycin A nor by lysosomotropic amines such as methylamine. Vacuoles harbouring

Coxiella burnetii, but not *C. trachomatis*, incorporate fluorescent labelled fluid-phase markers, indicating trafficking between the rickettsial vacuole and the endocytic pathway. The late endosomal-prelysosomal marker, the cation-independent mannose-6-phosphate receptor, was not detectable in the vacuolar membranes encompassing either organism. However, the lysosomal enzymes acid phosphatase and cathepsin D and the lysosomal glycoproteins LAMP-1 and LAMP-2 could be localized to the rickettsial endosome but not to its chlamydial counterpart. Endocytosed chlamydiae were transported to the Golgi region and acquired sphingolipids from the host within a few hours following infection, utilizing an active, energy and temperature-dependent, vesicle-mediated process. Lipid vesicles derived from the Golgi apparatus, but not the plasma membrane, were identified in the process of fusing to the chlamydial inclusion membrane, although the latter did not contain Golgi-specific glycoproteins. Collectively, these data suggest that *Coxiella burnetii* resides in a classic phagolysosomal vacuole, whereas *C. trachomatis* replicates within a unique non-acidified vacuole that is disconnected from endosome-lysosome trafficking but which interrupts a host exocytic pathway to receive sphingolipids in transit from the Golgi apparatus to the plasma membrane (Hackstadt *et al.*, 1996; Heinzen *et al.*, 1996).

Endosomes containing *C. trachomatis* tend to fuse with themselves to form one large inclusion, whereas endosomes containing *C. psittaci* remain distinct. Indeed, in co-infected cells, endosomes containing *C. psittaci* do not fuse with endosomes containing *C. trachomatis* (Matsumoto *et al.*, 1991). Chlamydial endosomes also escape fusion with cellular lysosomes (Matsumoto *et al.*, 1991). The mechanisms of these fusion and endosome-sorting events are unknown but the host cell cytoskeleton clearly plays a role which is serovar-dependent in the case of *C. trachomatis*. Thus, disruption of HEC-1B (epithelial) and McCoy (fibroblast) cell microfilaments with cytochalasin D markedly reduced serovar E, but not serovar L2, infection (Schramm & Wyrick, 1995). Conversely, cytochalasin D treatment, as well as microtubule disruption with colchicine or nocodazole, had no effect on serovar E inclusion development, whereas they resulted in the initial formation, during early and mid-phase development, of multiple serovar L2 inclusions per cell which subsequently fused to form one large inclusion in the absence of an intact cytoskeleton. Clathrin is probably also involved in the redistribution of chlamydial endosomes (Majeed & Kihlström, 1991), as well as calcium, and the selective translocation of annexins ii, iv and v (Majeed *et al.*, 1994), a group of host calcium- and phospholipid-binding proteins.

The genetic and metabolic events associated with the intracellular replication of chlamydial RBs and their differentiation into EBs have been reviewed in detail elsewhere (McClarty, 1994). A single RB often gives rise to more than one EB (Ward, 1988), and production of EBs requires the controlled up-regulation of MOMP expression by the *omp1* gene and its re-condensation onto histone protein. Differentiation of an RB from the original EB involves an increase in diameter from 0.3 μm to approximately 1 μm, reduction of S-S links in the outer membrane, unravelling of DNA in the condensed nucleoid and the appearance of granular ribosomes with their associated RNA/protein synthesis. Early proteins synthesized include chlamydial ribosomal protein S1, and the heat-shock proteins GroEL and DnaK (Lundemose *et al.*, 1990). Clearly this process requires the sequential regulation of chlamydial gene expression. In chlamydiae, only one sigma factor, namely σ^{66}, has been identified (Engel *et al.*, 1990), which is analogous to *E. coli* sigma factor σ^{70}, but others may exist. Chlamydial promoters have been mapped by primer extension or S1 exonuclease analyses for several genes and appear not to have strong similarities to each other or to the -10 and -35 consensus sequences recognized by *E. coli* σ^{70}. However it has been demonstrated that *in vitro* transcription of some chlamydial genes in a cell-free system required σ^{66} (Douglas, Saxena & Hatch, 1994; Fahr *et al.*, 1995).

Two basic, DNA-binding proteins termed Hc1 and Hc2, with molecular masses of 17 and 25.7 kDa and coded by the *hct1* and *hct2* genes, respectively, have also been identified (Pedersen, Birkelund & Christiansen, 1996*a*; Pedersen *et al.*, 1996*b*). Both proteins have close amino acid sequence identity with eukaryotic histone protein H1. Protein Hc1 is expressed concomitantly with the transformation of RBs to EBs; both the protein's presence in isolated EB nucleoids and its ability to bind DNA on blotting indicate that this protein plays a key role in condensation of the chlamydial nucleoid. Electron microscopic studies indicate that the Hc1 protein alone has the capability of inducing condensation of the chlamydial chromosome and that it is able to bind circular or linearized DNA. The Hc1-DNA complexes consist of spherical aggregates of looped and kinked DNA showing that the protein is able to modulate DNA topology and, thereby, to regulate gene expression (Barry, Hayes & Hackstadt, 1992; Barry, Brickman & Hackstadt, 1993; Christiansen *et al.*, 1993; Hackstadt *et al.*, 1993). Whether external signals in the micro-environment of the host cell trigger alterations in chlamydial genomic topology via a chlamydial sensory/regulatory system remains to be determined.

Within the endosome, chlamydial replication is independent of Golgi function as it is not inhibited by Brefeldin A (Stephens, 1994), even though some important chlamydial proteins, including MOMP, are glycosylated (Swanson & Kuo, 1994). Nevertheless, there is evidence that *C. psittaci* can modify the endosomal (inclusion) membrane. Thus, when an expression library of *C. psittaci* DNA was screened with hyperimmune antisera, clones were identified which were only recognized by convalescent sera from infected animals, but not by antisera to purified EBs. Two open-reading frames were identified, one of which coded for a 39 kDa protein which was subsequently expressed in *E. coli*. Antiserum to the expressed protein reacted on immunoblotting with lysates of infected cells, or with RBs, but reacted very weakly with lysates of purified EBs. Fluorescence microscopy indicated that the 39 kDa protein was localized in the inclusion membrane of infected HeLa cells and in structures that extended into the cytoplasm and over the nucleus of infected cells. This protein has been designated IncA (inclusion membrane protein A) (Rockey, Heinzen & Hackstadt, 1995) but its function is presently unknown. Differentiation of infectious EBs from the RBs and their release completes the chlamydial replication cycle which, *in vitro*, takes some 40–72 h.

Incomplete chlamydial development and its significance in clinical disease

The developmental cycle outlined above constitutes a productive cycle of infection by which chlamydial numbers increase as a result of infectious EBs being released. This is the norm as judged by the frequent recovery of infectious EBs from patients using cell culture. However, it has been known for a long time that *C. trachomatis* antigen can be detected by immunofluorescence in the eyes of many individuals who do not exhibit current signs of active trachoma (Mabey, Bailey & Hurtin, 1992*a*); indeed a hallmark of active trachoma is that infectious EBs are shed intermittently rather than continuously. Laboratory studies on mouse L cells indicate that incomplete chlamydial replication also occurs, resulting in a state of persistent infection. These infections could persist for many months *in vitro*, being characterized by large, periodic and inversely related fluctuations in host cell and chlamydial numbers. Cells persistently infected with *C. psittaci* but not *C. trachomatis* were resistant to attachment and super-infection by exogenous chlamydiae and showed minor alteration in surface exposed plasma membrane proteins. The chlamydiae persisted in immature cytoplasmic inclusions as small, aberrant, RB-like forms which were capable of being transferred at host cell division, although the ability of the host cells to replicate was impaired. Transfer from overt to covert infection was a penicillin-sensitive process dependent on host cell density and nutrition (Moulder, Zeichner & Levy, 1982). Other factors possibly favouring persistent infection included the action of penicillin, gamma interferon (Beatty, Byrne & Morrison, 1993), tumour necrosis factor (Beatty, Byrne & Morrison, 1994*b*) and deprivation of the amino acids tryptophan and cysteine essential for chlamydial replication (Coles & Pearce, 1987; Coles *et al.*, 1993). Isolates of *C. pecorum* from sheep faeces spontaneously produce persistent infection of cell cultures *in vitro*, failing to produce inclusions on serial passage in cycloheximide-treated monolayers for several weeks (Philips & Clarkson, 1995).

There is little evidence *in vivo* for the clinical significance of persistent, incomplete, chlamydial infection. It is clear both ocular and genital chlamydial infections persist for months in the absence of treatment. These infections are often asymptomatic or, as in the case of trachoma, are characterized by intermittent periods of clinical activity and associated chlamydial shedding. In general, chlamydial activity is thought to be greatest in host cells which are themselves actively growing and dividing. Thus if mucosal cell turnover is increased by trauma or coincident infection, quiescent chlamydial infection might be activated (Richmond, 1985; Campbell *et al.*, 1988). Support for this hypothesis came from the observations that the isolation of *C. trachomatis* from the genital tracts of female contacts of men with gonococcal urethritis was dependent not only on the presence of chlamydiae in the male partner but also on actual infection by *Neisseria gonorrhoeae* in the woman herself (Oriel & Ridgway, 1982; Katz, Batteiger & Jones, 1987; Batteiger *et al.*, 1989). This could be explained either by the reactivation of low level, asymptomatic, deep-seated chlamydial infection, or by the triggering of productive infection in cells latently infected with *Chlamydia*. Clinically, such infections would be characterized by positive assays for chlamydial antigen or DNA but only intermittent isolation of infectious EBs by tissue culture. However, interpretation of such findings will inevitably be confounded by the greatly different sensitivities of laboratory diagnostic tests themselves and by the fact that microbiological indices of chlamydial infection do not coincide in time with clinical signs (Ward *et al.*, 1990).

Studies using *in situ* hybridization do suggest that chlamydial DNA may persist in ocular tissue (Cosgrove *et al.*, 1992) and in a high proportion of fallopian tube tissue from women with presumed post infection infertility following tubal occlusion (Campbell *et al.*, 1993; Patton *et al.*, 1994). Long-term persistence of viable chlamydiae in chronic infection is suggested by evidence of chlamydial infection in the joints of patients with reactive arthritis or undifferentiated oligoarthritis (Bas *et al.*, 1995), by the presence of *C. pneumoniae* in coronary arteries of patients with heart disease (see below) and, perhaps most convincingly, by reports of *Chlamydia*-specific mRNA transcripts in the joints of patients with reactive arthritis (Beutler *et al.*, 1995; Gerard *et al.*, 1995). In the latter case there is evidence that, just as observed for gamma-interferon (Ifnγ)-induced delayed chlamydial infection *in vitro*, there is up-regulation of mRNA transcripts for chlamydial Hsp60 and down-regulation of transcripts for MOMP. It is not known whether a corresponding altered translation of chlamydial antigen *in vivo* actually plays a role in the immunopathology of chlamydial disease.

CHLAMYDIAL GENOMIC SEQUENCING

Chlamydiae have the genomic capacity to encode some six hundred proteins but, until recently, very few of these proteins had been identified. Moreover, although chlamydiae must have complex mechanisms of gene regulation related to the developmental cycle, very little is known about these mechanisms due to the absence of methods for the genetic manipulation of chlamydiae. However, the small genomic size of chlamydiae, their medical and veterinary importance and their unique biology make them an obvious candidate for gene assignment via genomic sequencing. As this manuscript was being revised, the first preliminary gene assignments from genomic sequencing of *C. trachomatis* serovar D (genome size 1045 kilobases) and from partial sequencing of *C. trachomatis* serovar L2 were presented by Dr R. Stephens at the Third Meeting of the European Society for *Chlamydia* Research in Vienna in September, 1996. This important project is a collaboration between the *Chlamydia* laboratory at Berkeley led by Dr R. Stephens and the Stanford Genome Sequencing Centre led by Dr R. Davis. The sequencing strategy involves random shearing of the DNA giving an eight- to ten-fold coverage of the genome (approximately 10 million bases) followed by the construction of an M13 library of fragments of approximately 1500 base pairs. Sequences with homologous regions are associated into bins and ultimately into contiguous sequence. Gaps between these contiguous regions will be filled by PCR-linking and positional assignments on the chlamydial chromosome are being determined by hybridisation to both bacteriophage λ and cosmid libraries. It is planned to amplify and clone each open reading frame into the prokaryotic expression vector pET and a derivative of the eukaryotic expression vector pCDNA1. At the time of presentation, approximately 9800 high-quality sequences were available and some 468 kilobase pairs had been assembled, corresponding to almost half the genome. From these, preliminary gene assignments had been made for approximately 260 chlamydial genes following Blast-X or -N searches of Genbase. These gene assignments and ultimately the corresponding sequences are being placed on the world wide web (http://violet.berkeley.edu:4231) to which the reader is referred for updated information. Some surprises were: (i) the presence of a large number of genes involved in muramic acid biosynthesis together with the LytB penicillin tolerance protein, even though chlamydiae are thought not to have peptidoglycan; (ii) the presence of homologues of the genes encoding cytoplasmic axial filament protein (CafA) and rod shape determining protein (MreB/RodA); and (iii) the presence of the gene encoding regulatory protein PilR. Of particular interest was the presence of homologues of the *Bacillus subtilis* sporulation regulator proteins SpoIIIE, SpoIIIEB and SpoVE coupled with the absence of sigma factors other than σ^{66}. This author has previously suggested that a control process analogous to sporulation in *Bacillus* species might be involved in the control of chlamydial differentiation (Ward, 1986).

INTERACTIONS OF CHLAMYDIAE WITH THE IMMUNE SYSTEM

Chlamydiae pose a major challenge to the immune system. As obligate intracellular pathogens, most of the chlamydial life cycle is shrouded from antibody-based immune effector systems. The superficial location of chlamydial infections of ocular and genital tissue means that chlamydiae largely have to be dealt with by the local, mucosal immune system. Moreover, some of the most serious sequelae of chlamydial infection, such as the scarring and fibrosis associated with severe trachoma or the infertility secondary to scarring obstruction of the fallopian tubes in chlamydial pelvic inflammatory disease, are thought to be driven as a pathological by-product of the host's immune response. Knowledge of the interaction of chlamydiae with the immune system has largely been driven by the desire to develop chlamydial vaccines.

The evidence for protective immunity

Multiple or repeat infections with *C. trachomatis* are common, suggesting that there is little natural immunity to chlamydial infections. However, in individuals repeatedly exposed to chlamydial infection, there is evidence that acquired immunity develops. Firstly, active trachoma is almost entirely a disease of school-age children, with chlamydiae rarely isolated from individuals over 15 years old (Treharne, 1985). Secondly, studies in the 1960s and 1970s to prevent trachoma by vaccination with crude, whole preparations of *C. trachomatis* reduced the incidence of trachoma in some studies and the numbers of inclusions in others, engendering short-lived protection ranging from 1.5 to 3 years (Sowa *et al.*, 1969). Blind volunteers experimentally challenged with *C. trachomatis* were more likely to have an improvement of experimental chlamydial eye infection if they had been inoculated with formalin-treated EBs rather than placebo (Grayston, Woolridge & Wang, 1962). Immunization of non-human primates with high dose, monovalent vaccines containing at least 10^9 purified, inactivated EBs in oil emulsion also conferred short-term, serovar-specific protection to ocular infection (Wang, Grayston & Alexander, 1967; Grayston *et al.*, 1971) which was associated with ocular antibody. Thirdly, in genital tract infections with *C. trachomatis*, chlamydial isolation rates decrease significantly with age; this could not all be ascribed to diminished sexual activity (Taylor-Robinson & Ward, 1989). Consistent with this, the recovery of viable chlamydiae from men with non-gonococcal urethritis (NGU) is lower if they have had an infection of NGU within the prior six months than if they have not (Jones & Batteiger, 1986). High levels of serum IgG antibody have also been associated with protection from subsequent salpingitis in women undergoing therapeutic abortion (Brunham *et al.*, 1987). Circulating IgG transudated across endometrial gland cells may be a significant source of local antibody in the genital tract. However, involvement of a common mucosal immune system in immunity to ocular or genital tract infections or both is suggested by the fact that in guinea-pigs (Nichols, Murray & Nisson, 1978), mice (LaScolea *et al.*, 1989) and cynomolgous monkeys (Whittum-Hudson, Prendergast & Taylor, 1986), oral immunization gives partial protection against subsequent challenge.

Taken together, the evidence suggests that weak, possibly serovar-specific immunity to *C. trachomatis* infection exists, which is increased as a result of chronic infection or repeated exposure to infection.

In the case of ovine abortion in sheep due to *C. psittaci*, a crude vaccine consisting of inactivated whole chlamydiae has been in use for many years, although its efficacy is uncertain. More recently, an attenuated live vaccine consisting of a temperature-sensitive strain has been developed and is reportedly effective (Rodolakis, 1995). However, it seems likely that protection against ovine abortion in sheep is relatively easier to achieve than protection against localized human infections with *C. trachomatis* because, in sheep, there is likely to be systemic transfer of chlamydiae through the circulation resulting in an excellent opportunity for exposure to the immune system prior to the localization of chlamydiae in the placenta.

The molecular and cellular basis of acquired immunity

The obligate intracellular replication of chlamydiae means that, for most of their life cycle, they will be inaccessible to antibody. Thus, the simplest model for immunity to chlamydial infections is that neutralizing antibody will be capable of protecting against initial chlamydial colonization by blocking attachment and invasion of EBs, but thereafter, cell-mediated immunity will be necessary to control established infection.

The molecular basis of chlamydial neutralization by antibody is now broadly understood, although the relative importance of continuous versus discontinuous epitopes is uncertain. In *C. trachomatis*, MOMP is the immunodominant protein at the surface of the infectious EB, forming the basis of the serological division of *C. trachomatis* into some 18 serovars. Comparison of inferred amino acid sequences of MOMP for the different serovars of *C. trachomatis* shows that the protein consists of five segments of conserved amino acid sequence interspersed with four segments of variable sequence (Yuan *et al.*, 1989), VSI to VSIV. Peptide mapping of MOMP B-cell epitopes shows that the immunodominant serovar specific epitopes responsible for short-term, type-specific immunity are usually located on VSI or VSII. The individual serovars may be classified serologically into distinct subgroups; related serovars share one or more subspecies-specific epitopes located in a surface exposed region of VSIV. Monoclonal antibodies to these surface-exposed, serovar- or group-specific epitopes neutralize *C. trachomatis* infectivity for cell culture, and *in vivo*, for the primate eye (Zhang, Y.-X. *et al.*, 1987). These protective B-cell epitopes have been defined to single amino acid resolution by peptide mapping (Conlan *et al.*, 1988; Zhong & Brunham, 1990, 1991; Zhong, Reid & Brunham, 1990).

The genetic equivalent of a serovar is termed a genovar, based on the *omp1* gene which encodes MOMP. Within individual genovars, variants arise by single base mutation which, if variation occurs at a dominant B-cell epitope, will result in a serologically distinct variant (that is, antigenic drift). Additionally, there is convincing evidence that recombination events involving the gene encoding MOMP also occur (that is, antigenic shift) giving rise to organisms which fail to bind neutralizing, serovar-specific antibody (Hayes *et al.*, 1994). The simplest explanation was that recombination arose within fused inclusions resulting from mixed serovar L1 and L2 infection. Interestingly, a gene analogous to the *recA* gene-mediating recombination in *E. coli* has been recognized in *C. trachomatis* (Zhang, D.-J. *et al.*, 1995). Clearly the ability of chlamydiae to evade neutralizing antibody directed against MOMP is going to depend on the tempo of antigenic change about which, as yet, there is insufficient information.

In STD clinics in Canada and among female sex workers in Nairobi, substantial variation within MOMP involving both point mutation (antigenic drift) and putative recombinational changes involving larger gene segments (antigenic shift) was observed (Brunham *et al.*, 1994). To date, antigenic shift appears to be limited to strains within the same serogroup. Among female sex workers, 18 different *C. trachomatis omp1* genotypes were observed, with the allelic composition of the *C. trachomatis* population changing significantly over time. Of 19 re-infections occurring 6 months or more apart, 17 were with different *omp1* genotypes, indicating a reduced rate of same-genotype re-infection consistent with the occurrence of strain- or variant-specific immunity (Brunham *et al.*, 1996). In contrast, in a complete cross-sectional study of two Gambian villages with endemic trachoma with a combined population of some 1300 people, 200 cases of infection with *C. trachomatis* were identified by PCR-sequencing, almost entirely due to organisms of serovar A or B. However, relatively few variants were present and there was no evidence for recombination events (Hayes *et al.*, 1990, 1992). Perhaps in these rural communities the relative lack of variation might indicate that many of the endemic trachoma agents were clonally related. Antigenic variation within MOMP in *C. pneumoniae* appears much more limited. Geographically distinct isolates of *C. pneumoniae* have essentially the same *omp1* gene sequence (Carter *et al.*, 1991; Melgosa, Kuo & Campbell, 1991), reflecting the fact that MOMP is less immunodominant in this species and perhaps not under the same immune pressure.

Antibody affords only limited resistance to chlamydial colonization as it can easily be overwhelmed. Thus, when monoclonal antibodies against neutralizing epitopes of MOMP were delivered into the serum and secretions of mice using the back-pack syngeneic hybridoma tumour system, it was found that both IgA and IgG antibodies significantly reduced infection following vaginal challenge with 5 ID_{50} of *C. trachomatis* mouse pneumonitis agent, but not with 10- or 100-fold higher challenge doses. There was a significant reduction in the severity of chlamydiae-induced inflammation in the oviducts of IgA- or IgG-treated animals and a reduced number of chlamydial inclusions in the oviduct epithelium (Cotter *et al.*, 1995). Thus, neutralizing antibody gave low level protection against chlamydial colonization which reduced disease severity but had only a marginal effect on vaginal shedding, presumably because of the intracellular location of established chlamydial infection.

In contrast, cell-mediated immunity appears crucial for resolving established chlamydial infection. In trachoma, cell-mediated immune responses to chlamydial and common recall antigens were measured in 26 subjects whose clinical signs of disease persisted over 6 months of follow-up and in 21 subjects whose clinical signs resolved spontaneously. Seven-day lymphocyte proliferative responses to chlamydial but not common recall antigens were significantly greater in subjects whose disease resolved spontaneously (Bailey *et al.*, 1995). Direct evidence of the importance of cell-mediated immunity came from the guinea-pig model of chlamydial infection with *C. psittaci*, in which susceptible animals could be made resistant by the passive transfer of chlamydiae-reactive T-cell clones (Rank *et al.*, 1989). Moreover, susceptibility to vaginal infection correlated with a decrease in antigen-specific T-cells in the genital tract (Igietseme & Rank, 1991). Replication of the mouse pneumonitis agent of *C. trachomatis* in polarized murine epithelial cells was inhibited by mouse pneumonitis-specific T-cell lines, probably by the action of short range cytokines (Igietseme *et al.*, 1994*b*). The most likely candidate is Ifnγ, as this cytokine resolved chronic infection with the mouse pneumonitis strain of *C. trachomatis* in nude mice (Rank *et al.*, 1992), while Ifnγ-producing T-helper 1 cells were the dominant $CD4^+$ T-cell response to intravaginal infection with mouse pneumonitis agent.

Protection of conventional mice against challenge with either *C. trachomatis* mouse pneumonitis agent or serovar E, or with the GPIC strain of *C. psittaci* has been related to the production of Ifnγ or tumour necrosis factor alpha (Igietseme *et al.*, 1994*a*). In primary murine pneumonia due to the mouse pneumonitis agent, depletion of $CD4^+$ or $CD8^+$ cells caused a significant increase in organism burden in the lungs. The depletion of $CD4^+$ cells led to a significant decrease in antibody and Ifnγ production and an increase in mortality; depletion of $CD8^+$ T-cells did not. Knock-out mice defective in $CD8^+$ T-cell function showed a significant increase in organism burden and mortality, but surviving mice were still able to clear the infection by day 34. It was considered that both $CD4^+$ and $CD8^+$ cells played a protective role in this model, but that $CD4^+$ cells were the most important (Magee *et al.*, 1995). Consistent with this, donor $CD4^+$ cells, but not $CD8^+$ cells obtained from mice after resolution of primary infection or after secondary challenge were effective in transferring significant protective immunity to naive animals (Su & Caldwell, 1995). Protection was relatively long lived as resting $CD4^+$ donor cells obtained four months after resolution of primary infection were still effective.

Further evidence of the importance of cell-mediated immunity in limiting chlamydial infection comes from experiments in gene-knock-out mice, discriminating between the roles of cytokines themselves, notably chlamydia-static Ifnγ and T-cell help for cognate B-cells. In mice $CD4^+$ T-cells develop from a common precursor into two functionally different, but reciprocally regulated subsets producing distinct cytokines. T-helper 1 cells produce IL-2, Ifnγ and lymphotoxin, whereas T-helper 2 cells produce IL-4, IL-5, IL-6, IL-10 and IL-13. T-helper 1 cells are involved in cell-mediated immunity and their differentiation is augmented by macrophage-derived IL-12 or Ifnγ and suppressed by IL-4. In contrast, for T-helper 2 cells, IL-4 is required for differentiation while Ifnγ is suppressive. Significantly higher levels of Ifnγ- as opposed to IL-4-secreting cells have been reported in the draining lymph nodes of the genital tract of mice three weeks after vaginal infection with the mouse pneumonitis agent, suggesting a dominant T-helper 1 response (Cain & Rank, 1995). However, others reported that protective $CD4^+$ T-cells recovered from mice following resolution of primary genital tract infection secreted IL-2, Ifnγ and IL-6 on restimulation *in vivo*, cytokine patterns characteristic of both T-helper 1 and T-helper 2 type responses (Su & Caldwell, 1995). Importantly, resting T-cells obtained from mice four months after resolution of a primary infection still conferred significant levels of protective immunity to naive mice. Taken together, these data suggest that T-helper 1 cells are crucial in the resolution of established chlamydial infection. This would explain why, although $CD8^+$ cytotoxic activity has been demonstrated in the mouse model (Starnbach, Bevan & Lampe, 1994), in gene-knock-out mice MHC Class II-restricted T-cell responses were necessary for the development of protective immunity whereas MHC Class I deficient mice resolved genital tract infection normally (Morrison, Feilzar & Tumas, 1995). This implied that the critical event in $CD4^+$ MHC Class II deficient animals was the inability to present antigen to T-helper cells, although other factors could not be excluded.

However, a criticism of these studies is that they were confounded by having investigated immune protection, including the role of antibody, in animals with a functioning Ifnγ system which is itself protective. Nils Lycke and colleagues (Johanson, Ward & Lycke, 1996; unpublished data) explored the development of natural immunity in progesterone-treated, wild-type C57BL or Ifnγ-receptor-deficient mice following vaginal infection with *C. trachomatis* serovar D. The Ifnγ-receptor-deficient mice were able to produce the Ifnγ characteristic of T-helper 1 responses but were unable to utilize it because of the absence of the cellular Ifnγ receptor. An overwhelmingly large infection and challenge dose of 100 ID_{50} of chlamydiae was used so that only strong immune responses associated with significant protection would be apparent. The Ifnγ-receptor-deficient mice compared with wild-type exhibited a much more severe ascending infection of prolonged duration but mounted augmented IgG and IgA responses in the genital tract to whole chlamydiae and to MOMP. IgG subclass analysis indicated that the response was predominantly of the T-helper 2 type, whereas in wild-type mice it was T-helper 1 regulated. Receptor-deficient mice showed a strong sub- and inter-epithelial $CD4^+$ T-cell response in the genital tract with a dominance of IL-4-producing cells but lacked the focalized clusters of $CD4^+$ T-cells of the wild-type.

Ifnγ might restrain chlamydial infection either by up-regulating indoleamine 2,3-deoxygenase expression (Hissong *et al.*, 1995) leading to decyclization and cellular depletion of tryptophan, an amino acid essential for chlamydial growth (Beatty *et al.*, 1994*a*), or by the up-regulation of inducible nitric oxide synthase and nitric oxide in macrophages. *In vitro*, chlamydiae-infected macrophages from the wild-type mice developed few chlamydial inclusions and they produced significant amounts of nitric oxide in the presence of Ifnγ. In contrast, macrophages from the Ifnγ-receptor-deficient mice contained many chlamydial inclusions but were unable to produce significant amounts of nitric oxide, even in the presence of exogenous Ifnγ (Johanson *et al.*, 1996). These results indicated again that antibody to MOMP was incapable of preventing overwhelming genital tract infection and showed that functional Ifnγ, presumably largely produced by T-helper 1 cells, was crucial for the resolution of ascendant *C. trachomatis* infection in the mouse female genital tract.

Cytotoxic T-cells may also play an important role in chlamydial infection with respect to protection or immunopathology. Intracellular pathogens such as viruses are able to present antigens via the endogenous pathway to MHC class I. However, chlamydiae are confined within an endosomal vacuole, so that it cannot be assumed that their antigens are similarly presented. Lysis of chlamydiae-infected epithelial cells by cytotoxic $CD8^+$ T-cells has been demonstrated, although only after transfection-induced increased expression of ICAM-1. Specific cytolysis could be eliminated by removal of the $CD8^+$ T-cells, by addition of the Golgi inhibitor Brefeldin A, or by blockade of host cell protein synthesis with cycloheximide. Blockade of the exogenous pathway of antigen processing with the receptor-mediated endocytosis inhibitors chloroquine or NH_4Cl had only minor effect (Beatty & Stephens, 1994). However, the model used was somewhat artificial. In a different study, a $CD8^+$ cytotoxic T-cell line derived from mice infected intraperitoneally with *C. trachomatis* serovar L2 was also shown to be specific and cytolytic for chlamydiae-infected cells, the peptide epitope being presented by infected cells in the classic MHC class I molecule H-2 L(d). Adoptive transfer of this $CD8^+$ cell line into an infected mouse was protective and required Ifnγ (Starnbach *et al.*, 1994). Cytotoxic T-lymphocytes in mice were also primed by infection with *C. trachomatis* serovar L2 *in utero* or in the ovarian bursa, or by infection with a human urogenital serovar D isolate. The resulting murine cytotoxic T-cell lines were broadly protective, lysing target cells infected with *C. trachomatis* serovars B, C, D, F, J, K, L2 or

L3 (Starnbach, Bevan & Lampe, 1995). The clinical significance of $CD8^+$ cytolysis is likely to be particularly marked in chlamydial infections such as reactive arthritis where there is a strong association with a particular MHC Class I haplotype, in this case HLA-B27 antigen. However, the evidence from animal models is that $CD4^+$ cells are more important for protection than $CD8^+$ cells, although the latter may be involved in immunopathology.

Immunopathology

If untreated, chlamydial infections tend to be long-term and recurrent, providing an opportunity for chronic host stimulation with chlamydial antigens. The chlamydial 12, 60 and 75 kDa heat-shock proteins (Hsps) are partially homologous with their *E. coli* counterparts GroEL, GroES and DnaK and with the related human mitochondrial proteins Cpn10, Hsp60 and Hsp70 and, as in mycobacterial infection, have been implicated in immunopathological damage. Presumably the chlamydial Hsps, like their human and *E. coli* counterparts, function as chaperonins, involved in the ATP-dependent renaturation and refolding of proteins damaged by proteolysis, heat or oxidative stress, as well as targeting proteins or peptide fragments to biological membranes (Hightower, 1991). Chlamydial Hsp60 and Hsp70 are early proteins, expressed 2 to 26 h post-infection (Lundemose *et al.*, 1990). Chlamydial Hsp70, like human Hsp70, is thought to confer partial resistance to the cytotoxic effects of tumour necrosis factor-alpha (TNFα) (Jaattela & Wissing, 1993; Lehtinen & Paavonen, 1994). The ratio of Hsp60 to MOMP increases dramatically in chlamydiae-infected, Ifnγ-treated, host cells (Beatty *et al.*, 1994*b*) and this has been mirrored in the mRNA transcripts produced in the synovium of patients with reactive arthritis (Beutler *et al.*, 1995; Gerard *et al.*, 1995). Reduced expression of MOMP and increased expression of Hsp60 in individuals persistently, recurrently or chronically infected by chlamydiae might contribute substantially to the delayed hypersensitivity type responses associated with the tubal or conjunctival scarring characteristic of pelvic inflammatory disease and trachoma, respectively. The B-cell epitopes on Hsp60 have been mapped (Paavonen *et al.*, 1994; Yi, Zhong & Brunham, 1993) and consisted of chlamydial species- and genus-specific epitopes, plus more broadly cross-reactive epitopes common to mycobacterial Hsp65. Human antibodies often recognize these cross-reactive epitopes and autoantibodies have been demonstrated to chlamydial Hsp60 and its human counterparts in infants with chlamydial pneumonitis (Paavonen *et al.*, 1994). However, although antibody to chlamydial Hsps have been associated with the damaging sequelae of chlamydial infection such as ectopic pregnancy (Brunham *et al.*, 1992), there is no clear indication that antibody to Hsp60 is disease specific (Lehtinen & Paavonen, 1994) rather than related to chronicity.

In contrast, there is increasing evidence that cell-mediated immune responses to chlamydial Hsps may be important. Chlamydial infections are characterized by plasma cell infiltration and the production of lymphoid follicles. Ifnγ production by T-helper 1 cells activates cellular immunity, and alters chlamydial growth and differentiation with up-regulation of chlamydial Hsp production in infected cells. In the guinea-pig model, ocular application of chlamydial Hsp60 in Triton X-100 elicited local monocytic infiltration (Morrison, Lyng & Caldwell, 1989). Lymphocyte proliferative responses to Hsp60 were detected in 50% of 18 women with laparoscopically verified pelvic inflammatory disease, in none of 10 women with chlamydial cervicitis and in 7% of 42 healthy controls (Witkin *et al.*, 1993). Not surprisingly, chlamydial Hsp60, as well as MOMP and other outer membrane antigens, stimulated proliferation of $CD4^+$ T-helper cells (Beatty & Stephens, 1992). It is likely that chlamydial Hsp60, like the mycobacterial Hsps, will be involved in the induction of T-helper 1 responses. The resulting Ifnγ, although chlamydiastatic and capable of recruiting activated immune cells, might also, by up-regulating Hsp expression, be immunopathological along with other proinflammatory cytokines. Chlamydial interaction with T-lymphocytes (Fitzpatrick *et al.*, 1991) and macrophages and with epithelial cells results in the up-regulation of the proinflammatory cytokines including Ifnγ, IL-6 and IL-8.

In peripheral blood mononuclear cells, *C. trachomatis* lipopolysaccharide (LPS) can persist for up to 14 days (Schmitz *et al.*, 1993) even though a productive infection may not be established. Both chlamydial LPS and chlamydial EBs induce the release of TNFα from whole blood. Chlamydial EBs also induced the translocation of nuclear factor κB in a Chinese hamster ovary fibroblast cell line transfected with the CD14 LPS receptor. This could be blocked with LPS antagonists, suggesting that chlamydial LPS interaction with CD14 is the main inducer of proinflammatory cytokines. Chlamydial LPS was roughly 100-fold less potent than gonococcal or *Salmonella minnesota* LPS at inducing proinflammatory cytokines, indicating that it is only a weak inducer of inflammation (Ingalls *et al.*, 1995). Interestingly, chlamydial infection is more likely than gonococcal infection to be asymptomatic.

Both the force of infection and genotypic factors appear to be related to chlamydiae-induced immunopathology. In women, each episode of pelvic inflammatory disease roughly doubled the risk of permanent tubal damage (Lehtinen & Paavonen, 1994) irrespective of whether infection was silent or overt (Patton *et al.*, 1989). Similarly in trachoma, recurrent infection was also associated with scarring sequelae (Grayston *et al.*, 1985). Sikhs migrating to British Columbia from a trachoma-endemic area of the Punjab showed an absence of otherwise expected severe scarring sequelae, presumably because they were no longer exposed to re-infection (Detels, Alexander & Dhir, 1966). In general, severe ocular scarring in trachoma-endemic areas has been associated with a high 'force of infection' or with severe trachoma occurring at a young age, both factors being inter-related. In monkeys, repeated inoculations with *C. trachomatis* were necessary to produce pelvic adhesions, tubal scarring and occlusion (Patton *et al.*, 1990) or corneal pannus and conjunctival scarring (Taylor *et al.*, 1982; Grayston *et al.*, 1985), while mice sensitized by immunization with either a crude chlamydial extract of the mouse pneumonitis agent (Blander & Amortegui, 1994) or Hsp60 (Blander, Amortegui & Wagar, 1994) suffered more severe inflammation when later challenged intravaginally with live organisms.

Cytokines likely to be involved in immunopathology include Ifnγ, TNFα and transforming growth factor-beta (TGFβ). Active TGFβ, which has been associated with fibrosis and scarring, was identified in the broncho-alveolar lavages of mice infected with the *C. trachomatis* mouse pneumonitis agent (Williams *et al.*, 1996), peaking two days post-infection. High levels of TNFα have been detected in the genital tract secretions of guinea-pigs infected with the *C. psittaci* GPIC agent. The intensity of the TNFα response was generally proportional to the intensity of infection, with high levels being coincident with a marked neutrophil influx as might be expected (Darville *et al.*, 1995).

A contributory role for host factors is indicated by the fact that individuals with moderate or severe inflammatory trachoma at one survey were more likely to have similar inflammatory changes at a previous or subsequent survey (Odds ratio 14.9, 95% confidence interval 3.9–68.0) (Mabey *et al.*, 1992*b*) than were those with milder disease. HLA-A28, an HLA-Class I antigen, occurred significantly more frequently in rural Gambians with moderate to severe ocular scarring than in controls with mild or absent disease. This association was due to the HLA-A*6802 allele. Polymorphisms related to ocular scarring were also observed in the gene encoding TNFα; individuals with severe trachoma tended to have higher levels of TNFα in their tear fluid (Conway *et al.*, 1996).

VACCINE DEVELOPMENT

In the 1960s and early 1970s a number of attempts were made to prevent trachoma by empirical vaccination with crude, whole, organisms (see for example Grayston *et al.*, 1962). The diversity of vaccines and test conditions used makes comparison of the sometimes conflicting reports of these studies difficult. However, protection, where elicited, was usually of short duration. Moreover, under certain circumstances vaccination increased both the attack rate of subsequent naturally acquired trachoma and the severity of the response to artificial challenge (Sowa *et al.*, 1969). In the third Gambian trial, using an aqueous vaccine of semi-purified 'fast killing' variants of *C. trachomatis* SA-2 and ASGH (now regarded as probably LGV agents), the proportion of vaccinated children progressing to scarring trachoma was less than in the unvaccinated controls but, paradoxically, the average severity of the disease in terms of clinical score was greater (Sowa *et al.*, 1969). This is arguably the first known example, before measles or respiratory syncytial virus vaccines, of an experimental vaccine apparently sensitizing individuals to more severe disease. In sheep, a live attenuated *C. psittaci* vaccine has had modest success in France in protecting against the ewe abortion agent (Rodolakis & Bernard, 1984; Rodolakis, 1995). However, ewe abortion is a special case of a systemic infection which subsequently localizes in the placental horns. In humans it would be considered unacceptable nowadays to use whole organism chlamydial vaccines. Attention has, therefore, focused on the use of purified or recombinant chlamydial antigens for vaccine development, particularly MOMP, as this antigen is immunodominant at the surface of the infectious EB and is capable of eliciting neutralizing antibody.

In C3H mice, parenteral immunization with a purified, heterologous, recombinant MOMP (rMOMP) preparation reduced the proportion of animals developing severe salpingitis following intra-uterine challenge by 77% compared with mock-immunized controls, but failed to reduce chlamydial colonization of the lower genital tract. In contrast, mice immunized with rMOMP directly into the Peyer's patches to stimulate mucosal immunity shed fewer chlamydiae from the vagina than controls, but showed little reduction in oviduct damage. Immunization with rMOMP via the presacral space, a route previously shown to stimulate mucosal immunity in the genital tract, produced high levels of circulating anti-rMOMP IgG but only traces of anti-rMOMP antibody in vaginal secretions. The preliminary conclusions drawn from this study were: (i) rMOMP does have some protective effect on both chlamydial colonization and subsequent tubal damage; (ii) heterotypic protection is feasible; (iii) there was no evidence that vaccination with rMOMP showed the same tendency as vaccination with viable organisms of increasing disease severity or frequency on rechallenge; and (iv) mucosal and parenteral immunization appeared to have distinct effects on chlamydial colonization and tubal damage, respectively (Tuffrey *et al.*, 1992).

In an alternative approach designed to elicit effective mucosal immunity, a neutralizing epitope VAGLEK from the serovar A-specific region of MOMP was cloned into antigenic site 1 of poliovirus type 1 Mahoney. The viable hybrid inoculated into rabbits induced strong neutralizing antibody responses to both poliovirus and to VAGLEK, some 10- to 100-fold greater than could be achieved by vaccination with MOMP or with VAGLEK alone (Murdin *et al.*, 1993). Recently, similar poliovirus hybrids have been constructed which incorporate serovar-, species- or subspecies-specific epitopes from variable sequence I or IV. Antisera to these hybrids strongly neutralized eight of the 12 oculogenital *C. trachomatis* serovars (Murdin *et al.*, 1995). The same group have also tested in mice and non-human primates the immunogenicity of a large synthetic oligopeptide consisting of the T-helper cell epitope A8 and B-cell epitopes from variable sequence 4. Six of eight H-2 congenic mouse strains immunized with the peptide produced high-titre IgG antibodies against the variable sequence 4 species-specific epitope LNPTIAG. These antibodies also reacted with intact *C. trachomatis* elementary bodies (EBs) by ELISA and neutralized chlamydial infectivity in a subspecies-specific manner *in vitro*. The immunogenicity of this oligopeptide in different strains of mice disparate at the H-2 locus, its immunogenicity in non-human primates, and its ability to target cross-reactive neutralizing antibody responses against multiple *C. trachomatis* serovars are encouraging. However, overall, the protection so far achieved with vaccines based on MOMP has been disappointing.

A key problem for vaccine development based on neutralizing antibody is that there is little information on how to target and sustain such antibody at the superficial sites in the ocular sac or genital tract where these infections occur. Furthermore, it is likely that the most effective neutralizing antibody will be directed against conformational epitopes whose three-dimensional structure will be difficult to mimic in a recombinant or peptide vaccine. This problem has been recognized for *Chlamydia* in studies aimed at eliciting conformationally relevant chlamydial antibodies using either mimotopic peptides (Yxfeldt *et al.*, 1994) or phage display techniques (Zhong *et al.*, 1994). T-cell epitopes, on the other hand, are often primary sequence dependent and effective local T-cell priming might be elicited by systemic immunization. These factors, together with realization of the crucial role of cell-mediated immunity and Ifnγ in resistance to chlamydial infection, have lead to a re-think of chlamydial vaccine development. One approach is to use a DNA vaccine. At a meeting on trachoma vaccines sponsored by the Edna McConnell Clark Foundation in New Orleans in May 1996, Brunham and colleagues reported convincing evidence of successful protection of mice from lethal lung challenge with the *C. trachomatis* mouse pneumonitis agent using a DNA-based vaccine. This vaccine consisted of the full length *omp1* gene, encoding expression of the mouse pneumonitis agent MOMP, in a cytomegalovirus-based plasmid capable of replicating in mouse cells. Protection was measured both by diminution of the burden of viable organisms in mouse lungs and by the reduction in the weight loss of mice caused by the infection. The DNA vaccine generated little antibody response to the MOMP and showed only low levels of expression in muscle cells. However, it generated a marked delayed-type hypersensitivity response as shown by mouse foot-pad swelling and there was evidence of a major participation of Ifnγ-producing T-helper 1 cells. These results are very encouraging for chlamydial vaccine development, being the first time such significant protection has been demonstrated in an animal model and providing key confirmation that MOMP generates protective cell-mediated immunity. In the latter context, a dodecameric peptide beginning with the conserved sequence TINKP adjacent to the N-terminus of VSIII was able to stimulate primary proliferative responses in virtually all human volunteers after presentation by dendritic cells, indicating that it functioned as an MHC Class II promiscuous peptide (Stagg *et al.*, 1993). Moreover, a single intradermal injection of this peptide in mice was sufficient to induce T-cell-mediated responses and partial protection against salpingitis in susceptible C3H mice following intra-uterine challenge with *C. trachomatis* (Knight *et al.*, 1995).

Overall, vaccines inducing cell-mediated immunity should be capable of eradicating established chlamydial infection and will have a public health impact by reducing shedding. However, safety concerns over the use of DNA vaccines for a non-life-threatening condition have to be taken into account. Likely next steps are: (i) to determine whether the vaccine construct is also protective against chlamydial challenge in the murine genital tract; this cannot be assumed as the lungs may be more accessible to the systemic immune response than the genital tract; (ii) to determine how long the protection can be sustained; (iii) to evaluate whether the vaccinated mice develop immunopathological complications when compared with controls; and (iv) to perform toxicity studies in higher animals. It seems likely that neutralizing antibody, although easily overwhelmed, is probably necessary for initial resistance to chlamydial colonization, so attempts to generate vaccines eliciting neutralizing antibody must continue.

CONCLUSIONS

The genus *Chlamydia* is one of the few groups of obligately intracellular bacterial pathogens. At the molecular level, comparatively little is known concerning host–pathogen interactions. To a large extent this is attributable to the particular difficulties of working with these organisms and the absence of any system for their genomic manipulation. By contrast, substantial progress has been made in elucidating chlamydial immunochemistry and the immunology of chlamydial infection. The *Chlamydia* cause a variety of important human and veterinary infections and experiments in animal models clearly indicate that the cell-mediated immune system and Ifnγ play an important role in their control. The prospect for chlamydial vaccine development is now particularly encouraging.

ACKNOWLEDGEMENTS

I thank Drs Art Andersen and Karin Everett for providing Fig. 1 and for discussions on chlamydial taxonomy. Chlamydial research in Southampton is supported by the Medical Research Council, the World Health Organisation Programme in Human Reproduction, the European Community and the Swedish International Development Agency, to all of whom I am most grateful.

REFERENCES

Bailey, R. L., Holland, M. J., Whittle, H. C. & Mabey, D. C. W. (1995). Subjects recovering from human ocular chlamydial infection have enhanced lymphoproliferative responses to chlamydial antigens compared with those of persistently diseased controls. *Infection and Immunity*, **63**, 389–92.

Barry, C. E., Hayes, S. F. & Hackstadt, T. (1992). Nucleoid condensation in *Escherichia coli* that express a chlamydial histone homolog. *Science*, **256**, 377–9.

Barry, C. E., Brickman, T. J. & Hackstadt, T. (1993). Hc1-mediated effects on DNA structure: a potential regulator of chlamydial development. *Molecular Microbiology*, **9**, 273–83.

Bas, S., Griffais, R., Kvien, T. K., Glennas, A., Melby, K. & Vischer, T. L. (1995). Amplification of plasmid and chromosome *Chlamydia* DNA in synovial fluid of patients with reactive arthritis and undifferentiated seronegative oligoarthropathies. *Arthritis and Rheumatism*, **38**, 1005–13.

Batteiger, B. E., Fraiz, J., Newhall, W. J., Katz, B. P. & Jones, R. B. (1989). Association of recurrent chlamydial infection with gonorrhea. *Journal of Infectious Diseases*, **159**, 661–9.

Bavoil, P., Ohlin, A. & Schachter, J. (1984). Role of disulfide bonding in outer membrane structure and permeability in *Chlamydia trachomatis*. *Infection and Immunity*, **44**, 479–85.

Beatty, P. R. & Stephens, R. S. (1992). Identification of *Chlamydia trachomatis* antigens by use of murine T-cell lines. *Infection and Immunity*, **60**, 4598–603.

Beatty, P. R. & Stephens, R. S. (1994). CD8(+) T-lymphocyte-mediated lysis of *Chlamydia*-infected L cells using an endogenous antigen pathway. *Journal of Immunology*, **153**, 4588–95.

Beatty, W. L., Byrne, G. I. & Morrison, R. P. (1993). Morphologic and antigenic characterization of interferon gamma-mediated persistent *Chlamydia trachomatis* infection *in vitro*. *Proceedings of the National Academy of Sciences, USA*, **90**, 3998–4002.

Beatty, W. L., Belanger, T. A., Desai, A. A., Morrison, R. P. & Byrne, G. I. (1994*a*). Tryptophan depletion as a mechanism of gamma interferon-mediated chlamydial persistence. *Infection and Immunity*, **62**, 3705–11.

Beatty, W. L., Byrne, G. I. & Morrison, R. P. (1994*b*). Repeated and persistent infection with *Chlamydia* and the development of chronic inflammation and disease. *Trends in Microbiology*, **2**, 94–8.

Beutler, A. M., Schumacher, H. R., Whittum-Hudson, J. A., Salameh, W. A. & Hudson, A. P. (1995). *In-situ* hybridization for detection of inapparent infection with *Chlamydia trachomatis* in synovial tissue of a patient with Reiter's syndrome. *American Journal of the Medical Sciences*, **310**, 206–13.

Birkelund, S., Johnsen, H. & Christiansen, G. (1994). *Chlamydia trachomatis* serovar L2 induces protein tyrosine phosphorylation during uptake by HeLa cells. *Infection and Immunity*, **62**, 4900–8.

Blander, S. J. & Amortegui, A. J. (1994). Mice immunized with a chlamydial extract have no increase in early protective immunity despite increased inflammation following genital infection by the mouse pneumonitis agent of *Chlamydia trachomatis*. *Infection and Immunity*, **62**, 3617–24.

Blander, S. J., Amortegui, A. & Wagar, E. (1994). Mice sensitized to the 60 kDa heat-shock protein have increased inflammation after genital chlamydial infection. *Clinical Research*, **42**, 150.

Brunham, R. C., Peeling, R., Maclean, I., McDowell, J., Persson, K. & Osser, S. (1987). Postabortal *Chlamydia trachomatis* salpingitis: correlating risk with antigen-specific serological responses and with neutralization. *Journal of Infectious Diseases*, **155**, 749–55.

Brunham, R. C., Peeling, R., Maclean, I., Kosseim, M. L. & Paraskevas, M. (1992). *Chlamydia trachomatis*-associated ectopic pregnancy: serologic and histologic correlates. *Journal of Infectious Diseases*, **165**, 1076–81.

Brunham, R. C., Yang, C., Maclean, I., Kimani, J., Maitha, G. & Plummer, F. (1994). *Chlamydia trachomatis* from individuals in a sexually transmitted disease core group exhibit frequent sequence variation in the major outer membrane protein (*omp1*) gene. *Journal of Clinical Investigation*, **94**, 458–63.

Brunham, R. C., Kimani, J., Bwayo, J., Maitha, G., Maclean, I., Yang, C., Shen, C., Roman, S., Nagelkerke, N. J. D., Cheang, M. & Plummer, F. A. (1996). The epidemiology of *Chlamydia trachomatis* within a sexually-transmitted diseases core group. *Journal of Infectious Diseases*, **173**, 950–6.

Buchan, H., Vessey, M., Goldacre, M. & Fairweather, J. (1993). Morbidity following pelvic inflammatory disease. *British Journal of Obstetrics and Gynaecology*, **100**, 558–62.

Cain, T. K. & Rank, R. G. (1995). Local Th1-like responses are induced by intravaginal infection of mice with the mouse pneumonitis biovar of *Chlamydia trachomatis*. *Infection and Immunity*, **63**, 1784–89.

Campbell, L. A., Patton, D. L., Moore, D. E., Cappuccio, A. L., Mueller, B. A. & Wang, S.-P. (1993). Detection of *Chlamydia trachomatis* deoxyribonucleic acid in women with tubal infertility. *Fertility and Sterility*, **59**, 45–50.

Campbell, S., Richmond, S. J., Haynes, P., Gump, D., Yates, P. & Allen, T. D. (1988). An *in vitro* model of *Chlamydia trachomatis* infection in the regenerative phase of the human endometrial cycle. *Journal of General Microbiology*, **134**, 2077–87.

Carter, M. W., Al-Mahdawi, S. A. H., Giles, I. G., Treharne, J. D., Ward, M. E. & Clarke, I. N. (1991). Nucleotide sequence and taxonomic value of the major outer membrane protein gene of *Chlamydia pneumoniae* IOL 207. *Journal of General Microbiology*, **137**, 465–75.

Chang, J. J., Leonard, K. & Arad, T. (1982). Structural studies of the outer envelope of *Chlamydia trachomatis* by electron microscopy. *Journal of Molecular Biology*, **161**, 579–90.

Christiansen, G., Pedersen, L. B., Koehler, J. E., Lundemose, A. G. & Birkelund, S. (1993). Interaction between the *Chlamydia trachomatis* histone H1-like protein (Hc1) and DNA. *Journal of Bacteriology*, **175**, 1785–95.

Coles, A. M. & Pearce, J. H. (1987). Regulation of *Chlamydia psittaci* (strain guinea pig inclusion conjunctivitis) growth in McCoy cells by amino acid antagonism. *Journal of General Microbiology*, **133**, 701–8.

Coles, A. M., Reynolds, D. J., Harper, A., Devitt, A. & Pearce, J. H. (1993). Low nutrient induction of abnormal chlamydial development – a novel component of chlamydial pathogenesis. *FEMS Microbiology Letters*, **106**, 193–200.

Conlan, J. W., Clarke, I. N. & Ward, M. E. (1988). Epitope mapping with solid phase peptides: identification of type-, subspecies-, species- and genus-reactive antibody binding domains on the major outer membrane protein of *Chlamydia trachomatis*. *Molecular Microbiology*, **2**, 673–9.

Conway, D. J., Holland, M. J., Bailey, R. L., Mahdi, S. & Mabey, D. C. W. (1996). Scarring trachoma is associated with polymorphism in the *TNF-α* gene promoter and with elevated TNF-α in tear fluid. *FASEB Journal, Abstracts*, **10**, Abstract no. 107, A1185.

Cosgrove, P. A., Patton, D. L., Tahija, S., Campbell, L., Kuo, C-C. & Cappuccio, A. L. (1992). *In situ* DNA hybridization detection of *Chlamydia trachomatis* in chronic ocular infection (stage-iv trachoma). *Investigative Ophthalmology and Visual Science*, **33**, 848.

Cotter, T. W., Meng, Q., Shen, Z.-L., Zhang, Y-X., Su, H. & Caldwell, H. D. (1995). Protective efficacy of major outer membrane protein-specific immunoglobulin A (IgA) and IgG monoclonal antibodies in a murine model of *Chlamydia trachomatis* genital tract infection. *Infection and Immunity*, **63**, 4704–14.

Darville, T., Laffoon, K. K., Kishen, L. R. & Rank, R. G. (1995). Tumor necrosis factor alpha activity in genital tract secretions of guinea pigs infected with chlamydiae. *Infection and Immunity*, **63**, 4675–81.

Detels, R., Alexander, E.R. & Dhir, S.P. (1966). Trachoma in Punjabi Indians in British Columbia: a prevalence study with comparisons to India. *American Journal of Epidemiology*, **84**, 81–91.

Douglas, A. L., Saxena, N. K. & Hatch, T. P. (1994). Enhancement of *in vitro* transcription by addition of cloned, overexpressed major sigma factor of *Chlamydia psittaci* 6BC. *Journal of Bacteriology*, **176**, 3033–9.

Engel, J. N., Pollack, J., Malik, F. & Ganem, D. (1990). Cloning and characterization of RNA polymerase core subunits of *Chlamydia trachomatis* by using the polymerase chain reaction. *Journal of Bacteriology*, **172**, 5732–41.

Fahr, M. J., Douglas, A. L., Xia, W. & Hatch, T. P. (1995). Characterization of late gene promoters of *Chlamydia trachomatis*. *Journal of Bacteriology*, **177**, 4252–60.

Fitch, W. M., Peterson, E. M. & de la Maza, L. M. (1993). Phylogenetic analysis of the outer membrane protein genes of chlamydiae, and its implication for vaccine development. *Molecular Biology and Evolution*, **10**, 892–913.

Fitzpatrick, D. R., Wie, J., Webb, D., Bonfiglioli, R., Gardner, I. D., Mathews, J. D. & Bielefeldtohmann, H. (1991). Preferential binding of *Chlamydia trachomatis* to subsets of human lymphocytes and induction of interleukin-6 and interferon-gamma. *Immunology and Cell Biology*, **69**, 337–48.

Fukushi, H. & Hirai, K. (1992). Proposal of *Chlamydia pecorum* sp. nov. for *Chlamydia* strains derived from ruminants. *International Journal of Systematic Bacteriology*, **42**, 306–8.

Fukushi, H. & Hirai, K. (1993). *Chlamydia pecorum*: the 4th species of genus *Chlamydia*. *Microbiology and Immunology*, **37**, 515–22.

Gerard, H. C., Branigan, P. J., Schumacher, H. R. & Hudson, A. P. (1995). Inapparently infecting *Chlamydia trachomatis* in the synovia of Reiter's syndrome (rs) reactive arthritis (rea) patients are viable. *Arthritis and Rheumatism*, **38**, R 24.

Grayston, J. T., Woolridge, R. L. & Wang, S-P. (1962). Trachoma vaccine studies on Taiwan. *Annals of the New York Academy of Sciences*, **98**, 1615–20.

Grayston, J. T., Kim, K. S. W., Alexander, E. R. & Wang, S-P. (1971). Protective studies in monkeys with trivalent and monovalent trachoma vaccines. In Nichols, R. L., ed. *Trachoma and Related Disorders*, pp. 377–85. Amsterdam: Excerpta Medica.

Grayston, J. T., Wang, S-P., Yeh, L-J. & Kuo, C-C. (1985). Importance of reinfection in the pathogenesis of trachoma. *Reviews of Infectious Diseases*, **7**, 717–25.

Grayston, J. T., Kuo, C.-C., Campbell, L. A. & Wang, S-P. (1989). *Chlamydia pneumoniae* sp-nov for *Chlamydia* sp. strain TWAR. *International Journal of Systematic Bacteriology*, **39**, 88–90.

Grayston, J. T., Aldous, M. B., Easton, A., Wang, S-P., Kuo, C-C., Campbell, L. A. & Altman, J. (1993). Evidence that *Chlamydia pneumoniae* causes pneumonia and bronchitis. *Journal of Infectious Diseases*, **168**, 1231–5.

Hackstadt, T., Brickman, T. J., Barry, C. E. & Sager, J. (1993). Diversity in the *Chlamydia trachomatis* histone homolog Hc2. *Gene*, **132**, 137–41.

Hackstadt, T., Rockey, D. D., Heinzen, R. A. & Scidmore, M. A. (1996). *Chlamydia trachomatis* interrupts an exocytic pathway to acquire endogenously synthesized sphingomyelin in transit from the Golgi apparatus to the plasma membrane. *EMBO Journal*, **15**, 964–77.

Hayes, L. J., Pickett, M. A., Conlan, J. W., Ferris, S., Everson, J. S., Ward, M. E. & Clarke, I. N. (1990). The major outer-membrane proteins of *Chlamydia trachomatis* serovar A and serovar B: intra-serovar amino acid changes do not alter specificities of serovar- and C subspecies-reactive antibody-binding domains. *Journal of General Microbiology*, **136**, 1559–66.

Hayes, L. J., Bailey, R. L., Mabey, D. C. W., Clarke, I. N., Pickett, M. A., Watt, P. J. & Ward, M. E. (1992). Genotyping of *Chlamydia trachomatis* from a trachoma-endemic village in The Gambia by a nested polymerase chain reaction: identification of strain variants. *Journal of Infectious Diseases*, **166**, 1173–7.

Hayes, L. J., Yearsley, P., Treharne, J. D., Ballard, R. A., Fehler, G. H. & Ward, M. E. (1994). Evidence for naturally occurring recombination in the gene encoding the major outer membrane protein of lymphogranuloma venereum isolates of *Chlamydia trachomatis*. *Infection and Immunity*, **62**, 5659–63.

Heinzen, R. A., Scidmore, M. A., Rockey, D. D. & Hackstadt, T. (1996). Differential interaction with endocytic and exocytic pathways distinguish parasitophorous vacuoles of *Coxiella burnetii* and *Chlamydia trachomatis*. *Infection and Immunity*, **64**, 796–809.

Hightower, L. E. (1991). Heat shock, stress proteins, chaperones and proteotoxicity. *Cell*, **56**, 191–7.

Hissong, B. D., Byrne, G., Padilla, M. L. & Carlin, J. M. (1995). Up-regulation of interferon-induced indoleamine 2,3-dioxygenase in human macrophage cultures by lipopolysaccharide, muramyl tripeptide, and interleukin-1. *Cellular Immunology*, **160**, 264–9.

Hsia, R.-C., Pannekoek, Y., Ingerowski, E. & Bavoil, P. M. (1996). Type III secretion in *Chlamydia psittaci* GPIC. In Stary, A., ed. *Proceedings of the Third Meeting of the European Society for* Chlamydia *Research*, p. 51. Bologna: Societa Editrice Esculapio.

Igietseme, J. U. & Rank, R. G. (1991). Susceptibility to reinfection after a primary chlamydial genital-infection is associated with a decrease of antigen-specific T cells in the genital tract. *Infection and Immunity*, **59**, 1346–51.

Igietseme, J. U., Magee, D. M., Williams, D. M. & Rank, R. G. (1994*a*). Role for $CD8^+$ T cells in antichlamydial immunity defined by chlamydia-specific T-lymphocyte clones. *Infection and Immunity*, **62**, 5195–7.

Igietseme, J. U., Wyrick, P. B., Goyeau, D. & Rank, R. G. (1994*b*). An *in vitro* model for immune control of chlamydial growth in polarized epithelial cells. *Infection and Immunity*, **62**, 3528–35.

Ingalls, R. R., Rice, P. A., Qureshi, N., Takayama, K., Lin, J. S. & Golenbock, D. T. (1995). The inflammatory cytokine response to *Chlamydia trachomatis* infection is endotoxin mediated. *Infection and Immunity*, **63**, 3125–30.

Jaattela, M. & Wissing, D. (1993). Heat shock proteins protect cells from monocyte cytotoxicity: possible mechanisms of self protection. *Journal of Experimental Medicine*, **177**, 231–6.

Johansson, M., Ward, M. & Lycke, N. (1996). Natural infection with *Chlamydia trachomatis* fails to induce protective immunity in CD4- and IFN-γR deficient mice. *FASEB Journal, Abstracts*, **10**, Abstract no. 122, A1020.

Jones, R. B. & Batteiger, B. E. (1986). Human immune response to *Chlamydia trachomatis* infections. In Oriel, D., Ridgway, G., Schachter, J., Taylor-Robinson,

D. & Ward, M., eds. *Chlamydial Infections*, pp. 423–32. Cambridge: Cambridge University Press.

Kahane, S., Gonen, R., Sayada, C., Elion, J. & Friedman, M.G. (1993). Description and partial characterization of a new chlamydia-like microorganism. *FEMS Microbiology Letters*, **109**, 329–33.

Kahane, S., Metzer, E. & Friedman, M. G. (1995). Evidence that the novel microorganism Z may belong to a new genus in the Family *Chlamydiaceae*. *FEMS Microbiology Letters*, **126**, 203–7.

Katz, B. P., Batteiger, B. E. & Jones, R. B. (1987). Effect of prior sexually transmitted disease on the isolation of *Chlamydia trachomatis*. *Sexually Transmitted Diseases*, **14**, 160–4.

Knight, S. C., Iqball, S., Woods, C., Stagg, A., Ward, M. E. & Tuffrey, M. (1995). A peptide of *Chlamydia trachomatis* shown to be a primary T cell epitope *in vitro* induces cell-mediated immunity *in vivo*. *Immunology*, **85**, 8–15.

Kuo, C-C., Wang, S-P. & Grayston, J. T. (1973). Effect of polycations, polyanions and neuraminidase on the infectivity of trachoma inclusion conjunctivitis and lymphogranuloma venereum organisms in HeLa cells. Sialic acid residues as possible receptors for trachoma inclusion conjunctivitis. *Infection and Immunity*, **8**, 74–9.

LaScolea, L. J., Zui-duan, C., Kopti, S., Fisher, J. & Ogra, P. L. (1989). Prevention of pulmonary and genital *Chlamydia* infection by oral immunization. In Meheus, A. & Spier, R.E., eds. *Vaccines for Sexually Transmitted Diseases*, pp. 86–91. London: Butterworths.

Lehtinen, M. & Paavonen, J. (1994). Heat shock proteins in the immunopathogenesis of chlamydial pelvic inflammatory disease. In Orfila, J., Byre, G. I., Chernesky, M. A., Grayston, J. T., Jones, R. B., Ridgway, G. L., Saikku, P., Schachter, J., Stamm, W. E. & Stephens, R. S., eds. *Chlamydial infections*, pp. 599–610. Bologna: Societa Editrice Esculapio.

Lundemose, A. G., Birkelund, S., Larsen, P. M., Fey, S. J. & Christiansen, G. (1990). Characterization and identification of early proteins in *Chlamydia trachomatis* serovar L2 by two-dimensional gel electrophoresis. *Infection and Immunity*, **58**, 2478–86.

Mabey, D. C. W., Bailey, R. L. & Hutin, Y. J. F. (1992*a*). The epidemiology and pathogenesis of trachoma. *Reviews in Medical Microbiology*, **3**, 112–9.

Mabey, D. C. W., Bailey, R. L., Ward, M. E. & Whittle, H. C. (1992*b*). A longitudinal study of trachoma in a Gambian village – implications concerning the pathogenesis of chlamydial infection. *Epidemiology and Infection*, **108**, 343–51.

Magee, D. M., Williams, D. M., Smith, J. G., Bleicker, C. A., Grubbs, B. G., Schachter, J. & Rank, R. G. (1995). Role of CD8 T cells in primary *Chlamydia* infection. *Infection and Immunity*, **63**, 516–21.

Majeed, M. & Kihlström, E. (1991). Mobilization of F-actin and clathrin during redistribution of *Chlamydia trachomatis* to an intracellular site in eukaryotic cells. *Infection and Immunity*, **59**, 4465–72.

Majeed, M., Ernst, J. D., Magnusson, K-E., Kihlström, E. & Stendahl, O. (1994). Selective translocation of annexins during intracellular redistribution of *Chlamydia trachomatis* in HeLa and McCoy cells. *Infection and Immunity*, **62**, 126–34.

Matsumoto, A., Bessho, H., Uehira, K. & Suda, T. (1991). Morphological studies of the association of mitochondria with chlamydial inclusions and the fusion of chlamydial inclusions. *Journal of Electron Microscopy*, **40**, 356–63.

McClarty, G. (1994). Chlamydiae and the biochemistry of intracellular parasitism. *Trends in Microbiology*, **2**, 157–64.

Melgosa, M. P., Kuo, C-C. & Campbell, L. A. (1991). Sequence analysis of the major outer membrane protein gene of *Chlamydia pneumoniae*. *Infection and Immunity*, **59**, 2195–9.

Moorman, D. R., Sixbey, J. W. & Wyrick, P. B. (1986). Interaction of *Chlamydia trachomatis* with human genital epithelium in culture. *Journal of General Microbiology*, **132**, 1055–67.

Morrison, R. P., Lyng, K. & Caldwell, H. D. (1989). Chlamydial disease pathogenesis: ocular hypersensitivity elicited by a genus-specific 57 kd protein. *Journal of Experimental Medicine*, **169**, 663–75.

Morrison, R. P., Feilzer, K. & Tumas, D. B. (1995). Gene knockout mice establish a primary protective role for major histocompatibility complex class II-restricted responses in *Chlamydia trachomatis* genital tract infection. *Infection and Immunity*, **63**, 4661–8.

Moulder, J. W., Zeichner, S. L. & Levy, N. J. (1982). Association between resistance to superinfection and patterns of surface protein labelling in mouse fibroblasts (L cells) persistently infected with *Chlamydia psittaci*. *Infection and Immunity*, **35**, 834–9.

Moulder, J. W. (1988). Characteristics of chlamydiae. In Barron, A. L., ed. *Microbiology of* Chlamydia, pp. 3–19. Boca Raton, Florida: CRC Press.

Murdin, A. D., Su, H., Manning, D. S., Klein, M. H., Parnell, M. J. & Caldwell, H. D. (1993). A poliovirus hybrid expressing a neutralization epitope from the major outer membrane protein of *Chlamydia trachomatis* is highly immunogenic. *Infection and Immunity*, **61**, 4406–14.

Murdin, A. D., Su, H., Klein, M. H. & Caldwell, H. D. (1995). Poliovirus hybrids expressing neutralization epitopes from variable domains I and IV of the major outer membrane protein of *Chlamydia trachomatis* elicit broadly cross-reactive *C. trachomatis*-neutralizing antibodies. *Infection and Immunity*, **63**, 1116–21.

Nichols, R. L., Murray, E. S. & Nisson, P. E. (1978). Use of enteric vaccines in protection against chlamydial infections of the genital tract and the eye of guinea pigs. *Journal of Infectious Diseases*, **138**, 742–7.

Oriel, J. D. & Ridgway, G. L. (1982). Studies of the epidemiology of chlamydial infection of the human genital tract. In Mårdh, P.-A., ed. *Chlamydial Infections*, pp. 425–8. Amsterdam: Elsevier Biomedical Press.

Paavonen, J., Lähdeaho, M. L., Puolakkainen, M., Mäki, M., Parkkonen, P. & Lehtinen, M. (1994). Antibody-response to B cell epitopes of *Chlamydia trachomatis* 60-kDa heat-shock protein and corresponding mycobacterial and human peptides in infants with chlamydial pneumonitis. *Journal of Infectious Diseases*, **169**, 908–11.

Patton, D. L., Moore, D. E., Spadoni, L. R., Soules, M. R., Halbert, S. A. & Wang, S.-P. (1989). A comparison of the fallopian tube's response to overt and silent salpingitis. *Obstetrics and Gynecology*, **73**, 622–30.

Patton, D. L., Wolnerhanssen, P., Cosgrove, S. J. & Holmes, K. K. (1990). The effects of *Chlamydia trachomatis* on the female reproductive tract of the *Macaca nemestrina* after a single tubal challenge following repeated cervical inoculations. *Obstetrics and Gynecology*, **76**, 643–50.

Patton, D. L., Askienazy-Elbhar, M., Henry-Suchet, J., Campbell, L. A., Cappuccio, A., Tannous, W., Wang, S-P. & Kuo, C-C. (1994). Detection of *Chlamydia trachomatis* in fallopian tube tissue in women with postinfectious tubal infertility. *American Journal of Obstetrics and Gynecology*, **171**, 95–101.

Pedersen, L. B., Birkelund, S. & Christiansen, G. (1996*a*). Purification of recombinant *Chlamydia trachomatis* histone H1-like protein Hc2, and comparative functional analysis of Hc2 and Hc1. *Molecular Microbiology*, **20**, 295–311.

Pedersen, L. B., Birkelund, S., Holm, A., Østergaard, S. & Christiansen, G. (1996*b*).

The 18- kilodalton *Chlamydia trachomatis* histone H1-like protein (Hc1) contains a potential N-terminal dimerization site and a C-terminal nucleic acid-binding domain. *Journal of Bacteriology*, **178**, 994–1002.
Philips, H. L. & Clarkson, M. J. (1995). Spontaneous change from overt to covert infection of *Chlamydia pecorum* in cycloheximide-treated mouse McCoy cells. *Infection and Immunity*, **63**, 3729–30.
Rank, R. G., Soderberg, L. S. F., Sanders, M. M. & Batteiger, B. E. (1989). Role of cell-mediated immunity in the resolution of secondary chlamydial genital infection in guinea-pigs infected with the agent of guinea pig inclusion conjunctivitis. *Infection and Immunity*, **57**, 706–10.
Rank, R. G., Ramsey, K. H., Pack, E. A. & Williams, D. M. (1992). Effect of gamma interferon on resolution of murine chlamydial genital infection. *Infection and Immunity*, **60**, 4427–29.
Raulston, J. E. (1995). Chlamydial envelope components and pathogen host cell interactions. *Molecular Microbiology*, **15**, 607–16.
Raulston, J. E., Davis, C. H., Schmiel, D. H., Morgan, M. W. & Wyrick, P. B. (1993). Molecular characterization and outer membrane association of a *Chlamydia trachomatis* protein related to the Hsp70 family of proteins. *Journal of Biological Chemistry*, **268**, 23139–47.
Reynolds, D. J. & Pearce, J. H. (1993). Endocytic mechanisms utilised by chlamydiae and their influence on induction of productive infection. *Infection and Immunity*, **59**, 3033–9.
Richmond, S. J. (1985). Division and transmission of inclusions of *Chlamydia trachomatis* in replicating McCoy cell monolayers. *FEMS Microbiology Letters*, **29**, 49–52.
Rockey, D. D., Heinzen, R. A. & Hackstadt, T. (1995). Cloning and characterization of a *Chlamydia psittaci* gene coding for a protein localized in the inclusion membrane of infected cells. *Molecular Microbiology*, **15**, 617–26.
Rodolakis, A. (1995). Vaccination against chlamydial abortion in livestock. *Veterinary Research*, **26**, 212–3.
Rodolakis, A. & Bernard, F. (1984). Vaccination with temperature-sensitive mutant of *Chlamydia psittaci* against enzootic abortion of ewes. *Veterinary Record*, **114**, 193–4.
Schmitz, E., Nettelnbreker, E., Zeidler, H., Hammer, M., Manor, E. & Wollenhaupt, J. (1993). Intracellular persistence of chlamydial major outer-membrane protein, lipopolysaccharide and ribosomal RNA after non-productive infection of human monocytes with *Chlamydia trachomatis* serovar K. *Journal of Medical Microbiology*, **38**, 278–85.
Schramm, N. & Wyrick, P. B. (1995). Cytoskeletal requirements in *Chlamydia trachomatis* infection of host cells. *Infection and Immunity*, **63**, 324–32.
Sowa, S., Sowa, J., Collier, L. H. & Blyth, W. A. (1969). Trachoma vaccine field trials in The Gambia. *Journal of Hygiene (Cambridge)*, **67**, 699–717.
Stagg, A. J., Elsley, W. A. J., Pickett, M. A., Ward, M. E. & Knight, S. C. (1993). Primary human T-cell responses to the major outer membrane protein of *Chlamydia trachomatis*. *Immunology*, **79**, 1–9.
Starnbach, M. N., Bevan, M. J. & Lampe, M. F. (1994). Protective cytotoxic T lymphocytes are induced during murine infection with *Chlamydia trachomatis*. *Journal of Immunology*, **153**, 5183–9.
Starnbach, M. N., Bevan, M. J. & Lampe, M. F. (1995). Murine cytotoxic T lymphocytes induced following *Chlamydia trachomatis* intraperitoneal or genital tract infection respond to cells infected with multiple serovars. *Infection and Immunity*, **63**, 3527–30.

Stephens, R. S. (1994). Cell biology of *Chlamydia* infection. In Orfila, J., Byre, G. I., Chernesky, M. A., Grayston, J. T., Jones, R. B., Ridgway, G. L., Saikku, P., Schachter, J., Stamm, W. E. & Stephens, R. S., eds. *Chlamydial Infections*, pp. 377–86. Bologna: Societa Editrice Esculapio.

Su, H. & Caldwell, H. D. (1995). $CD4^+$ T cells play a significant role in adoptive immunity to *Chlamydia trachomatis* infection of the mouse genital tract. *Infection and Immunity*, **63**, 3302–8.

Su, H., Zhang, Y-X., Barrera, O., Watkins, N. G. & Caldwell, H. D. (1988). Differential effect of trypsin on infectivity of *Chlamydia trachomatis*: loss of infectivity requires cleavage of major outer membrane protein variable domains II and domain IV. *Infection and Immunity*, **56**, 2094–100.

Swanson, A. F. & Kuo, C-C. (1994). Binding of the glycan of the major outer membrane protein of *Chlamydia trachomatis* to HeLa cells. *Infection and Immunity*, **62**, 24–8.

Tam, J. E., Davis, C. H. & Wyrick, P. B. (1994). Expression of recombinant DNA introduced into *Chlamydia trachomatis* by electroporation. *Canadian Journal of Microbiology*, **40**, 583–91.

Taylor, H. R., Johnson, S. L., Prendergast, R. A., Schachter, J., Dawson, C. R. & Silverstein, A. M. (1982). An animal model of trachoma. II. The importance of repeated infection. *Investigative Ophthalmology and Visual Science*, **23**, 507–15.

Taylor-Robinson, D. & Ward, M. E. (1989). Immunity to chlamydial infections and the outlook for vaccination. In Meheus, A. & Spier, R. E., eds. *Vaccines for Sexually Transmitted Diseases*, pp. 67–85. London: Butterworths.

Ting, L.-M., Hsia, R.-C., Haidaris, C. G. & Bavoil, P. M. (1995). Interaction of outer envelope proteins of *Chlamydia psittaci* GPIC with the HeLa cell surface. *Infection and Immunity*, **63**, 3600–8.

Treharne, J. D. (1985). The community epidemiology of trachoma. *Reviews of Infectious Diseases*, **7**, 760–4.

Tuffrey, M., Alexander, F., Conlan, W., Woods, C. & Ward, M. E. (1992). Heterotypic protection of mice against chlamydial salpingitis and colonization of the lower genital tract with a human serovar F isolate of *Chlamydia trachomatis* by prior immunization with recombinant serovar L1 major outer-membrane protein. *Journal of General Microbiology*, **138**, 1707–15.

van Putten, J. P. M. & Paul, S. M. (1995). Binding of syndecan-like cell surface proteoglycan receptors is required for *Neisseria gonorrhoeae* entry into human mucosal cells. *EMBO Journal*, **14**, 2144–54.

Wang, S.-P., Grayston, J. T. & Alexander, E. R. (1967). Trachoma vaccine studies in monkeys. *American Journal of Ophthalmology*, **63**, 1615–20.

Ward, M. E. (1986). Outstanding problems in chlamydial cell biology. In Oriel, D., Ridgway, G., Schachter, J., Taylor-Robinson, D. & Ward, M., eds. *Chlamydial Infections*, pp. 3–14. Cambridge: Cambridge University Press.

Ward, M. E. (1988). The chlamydial developmental cycle. In Barron, A. L., ed. *Microbiology of* Chlamydia, pp. 71–95. Boca Raton, Florida: CRC Press.

Ward, M. E. (1995). The immunobiology and immunopathology of chlamydial infections. *APMIS*, **103**, 769–96.

Ward, M. E., Bailey, R., Lesley, A., Kajbaf, M., Robertson, J. & Mabey, D. (1990). Persisting inapparent chlamydial infection in a trachoma endemic community in The Gambia. *Scandinavian Journal of Infectious Diseases*, **Suppl. 69**, 137–48.

Weisburg, W. G., Barns, S. M., Pelletier, D. A. & Lane, D. J. (1991). 16S ribosomal DNA amplification for phylogenetic study. *Journal of Bacteriology*, **173**, 697–703.

Whittum-Hudson, J. A., Prendergast, R. A. & Taylor, H. R. (1986). Effects of oral preimmunization on chlamydial eye infection. In Oriel, D., Ridgway, G.,

Schachter, J., Taylor-Robinson, D. & Ward, M., eds. *Chlamydial Infections*, pp. 469–72. Cambridge: Cambridge University Press.

Williams, D. M., Grubbs, B. G., Parksnyder, S., Rank, R. G. & Bonewald, L. F. (1996). Activation of latent transforming growth-factor-beta during *Chlamydia trachomatis*-induced murine pneumonia. *Research in Microbiology*, **147**, 251–62.

Witkin, S. S., Jeremias, J., Toth, M. & Ledger, W. J. (1993). Cell-mediated immune response to the recombinant 57-kDa heat-shock protein of *Chlamydia trachomatis* in women with salpingitis. *Journal of Infectious Diseases*, **167**, 1379–83.

Wyrick, P. B., Choong, J., Davis, C. H., Knight, S. T., Royal, M. O., Maslow, A. S. & Bagnell, C. (1989). Entry of genital *Chlamydia trachomatis* into polarized human epithelial cells. *Infection and Immunity*, **57**, 2378–89.

Wyrick, P. B., Davis, C. H., Knight, S. T., Choong, J., Raulston, J. E. & Schramm, N. (1993). An *in vitro* human epithelial cell culture system for studying the pathogenesis of *Chlamydia trachomatis*. *Sexually Transmitted Diseases*, **20**, 248–56.

Yi, Y., Zhong, G. & Brunham, R. C. (1993). Continuous B-cell epitopes in *Chlamydia trachomatis* heat shock protein 60. *Infection and Immunity*, **61**, 1117–20.

Yuan, Y., Zhang, Y-X., Watkins, N. G. & Caldwell, H. D. (1989). Nucleotide and deduced amino acid sequences for the four variable domains of the major outer membrane proteins of the 15 *Chlamydia trachomatis* serovars. *Infection and Immunity*, **57**, 1040–9.

Yxfeldt, G., Fröman, G., Mårdh, P-A. & Ward, M. E. (1994). Reactivity of antibodies to heteroclitic peptides based on the *Chlamydia trachomatis* major outer-membrane protein. *Microbiology*, **140**, 815–21.

Zahn, I., Szeredi, L., Schiller, I., Kunz, U. S., Burgi, E., Guscetti, F., Heinen, E., Corboz, L., Sydler, T. & Pospischil, A. (1995). Immunohistological determination of *Chlamydia psittaci*/*Chlamydia pecorum* and *C. trachomatis* in the piglet gut. *Journal of Veterinary Medicine Series B-Infectious Diseases and Veterinary Public Health*, **42**, 266–76.

Zaretzky, F. R., Pearce-Pratt, R. & Phillips, D. M. (1995). Sulfated polyanions block *Chlamydia trachomatis* infection of cervix-derived human epithelia. *Infection and Immunity*, **63**, 3520–6.

Zhang, D.-J., Fan, H., McClarty, G. & Brunham, R. C. (1995). Identification of the *Chlamydia trachomatis* RecA-encoding gene. *Infection and Immunity*, **63**, 676–80.

Zhang, J. P. & Stephens, R. S. (1992). Mechanism of *C. trachomatis* attachment to eukaryotic host cells. *Cell*, **69**, 861–9.

Zhang, Y.-X., Stewart, S., Joseph, T., Taylor, H. R. & Caldwell, H. D. (1987). Protective monoclonal antibodies recognize epitopes located on the major outer membrane protein of *Chlamydia trachomatis*. *Journal of Immunology*, **138**, 575–81.

Zhong, G. & Brunham, R. C. (1990). Immunoaccessible peptide sequences of the major outer membrane protein from *Chlamydia trachomatis* serovar C. *Infection and Immunity*, **58**, 3438–41.

Zhong, G. & Brunham, R. C. (1991). Antigenic determinants of the chlamydial major outer membrane protein resolved at a single amino acid level. *Infection and Immunity*, **59**, 1141–7.

Zhong, G., Reid, R. E. & Brunham, R. C. (1990). Mapping antigenic sites on the major outer membrane protein of *Chlamydia trachomatis* with synthetic peptides. *Infection and Immunity*, **58**, 1450–5.

Zhong, G., Smith, G. P., Berry, J. & Brunham, R. C. (1994). Conformational mimicry of a chlamydial neutralization epitope on filamentous phage. *Journal of Biological Chemistry*, **269**, 24183–8.

CONTROL OF TYPE 1 FIMBRIAL EXPRESSION BY A RANDOM GENETIC SWITCH IN *ESCHERICHIA COLI*

C. J. DORMAN, N. C. NOLAN AND S. G. J. SMITH

Department of Microbiology, Moyne Institute of Preventive Medicine, University of Dublin, Trinity College, Dublin 2, Republic of Ireland

INTRODUCTION

Pathogenic bacteria often express particular traits which render them virulent in their interactions with the host. These traits are termed virulence factors and well-studied examples include adhesins which attach the bacterium to its host, toxins which damage the host, and extracellular polysaccharide capsules which hide the bacterium from the host defences. The genes which encode these traits are known collectively as virulence genes.

In general, virulence factors are not expressed constitutively by bacteria. Instead, their expression is regulated to ensure that this occurs only under appropriate conditions. If expression of a virulence gene is required only while the bacterium is in intimate association with the host, then an environmental cue supplied by the host is often used to trigger expression. Changes in physical parameters of the environment, such as temperature, pH or osmolarity, are good examples of cues of this sort. Chemical signals are also used, such as the presence or absence of particular amino acids or iron.

By evolving mechanisms which regulate virulence gene expression tightly, bacteria avoid wasteful expression under inappropriate circumstances. Many virulence factors, such as outer membrane proteins and fimbrial adhesins, are located on the outer surface of the bacterium and are immunogenic. Inappropriate expression within the host may trigger the host defences, resulting in the death of the bacterium. Regulation through mechanisms which alter gene expression in every cell in the bacterial population at the same time ('stereotypic' control) appears to be a successful solution to the metabolically wasteful expression of virulence factors out of context. The same strategy seems less useful in helping to overcome the difficulties with the host defences. In this case, bacteria employ mechanisms which either vary the structure of the virulence factor so that a primed defence system can no longer detect it (antigenic variation) or vary virulence factor expression at random in different cells in the population (phase variation) through stochastic mechanisms. With phase variation in operation, the population becomes phenotypically mixed from the point of view of the virulence trait concerned. If host defences are primed to kill cells which are in the expressing mode, those in the non-expressing mode will survive and can produce expressing variants at a later time, possibly at another site in the host. This subsequent switch to expression by a small subset of bacteria represents another trial of the environment. If the trial is successful, the participating bacteria will be rewarded by being permitted to proliferate, while those in the alternative or non-expressing phase will be at a disadvantage. Bacterial populations are highly sophisticated in their interactions with the host, employing sterotypic and stochastic systems simultaneously to maximize the probability that the population will be successful in its attempt at colonization. Since this success may result in disease for the host, it is important to understand the molecular mechanisms which underly the regulatory processes. Type 1 fimbriae in *Escherichia coli* are just one of a number of model systems which are currently the subject of intensive research aimed at assisting this understanding.

Fimbriae contribute to host–pathogen interactions by promoting adhesion of bacteria to host cells (Smyth & Smith, 1992). Type 1 fimbriae are characterized by having the ability to agglutinate erythrocytes from several species in a manner which is prevented by co-incubation with D-mannose (Klemm, 1994). Their *in vivo* target is thought to be basement membrane carbohydrate (Kukkonen *et al.*, 1993). Most clinical isolates of *Escherichia coli* express type 1 fimbriae and many also express mannose-insensitive fimbriae. Although controversy still surrounds the precise role of type 1 fimbriae in some infections (Orndorff & Bloch, 1990), they have been described as enhancing the communicability of bacteria between infected and uninfected hosts, contributing to bacterial virulence during urinary tract infections in animal models, being involved in intestinal infections and colonization of the large bowel, and down-regulation of their expression has been associated with the establishment of peritoneal *E. coli* infection (Smyth, 1986; Schaeffer *et al.*, 1987; Orndorff & Bloch, 1990; Bloch, Stocker & Orndorff, 1992; May *et al.*, 1993; Smyth, Marron & Smith, 1994). Non-pathogenic strains of *E. coli* K-12 express only type 1 fimbriae and this has made the type 1 system of K-12 strains an attractive subject for fimbrial research (Eisenstein, 1987). One of the most attractive features of type 1 fimbrial expression in *E. coli* is that it is subject to stochastic control mechanisms.

PHASE VARIATION

Expression of type 1 fimbriae is phase variable, oscillating between an expressed (Phase ON) and a non-expressed (Phase OFF) state (Eisenstein, 1987). Switching between the ON and OFF phases occurs at a frequency of approximately 5×10^{-2} per cell per generation. These phases reflect expression of the *fimA* gene, which encodes the fimbrial subunit protein. It should be noted, however, that a poorly characterized phase-variation mechanism has been reported which does not appear to depend on variations in *fimA* transcription (McClain *et al*., 1993).

Fimbriae are strongly immunogenic and it has been rationalised that phase variation permits the bacterial population to include both fimbriate (adhesive) and non-fimbriate (lacking the fimbrial antigen) members at any one time. The environment then selects the bacteria with the phase which best promotes survival. Phase variation operates at random in the bacterial population. One cannot predict in advance which Phase-ON cell will undergo the ON-to-OFF switch or which Phase-OFF cell, the OFF-to-ON switch. This is a very different situation from stereotypic adaptive processes in which the entire population responds in unison to a cue from the environment, as happens, for example, when a heat shock is administered (Smyth, 1986; Robertson & Meyer, 1992; Dybvig, 1993; Moxon *et al*., 1994; Smyth *et al*., 1994; Dorman, 1995; Saunders, 1995).

SITE-SPECIFIC RECOMBINATION

In *E. coli* K-12 strains, transcription of the *fimA* gene depends on a promoter which is carried on an invertible DNA segment (Abraham *et al*., 1985). The open reading frame of the *fimA* gene lies outside, but immediately adjacent to, the invertible segment. Inversion requires the product of the *fimB* gene, which is located nearby (Fig. 1) (Klemm, 1986). This weakly transcribed gene codes for a protein showing significant amino acid sequence similarity to the integrase family of site-specific recombinases (Dorman & Higgins, 1987; Eisenstein *et al*., 1987). The matching amino acid residues are found in the carboxy-terminal domain of the FimB protein and include amino acids known to be involved in the site-specific recombination activity of integrase family members (Fig. 2) (Argos *et al*., 1986; Blakely & Sherratt, 1996). Genetic analysis has shown that the FimB protein can catalyse the inversion of the *fimA* promoter element in the ON-to-OFF and the OFF-to-ON directions with equal efficiency. Lying between the *fimB* gene and the invertible element is the *fimE* gene (Klemm, 1986). This also encodes a recombinase of the integrase family (Figs. 1 and 2) but in many *E. coli* K-12 strains this gene is found in a mutated form. For example, strain CSH50 and its derivatives, in which much of the *fim* locus research has been performed, possess an IS*1* insertion mutation in the *fimE* gene. Active copies of the *fimE* gene are found in other K-12 strains such as strain W3110 (from which the widely used Kohara library of *E. coli* genes was derived) and in clinical isolates (Kohara, Akiyama & Isono, 1987; Blomfield *et al*., 1991). The FimE protein is 40% identical to FimB in its amino acid sequence, yet it has a distinct effect on inversion of the *fimA* gene promoter segment (Klemm, 1986; McClain, Blomfield & Eisenstein, 1991). Unlike the FimB protein which inverts the *fimA* promoter element in either direction with approximately equal facility, the FimE protein has a strong preference for inversion in the

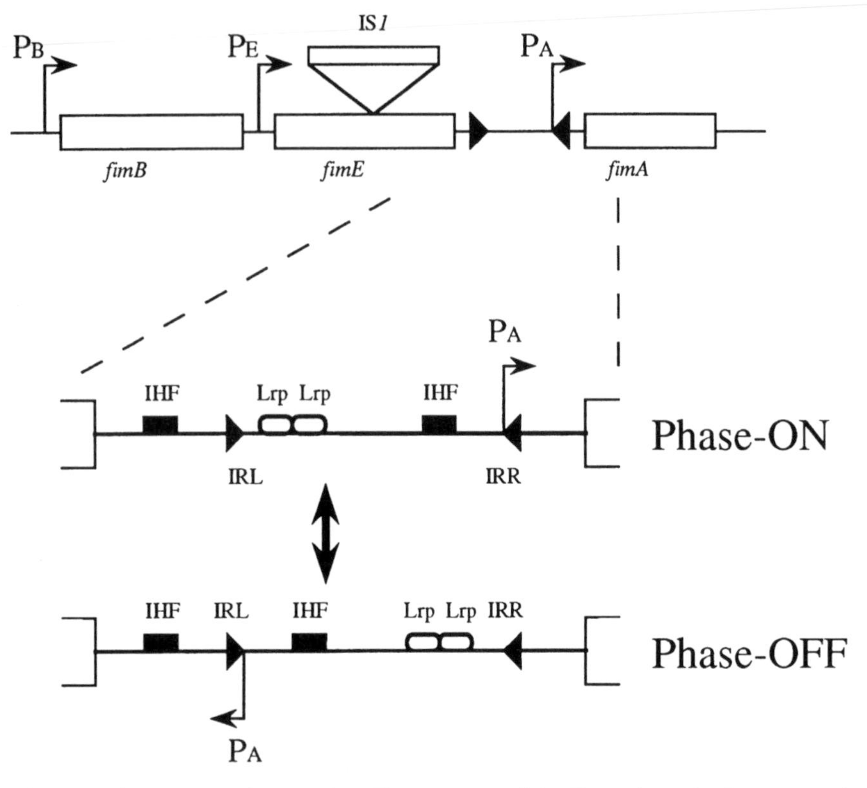

Fig. 1. The *fimA* gene regulatory region. The upper part of the Figure shows the *fimA* gene and the regulatory genes *fimB* and *fimE*. Each gene has its own promoter. In several *E. coli* K-12 strains, the *fimE* gene is inactive due to interruption by an insertion sequence. An expanded view of the *fimA* promoter (PA) region is shown in the lower parts of the Figure. When PA is directed to the right (towards the *fimA* open-reading frame), the cell is Phase-ON; when it is directed away from the *fimA* gene (as in the bottom portion of the Figure), the cell is Phase-OFF. The positions of the 9 bp inverted repeats are represented by the filled triangles at the left (IRL) and right (IRR) ends of the 314 bp invertible element. Binding sites (two each) for the accessory proteins IHF and Lrp are also shown. The diagram is not drawn to scale.

```
138-IHPHMLRHSCGFALANMGIDTRLIQDYLGHRNIRHTVWYTASNAG-182  FimB
133-THPHMLRHACGYELAERGADTRLIQDYLGHRNIRHTVRYTASNAA-177  FimE
    ***H*LRH**A**L***G*********LGH*****T**Y*H****     Consensus
       Δ  Δ                               Δ
```

Fig. 2. The FimB and FimE proteins show amino acid sequence identity to integrase proteins. The amino acid sequences of the carboxy-termini of the FimB protein (residues 138 to 182) and the FimE protein (residues 133 to 177) are shown. The consensus sequence for the catalytic domain of the integrase family of site-specific recombinases is shown below, with the three amino acids which participate in the recombination reaction shown by the arrowheads. All three of these amino acids, together with eight of the remaining ten consensus sequence residues, are present in both the FimB and FimE proteins.

ON-to-OFF direction (McClain *et al.*, 1991; Gally, Leathart & Blomfield, 1996). Presumably, in cells expressing both the *fimB* and *fimE* genes, there is an antagonism between their products as they attempt to invert their DNA substrate.

The invertible DNA element (the *fim* switch) is 314 bp long and is flanked by 9 bp inverted repeats (Abraham *et al.*, 1985). The FimB protein has been shown to bind to these repeats *in vivo* (Dove & Dorman, 1996) and both the FimB and FimE proteins bind to these repeats *in vitro* (Gally *et al.*, 1996). Crude lysates from cells over-producing FimB or FimE have been used to show that both recombinases bind to sites that flank and overlap the inverted repeat at the left (IRL) and the right (IRR) of the *fim* switch (Gally *et al.*, 1996). By analogy with related site-specific recombination systems, the FimB protein is thought to recombine these repeats with one another via an asymmetric cleavage of the DNA within the repeats. This is followed by the formation of a synaptic complex in which the protein is covalently bound via a tyrosine residue to the cleaved DNA. (This tyrosine residue is conserved throughout the integrase family (Fig. 2) and when it is altered to phenylalanine by site-directed mutagenesis, FimB ceased to be capable of catalysing recombination (S. G. J. Smith & C. J. Dorman, unpublished data).) Next, strand exchange and resolution produce the recombination product (for review see Blakely & Sherratt, 1996). Genetic analyses have revealed that, in addition to the FimB protein, protein co-factors are required for normal control of *fimA* gene expression (see below).

PROTEIN CO-FACTORS AND *fimA* EXPRESSION

Protein H-NS

Orndorff and coworkers identified the *pilG* locus at 27 minutes on the *E. coli* chromosome as encoding a *trans*-acting repressor of the *fim* switch (Spears, Schauer & Orndorff, 1986). This locus is now known to be allelic with the *hns* locus, encoding the bacterial chromatin-associated protein H-NS (*h*istone-like *n*ucleoid *s*tructuring protein) (Higgins *et al.*, 1988; Kawula & Orndorff, 1991). This is an abundant (20 000 copies per cell), heat-stable protein and has a molecular mass of 15.6 kDa (Lammi *et al.*, 1984; Spassky *et al.*, 1984). The H-NS protein is neutral under native conditions but under denaturing conditions migrates with a pI of 5.6 (van Bogelan *et al.*, 1992). It contains many acidic residues, in contrast to eukaryotic histones, which are basic proteins. The H-NS protein is located in the bacterial nucleoid, where it is assumed to be bound to DNA (it can also bind to RNA with low affinity) (Varshavsky *et al.*, 1977; Friedrich *et al.*, 1988; Durrenberger *et al.*, 1991). It is thought to form oligomers but the form in which the H-NS protein binds to DNA has not been determined unambiguously (for review see Ussery *et al.*, 1994).

Binding of the H-NS protein to double-stranded DNA is relatively non-specific; there is no consensus DNA sequence for H-NS binding sites. Binding is influenced, however, by DNA structure, with the H-NS protein showing a marked preference for binding to sequences which are curved intrinsically (Yamada *et al.*, 1991). The feature of curved DNA which makes it attractive to the H-NS protein for binding is presently unknown. Curved sequences are associated with promoters which the H-NS protein regulates, including the promoter of its own gene, *hns* (Dersch, Schmidt & Bremer, 1993; Ueguchi, Kakeda & Mizuno, 1993). Binding to this curved sequence allows the H-NS protein to autoregulate transcription of the *hns* gene. Transcription is activated by DNA synthesis and ceases when the cell enters stationary phase, probably reflecting a need to maintain an appropriate intracellular H-NS protein:DNA ratio (Free & Dorman, 1995).

Mutations in the *hns* gene alter the topology of the DNA in the cell and purified H-NS protein can constrain supercoils *in vitro* (Higgins *et al.*, 1988; Tupper *et al.*, 1994). The *in vitro* interaction depends on the protein-DNA ratio and on the ionic strength of the buffer. Changes in the physical environment of the DNA in the cell, caused in response to alterations in environmental conditions, may allow the H-NS protein to interact with DNA differently under different growth conditions. This could account for the role the H-NS protein appears to play in regulating the transcriptional profile of the cell.

Many genes show altered expression in *hns* mutants. The overall picture is one in which the H-NS protein acts as a negative factor in controlling transcription, although there may be exceptions to this (Bertin *et al.*, 1990; Yamada *et al.*, 1991; Ussery *et al.*, 1994). The only common feature of these genes is that they contribute in a general sense to the adaptation of the bacterium to alterations in its environment. This broad category includes genes which assist pathogens during interactions with the host, such as the *virB* gene of *Shigella flexneri* which encodes a positive regulator of structural genes that express factors which promote invasion of human cells (Tobe *et al.*, 1993).

There is no agreement yet on the mechanism by which the H-NS protein influences transcription. It may simply exclude RNA polymerase from the promoter or it may inhibit transcription initiation by altering the local topology of the DNA (Ueguchi & Mizuno, 1993; Ussery *et al.*, 1994). These possibilities are not mutually exclusive.

In the case of *fim* gene expression, *hns* mutations result in a 100-fold increase in the rate of inversion of the *fimA* promoter segment (Spears *et al.*, 1986; Higgins *et al.*, 1988; May *et al.*, 1990; Kawula & Orndorff, 1991). It is possible that the structure of the nucleoprotein complex required for inversion is altered in the absence of the H-NS protein. This might involve a change to the topology of the DNA at the inverting segment which makes it an improved substrate for the recombinases to act on. The binding of the recombinases may be better in the absence of the H-NS protein or the binding of accessory factors with a positive role in promoting inversion might be improved or both. The H-NS protein may influence the supply of recombinases, through its action at the promoters of the genes which code for them. For example, it has been demonstrated that the promoters for both the *fimB* and *fimE* genes are regulated negatively by the H-NS protein (Olsen & Klemm, 1994; S. L. Dove & C. J. Dorman, unpublished data). Transcription of these genes is elevated in *hns* mutants at both 30°C and 37°C, although the effect is stronger at the lower temperature (Olsen & Klemm, 1994). In addition to its effect on the DNA inversion system, the H-NS protein regulates the *fimA* gene promoter negatively, particularly at 30°C (Dorman & Ní Bhriain, 1992). Thus, the H-NS protein has an input into *fimA* expression at several points. However, it is not yet clear which inputs are the most significant physiologically. It is interesting to note that its role has been linked to growth temperature, since the H-NS protein has been shown to regulate other systems subject to temperature control, such as the invasion genes of *Shigella flexneri* and colonization antigens in enterotoxigenic *E. coli* (Hromockji, Tucker & Maurelli, 1992; Jordi *et al.*, 1992; Porter & Dorman, 1994). The H-NS protein may serve to make *fim* gene expression thermally responsive. Given the environments in which *E. coli* may or may not benefit from fimbrial expression, placing *fim* gene expression at least partly under thermal control might be very significant.

Protein StpA

E. coli possesses an H-NS homologue known as the StpA protein (Zhang *et al.*, 1996). The proteins are 58% identical in amino acid sequence and are similar in size. The biological function of the StpA protein is still unclear; its best characterized role *in vivo* is in stimulating self-splicing of bacteriophage T4 thymidylate synthase mRNA (Zhang *et al.*, 1995). Given the strong similarity of the StpA and H-NS proteins, it has been hypothesised that one protein may be capable of influencing genes whose expression is regulated by the other (Zhang *et al.*, 1996). The StpA protein has a poorly characterized negative effect on inversion of the *fim* switch (A. Free & C. J. Dorman, unpublished data). Given the structural similarity of the StpA and H-NS proteins and the role played by the latter in inhibiting this process, this is perhaps unsurprising. The StpA protein may influence *fim* gene expression by acting alone, or it may exert its effects by forming heteromeric structures with the H-NS protein, as has been suggested to explain its role in other systems (Ussery *et al.*, 1994; Williams, Rimsky & Buc, 1996).

Protein IHF

The integration host factor (IHF) has been shown by genetic methods to be essential for normal expression of the *fimA* gene in *E. coli* (Dorman & Higgins, 1987; Eisenstein *et al.*, 1987). The IHF protein is another nucleoid-associated protein, but it differs from the H-NS protein in many respects, not least in being sequence-specific in its interactions with DNA. Having been discovered initially as an important contributor to the life cycle of bacteriophage lambda, the IHF protein is now recognised as playing numerous roles in the cell, influencing (*inter alia*) transcription, transposition, site-specific recombination, and plasmid stability (Miller & Friedman, 1980; Nash & Robertson, 1981; Friedman, 1988; Freundlich *et al.*, 1992; Goosen & van de Putte, 1995).

The IHF protein is heterodimeric and is composed of IHFα (11.35 kDa) and IHFβ (10.65 kDa) subunits, encoded by the unlinked genes *ihfA* and *ihfB*, previously known as *himA* and *himD* (or *hip*), respectively (Miller, 1984; Flamm & Weisberg, 1985). When the IHF protein interacts with its binding site in DNA, the DNA becomes bent by up to 140° (Kosturko, Daub & Murialdo, 1989). In this way, the IHF protein serves as a site-specific architectural element in the genome, directing the pathway of the DNA either side of the binding site. This DNA-bending function is crucial to the contribution of the IHF protein to bacteriophage lambda integration; here, the IHF protein assists in the assembly of the nucleoprotein complex known as the intasome (Goodman & Nash, 1989; Snyder, Thompson & Landy, 1989). It is likely that bending of DNA is the key role performed by the IHF protein in most of the systems where it plays a part.

The intracellular concentration of the IHF protein is variable and these fluctuating protein levels may be physiologically significant (Bushman *et al.*, 1985; Aviv *et al.*, 1994). During exponential growth, IHF levels in *E. coli* are low, but as the culture enters stationary phase, the levels rise. Thus, events subject to control by the IHF protein may be more likely to occur at some phases of growth than at others. Transcription of the *ihfA* and *ihfB* genes is under complex control. Roles have been identified for the stationary-phase-specific sigma factor, RpoS, for the alarmone guanosine tetraphosphate and for autoregulation by the IHF protein itself in the control of the *ihf* genes (Aviv *et al.*, 1994).

In IHF mutants, inversion of the *fimA* promoter segment ceases completely (Dorman & Higgins, 1987; Eisenstein *et al.*, 1987). Two matches to the consensus sequence for IHF binding sites are located at the invertible element, one within the element itself and the other to its left (Fig. 1). Specific binding of the IHF protein to both sites has been demonstrated *in vitro* by DNase I protection studies and band-shift analysis (S. L. Dove & C. J. Dorman, unpublished data). By analogy with the lambda integration system, it is likely that the IHF protein organizes the architecture of a nucleoprotein complex required for inversion of the *fim* switch. The purpose of this complex would be to bring together the 9 bp inverted repeats with their bound recombinases in order to facilitate recombination. Presumably, in the absence of the IHF protein, the probability of this complex forming approaches zero and inversion does not occur.

In addition to its role in facilitating DNA inversion, the IHF protein also influences the potency of the *fimA* promoter. In mutants deficient in IHF expression, the level of transcription from the *fimA* promoter is reduced by about 7-fold (Dorman & Higgins, 1987). The IHF protein has been identified as a positive factor in the control of transcription from other promoters (Goosen & van de Putte, 1995) and it is possible that it helps to activate *fimA* gene transcription via an interaction between IHF protein bound to the site internal to the invertible element and the promoter (Fig. 1). By analogy with other systems, this could involve DNA bending or direct contact between the IHF protein and bound RNA polymerase or both.

Protein LRP

In addition to the IHF protein, the leucine-responsive regulatory protein (LRP) is a positive factor in controlling *fimA* gene expression (Blomfield *et al.*, 1993; Gally, Rucker & Blomfield, 1994). The LRP protein is a basic (pI = 9.3) protein composed of two identical subunits of 18.8 kDa each, present in about 3000 copies per cell (Calvo & Matthews, 1994; Newman & Lin, 1995). Comparisons of the nucleotide sequences of the sites to which the LRP protein is known to bind at a number of genetic loci has allowed a consensus sequence to be described. Thus, the LRP protein resembles protein IHF rather than H-NS in being a sequence-specific DNA binding protein. Like protein IHF, the LRP protein bends DNA. The angle for LRP-induced bending has been estimated to be 52° for single sites and 135° for two adjacent sites (Wang & Calvo, 1993). The catalogue of genes known to be subject to regulation by the LRP protein is large and is still growing. In addition to the *fim* operon (type 1), this list includes the fimbrial systems encoded by *daa* (F1845), *fae* (K88), *fan* (K99), *pap* (P) and *sfa* (S) (Braaten *et al.*, 1992; Huisman *et al.*, 1994; van der Woude & Low, 1994).

Individual systems can be regulated by the LRP protein positively or negatively and this control may or may not be modulated by L-leucine. The LRP protein has a leucine-responsive domain within its carboxy terminus (Platko & Calvo, 1993). In cases where there is a leucine effect, this can accentuate or antagonise the input from protein LRP. Thus, the LRP protein is a sophisticated regulator and its role in each system must be studied without preconceived ideas about the input which it is likely to make. In the case of *fimA* gene expression, the input from the LRP protein is positive and this is accentuated by leucine, and also by the amino acids alanine, isoleucine and valine (Blomfield *et al.*, 1993; Gally *et al.*, 1993, 1994). Two binding sites for the LRP protein have been identified at the *fim* invertible sequence (Fig. 1). Mutating these sites prevents binding of the LRP protein *in vitro* and slows the rate of inversion *in vivo* (Gally *et al.*, 1994). It is likely that the LRP protein contributes to the formation of the nucleoprotein complex which promotes site-specific recombination at the *fim* locus and that this contribution is modulated by the composition of the growth medium.

RECOMBINASE OVER-EXPRESSION AND ACCESSORY PROTEIN REDUNDANCY

Increasing the number of copies of the functional *fimB* gene in the cell accelerates inversion of the *fimA* promoter segment. It also permits inversion to occur at normal rates in the absence of either the IHF or LRP proteins; in mutants deficient in both the IHF and LRP proteins, multiple copies of the *fimB* gene restore a low level of phase-variable *fimA* gene expression (Dove & Dorman, 1996). These data suggest that the FimB protein can interact with the invertible sequence in novel ways when over-expressed, rendering the normally essential accessory proteins individually unnecessary. This is confirmed by studying FimB protein interactions with DNA *in vivo* using dimethyl sulphate-mediated DNA methylation to investigate protection patterns in cells harbouring single and multiple copies of the *fimB* gene. Interactions between the FimB protein and the inverted repeats is detectable in both types of cells but the degree of interaction is more pronounced when the recombinase is over-expressed (Dove & Dorman, 1996). In addition, novel patterns of interaction are seen at the right-hand inverted repeat when the FimB protein is over-expressed. In Phase-ON cells, this is the inverted repeat which lies adjacent to the *fimA* promoter and transcription initiating at this promoter must traverse this inverted repeat. When the FimB protein is over-expressed, *fimA* transcription is strongly repressed (by about ten-fold), which is consistent with a negative effect on transcription initiation due to FimB protein binding at the IRR site (Dove & Dorman, 1996). These data suggest that there may be an interplay between transcription and recombination in this system. This hypothesis is supported by data from other studies (see below).

DNA SUPERCOILING AND FIM INVERSION

The involvement of the H-NS, IHF and LRP proteins in *fim* gene regulation raises the possibility that other aspects of DNA topology, such as supercoiling, may also influence this system. Supercoiling levels in bacterial DNA are set by the activities of DNA topoisomerases, with gyrase introducing negative supercoils and topoisomerase I removing them (Dorman, 1995). The degree of supercoiling seen at any time is a reflection of the physiological state of the cell, which is influenced in turn by the physical and chemical make-up of the external environment. Variations in supercoiling have been shown to occur in response to changes in temperature, oxygen availability, osmotic stress and other parameters (for review see Dorman, 1995).

Supercoiling can vary at a local level (over distances of a few hundred base pairs) due to the unwinding of the DNA duplex during movement of complexes involved in processes such as DNA replication or transcription. The moving complex leaves a domain of underwound or negatively supercoiled DNA in its wake and creates a relaxed (or even positively supercoiled) domain in front (Liu & Wang, 1987; Wu *et al.*, 1988; Tsao, Wu & Liu, 1989). These domains become substrates for gyrase (relaxes positive supercoils) and topoisomerase I (relaxes negative supercoils) and are thought to have a transient existence under normal circumstances. However, if the physiological state of the cell impairs topoisomerase activity, the domains may linger and exert a biological effect. This may be particularly relevant in the case of DNA gyrase. Local supercoiling generated by transcription has been studied in a number of systems and shown to have profound effects on promoter function and site-specific recombination (Figuero & Bossi, 1988; Chen *et al.*, 1992; Rahmouni & Wells, 1992; Dröge, 1993; Wang & Dröge, 1996).

In cells harbouring null mutations in the *topA* gene, which encodes DNA topoisomerase I, inversion of the *fimA* promoter segment ceases (Dove & Dorman, 1994). These mutants have levels of negative supercoiling which exceed those of wild-type cells, because they have lost the DNA relaxing activity of topoisomerase I. Inversion is not reinstated when supercoiling levels are restored to values found in wild-type cells by the acquisition by the *topA* mutant strains of compensatory mutations. Normal rates of inversion of the *fim* switch correlate with the presence or absence of topoisomerase I and not with the global level of supercoiling in the cell. These data suggest that topoisomerase I plays a role in the vicinity of the inverting DNA segment itself and, in the *topA* mutant, this role is not fulfilled by any other cellular component. The most likely explanation is that topoisomerase I is required to relax a domain of negatively supercoiled DNA which is specific to the inverting DNA segment or its immediate environment.

Inhibiting DNA gyrase with the antibiotic novobiocin results in a general relaxation of DNA. Loss of some gyrase activity (complete loss is lethal) leads to a phase-specific effect on inversion of the *fimA* promoter segment (Dove & Dorman, 1994). Phase-ON cells are unaffected by gyrase inhibition in terms of the orientation of the invertible element; Phase-OFF cells are affected strongly and invert the element to the ON orientation. This suggests that gyrase plays a key role in Phase-OFF cells which it is not required to fulfil when cells are Phase-ON. The simplest interpretation of the data is that a substrate for gyrase exists in Phase-OFF cells which is absent in Phase-ON cells. A requirement for gyrase to relax a domain of

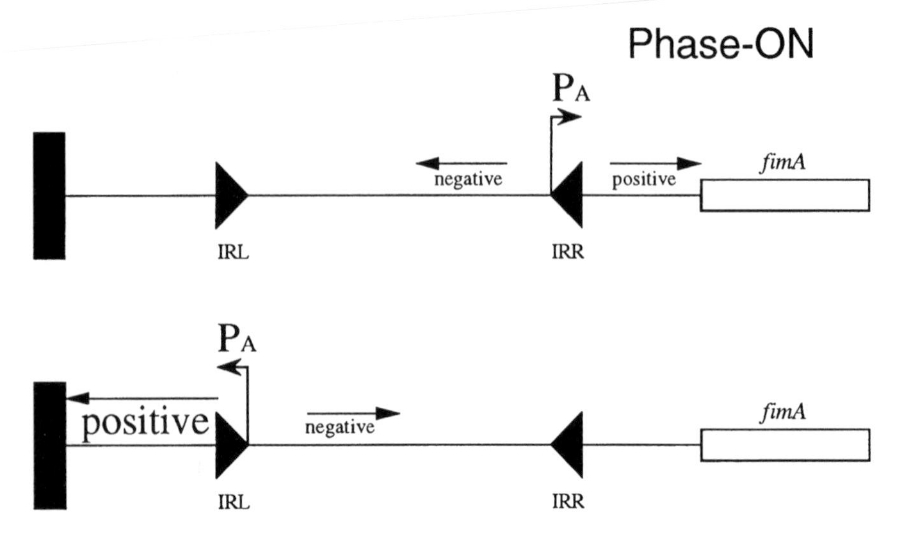

Fig. 3. Domains of differentially supercoiled DNA at the *fim* switch. The inverted repeats flanking the invertible element are represented by the filled triangles. The *fimA* promoter is denoted by the angled arrow labelled 'PA'. In Phase-ON, this promoter is directed towards the *fimA* gene; in Phase-OFF, it is directed away. Transcriptional activity generates domains of supercoiled DNA behind and ahead of the moving transcription complex. The horizontal arrows show the directions in which these domains move as they dissipate by rotation of the DNA helix. The domains of negative supercoiling behind the promoter are of equal magitude in both phases. Positive supercoils generated ahead of the promoter in Phase-ON are hypothesised to dissipate by rotational diffusion, whereas those generated in Phase-OFF become trapped by an unidentified barrier to rotational diffusion (the filled rectangle at left). In circumstances where DNA gyrase is unable to relax the positive supercoils, the two phases contain domains of DNA which are positively supercoiled to different extents.

positively supercoiled DNA generated only in Phase-OFF cells would offer an explanation. This domain could be generated by the activity of the *fimA* promoter, which is equally active in both Phase-ON and Phase-OFF cells (S. L. Dove & C. J. Dorman, unpublished data). However, a supercoil-trapping mechanism which operates only in Phase-OFF would give the system the requisite asymmetry to account for the experimental data (Fig. 3). This supercoil trapping system could be a complex similar to the nuclear scaffolding of eukaryotes, perhaps involving the IHF binding site to the left of the left-hand inverted repeat (Figs. 2 and 3). Matrix association regions (MAR sites) perform an analogous function in eukaryotes. These are AT-rich regions of DNA to which the type II topoisomerase known as topo II binds (interestingly, bacterial gyrase is also a type II topoisomerase). MAR sites which block rotational diffusion of positive supercoils occur in association with immunoglobulin genes, and transcriptionally induced variations in supercoiling contribute to the recombination machinery which underlies immunoglobulin heavy chain class switching (Cockerill & Garrard, 1986; Schwedler, Jäck & Wabl, 1990; Lee & Garrard, 1991; Leung & Maizels, 1992; Xu *et al.*, 1993). This raises the intriguing possibility that host and pathogen may have evolved similar mechanisms for varying the expression of defence and virulence factors, respectively.

The result of supercoil trapping in the vicinity of the *fim* switch is that the inverted repeats lie in domains of DNA supercoiled to different extents in Phase-ON and Phase-OFF cells (Fig. 3), perhaps altering their suitability as substrates for site-specific recombination. This only becomes relevant when gyrase activity is impaired, either through treatment with an inhibitory drug or through a change in cellular physiology.

ENVIRONMENTAL INFLUENCES

The expression of type 1 fimbriae is most likely to benefit the bacterium when it occurs during colonization of the host. The benefits of phase-variable expression can be rationalized on the basis of the risk which expression poses in the presence of the host immune system and the benefits which the bacterium receives when the fimbriae are expressed in an appropriate niche within the host. Attracting the attention of the host defences to surface features such as fimbriae may result in the death of the bacterium whereas expressing the fimbriae appropriately in time and space will enhance the survival of the bacterium within the host (Klemm, 1994).

It can be seen from the foregoing discussion that *fim* gene expression is subject to complex regulation. There are several possibilities for environmental signals to influence the process. Regulating the supply of recombinase protein might be a useful way to control the rate of DNA inversion. The *fimB* promoter is influenced strongly by variations in the supercoiling of chromosomal DNA, becoming stronger as the DNA becomes relaxed (Dove & Dorman, 1994). The H-NS protein represses *fimB* (and *fimE*) gene transcription (Olsen & Klemm, 1994; Dove & Dorman, 1997). Transcription of the *fimB* gene is also regulated negatively by the stationary-phase-specific sigma factor, RpoS, with repression being exerted as cells enter stationary phase (Dove, Smith & Dorman, 1997). The mechanism by which the RpoS sigma factor influences *fimB* gene transcription is unknown and may be indirect. Although usually playing a positive role in transcriptional control in stationary phase cultures, a negative role for RpoS in gene regulation has been reported before (Nystrom, 1994; O'Neal *et al.*, 1994; Xu & Johnson, 1995). FimB expression is also controlled post-transcriptionally; in *E. coli* K-12 strains, the *fimB* gene contains five codons for the rare $tRNA_5^{Leu}$ encoded by the *leuX* gene (Klemm, 1986). The levels of the *leuX* gene product are thought to govern the rate at which *fimB* mRNA is translated (Newman *et al.*, 1994; Ritter *et al.*, 1995). Growth on solid or in liquid medium can also influence expression of the *fim* recombinase genes,

Table 1. *Properties of factors which affect* fim *gene expression*

Regulatory factor	Description	Effect on *fim* expression	Specific to *fim* gene regulation?
FimB	Recombinase	Promotes ON-to-OFF and OFF-to-ON inversion	Yes
FimE	Recombinase	Promotes ON-to-OFF inversion	Yes
IHF	Bends DNA	Helps promote inversion; enhances *fimA* promoter	No
LRP	Bends DNA	Helps promote inversion	No
H-NS	Binds curved DNA	Inhibits inversion; inhibits *fimA*, *fimB*, *fimE* promoters	No
StpA	Homologue of H-NS	Inhibits inversion	No
DNA topo-isomerase I	Relaxes negatively supercoiled DNA	Facilitates ON-to-OFF and OFF-to-ON inversion	No
DNA gyrase	Negatively supercoils DNA; relaxes positive supercoils	Retards OFF-to-ON inversion	No
$tRNA_3^{Leu}$	Rare tRNA	Regulates *fimB* mRNA translation	No

with some strains of *E. coli* showing differential transcription of the *fimB* and *fimE* genes under these two growth conditions (Schwan, Seifert & Duncan, 1992).

Temperature modulates the ability of the recombinases to invert the *fimA* promoter segment. At low temperatures, the FimE protein is more effective at switching expression to Phase-OFF than at 37°C; conversely, FimB activity increases with temperature, and has an optimum at or just above 37°C (Gally *et al.*, 1993). The *fimA* promoter is also sensitive to temperature and is more active at 37°C than at 30°C. This promoter is also subject to repression by the H-NS protein (as are the *fimB* and *fimE* promoters) (Dorman & Ní Bhriain, 1992). The RpoS sigma factor represses transcription of the *fimA* gene, in addition to that of the *fimB* gene (the *fimE* gene has not been tested), as the cell enters stationary phase.

Clearly, *fimA* gene expression is under complex, multifactorial control (Table 1). Several of the factors which influence its expression are responsive to changes in the physiology of the cell or the external environment or both. In addition to conventional positive and negative regulation, the *fimA* gene is controlled in a stochastic manner via phase variation. The latter determines whether or not the gene is capable of being expressed while the former ensures that, if it is, it will be expressed optimally for the environment in which the bacterium finds itself at that particular time. The regulatory mechanisms governing expression of the *fim* genes illustrate clearly that stereotypic and stochastic control mechanisms do not operate in isolation but can combine to influence the expression of one system. Still lacking at the time of writing is a complete, detailed picture of how these diverse regulatory inputs come together at the molecular level to optimize *fim* gene expression.

ACKNOWLEDGEMENTS

We thank Cyril J. Smyth for a critical reading of the manuscript and for many helpful suggestions. Work on *fim* gene expression is supported in our laboratory by Forbairt Grant SC/96/301 (Republic of Ireland) and by Wellcome Trust Grant 046233/Z/95/Z (United Kingdom).

REFERENCES

Abraham, J. M., Freitag, C. S., Clements, J. R. & Eisenstein, B. I. (1985). An invertible element of DNA controls phase variation of type 1 fimbriae of *Escherichia coli*. *Proceedings of the National Acadamy of Sciences, USA*, **82**, 5724–7.

Argos, P., Landy, A., Abremski, K., Egan, J. B., Haggard-Ljungquist, E., Hoess, R. H., Khan, M. L., Kalionis, B., Narayana, S. V. L., Pierson III, L. S., Sternberg, N. & Leong, J. M. (1986). The integrase family of site-specific recombinases: regional similarities and global diversity. *EMBO Journal*, **5**, 433–40.

Aviv, M., Giladi, H., Schreiber, G., Oppenheim, A. B. & Glaser, G. (1994). Expression of the genes coding for the *Escherichia coli* integration host factor are controlled by growth phase, *rpoS*, ppGpp and by autoregulation. *Molecular Microbiology*, **14**, 1021–31.

Bertin, P., Lejeune, P., Laurent, W. C. & Danchin, A. (1990). Mutations in *bglY*, the structural gene for the DNA-binding protein H1, affect expression of several *Escherichia coli* genes. *Biochimie*, **72**, 889–91.

Blakely, G. W. & Sherratt, D. J. (1996). *Cis* and *trans* in site-specific recombination. *Molecular Microbiology*, **20**, 234–7.

Bloch, C. A., Stocker, B. A. & Orndorff, P. E. (1992). A key role for type 1 pili in enterobacterial communicability. *Molecular Microbiology*, **6**, 697–701.

Blomfield, I. C., Calie, P. J., Eberhardt, K. J., McClain, M. S. & Eisenstein, B. I. (1993). LRP stimulates phase variation of type 1 fimbriation in *Escherichia coli* K-12. *Journal of Bacteriology*, **175**, 27–36.

Blomfield, I. C., McClain, M. S., Princ, J. A., Calie, P. J. & Eisenstein, B. I. (1991). Type 1 fimbriation and *fimE* mutants of *Escherichia coli* K-12. *Journal of Bacteriology*, **173**, 5298–307.

Braaten, B. A., Platko, J. V., van der Woude, M., de Graaf, F. K., Calvo, J. M. & Low, D. A. (1992). Leucine responsive regulatory protein (Lrp) controls the expression of both the Pap and Fan pili operons in *Escherichia coli*. *Proceedings of the National Academy of Sciences, USA*, **89**, 4250–4.

Bushman, W., Thompson, J. F., Vargas, L. & Landy, A. (1985). Control of directionality in lambda site specific recombinations. *Science*, **230**, 906–11.

Calvo, J. M. & Matthews, R. G. (1994). The leucine-responsive regulatory protein, a global regulator of metabolism in *Escherichia coli*. *Microbiological Reviews*, **58**, 466–90.

Chen, D., Bowater, R., Dorman, C. J. & Lilley, D. M. J. (1992). Activity of a plasmid-borne *leu500* promoter depends on the transcription of an adjacent gene. *Proceedings of the National Academy of Sciences, USA*, **89**, 8784–8.

Cockerill, P. N. & Garrard, W. T. (1986). Chromosomal loop anchorage of the kappa

immunoglobulin gene occurs next to the enhancer in a region containing topoisomerase II sites. *Cell* **44**, 273–82.
Dersch, P., Schmidt, K. & Bremer, E. (1993). Synthesis of the *Escherichia coli* K-12 nucleoid-associated DNA-binding protein H-NS is subjected to growth-phase control and autoregulation. *Molecular Microbiology*, **8**, 875–89.
Dorman, C. J. (1995). DNA topology and the global control of bacterial gene expression: implications for the regulation of virulence gene expression. *Microbiology*, **141**, 1271–80.
Dorman, C. J. & Higgins, C. F. (1987). Fimbrial phase variation in *Escherichia coli*: dependence on integration host factor and homologies with other site-specific recombinases. *Journal of Bacteriology*, **169**, 3840–3.
Dorman, C. J. & Ní Bhriain, N. (1992). Thermal regulation of *fimA*, the *Escherichia coli* gene coding for the type 1 fimbrial subunit protein. *FEMS Microbiology Letters*, **99**, 125–30.
Dove, S. L. & Dorman, C. J. (1994). The site-specific recombination system regulating expression of the type 1 fimbrial subunit gene of *Escherichia coli* is sensitive to changes in DNA supercoiling. *Molecular Microbiology*, **14**, 975–88.
Dove, S. L. & Dorman, C. J. (1996). Multicopy *fimB* gene expression in *Escherichia coli*: binding to inverted repeats *in vivo*, effect on *fimA* gene transcription and DNA inversion. *Molecular Microbiology*, **21**, 1161–73.
Dove, S. L., Smith, S. G. J. & Dorman, C. J. (1997). Control of *Escherichia coli* type 1 fimbrial gene expression in stationary phase: a negative role for RpoS. *Molecular and General Genetics*, in press.
Dröge, P. (1993). Transcription-driven site-specific DNA recombination *in vitro*. *Proceedings of the National Academy of Sciences, USA*, **90**, 2759–63.
Durrenberger, M., La Teana, A., Citro, G., Venanzi, F., Gualerzi, C. O. & Pon, C. L. (1991). *Escherichia coli* DNA-binding protein H-NS is localised in the nucleoid. *Research in Microbiology*, **142**, 373–80.
Dybvig, K. (1993). DNA rearrangements and phenotypic switching in prokaryotes. *Molecular Microbiology*, **10**, 465–71.
Eisenstein, B. I. (1987). Fimbriae. In Neidhardt, F. C., Ingraham, J. L., Low, K. B., Magasanik, B., Schaechter, M. & Umbarger, H. E., eds. Escherichia coli *and* Salmonella typhimurium: *Cellular and Molecular Biology*, pp. 84–90. Washington, DC: American Society for Microbiology.
Eisenstein, B. I., Sweet, D., Vaughn, V. & Friedman, D. I. (1987). Integration host factor is required for the DNA inversion event that controls phase variation in *Escherichia coli*. *Proceedings of the National Academy of Sciences, USA*, **84**, 6506–10.
Figuero, N. & Bossi, L. (1988). Transcription induces gyration of the DNA template in *Escherichia coli*. *Proceedings of the National Academy of Sciences, USA*, **85**, 9416–20.
Flamm, E. L. & Weisberg, R. A. (1985). Primary structure of the *hip* gene of *Escherichia coli* and of its product, the β subunit of integration host factor. *Journal of Molecular Biology*, **183**, 117–28.
Free, A. & Dorman, C. J. (1995). Coupling of *Escherichia coli hns* mRNA levels to DNA synthesis by autoregulation: implications for growth phase control. *Molecular Microbiology*, **18**, 101–13.
Freundlich, M., Ramani, N., Mathew, E., Sirko, A. & Tsui, P. (1992). The role of integration host factor in gene expression in *Escherichia coli*. *Molecular Microbiology*, **6**, 2557–63.
Friedman, D. I. (1988). Integration host factor: a protein for all reasons. *Cell*, **55**, 545–54.

Friedrich, K., Gualerzi, C. O., Lammi, M., Losso, M. A. & Pon, C. L. (1988). Proteins from the prokaryotic nucleoid. Interaction of nucleic acids with the 15 kDa *Escherichia coli* histone-like protein H-NS. *FEBS Letters*, **229**, 197–202.

Gally, D. L., Bogan, J. A., Eisenstein, B. I. & Blomfield, I. C. (1993). Environmental regulation of type 1 fimbrial phase variation in *Escherichia coli* K-12: effects of temperature and media. *Journal of Bacteriology*, **175**, 6186–93.

Gally, D. L., Rucker, T. J. & Blomfield, I. C. (1994). The leucine-responsive regulatory protein binds to the *fim* switch to control phase variation of type 1 fimbrial expression in *Escherichia coli* K-12. *Journal of Bacteriology*, **176**, 5665–72.

Gally, D. L., Leathart, J. & Blomfield, J. C. (1996). Interaction of FimB and FimE with the *fim* switch that controls the phase variation of type 1 fimbriae in *Escherichia coli* K-12. *Molecular Microbiology*, **21**, 725–38.

Goodman, S. D. & Nash, H. A. (1989). Functional replacement of a protein-induced bend in a DNA recombination site. *Nature*, **341**, 251–4.

Goosen, N. & van de Putte, P. (1995). The regulation of transcription initiation by integration host factor. *Molecular Microbiology*, **16**, 1–7.

Higgins, C. F., Dorman, C. J., Stirling, D. A., Waddell, L., Booth, I. R., May, G. & Bremer, E. (1988). A physiological role for DNA supercoiling in the osmotic regulation of gene expression in *S. typhimurium* and *E. coli*. *Cell*, **52**, 569–84.

Hromockji, A. E., Tucker, S. C. & Maurelli, A. T. (1992). Temperature regulation of *Shigella* virulence: identification of the repressor gene *virR*, an analogue of *hns*, and partial complementation by tyrosyl transfer RNA ($tRNA_1^{tyr}$). *Molecular Microbiology*, **6**, 2113–24.

Huisman, T. T., Bakker, D., Klaasen, P. & de Graaf, F. K. (1994). Leucine-responsive regulatory protein, IS*1* insertions, and the negative regulator FaeA control the expression of the *fae* (K88) operon in *Escherichia coli*. *Molecular Microbiology*, **11**, 525–36.

Jordi, B. J. A. M., Dagberg, B., de Haan, A. A. M., van der Zeijst, B. A. M., Gaastra, W. & Uhlin, B. E. (1992). The positive regulator CfaD overcomes the repression mediated by histone-like protein H-NS (H1) in the CFA/I fimbrial operon of *Escherichia coli*. *EMBO Journal*, **11**, 2627–32.

Kawula, T. H. & Orndorff, P. E. (1991). Rapid site-specific DNA inversion in *Escherichia coli* mutants lacking the histonelike protein H-NS. *Journal of Bacteriology*, **173**, 4116–23.

Klemm, P. (1986). Two regulatory *fim* genes, *fimB* and *fimE*, control the phase variation of type 1 fimbriae in *Escherichia coli*. *EMBO Journal*, **5**, 1389–93.

Klemm, P. (1994). Type 1 fimbriae of *Escherichia coli*. In Klemm, P., ed. *Fimbriae: Adhesion, Genetics, Biogenesis, and Vaccines*, pp. 9–26. London, CRC Press.

Kohara, Y., Akiyama, K. & Isono, K. (1987). The physical map of the whole *E. coli* chromosome: application of a new strategy for rapid analysis and sorting of a large genomic library. *Cell*, **50**, 495–508.

Kosturko, L. D., Daub, E. & Murialdo, H. (1989). The interaction of *E. coli* integration host factor and λ *cos* DNA: multiple complex formation and protein-induced bending. *Nucleic Acids Research*, **17**, 317–34.

Kukkonen, M., Raunio, T., Virkola, R., Lähteenmäki, K., Mäkelä, P. H., Klemm, P., Clegg, S. & Korhonen, T. K. (1993). Basement membrane carbohydrate as a target for bacterial adhesion: binding of type 1 fimbriae of *Salmonella enterica* and *Escherichia coli* to laminin. *Molecular Microbiology*, **7**, 229–37.

Lammi, M., Paci, M., Pon, C. L., Lesso, M. A., Miano, A., Pawlik, R. T. Gianfranceschi, G. L. & Gualerzi, C. O. (1984). Proteins from the prokaryotic nucleoid: biochemical and ^{1}H-NMR studies on three bacterial histone-like proteins. *Advances in Experimental Medical Biology*, **179**, 467–77.

Lee, M.S. & Garrard, W. T. (1991). Transcription-induced nucleosome 'splitting': an underlying structure for DNase sensitive chromatin. *EMBO Journal*, **10**, 607–15.

Leung, H. & Maizels, N. (1992). Transcriptional regulatory elements stimulate recombination in extrachromosomal substrates carrying immunoglobulin switch-region sequences. *Proceedings of the National Academy of Sciences, USA*, **89**, 4154–8.

Liu, L. F. & Wang, J. C. (1987). Supercoiling of the DNA template during transcription. *Proceedings of the National Academy of Sciences, USA*, **84**, 7024–7.

May, A. K., Bloch, C. A., Sawyer, R. G., Spengler, M. D. & Pruett, T. L. (1993). Enhanced virulence of *Escherichia coli* bearing a site-targeted mutation in the major structural subunit of type 1 fimbriae. *Infection and Immunity*, **61**, 1667–73.

May, G., Dersch, P., Haardt, M., Middendorf, A. & Bremer, E. (1990). The *osmZ (bglY)* gene encodes the DNA binding protein H-NS (H1a), a component of the *Escherichia coli* K-12 nucleoid. *Molecular and General Genetics*, **224**, 81–90.

McClain, M. S., Blomfield, I. C., Eberhardt, K. J. & Eisenstein, B. I. (1993). Inversion-independent phase variation of type 1 fimbriae in *Escherichia coli. Journal of Bacteriology*, **175**, 4335–44.

McClain, M. S., Blomfield, I. C. & Eisenstein, B. I. (1991). Roles of *fimB* and *fimE* in site-specific DNA inversion associated with phase variation of type 1 fimbriae in *Escherichia coli. Journal of Bacteriology*, **173**, 5308–14.

Miller, H. I. (1984). Primary structure of the *himA* gene of *Escherichia coli*: homology with DNA-binding protein HU and association with the phenylalanine-tRNA synthetase operon. *Cold Spring Harbor Symposia on Quantitative Biology*, **49**, 691–8.

Miller, H. I. & Friedman, D. I. (1980). An *E. coli* gene product required for λ site-specific recombination. *Cell*, **20**, 711–19.

Moxon, E. R., Rainey, P. B., Nowak, M. A. & Lenski, R. E. (1994). Adaptive evolution of highly mutable loci in pathogenic bacteria. *Current Biology*, **4**, 24–33.

Nash, H. & Robertson, C. A. (1981). Purification and properties of the *E. coli* protein factor required for lambda integrative recombination. *Journal of Biological Chemistry*, **256**, 9246–53.

Newman, E. B. & Lin, R. (1995). Leucine-responsive regulatory protein: a global regulator of gene expression in *Escherichia coli. Annual Review of Microbiology*, **49**, 747–75.

Newman, J. V., Burghoff, R. L., Pallesen, L., Krogfelt, K. A., Kristensen, C. S., Laux, D. C. & Cohen, P. S. (1994). Stimulation of *Escherichia coli* F-18Col$^-$ type 1 fimbriae synthesis by *leuX. FEMS Microbiology Letters*, **122**, 281–7.

Nystrom, T. (1994). Role of guanosine tetraphosphate in gene expression and the survival of glucose or seryl-tRNA starved cells of *Escherichia coli. Molecular and General Genetics*, **245**, 355–62.

Olsen, P. B. & Klemm, P. (1994). Localization of promoters in the *fim* gene cluster and the effect of H-NS on the transcription of *fimB* and *fimE. FEMS Microbiology Letters*, **116**, 95–100.

O'Neal, C. R., Gabriel, W. M., Turk, A. K., Libby, S. J., Fang, F. C. & Spector, M. P. (1994). RpoS is necessary for both positive and negative regulation of starvation survival genes during phosphate, carbon, and nitrogen starvation in *Salmonella typhimurium. Journal of Bacteriology*, **176**, 4610–16.

Orndorff, P. E. & Bloch, C. A. (1990). The role of type I pili in the pathogenesis of *Escherichia coli* infections: a short review and some new ideas. *Microbial Pathogenesis*, **9**, 75–9.

Platko, J. V. & Calvo, J. M. (1993). Mutations affecting the ability of *Escherichia coli*

Lrp to bind DNA, activate transcription, or respond to leucine. *Journal of Bacteriology*, **175** 1110–17.

Porter, M. E. & Dorman, C. J. (1994). A role for H-NS in the thermo-osmotic regulation of virulence gene expression in *Shigella flexneri*. *Journal of Bacteriology*, **176**, 4187–91.

Rahmouni, A. R. & Wells, R. D. (1992). Direct evidence for the effect of transcription on local DNA supercoiling *in vivo*. *Journal of Molecular Biology*, **223**, 131–44.

Ritter, A., Blum, G., Emböd y, L., Kereni, M., Böck, A., Neuhierl, B., Rabsch, W., Scheutz, F. & Hacker, J. (1995). tRNA genes and pathogenicity islands: influence on virulence and metabolic properties of uropathogenic *Escherichia coli*. *Molecular Microbiology*, **17**, 109–21.

Robertson, B. D. & Meyer, T. F. (1992). Genetic variation in pathogenic bacteria. *Trends in Genetics*, **8**, 422–7.

Saunders, J. R. (1995). Population genetics of phase variable antigens. In Baumberg, S., Young, J. P. W., Wellington, E. M. H. & Saunders, J. R., eds. *Population Genetics of Bacteria*, Symposium Volume no. 52 of the Society for General Microbiology, pp. 247–68. Cambridge: Cambridge University Press.

Schaeffer, A. J., Schwan, W. R., Hultgren, S. J. & Duncan, J. L. (1987). Relationship of type 1 pilus expression in *Escherichia coli* to ascending urinary tract infections in mice. *Infection and Immunity*, **55**, 373–80.

Schwan, W. R., Seifert, H. S. & Duncan, J. L. (1992). Growth conditions mediate differential transcription of *fim* genes involved in phase variation of type I pili. *Journal of Bacteriology*, **174**, 2367–75.

Schwedler, U., Jäck, H.-M. & Wabl, M. (1990). Circular DNA is a product of the immunoglobulin class switch rearrangement. *Nature*, **345**, 452–6.

Smyth, C. J. (1986). Fimbrial variation in *Escherichia coli*. In Birkbeck, T. H. & Penn, C. W., eds. *Antigenic Variation in Infectious Diseases*, Special publications of the Society for General Microbiology, Volume no. 19, pp. 95–125. Oxford: IRL Press.

Smyth, C. J., Marron, M. & Smith, S. G. J. (1994). Fimbriae of *Escherichia coli*. In Gyles, C. L., ed. Escherichia coli *in Domestic Animals and Humans*, pp. 399–435. Wallingford: CAB International.

Smyth, C. J. & Smith, S. G. J. (1992). Bacterial fimbriae: variation and regulatory mechanisms. In Hormaeche, C. E., Penn, C. W. & Smyth, C. J., eds. *Molecular Biology of Bacterial Infection: Current Status and Future Perspectives*, Symposium Volume no. 49 of the Society for General Microbiology, pp. 267–97. Cambridge: Cambridge University Press.

Snyder, U. K., Thompson, J. F. & Landy, A. (1989). Phasing of protein-induced DNA bends in a recombination complex. *Nature*, **341**, 255–7.

Spassky, A., Rimsky, S., Gareau, H. & Buc, H. (1984). H1a, an *E. coli* DNA-binding protein which accumulates in stationary phase, strongly compacts DNA *in vitro*. *Nucleic Acids Research*, **12**, 5321–40.

Spears, P. A., Schauer, D. A. & Orndorff, P. E. (1986). Metastable regulation of type 1 piliation in *Escherichia coli* and isolation and characterization of a phenotypically stable mutant. *Journal of Bacteriology*, **168**, 179–85.

Tobe, T., Yoshikawa, M., Mizuno, T. & Sasakawa, C. (1993). Transcriptional control of the invasion regulatory gene *virB* of *Shigella flexneri*: activation by VirF and repression by H-NS. *Journal of Bacteriology*, **175**, 6142–9.

Tsao, Y.-P., Wu, H.-Y. & Liu, L. F. (1989). Transcription-driven supercoiling of DNA: direct biochemical evidence from *in vitro* studies. *Cell*, **56**, 111–18.

Tupper, A. E., Owen-Hughes, T. A., Ussery, D. W., Santos, D. S., Ferguson, D. J. P., Sidebotham, J. M., Hinton, J. C. D. & Higgins, C. F. (1994). The chromatin-associated protein H-NS alters DNA topology *in vitro*. *EMBO Journal*, **13**, 258–68.

Ueguchi, C., Kakeda, M. & Mizuno, T. (1993). Autoregulatory expression of the *Escherichia coli hns* gene encoding a nucleoid protein: H-NS functions as a repressor of its own transcription. *Molecular and General Genetics*, **236**, 171–8.

Ueguchi, C. & Misuno, T. (1993). The *Escherichia coli* nucleoid protein H-NS functions directly as a transcriptional repressor. *EMBO Journal*, **12**, 1039–46.

Ussery, D. W., Hinton, J. C. D., Jordi, B. J. A. M., Granum, P. E., Seirafi, A., Stephen, R. J., Tupper, A. E., Berridge, G., Sidebotham, J. M. & Higgins, C. F. (1994). The chromatin-associated protein H-NS. *Biochimie*, **76**, 968–80.

van Bogelan, R. A., Sankar, P., Clark., R. L., Bogan, J. A. & Neidhardt, F. C. (1992). The gene-protein database of *Escherichia coli*. Edition 5. *Electrophoresis*, **13**, 1014–54.

van der Woude, M. W. & Low, D. A. (1994). Leucine responsive regulatory protein and deoxyadenosine methylase control the phase variation and expression of the *E. coli sfa* and *daa* pili operons. *Molecular Microbiology*, **11**, 605–18.

Varshavsky, A. J., Nedospasov, S. A., Bakayev, V. V., Bakayeva, T. G. & Georgiev, G. P. (1977). Histone-like proteins in the purified *Escherichia coli* deoxyribonucleoprotein. *Nucleic Acids Research*, **4**, 2725–45.

Wang, Q. & Calvo, J. M. (1993). Lrp, a major regulatory protein in *Escherichia coli*, bends DNA and can organise the assembly of a higher order nucleoprotein structure. *EMBO Journal*, **12**, 2495–501.

Wang, Z. & Dröge, P. (1996). Differential control of transcription-induced and overall DNA supercoiling by eukaryotic topoisomerases *in vitro*. *EMBO Journal*, **15**, 581–9.

Williams, R. M., Rimsky, S. & Buc, H. (1996). Probing the structure, function and interactions of the *Escherichia coli* H-NS and StpA proteins by using dominant negative derivatives. *Journal of Bacteriology*, **178**, 4335–4343.

Wu, H.-Y., Shyy, S., Wang, J. C. & Liu, L. F. (1988). Transcription generates positively and negatively supercoiled domains in the template. *Cell*, **53**, 433–40.

Xu, J. & Johnson, R. C. (1995). Identification of genes negatively regulated by Fis: Fis and RpoS comodulate growth-phase-dependent gene expression in *Escherichia coli*. *Journal of Bacteriology*, **177**, 938–47.

Xu, L., Gorham, B., Li, S. C., Bottaro, A., Alt, F. W. & Rothman, P. (1993). Replacement of germline ε promoter by gene targeting alters control of immunoglobulin heavy chain class switching. *Proceedings of the National Academy of Sciences, USA*, **90**, 3705–9.

Yamada, H., Yoshida, T., Tanaka, K., Sasakawa, C. & Mizuno, T. (1991). Molecular analysis of the *Escherichia coli hns* gene encoding a DNA-binding protein, which preferentially recognizes curved DNA sequences. *Molecular and General Genetics*, **230**, 332–6.

Zhang, A., Derbyshire, V., Galloway Salvo, J. L. & Belfort, M. (1995). *Escherichia coli* protein StpA stimulates self-splicing by promoting RNA assembly *in vitro*. *RNA*, **1**, 783–93.

Zhang, A., Rimsky, S., Reaban, M. E., Buc, H. & Belfort, M. (1996). *Escherichia coli* protein analogues StpA and H-NS: regulatory loops, similar and disparate effects on nucleic acid dynamics. *EMBO Journal*, **15**, 1340–9.

VIRUSES AND THE PROTEIN SYNTHESIS MACHINERY OF THE CELL: OFFENCE, DEFENCE AND DEPENDENCE

M. B. MATHEWS

Cold Spring Harbor Laboratory, PO Box 100, Cold Spring Harbor, New York, NY 11724, USA[1]

INTRODUCTION

Viruses lack the ability to conduct most metabolic and biosynthetic reactions, relying instead on the cells that they infect to supply the building blocks and much of the machinery required for viral replication. With respect to protein synthesis, this dependence is almost total since, with the exception of some tRNAs, viruses do not encode any part of the translational apparatus. Viruses do not simply coexist with the cellular protein synthetic machinery, however; rather, they co-opt it for their own purposes. Many of them exploit mechanisms and control circuits in ways that in some cases imitate cellular processes, and in others appear to be unique to viruses. Furthermore, some viruses usurp the translational system to the extent that viral protein synthesis predominates in the infected cell, or subvert it to neutralize cellular anti-viral defences. Such interactions can play an important part in determining the virulence of a viral infection. This chapter summarizes the principal interactions between viruses and the translation system of the cell.

General considerations

The site of viral genome replication within the cell (in the nuclear or the cytoplasmic compartment) is a fundamental characteristic that appears to be determined by the viruses' strategy for mRNA production. Viruses which depend on cellular transcription enzymes replicate in the nucleus. This group includes all the DNA viruses except for vaccinia, together with the retroviruses (whose genomes go through a chromosomally integrated DNA phase) and influenza virus (which pirates the capped 5′ end of nuclear mRNA precursors as primers for viral transcription). All other viruses replicate in the cytoplasm, which they are equipped to do by virtue of replicases that they encode or import in the virion. Vaccinia virus can

[1] *Present address*: Department of Biochemistry and Molecular Biology, UMDNJ – New Jersey Medical School, 185 South Orange Avenue, Newark, NJ 07103-2714, USA.

replicate in the cytoplasm because its virions contain viral enzymes for RNA synthesis and modification, making it self-sufficient for transcription. Among the RNA viruses (retroviruses and influenza viruses excepted), those whose genomes are plus strands (e.g. picornaviruses) generate the requisite enzymes directly by translation; those whose genomes are double-stranded (e.g. reovirus) or of negative polarity (e.g. vesicular stomatitis virus, VSV) package RNA-dependent RNA polymerases in their virions, permitting them to generate plus strands and mRNA. It seems that cytoplasmic replication confers an advantage, because no virus that is equipped for replication in this compartment has opted to replicate in the nucleus: possibly access to the protein synthesis machinery is the determining factor.

Many viruses, especially DNA viruses, generate products in an orderly fashion and temporal sequence which is determined largely through transcriptional controls, but in some cases by regulation at the translational level. Possibly the clearest examples of temporal switches operating at the translational level are afforded by RNA viruses whose genomes serve directly as mRNAs. For instance, synthesis of the replicase protein of the single-stranded RNA phages (f2, R17, MS2, Qβ, etc.), which begins shortly after infection, is curtailed well before the synthesis of the other cistrons has peaked (van Duin, 1988). The translation of replicase is first activated by translation of the coat protein cistron and then switched off by the coat protein itself. Protein synthesis in poliovirus-infected cells exemplifies a different kind of early/late transition. An early decline in amino acid incorporation is followed by a later upswing due to translation of the increasing number of viral templates. The initial inhibition of host protein synthesis is attributable to proteolysis of initiation factor eIF4G (previously p220), catalysed by a virus-coded enzyme whose primary role is to cleave the viral polyprotein (see below).

Since viruses are very diverse, their interactions with the translation system are correspondingly varied; consequently, a phenomenon apparent in one virus-cell combination does not always occur with a different, albeit closely related, virus or in another cell type. This inconsistency is particularly manifest in the phenomenon of host cell shutoff, whereby many viruses interfere with the production, maturation, or stability of cellular macromolecules in the later stages of a productive infection. Together with the increasing rate of synthesis of viral products, host shutoff contributes to the viral domination of cellular biosynthetic pathways that is seen in many infections. Although widespread, shutoff is not ubiquitous and is not a single phenomenon with a unique mechanism. Related viruses may employ different mechanisms to accomplish the same end, or even do without shutoff altogether, as illustrated by members of the picornavirus family. Shutoff usually requires viral protein synthesis but may be independent of it because the factor responsible is introduced by the infecting virions (e.g. vaccinia and herpesviruses). Moreover, shutoff is not tightly linked to the abundant synthesis of viral proteins: in poliovirus infected cells, shutoff precedes the appearance of high levels of viral proteins, whereas the reverse is seen with adenovirus. Finally, virus production is not necessarily more efficient when shutoff occurs than in comparable infections, with another cell line or virus strain, where it does not (see Kozak, 1986, 1992). Indeed, shutoff during adenovirus infection has less to do with the accumulation of viral proteins than with the release of completed virions from degenerating cells (Zhang & Schneider, 1994). In a few cases, the selective translation of viral mRNA has been traced to modifications of host cell translational components, as discussed below. In most cases, however, the picture remains more nebulous and mRNA competition has been held responsible for the phenomenon, possibly aided by ionic changes favouring viral mRNA translation (Nuss, Opperman & Koch, 1975; Alonso & Carrasco, 1981).

Regardless of the advantages that might accrue from monopolizing cellular resources, viruses need to keep their hosts alive and functioning so that the viral life cycle can be completed. Accordingly, many viruses produce proteins that promote cell functions, including translation (see below), and shutoff must be a delicately balanced affair. Moreover, infection does not lead inexorably to virus multiplication and cell death: other possible outcomes include abortive, persistent and latent infections and oncogenesis. Whether, and to what extent, a virus multiplies depends on its virulence and the permissivity of the cell, both of which are multifactorial properties operating at many levels. Correspondingly, there exists a spectrum of pathogenicity ranging from acute to chronic effects. The more subtle interactions of virus and cell are common but difficult to study from the perspective of translational control. In such infections, it is assumed that one or more permissivity factors are missing or limiting: in some cases, for example, in VSV infections (Schmidt, Gravel & Woodland, 1995), such factors appear to operate at the translational level. This should be a fertile field for investigation in the future.

Redeployment of the translational apparatus by viruses

Despite their dependence on the cellular apparatus for translating their mRNAs, viruses also appropriate parts of the host apparatus for other purposes. Some examples of the ways in which components of the translational machinery are co-opted by viruses and pressed into service in other functions are outlined here.

RNA phages

The enzyme that replicates the genome of bacteriophage Qβ and its relatives is composed of five subunits, one of which is the product of the viral replicase gene while four are host encoded (Kamen, 1975; van Duin, 1988). Three of the proteins contributed by the bacterial cell have long been identified with components of the translational system, namely, elongation factors EF1A and EF1B (formerly EF-T_u and EF-T_s) and the ribosomal protein S1. The fifth protein, HF-I is also ribosome associated and has recently been shown to be involved in translational initiation of at least one bacterial mRNA (Muffler, Fischer & Henegge-Aronis, 1996). Abduction of translational proteins by the phage seems to serve two purposes: it provides the replicase with necessary biochemical functions and also couples replication to translation.

Plant viruses

Many plant viral RNAs, including those of tobacco mosaic virus and cucumber mosaic virus, carry tRNA-like sequences at their 3′ ends. These can interact with tRNA-specific enzymes, the best known being aminoacyl-tRNA synthetases (Haenni, Joshi & Chapeville, 1982). The structure at the 3′ end of brome mosaic virus RNA is required for minus strand RNA replication as well as for charging with tyrosine (Dreher & Hall, 1988), raising the possibility that cellular enzymes of protein synthesis and tRNA metabolism are co-opted into the replicase complexes of these plant viruses also.

Epstein–Barr virus

Epstein–Barr virus is an oncogenic herpesvirus. The most abundant viral transcript found in human B lymphocytes transformed by this virus is a short non-coding RNA, EBER-1. EBER-1 is synthesized by RNA polymerase III and is found in association with two cellular proteins, the La antigen (Lerner *et al.*, 1981) and ribosomal protein L22 (Toczyski *et al.*, 1994), a component of the large ribosomal subunit. As much as half of the cells' L22 is present in these ribonucleoprotein particles, which seem to reside in the nucleoplasmic compartment. Although the significance of these particles is unknown, two suggestive observations have been made. First, EBER-1 can bind to, and inactivate, the protein kinase PKR, whose best known substrate is the initiation factor eIF2 (Clarke, Sharp & Clemens, 1990; see below); dominant negative mutants of PKR can transform cells and render them tumourigenic (Koromilas *et al.*, 1992; Meurs *et al.*, 1993; Clemens, 1996). Secondly, the gene for L22 is the target of a chromosomal translocation in some leukaemia patients (Nucifora *et al.*, 1993).

Retroviruses

The fourth example of redeployment involves RNA rather than protein as the cellular component. Host cell tRNA plays an essential role in the conversion of retroviral genomes into DNA by the viral enzyme reverse transcriptase (Coffin, 1996). Both the enzyme and the tRNA are associated with the nucleocapsid in the virus particle and the 3′ end of the tRNA serves as primer for the first step of reverse transcription. A second priming step results in the assembly of viral sequences that comprise the long terminal repeat (LTR), which performs numerous essential functions during the virus life cycle. Each virus (or viral subgroup) commandeers a specific tRNA for this purpose (for example, the primer for HIV-1 and other lentiviruses is a lysine acceptor), and at least five different tRNA species are thus exploited for retroviral replication.

Unorthodox viral translational tactics

Although they are wholly dependent on the host translational apparatus, viruses provide many exceptions to entrenched concepts in the field of translation. At first glance this may seem paradoxical because fundamental mechanisms of protein synthesis were in large measure established using viral systems, but as several of the unconventional gambits first recognized in viral systems have also been found to operate in cellular systems they may turn out to be less unorthodox than they now seem. None the less, it is difficult to avoid the impression that viruses employ 'tricks' more often than do their host cells, as part of a regulatory strategy or as a way to expand their coding capacity within a confined genome. These unorthodox tactics, depicted in Fig. 1, represent departures from the orderly and sequential readout of an mRNA by conventional scanning and decoding from the 5′ end. Signals that are ordinarily respected are bypassed or ignored, either in whole or in part; such behaviour is dictated by overriding or additional signals which specify deviations from the standard mechanism.

Internal ribosome entry

The entry of ribosomes to mRNAs at internal sites constitutes a radical departure from scanning (Kozak, 1989). It occurs in response to a signal known as the internal ribosome entry sequence (IRES) found in all picornaviruses as well as a growing number of other viruses and some cellular mRNAs (Jang *et al.*, 1988; Pelletier & Sonenberg, 1988; Ehrenfeld, 1996; Jackson, 1996). An IRES typically extends over some 500 nucleotides in the 5′ untranslated region (UTR) of the mRNA and involves higher-order structure. Cellular proteins recognize features of the IRES and are apparently sufficient for its function, although the poliovirus protease 2A plays a stimulatory role (Hambridge & Sarnow, 1992). Once it has become associated with the mRNA, the ribosome may initiate immediately after the IRES, or it may scan briefly to a nearby downstream initiation codon.

IRES-mediated initiation confers three potential advantages. First, ribosomes can ignore upstream sequences that are irrelevant to translation but play other roles in the viral life cycle, so that sequences containing AUG triplets and secondary structure can be tolerated in the 5′ end of these viral mRNAs. Second, the presence of a 5′ cap structure is irrelevant for IRES-

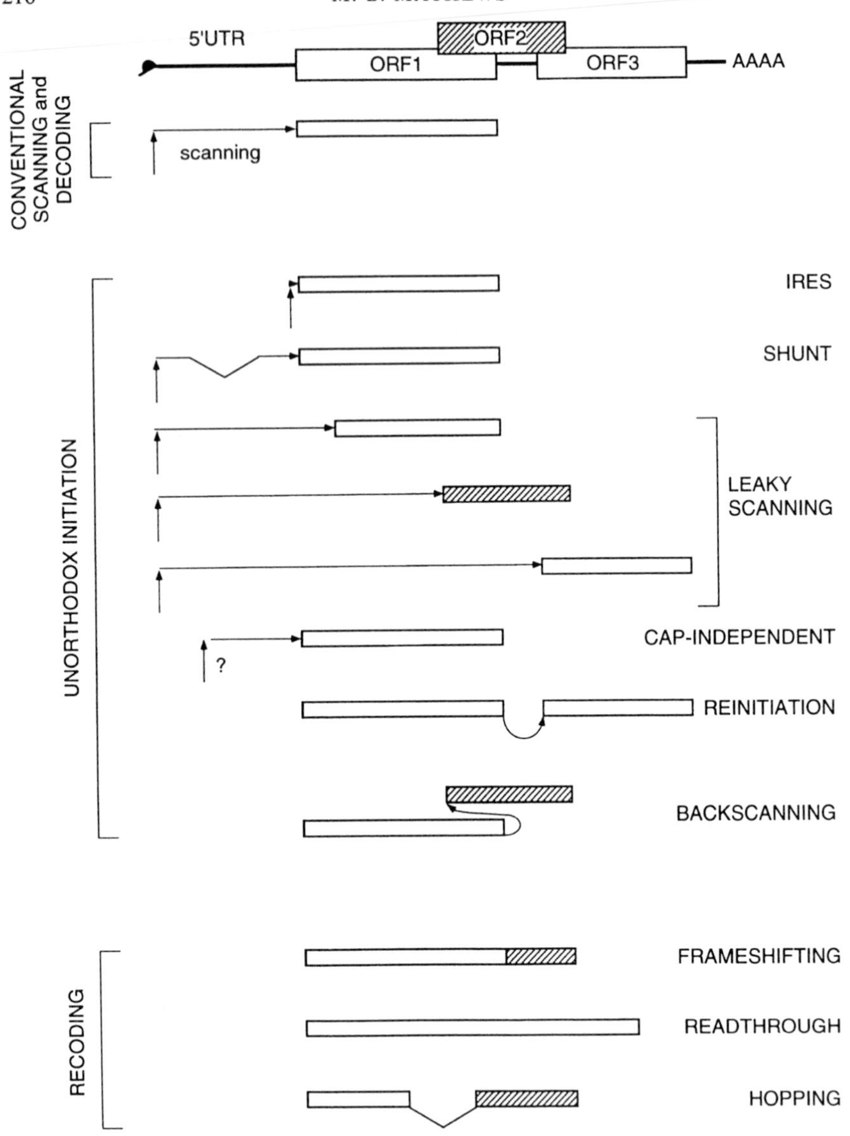

Fig. 1. Unconventional translational strategies. The top line represents a capped and poly-adenylated mRNA with a 5′ untranslated region (5′ UTR) and three open reading frames (ORFs 1–3). Various strategies for initiating protein synthesis on this mRNA and for decoding its information are depicted on the lines below. (From Mathews, 1996, with permission.)

mediated initiation, relieving the virus from the need to replicate in the nucleus or provide its own capping enzyme. Third, although the same complement of initiation factors is required, the IRES element frees the mRNA which contains it from dependence on the integrity of initiation factor eIF4G, providing the basis for host protein synthesis shutoff (see below).

Ribosome shunting

Another mechanism for avoiding primary and secondary structure in the 5′UTR is the ribosome shunt (Jackson, 1996). First characterized in cauliflower mosaic virus (CaMV) 35S RNA (Fütterer, Kiss-László & Hohn, 1993), shunting has recently been shown to operate on adenovirus RNAs containing the tripartite leader (Yueh & Schneider, 1996). Shunting allows ribosomes to scan for a short distance from the 5′ terminus of the mRNA, then skip several hundred nucleotides to a site near the 3′ border of the UTR without scanning through the intervening sequences. The determinants of the signal are not yet established, but, in CaMV, the 5′UTR has a considerable degree of secondary structure and the efficiency of shunting seems to depend on host factors. No viral protein is essential, but the product of the viral ORF VI gene, TAV, is stimulatory (Bonneville *et al.*, 1989). Like an IRES, the shunt permits ribosomes to avoid 5′ UTR sequences and structure (even large open reading frames (ORFs)) which are incompatible with conventional scanning. Unlike the IRES, however, the shunt would not be expected to alleviate the 5′ cap and eIF4G requirements.

Leaky scanning

The leaky scanning hypothesis accommodates mRNAs in which a downstream AUG is used for initiation in preference or in addition to the first (i.e. 5′ proximal) AUG. Up to 10% of eukaryotic mRNAs are in this class, and many of these encode two proteins which may be in the same or different reading frames (Geballe, 1996; Jackson, 1996). One of the principal features defining the initiation site on an mRNA is the context in which the initiation codon is embedded. The most favourable context is **ACC<u>AUG</u>G**, where the A at position −3 and the G at position +4 (relative to the A of the initiator AUG triplet, assigned +1) are the chief determinants of a strong initiation site (Kozak, 1989). In dicistronic mRNAs, the first AUG is often in a suboptimal context, suggesting that it functions as a weak initiator which would be bypassed at a substantial frequency according to the scanning model, allowing for initiation at a subsequent site downstream.

Leaky scanning is widely used in viruses where it presumably helps to economize on coding space and signals for transcription and RNA processing. In HIV-1, for example, the essential envelope protein (Env) is translated from mRNAs which contain an upstream ORF encoding the accessory protein Vpu in a different reading frame. To permit Env synthesis, the *vpu* initiation site is required to be weak (Schwartz, Felber & Pavlakis, 1992). When the two ORFs are in-frame with each other, the result is a nested pair of proteins with overlapping C-terminal sequences and related functions (e.g. SV40 coat proteins VP2 and VP3). On the other hand, when the two ORFs are in different reading frames, the resultant proteins need not be functionally related (e.g. HIV-1 Vpu and Env) although they sometimes are (e.g. adenovirus-5 E1B 19kD and 55kD proteins). Leaky scanning offers numerous regulatory possibilities: for example, in the reovirus S1 RNA ribosomes translating one ORF may impede those translating the other (Fajardo & Shatkin, 1990; Belli & Samuel, 1993).

Translation of uncapped RNA

The RNAs of some plant viruses, such as satellite tobacco necrosis virus (Wimmer *et al.*, 1968), and of the yeast virus L-A (Nemeroff & Bruenn, 1987; Masison *et al.*, 1995) are exceptional in lacking 5′ cap and 3′ poly A structures. Unlike picornavirus RNA, which is also not capped, they function without the benefit of an IRES element. In the case of L-A, this depends on interactions with products of the cellular SKI genes (see below). Judging from experiments with an artificial uncapped RNA encoding HIV-1 Tat, initiation involves a scanning mechanism despite the absence of a cap (Gunnery & Mathews, 1995).

Reinitiation of translation

Since scanning and 5′ caps are specific to eukaryotes, the foregoing events apply exclusively to viruses that infect eukaryotic cells. Reinitiation provides another means whereby two proteins can be made from a single mRNA. Although rare in eukaryotes, it is common in prokaryotes where internal ribosome entry, rather than scanning, is standard. Prokaryotic cellular and viral mRNAs frequently contain non-overlapping ORFs in tandem, and their ribosomes can generally gain access to all of the initiation sites within a prokaryotic polycistronic mRNA, allowing for several proteins to be translated independently. In phage ϕX174, for example, two proteins are translated from the same sequence using separate ribosome binding sites that give access to different reading frames (Ravetch, Model & Robertson, 1977). The individual sites are not necessarily utilized with equal efficiencies, however, and in several cases, the translation of a downstream cistron depends upon the translation of an upstream cistron (see below).

Evidence that reinitiation can be used by eukaryotes to gain access to downstream ORFs in a dicistronic mRNA is presently limited to work with artificially constructed gene fusions (Peabody, Subramani & Berg, 1986). In the natural eukaryotic cases that have been studied most intensively, the upstream ORFs are generally short and often appear to be regulatory in nature, typically exerting a negative influence over the translation of the downstream ORF (for review see Geballe, 1996; Hinnebusch, 1996). For example, the 22-residue peptide produced by translating an upstream ORF in the cytomegalovirus gp48 mRNA is thought to act by blocking the scanning process (Cao & Geballe, 1995). On the other hand, an upstream ORF in CaMV 35S RNA exerts a positive effect on the translation of downstream

ORFs which action requires the presence of the viral transactivator protein TAV (Fütterer & Hohn, 1992).

Translation frameshifting

During the decoding of some mRNAs, the advancing ribosome slips forward or back by one nucleotide, resulting in programmed +1 or −1 change of reading frame (Jacks & Varmus, 1985; Atkins and Gesteland, 1996). Such frameshifting events are common in retroviruses, for many of which a −1 shift is an essential event in reverse transcriptase synthesis. The shift takes place at a 'slippery' site, which is generally followed by a second element, either a pseudoknot or a hairpin positioned downstream, that presumably acts by causing ribosomes to pause at the shift site. The proportion of ribosomes changing reading frames is characteristic of each site, and is usually rather low; hence, the products comprise a majority of the conventionally decoded polypeptide and a minority of the 'recoded' form. In retroviruses, the Gag-Pol shifted product is usually about 5% of the unshifted Gag product. Most retroviruses translate *pol* this way, and similar events occur in coronaviruses and yeast L-A virus. There is little evidence that the proportion of frameshifting at a particular site is controlled, so the mechanism seems to be a device for increasing the coding capacity of the viral genome and for producing the two products in a fixed ratio. The latter may be of critical importance, since a small increase or decrease in the Gag to Gag-Pol ratio is detrimental to virus assembly and proliferation (Dinman & Wickner, 1992; Karacostas *et al.*, 1993).

Read-through

Suppression or read-through of stop codons is also used by many viruses to generate a C-terminally extended protein at a fixed ratio to the conventionally translated product. Examples are found in viruses infecting plants, bacteria, and mammals (e.g. the Moloney murine leukemia virus *gag-pol* protein) and rarely in cellular mRNAs (for review see Atkins & Gesteland, 1996).

Hopping

During the translation of T4 gene 60 mRNA, 50 nucleotides are bypassed. This extraordinary event is mediated by the nascent peptide and requires duplications flanking the bypassed sequence, as well as a stop codon (Weiss, Huang & Dunn, 1990).

Regulation of viral initiation site availability and utilization

Ribosome access to translational initiation sites can be regulated via RNA:RNA and RNA:protein interactions. Such mechanisms are common in bacteriophage systems, where they subserve autoregulatory roles. mRNA secondary structure, particularly in the region of the ribosome binding site, is a major determinant of translational efficiency in bacteria (deSmit & van Duin, 1990). The degree of secondary structure is subject to regulation by ribosomes (serving as activators of translation) and by *trans*-acting proteins (which generally function as translational repressors). Proteins binding in the vicinity of the initiation site can also influence ribosome binding directly or indirectly (Gold, 1988; McCarthy & Gualerzi, 1990). Both of these mechanisms are practised by phages to regulate the translation of their own genes.

Translational coupling

The initiation site is frequently engaged in higher-order structure that wholly or partly obscures an essential feature such as the AUG or the Shine–Dalgarno sequence, thereby restricting initiation. In some viral and cellular mRNAs, the cistrons are arranged such that the restriction is lifted by ribosomes traversing a different cistron (usually, but not necessarily, upstream). This mechanism is termed translational coupling; in essence, translating ribosomes act as derepressors. In the best-known case, the synthesis of the replicase protein of the RNA phages depends on translation of the coat protein cistron which lies upstream, as evidenced by the observation that an amber mutation early in the coat protein gene exerts a polar effect on replicase synthesis whereas a downstream amber mutation does not (Lodish & Zinder, 1966; Lodish, 1975). The effect is due to long-range base-pairing between a sequence in the coat protein cistron and nucleotides immediately upstream of the replicase AUG. The interaction restricts access to the replicase initiation site and is lifted by elongating ribosomes which disrupt the base-pairing.

More often, the interactions occur over a short range and involve terminating, rather than elongating, ribosomes. For example, the ORF encoding the lysis peptide of the RNA phages f2 and MS2 overlaps the 3′ end of the coat protein cistron and it has been proposed that lysis protein initiation makes use of ribosomes that have just finished synthesizing coat protein. In phage f1, where gene VII is translationally coupled to gene V, the gene VII initiation site is incapable of directly binding 30S ribosomal subunits; instead, these are supplied by the upstream gene V (Ivey-Hoyle & Steege, 1989, 1992). Similar cases are suspected in eukaryotic viruses but have not yet been subjected to such rigorous analysis: a possible occurrence in the reovirus S1 transcript was mentioned above; another is in parainfluenza virus P/C mRNA (Boeck *et al.*, 1992), and further candidates are discussed in Geballe (1996).

Repression and activation

Whereas translation of the coat protein cistron of the RNA phages is responsible for upregulating replicase synthesis, the coat protein itself subsequently downregulates replicase synthesis. Coat protein binds to a stem–loop structure that encompasses the replicase AUG initiation site, thereby excluding ribosomes and blocking initiation by obscuring both the Shine–Dalgarno sequence and the AUG codon (Witherell, Gott & Uhlenbeck, 1991). Similar principles apply to several translational repressors found in phage T4 which, in different ways, all interfere with the formation of initiation complexes on their mRNA targets (Gold, 1988). While most such mRNA:protein interactions known to date are negative, phage Mu furnishes a precedent for a positive translational effector. Translation of the Mom protein, a DNA modification enzyme, depends on the Com protein encoded by the same operon. The Mom initiation site is structured such that the AUG and part of the Shine–Dalgarno sequence are sequestered in a stem. Binding of the Com protein to a site upstream of the cryptic Mom initiation site causes a conformational change which enables ribosome access (Hattman *et al.*, 1991; Wulczyn & Kahmann, 1991).

No such translational activators or repressors have been definitively identified in eukaryotic viruses to date, but one candidate is the adenovirus 100K protein that accumulates to high levels in the cytoplasm at late times of infection. This protein is an RNA-binding protein, although no specificity has yet been demonstrated. A mutation in the protein leads to a defect in the translation of late viral mRNAs without deterring the translation of early viral mRNAs or the shutoff of host protein synthesis in the late phase (Hayes *et al.*, 1990). Positive translational functions have also been ascribed to two RNA-binding proteins of HIV-1, Tat (Cullen, 1986; SenGupta *et al.*, 1990) and Rev (Arrigo & Chen, 1991; D'Agostino *et al.*, 1992; Campbell *et al.*, 1994), as well as to CaMV TAV: these activities merit further investigation.

Modifications of the translational apparatus by viruses and host shutoff

Another class of regulatory mechanism controls the activity of cellular translational components and can influence the translation of both viral and cellular mRNAs. Such modifications, which are prominent with viruses that infect eukaryotes, play important roles in the shutoff of host cell protein synthesis. Shutoff at the translational level generally entails two events: the virus places itself at an advantage by imposing a limitation upon the translation system, concomitantly creating a bypass so that it escapes the limitation. Four kinds of virus-induced covalent modifications of translational components are presently known to take place during infection, three in eukaryotes and one in prokaryotes.

tRNA cleavage

The T-even phages encode tRNAs and engender a large number of alterations in the bacterial protein synthetic apparatus (Mosig & Eiserling, 1988): accordingly, tRNA cleavage reactions are the most conspicuous of the modifications that take place in T4-infected *E. coli*. The destruction of host tRNAs accentuates the dependence of protein synthesis on T4-derived mRNA. For example, the $tRNA^{Leu}$ species that recognizes the codon CUG, rare in T4 mRNAs, is cleaved soon after infection. Phage T4 encodes eight tRNAs which seem to be advantageous because they serve codons that are frequent in T4 mRNAs but rare in host mRNAs, or because they provide a bypass around a virus-induced lesion. Thus, in some *E. coli* strains, the cellular $tRNA^{Ile}$ is cleaved in its anticodon loop and is functionally replaced by the phage-encoded isoaccepting species. Such changes are part of a widespread alteration of the cellular machinery that takes place after T4 infection to the benefit of the phage. In contrast, the cleavage of $tRNA^{Lys}$, which is not covered by a phage gene, is part of a cellular defense mechanism (the *prr* exclusion system, discussed below).

eIF2 phosphorylation

Eukaryotic initiation factor 2, eIF2, serves to transport the initiator tRNA to the 40S ribosomal particle, in the form of a ternary complex (eIF2•GTP•Met-$tRNA_i$), and its activity is modulated by phosphorylation (Clemens, 1996; Merrick & Hershey, 1996; Trachsel, 1996). Eukaryotes possess three kinases that can inhibit initiation by phosphorylating eIF2, one of which is the double-stranded RNA-activated inhibitor (PKR, DAI, dsI or P1). This enzyme is widespread in higher cells and is intimately linked with the host response to viral infection. Phosphorylation of eIF2 curtails its function by trapping the GTP exchange factor (eIF2B) which is required for eIF2 to recycle, resulting in ternary complex depletion. Since eIF2B is less abundant than eIF2, phosphorylation of a fraction of the eIF2 ($\sim$30%) can sequester all of the recycling factor and lead to a complete block to protein synthesis. This outcome may contribute to the interferon-induced anti-viral response (see below). A lesser degree of eIF2 phosphorylation is observed in cells infected with many different viruses (see Kozak, 1986, 1992), and is believed to lead to host shutoff. Support for this idea has been drawn from the observation that host cell shutoff does not take place when PKR-deficient cells are infected with adenovirus (O'Malley *et al.*, 1989; Huang & Schneider, 1990). Although the basis for the implied translational selectivity is not clear, two possibilities have been entertained. The first suggests that viral mRNAs are intrinsically more efficient or abundant, and hence less sensitive to a reduction in the effective concentration of initiation factor; the second supposes that the inhibitory effect of eIF2 phosphorylation is compartmentalized in the cell, so that the host mRNAs are preferentially inhibited while the viral mRNAs are spared (Schneider, 1996).

eIF4E dephosphorylation and binding

The entry of mRNA into the initiation pathway is mediated by factors in the eIF4 group (Merrick & Hershey, 1996; Sonenberg, 1996). In many circumstances, the cap-binding protein, eIF4E, is the rate-limiting initiation factor. It binds to the cap structure at the 5′ end of viral and cellular mRNAs. It also complexes with two other factors, eIF4A and eIF4G, to form the cap-binding complex, eIF4F, which with eIF4B facilitates unwinding of secondary structure in the 5′ end of the mRNA. The eIF4F complex then catalyses the binding of the mRNA to the 40S ribosomal subunit. Phosphorylation of eIF4E correlates with its increased activity in the initiation pathway, possibly because of an elevated affinity for the 5′ cap and for other components of the eIF4F complex (Sonenberg, 1996). In cells infected with a number of viruses, including adeno- and influenza viruses, the extent of eIF4E phosphorylation falls (Huang & Schneider, 1991; Feigenblum & Schneider, 1993). Such dephosphorylation is believed to contribute to host cell shutoff by placing weak cellular mRNAs at a disadvantage when they are competing against strong viral mRNAs. The strong adenoviral and influenza virus mRNAs presumably have relatively little secondary structure in their 5′ UTRs and a correspondingly low requirement for eIF4 activity (Schneider, 1996; Katze, 1996). The activity of initiation factor eIF4E can also be modulated by a protein called 4E-BP1; in its under-phosphorylated form, 4E-BP1 is a repressor of eIF4E function (Sonenberg, 1996). Recent work has demonstrated that the phosphorylation of 4E-BP1 decreases concomitantly with the inhibition of host protein synthesis in encephalomyocarditis virus-infected cells, suggesting that translation of the uncapped viral RNA is favoured by diminished cap-binding activity (Gingras *et al.*, 1996).

eIF4G cleavage

eIF4G, the largest subunit of eIF4F, is cleaved into two fragments in cells infected with certain picornaviruses (Ehrenfeld, 1996; Jackson, 1996). Cleavage is mediated by proteolytic enzymes that process the primary viral translation product, protease 2A (in the enteroviruses and rhinoviruses) and protease L (in foot and mouth disease virus). Cleavage of eIF4G effectively separates it into two domains; an N-terminal part which interacts with eIF4E, and a C-terminal part which interacts with eIF4A and eIF3 (Lamphear *et al.*, 1995; Mader *et al.*, 1995). As a result, cap-dependent initiation is severely inhibited. Since the IRES-dependent initiation mechanism is unaffected or even facilitated (Ohlmann, Rau & Pain, 1995), eIF4G cleavage sets these picornavirus mRNAs at an advantage. The proteases of other picornavirus groups (the cardioviruses and hepatoviruses) do not cleave eIF4G, but the cardioviruses at least have contrived a different mechanism (4E-BP1 dephosphorylation) to achieve protein synthesis shutoff as outlined above.

Cellular defences and viral countermeasures

Higher organisms possess enzymes that act at the cellular level as first line defenses against viral infection. These include the eIF2 kinase PKR and a ribonuclease, RNase L, which both affect protein synthesis. They are expressed constitutively at a low level, from which they can moderate virus infection, and they play key roles in the anti-viral state established by interferon. In response, viruses generate products that attenuate or neutralize the impact of these defensive enzymes on translation. Although interferon is restricted to vertebrates, comparable systems must be widespread in nature since cellular defence mechanisms are emplaced in organisms as distant as *E. coli* and yeast; moreover, as described below, their viruses are also armed with appropriate countermeasures.

Translational inhibition in higher cells

The defence system composed of PKR and RNase L, together with 2′,5′ oligoadenylate synthetase (2–5A synthetase, a regulator of RNase L), is present in untreated, uninfected cells, but is neither fully mobilized in the absence of interferon induction nor fully armed without virus infection. Interferon synthesis is induced by virus infection, especially by RNA viruses, and by the exposure of cells to other stimuli such as dsRNA (Gilmour & Reich, 1995). The cytokine diffuses to adjacent cells and triggers the transcriptional activation of over 30 genes whose products establish an anti-viral state during which viral replication may be blocked at a number of levels. In many cases, infection follows a normal course up to, and including, the synthesis of viral mRNA, but this mRNA does not become stably associated with polysomes as a result of the actions of PKR and RNase L. Both of these enzymes are found at relatively low levels in many uninduced cells and their synthesis is induced by interferon at the transcriptional level (Laurent *et al.*, 1985; Zhou, Hassel & Silverman, 1993). The 2–5A synthetases are also highly inducible by interferon treatment (Kerr, Brown & Hovanessian, 1977).

PKR and RNase L are present in uninfected cells mainly in an inactive (latent) form, and both are activated by dsRNA although via different mechanisms, as depicted in Fig. 2. PKR activation occurs as a direct response to dsRNA, possibly of viral origin (Maran & Mathews, 1988; Gribaudo *et al.*, 1991), and is accompanied by autophosphorylation. The enzyme has two copies of a dsRNA-binding motif (dsRBM); the binding of dsRNA is thought to cause a conformational change in the enzyme which unmasks its kinase activity. Autophosphorylation then might allow PKR to phosphor-

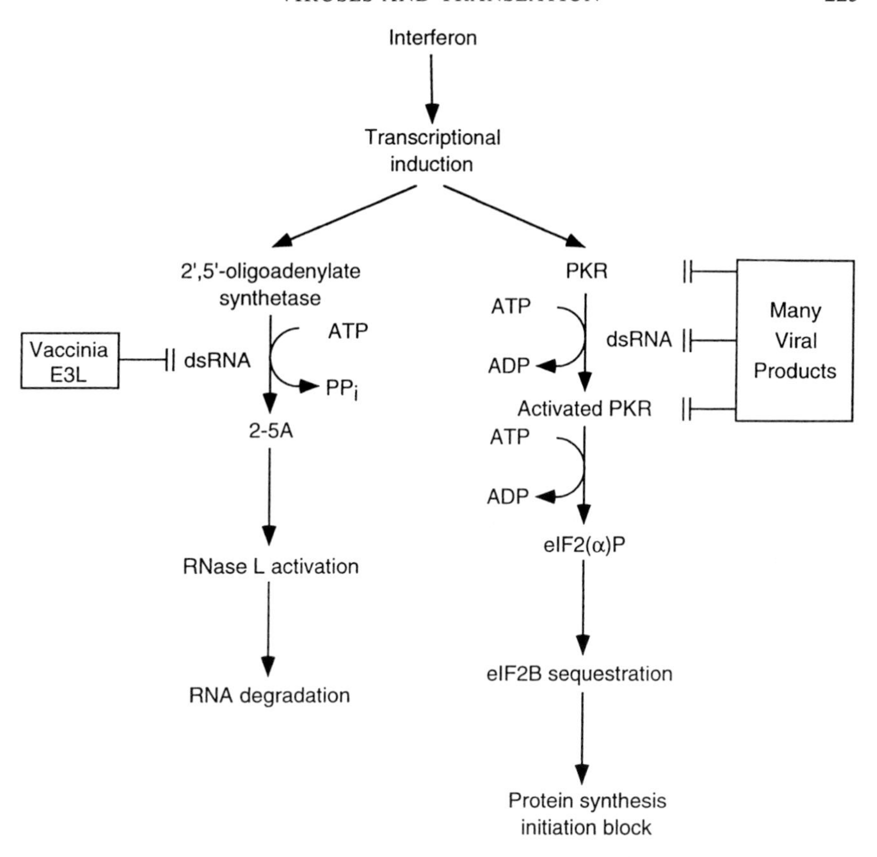

Fig. 2. Translational inhibition by interferon-induced pathways. The flowchart summarizes the pathways for induction and activation of PKR and RNase L, and the result of their activation. Also indicated are positions in the pathways where viral products can intercede to overcome translational inhibition. (From Mathews, 1996, with permission.)

ylate eIF2, resulting in the inhibition of initiation (Manche *et al.*, 1992; Green, Manche & Mathews, 1995; Clemens, 1996). The activation of RNase L occurs by a more indirect route. Although they do not appear to contain a dsRBM (Patel & Sen, 1992), the 2–5A synthetases are activated by dsRNA, producing a series of short, 2′–5′ linked oligoadenylates of the form pppA(pA)$_n$. These specifically activate RNase L which degrades RNA, chiefly by cutting at the 3′ side of UpUp and UpAp sequences.

The anti-viral actions of 2–5A synthetases are principally directed against picornaviruses and possibly vaccinia virus, whereas PKR affects a broader range of viruses (Samuel, 1988; Staeheli, 1990). For example, constitutive expression of 2–5A synthetase confers resistance to mengovirus but not to VSV or herpesvirus type 2 (Chebath *et al.*, 1987; Coccia *et al.*, 1990), whereas overexpression of PKR inhibits the replication of vaccinia and encephalomyocarditis viruses but not of VSV (Lee & Esteban, 1993; Meurs *et al.*, 1992).

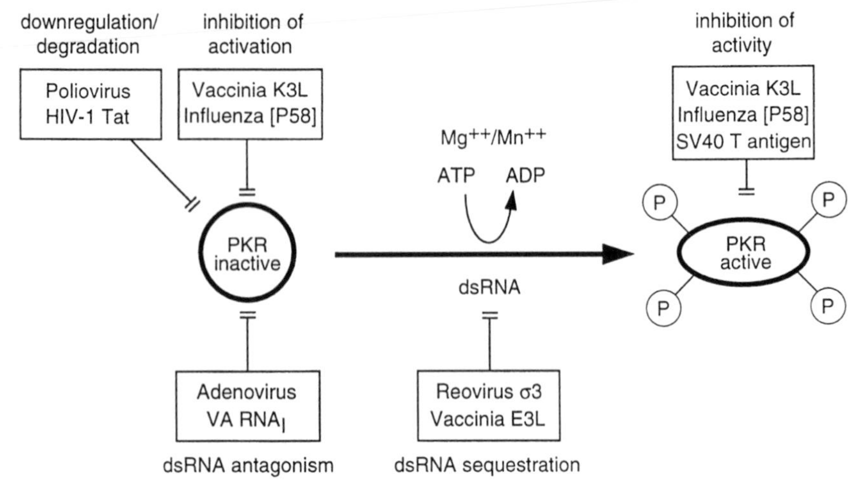

Fig. 3. The activation and inhibition of PKR. The activation of PKR and dsRNA-mediated autophosphorylation, and the interference with its activation and activity by viral products are shown schematically. (From Mathews, 1996, with permission.)

Persuasive support for the anti-viral roles of these enzymes can be inferred from the study of viral countermeasures (described below). For example, the absence of VA RNA$_I$ (an adenoviral PKR antagonist) is more deleterious to viral multiplication in interferon-treated cells than in untreated cells (Kitajewski *et al.*, 1986). Therefore, the uninduced levels of these enzymes confer partial resistance to virus infection, and interferon treatment enables cells to mount a more effective response.

Viral countermeasures against PKR

Most viruses induce interferon and many of them produce dsRNA that can activate PKR and the 2–5A synthetase/RNase L system. PKR in particular poses a serious threat to viral multiplication, judging by the number of mechanisms that viruses have elaborated to counteract its effects on protein synthesis. These countermeasures are diverse, involving viral RNAs and proteins as well as host proteins, and PKR function is inhibited at different levels as illustrated in Fig. 3 (Sonenberg, 1990; Mathews & Shenk, 1991; Mathews, 1993; Katze, 1995, 1996; Schneider, 1996).

Sequestration of dsRNA

Both vaccinia and reoviruses produce proteins that bind dsRNA, rendering it unavailable to activate PKR and 2–5A synthetase (Katze, 1996; Schneider, 1996). The vaccinia virus protein, E3L, is identical with a specific kinase inhibitory factor (SKIF) detected in infected cell extracts. Its dsRNA-binding capability is due to a single copy of the dsRBM that is found as a tandem repeat in PKR. Vaccinia virus mutants lacking E3L exhibit increased sensitivity to interferon, and cells infected with them display high activity of PKR and the 2–5A synthetase/RNase L system. The product of the reovirus S4 gene, $\sigma 3$, also binds dsRNA tightly although it lacks a dsRBM. The $\sigma 3$ protein inhibits PKR activation and can partially substitute for vaccinia E3L or adenovirus VA RNA_I; it also can counter the effects of interferon (Beattie *et al.*, 1995).

Inhibition by dsRNA analogues

The transcription of adenovirus by RNA polymerase III produces VA RNA_I, a small RNA that accumulates to high concentrations in the cytoplasm at late times of infection. It imitates dsRNA in that it interacts with the dsRNA-binding region of PKR, but instead of causing activation of the kinase, the binding of VA RNA prevents PKR activation. Critical features of the RNA molecule are a stem that ensures efficient binding to PKR, and a region of complex tertiary structure including a pseudoknot that blocks activation (Ma & Mathews, 1996). VA RNA_I enhances virus multiplication, confers interferon-resistance, and may be involved in host shutoff (Mathews & Shenk, 1991; Schneider, 1996). Other viruses also produce small RNAs that have the capacity to inhibit PKR, but their roles in virus infection remain to be established. The EBERs of Epstein–Bar virus, which resemble the VA RNAs in many respects, and HIV-1 TAR RNA are in this category.

Inhibition of kinase function

In addition to E3L, described above, vaccinia virus encodes a second protein that interacts with PKR. This protein, K3L, appears to act as a pseudosubstrate for the kinase. It blocks the activation of PKR as well as the activity of the kinase once activated, and has been shown to counteract the effects of PKR and interferon *in vivo* (Schneider, 1996). Influenza virus-infected cells also contain an inhibitor of PKR activation and activity called P58 (Katze, 1996). P58 is a cellular protein and its capacity to block PKR is masked in uninfected cells. SV40 T antigen antagonizes PKR activity, though not its activation (Rajan *et al.*, 1995). HIV-1 Tat inhibits PKR activation and also serves as a substrate for the enzyme (Maitra *et al.*, 1994; S. R. Brand, R. Kobayaski, M. B. Mathews, unpublished observations), but the physiological importance of this observation remains to be elucidated.

Downregulation of PKR levels

Two viruses reduce the concentration of the kinase. Poliovirus seems to destabilize PKR (Black *et al.*, 1989), as well as eIF4G (see above). Cells expressing the *tat* gene or infected with HIV-1 virus contain reduced amounts of PKR, although the mRNA for 2–5A synthetase is unaffected (Roy *et al.*, 1990).

Host–virus interactions in yeast and bacterial cells

Systems with interesting parallels to those induced by interferon have recently been uncovered in yeast and bacteria.

Bacteriophage T4 and the *prr* exclusion system

Some *E. coli* strains carry a non-essential operon, *prr*, which confers resistance to phage T4 infection by blocking protein synthesis at the elongation stage. The *prr* locus contains four ORFs, one of which (*prrC*) encodes a specific endonuclease called anticodon nuclease. This enzyme inactivates the cellular $tRNA^{Lys}$ by cleaving it at the 5′ border of its anticodon (Penner *et al.*, 1995; Shterman *et al.*, 1995). In uninfected bacteria, the nuclease is present in a latent state, masked by the products of the other three genes of the *prr* locus (*prr A, B* and *D*). It is unmasked by a short polypeptide produced by the phage T4 *stp* gene. Thus, the phage Stp polypeptide brings about activation of the cellular anticodon nuclease, resulting in tRNA cleavage, inhibition of late T4 protein synthesis, and aborting the infection. Phage T4 is equipped with countermeasures to meet this cellular response to infection. It encodes polynucleotide kinase and RNA ligase which repair the severed tRNA and restore the cell to translational competence. In formal terms, the *prr* system is analogous to the PKR system: a viral product activates a latent cellular enzyme, resulting in the modification of a component of the translation apparatus; moreover, in both cases viruses have elaborated products to neutralize the threats posed by the imminent shutdown of protein synthesis.

Yeast L-A virus and the SKI system

The SKI1 gene is one of a group of genes that regulates the 'superkiller' phenotype observed in yeast carrying the L-A virus. As noted above, the mRNA of this virus is uncapped and non-polyadenylated. Mutations in the SKI genes allow enhanced replication of L-A and of its satellite virus, M_1, which produces a secreted toxin (killer toxin). The functions of the SKI genes are exerted at least partly at the level of protein synthesis (Masison *et al.*, 1995). SKI1 is an essential gene, recently identified with *XRN1* whose product is a 5′–3′ exoribonuclease specific for uncapped RNA; mutants in this gene permit the accumulation of elevated concentrations of uncapped L-A virus mRNA (as well as uncapped cellular mRNA fragments). Thus, the wild-type SKI1 nuclease is an antiviral protein, as well as an enzyme in the normal RNA degradative pathway (Masison *et al.*, 1995). In the case of L-A virus, the countermeasure is mediated by the coat protein of L-A virus, Gag, which removes the cap structure from mRNA, leaving an uncapped mRNA (Blanc, Goyer & Sonenberg, 1992; Blanc *et al.*, 1994). The uncapped RNA can then serve as a substrate or 'decoy' for the exonuclease, diverting it away from the L-A mRNA which is therefore spared destruction. This scheme is reminiscent of the 2–5A synthetase/RNase L system, which is blocked by the vaccinia K3L protein (albeit by a different mechanism).

CONCLUDING REMARKS

Viruses exhibit ingenious and versatile strategies to regulate, modify, and exploit the host protein synthetic machinery, and to defend themselves against changes in it that compromise their replication. Further examples and details can be found in the recent review chapter (Mathews, 1996) and references cited therein. In their transactions with the cell, viruses apparently strive to capitalize on the full potential of the translational system, or perhaps even to increase its range. An appreciation of the underlying mechanisms is important because it enhances our understanding of viruses and of the translation process; furthermore, it may also prove relevant in a broader sphere since virus/host cell differences are potential points of therapeutic intervention.

ACKNOWLEDGEMENTS

The author is supported by grants from the National Cancer Institute and National Institute of Allergy and Infectious Diseases, NIH.

REFERENCES

Alonso, M. A. & Carrasco, L. (1981). Reversion by hypotonic medium of the shutoff of protein synthesis induced by encephalomyocarditis virus. *Journal of Virology*, **37**, 535.

Arrigo, S. J. & Chen, I. S. Y. (1991). Rev is necessary for translation but not cytoplasmic accumulation of HIV-1 vif, vpr, and env/vpu-2 RNAs. *Genes and Development*, **5**, 808–19.

Atkins, J. F. & Gesteland, R. F. (1996). Regulatory recoding. In Hershey, J. W. B., Mathews, M. B. and Sonenberg, N., eds. *Translational Control*, p. 653. Cold Spring Harbor Laboratory Press, Cold Spring Harbor, NY.

Beattie, E., Denzler, K. L., Tartaglia, J., Perkus, M. E., Paoletti, E. & Jacobs, B. L. (1995). Reversal of the interferon-sensitive phenotype of a vaccinia virus lacking E3L by expression of the reovirus S4 gene. *Journal of Virology*, **69**, 499–505.

Belli, B. A. & Samuel, C. E. (1993). Biosynthesis of reovirus-specified polypeptides: identification of regions of the bicistronic reovirus S1 mRNA that affect the efficiency of translation in animal cells. *Virology*, **193**, 16–27.

Black, T. L., Safer, B., Hovanessian, A. & Katze, M. (1989). The cellular 68,000-Mr protein kinase is highly autophosphorylated and activated yet significantly degraded during poliovirus infection: implications for translational regulation. *Journal of Virology*, **63**, 2244–51.

Blanc, A., Goyer, C. & Sonenberg, N. (1992). The coat protein of the yeast double-stranded RNA virus L-A attaches covalently to the cap structure of eukaryotic mRNA. *Molecular and Cellular Biology*, **12**, 3390–8.

Blanc, A., Ribas, J. C., Wickner, R. B. & Sonenberg, N. (1994). His-154 is involved in the linkage of the *Saccharomyces cerevisiae* L-A double-stranded RNA virus Gag protein to the cap structure of mRNAs and is essential for M_1 satellite virus expression. *Molecular and Cellular Biology*, **14**, 2664–74.

Boeck, R., Curran, Y., Matsuoka, R., Compans, R. & Kolakofsky, D. (1992). The parainfluenza virus type 1 P/C gene uses a very efficient GUG codon to start its C′ protein. *Journal of Virology*, **66**, 1765–8.

Bonneville, J. M., Sanfacon, H., Fütterer, J. & Hohn, T. (1989). Postranscriptional *trans*-activation in cauliflower mosaic virus. *Cell*, **59**, 1135–43.

Campbell, L. H., Borg, K. T., Haines, J. K., Moon, R. T., Schoenberg, D. R. & Arrigo, S. J. (1994). Human immunodeficiency virus type 1 rev is required *in vivo* for binding of poly(A)-binding protein to rev-dependent RNAs. *Virology*, **68**, 5433–8.

Cao, J. & Geballe, A. P. (1995). Translational inhibition by a human cytomegalovirus upstream open reading frame despite inefficient utilization of its AUG codon. *Journal of Virology*, **69**, 1030–6.

Chebath, J., Benech, P., Revel, M. & Vigneron, M. (1987). Constitutive expression of (2′–5′) oligo A synthetase confers resistance to picornavirus infection. *Nature*, **330**, 587–8.

Clarke, P. A., Sharp, N. A. & Clemens, M. J. (1990). Translational control by the Epstein–Barr virus small RNA EBER-1. Reversal of the double-stranded RNA-induced inhibition of protein synthesis in reticulocyte lysates. *European Journal of Biochemistry*, **193**, 635–41.

Clemens, M. J. (1996). Protein kinases that phosphorylate eIF2 and eIF2B, and their role in eukaryotic cell translational control. In Hershey, J. W. B., Mathews, M. B. & Sonenberg, N., eds. *Translational Control*, p. 139. Cold Spring Harbor Press, Cold Spring Harbor, NY.

Coccia, E. M., Romeo, G., Nissim, A., Marziali, G., Albertini, R., Affabris, E., Battistini, A., Fiorucci, G., Orsatti, R., Rossi, G. B. & Chebath, J. (1990). A full-length murine 2–5A synthetase cDNA transfected in NIH-3T3 cells impairs EMCV but not VSV replication. *Virology*, **179**, 228–33.

Coffin, J. M. (1996). Retroviridae: the viruses and their replication. In Fields, B. N., Knipe, D. M., Howley, P. M., Chanock, R. M., Melnick J. L., Monath, T. P., Roizman, B. & Straus, S. E., eds. *Fields Virology*, p. 1767. Lippincott-Raven Publishers, Philadelphia, USA.

Cullen, B. R. (1986). Trans-activation of human immunodeficiency virus occurs via a bimodal mechanism. *Cell*, **46**, 973–82.

D'Agostino, D. M., Felber, B. K., Harrison, J. E. & Pavlakis, G. (1992). The rev protein of human immunodeficiency virus type 1 promotes polysomal association and translation of *gag/pol* and *vpu/env* mRNAs. *Molecular and Cellular Biology*, **12**, 1375–86.

de Smit, M. H. & van Duin, J. (1990). Control of prokaryotic translational initiation by mRNA secondary structure. *Progress in Nucleic Acid Research and Molecular Biology*, **38**, 1–35.

Dinman, J. D. & Wickner, R. B. (1992). Ribosomal frameshifting efficiency and *gag/gag-pol* ratio are critical for yeast M1 double-stranded RNA virus propagation. *Journal of Virology*, **66**, 3669–76.

Dreher, T. W. & Hall, T. C. (1988). Mutational analysis of the sequence and structural requirements in brome mosaic virus RNA for minus strand promoter activity. *Journal for Molecular Biology*, **201**, 31–40.

Ehrenfeld, E. (1996). Initiation of translation by picornavirus RNAs. In Hershey, J. W. B., Mathews, M. B. & Sonenberg, N., eds. *Translational Control*, p. 549. Cold Spring Harbor Laboratory Press, Cold Spring Harbor, NY.

Fajardo, J. E. & Shatkin, A. J. (1990). Translation of bicistronic viral mRNA in transfected cells: regulation at the level of elongation. *Proceedings of the National Academy of Sciences, USA*, **87**, 328–32.

Feigenblum, D. & Schneider, R. J. (1992). Modification of eukaryotic initiation factor 4F during infection by influenza virus. *Journal of Virology*, **67**, 3027–35.

Fütterer, J., Kiss-László, Z. & Hohn, T. (1993). Nonlinear ribosome migration on cauliflower mosaic virus 35S RNA. *Cell*, **73**, 789–802.

Fütterer, J. & Hohn, T. (1992). Role of an upstream open reading frame in the translation of polycistronic mRNA in plant cells. *Nucleic Acids Research*, **20**, 3851–7.

Geballe, A. P. (1996). Translational control medicated by upstream AUG codons. In Hershey, J. W. B., Mathews, M. B. & Sonenberg, N., eds. *Translational Control*, p. 173. Cold Spring Harbor Laboratory Press, Cold Spring Harbor.

Gilmour, K. C. & Reich, N. C. (1995). Signal transduction and activation of gene transcription by interferons. *Gene Expression*, **5**, 1–18.

Gingras, A., Svitkin, Y., Belsham, G. J., Pause, A. & Sonenberg, N. (1996). Activation of the translational suppressor 4E-BP1 following infection with encephalomyocarditis virus and poliovirus. *Proceedings of the National Academy of Sciences*, **93**, 5578–83.

Gold, L. (1988). Posttranscriptional regulatory mechanisms in *Escherichia coli*. *Annual Review of Biochemistry*, **57**, 199–233.

Green, S. R., Manche, L. & Mathews, M. B. (1995). Two functionally distinct RNA binding motifs in the regulatory domain of the protein kinase DAI. *Molecular and Cellular Biology*, **15**, 358–64.

Gribaudo, G., Lembo, D., Cavallo, G., Landolfo, S. & Lengyel, P. (1991). Interferon action: Binding of viral RNA to the 40-kilodalton 2′-5′-oligoadenylate synthetase in interferon-treated HeLa cells infected with encephalomyocarditis virus. *Journal of Virology*, **65**, 1748–57.

Gunnery, S., Green, S. R. & Mathews, M. B. (1992). Tat-responsive region RNA of human immunodeficiency virus type 1 stimulates protein synthesis *in vivo* and *in vitro*: relationship between structure and function. *Proceedings of the National Academy of Sciences, USA*, **89**, 11557–61.

Gunnery, S. & Mathews, M. B. (1995). Functional mRNA can be generated by RNA polymerase III. *Molecular and Cellular Biology*, **15**, 3597–607.

Haenni, A., Joshi, S. & Chapeville, F. (1982). tRNA-like structures in the genomes of RNA viruses. *Progress in Nucleic Acid Research and Molecular Biology*, **27**, 101–2.

Hambridge, S. & Sarnow, P. (1992). Translational enhancement of the poliovirus 5′ noncoding region mediated by virus-encoded polypeptide 2A. *Proceedings of the National Academy of Sciences, USA*, **89**, 10272–6.

Hattman, S., Newman, L., Krishna Murthy, H. M. & Nagaraja, V. (1991). Com, the phage Mu *mom* translational activator, is a zinc-binding protein that binds specifically to its cognate mRNA. *Proceedings of the National Academy of Sciences, USA*, **88**, 10027–31.

Hayes, B. W., Telling, G. C., Myat, M. M., Williams, J. F. & Flint, S. J. (1990). The adenovirus L4 100-kilodalton protein is necessary for efficient translation of viral late mRNA species. *Journal of Virology*, **64**, 2732–42.

Hinnebusch, A. G. (1996). Translational control of GCN4: gene-specific regulation by phosphorylation of eIF2. In Hershey, J. W. B., Mathews, M. B. & Sonenberg, N., eds. *Translational Control*, p. 199. Cold Spring Harbor Laboratory Press, Cold Spring Harbor, NY, USA.

Huang, J. T. & Schneider, R. J. (1990). Adenovirus inhibition of cellular protein synthesis is prevented by the drug 2-aminopurine. *Proceedings of the National Academy of Sciences, USA*, **87**, 7115–19.

Huang, J. T. & Schneider, R. J. (1991). Adenovirus inhibition of cellular protein synthesis involves inactivation of cap binding protein. *Cell*, **65**, 271–80.

Ivey-Hoyle, M. & Steege, D. A. (1989). Translation of phage f1 gene VII occurs from an inherently defective initiation site made functional by coupling. *Journal for Molecular Biology*, **208**, 233–44.

Ivey-Hoyle, M. & Steege, D. A. (1992). Mutational analysis of an inherently defective translation initiation site. *Journal for Molecular Biology*, **224**, 1039–54.

Jacks, T. & Varmus, H. E. (1985). Expression of the rous sarcoma virus *pol* gene by ribosomal frameshifting. *Science*, **230**, 1237–42.

Jackson, R. J. (1996). A comparative view of initiation site selection mechanisms. In Hershey, J. W. B., Mathews, M. B. & Sonenberg, N., eds. *Translational Control*, p. 71. Cold Spring Harbor Laboratory Press, Cold Spring Harbor, NY.

Jang, S. K., Krausslich, H. G., Nicklin, M. J., Duke, G. M., Palmenberg, A. C. & Wimmer, E. (1988). A segment of the 5′ nontranslated region of encephalomyocarditis virus RNA directs internal entry of ribosome during *in vitro* translation. *Journal of Virology*, **62**, 2636–43.

Kamen, R. I. (1975). Structure and function of the Qb RNA replicase. In Zinder, N. D., ed. *RNA Phages*, p. 203.

Karacostas, V., Wolffe, E. J., Nagashima, K., Gonda, M. A. & Moss, B. (1993). Overexpression of the HIV-1 gag-pol polyprotein results in intracellular activation of HIV-1 protease and inhibition of assembly and budding of virus-like particles. *Virology*, **193**, 661–71.

Katze, M. G. (1995). Regulation of the interferon-induced PKR: can viruses cope. *Trends in Microbiology*, **3**, 75–8.

Katze, M. G. (1996). Translational control in cells infected with influenza virus and reovirus. In Hershey, J. W. B., Mathews, M. B. & Sonenberg, N., eds. *Translational Control*, p. 607. Cold Spring Harbor Laboratory Press, Cold Spring Harbor, NY.

Kerr, T. M., Brown, R. E. & Hovanessian, A. G. (1977). Nature of inhibitor of cell-free protein synthesis formed in response to interferon and double-stranded RNA. *Nature*, **268**, 540–2.

Kitajewski, J., Schneider, R. J., Safer, B., Munemitsu, S. M., Samuel, C. E., Thimmappaya, B. & Shenk, T. (1986). Adenovirus VAI RNA antagonizes the antiviral action of interferon by preventing activation of the interferon-induced eIF-2a kinase. *Cell*, **45**, 195–200.

Koromilas, A. E., Roy, S., Barber, G. N., Katze, M. G. & Sonenberg, N. (1992). Malignant transformation by a mutant of the IFN-inducible dsRNA-dependent protein kinase. *Science*, **257**, 1685–9.

Kozak, M. (1986). Point mutations define a sequence flanking the AUG initiator codon that modulates translation by eukaryotic ribosomes. *Cell*, **44**, 283–92.

Kozak, M. (1989). The scanning model for translation: an update. *Journal of Cell Biology*, **108**, 229–41.

Kozak, M. (1992). Regulation of translation in eukaryotic systems. *Annual Reviews in Cell Biology*, **8**, 197–225.

Lamphear, B. J., Kirchweger, R., Skern, T. & Rhoads, R. E. (1995). Mapping of functional domains in eIF4G with picornaviral proteases. Implications for cap-dependent and cap-independent translation initiation. *Journal of Biological Chemistry*, **270**, 21975–83.

Laurent, A. G., Krust, B., Galabru, J., Svab, J. & Hovanessian, A. G. (1985). Monoclonal antibodies to an interferon-induced M_r 68,000 protein and their use for the detection of double-stranded RNA-dependent protein kinase in human cells. *Proceedings of the National Academy of Sciences, USA*, **82**, 4341–5.

Lee, S. B. & Esteban, M. (1993).The interferon-induced double-stranded RNA-activated human p68 protein kinase inhibits the replication of vaccinia virus. *Virology*, **193**, 1037–41.

Lengyel, P. (1993). Tumor-suppressor genes: news about the interferon connection. *Proceedings of the National Academy of Sciences, USA*, **90**, 5893–5.

Lerner, M. R., Andrews, N. C., Miller, G. & Steitz, J. A. (1981). Two small RNAs encoded by Epstein–Barr virus and complexed with protein are precipitated by antibodies from patients with systemic lupus erythematosus. *Proceedings of the National Academy of Sciences, USA*, **78**, 805–9.

Lodish, H. F. (1975). Regulation of *in vitro* protein synthesis by bacteriophage RNA by RNA tertiary structure. In Zinder, N. D., ed. *RNA Phages*, p. 301.

Lodish, H. F. & Zinder, N. D. (1966). Mutants of the bacteriophage f2. VIII. Control mechanisms for phage-specific syntheses. *Journal of Molecular Biology*, **19**, 333–48.

Ma, Y. & Mathews, M. B. (1996). Structure, function, and evolution of adenovirus-associated RNA: a phylogenetic approach. *Journal of Virology*, **70**, 5083–99.

Mader, S., Lee, H., Pause, A. & Sonenberg, N. (1995). The translation initiation factor eIF-4E binds to a common motif shared by the translation factor eIF-4y and the translational repressors 4E-binding proteins. *Molecular & Cellular Biology*, **15**, 4990–7.

Maitra, R. K., McMillan, N. A. J., Desai, S., McSwiggen, J., Hovanessian, A. G., Sen, G., Williams, B. R. G. & Silverman, R. H. (1994). HIV-1 TAR RNA has an intrinsic ability to activate interferon-inducible enzymes. *Virology*, **204**, 823–7.

Manche, L., Green, S. R., Schmedt, C. & Mathews, M. B. (1992). Interactions between double-stranded RNA regulators and the protein kinase DAI. *Molecular and Cellular Biology*, **12**, 5238–48.

Maran, A. & Mathews, M. B. (1988). Characterization of the double-stranded RNA implicated in the inhibition of protein synthesis in cells infected with a mutant adenovirus defective for VA RNA_I. *Virology*, **164**, 106–13.

Masison, D. C., Blanc, A., Ribas, J. C., Carroll, K., Sonenberg, N. & Wickner, R. B. (1995). Decoying the cap mRNA degradation system by a double-stranded RNA virus and poly(A) mRNA surveillance by a yeast antiviral system. *Molecular and Cellular Biology*, **15**, 2763–71.

Mathews, M. B. (1993). Viral evasion of cellular defense mechanisms: regulation of the protein kinase DAI by RNA effectors. *Seminars in Virology*, **4**, 247–57.

Mathews, M. B. (1996). Interactions between viruses and the cellular machinery for protein synthesis. In Hershey, J. W. B., Mathews, M. B. & Sonenberg, N., eds. *Translational Control*, p. 505. Cold Spring Harbor Laboratory Press, Cold Spring Harbor, NY.

Mathews, M. B. & Shenk, T. (1991). Adenovirus virus-associated RNA and translational control. *Journal of Virology*, **65**, 5657–62.

McCarthy, J. E. G. & Gualerzi, C. (1990). Translational control of prokaryotic gene expression. *Trends in Genetics*, **6**, 78–85.

Merrick, W. C. & Hershey, J. W. B. (1996). The pathway and mechanism of eukaryotic protein synthesis. In Hershey, J. W. B., Mathews, M. B. & Sonenberg, N., eds. *Translational Control*, p. 31. Cold Spring Harbor Laboratory, Cold Spring Harbor, NY.

Meurs, E. F., Watanabe, Y., Kadereit, S., Barber, G. N., Katze, M. G., Chong, K., Williams, B. R. G. & Hovanessian, A. G. (1992). Constitutive expression of human double-stranded RNA-activated p68 kinase in murine cells mediates phosphorylation of eukaryotic initiation factor 2 and partial resistance to encephalomyocarditis virus growth. *Journal of Virology*, **66**, 5805–14.

Meurs, E. F., Galabru, J., Barber, G. N., Katze, M. G. & Hovanessian, A. G. (1993). Tumor suppressor function of the interferon-induced double-stranded RNA-activated protein kinase. *Proceedings of the National Academy of Sciences, USA*, **90**, 232–6.

Mosig, G. & Eiserling, F. (1988). Phage T4 structure and metabolism. In Calendar, R., ed. *The bacteriophages*, p. 521.

Muffler, A., Fischer, D. & Henegge-Aronis, R. (1996). The RNA-binding protein HF-I, known as a host factor for phage Qb RNA replication, is essential for rpoS translation in *Escherichia coli. Genes and Development*, **10**, 1143–51.

Nemeroff, M. E. & Bruenn, J. A. (1987). Initiation by the yeast viral transcriptase *in vitro*. *Journal of Biological Chemistry*, **262**, 6785–7.

Nucifora, G., Begy, C. R., Erickson, P., Drabkin, H. A. & Rowley, J. D. (1993). The 3;21 translocation in myelodysplasia results in a fusion between the *AML1* gene and the gene for EAP, a highly conserved protein associated with the epstein-barr virus small RNA EBER 1. *Proceedings of the National Academy of Sciences, USA*, **90**, 7784–8.

Nuss, D. L., Opperman H. & Koch, G. (1975). Selective blockage of initiation of host protein synthesis in RNA virus-infected cells. *Proceedings of the National Academy of Sciences, USA*, **72**, 1258.

O'Malley, R. P., Duncan, R. F., Hershey, J. W. B. & Mathews, M. B. (1989). Modification of protein synthesis initiation factors and shut-off of host protein synthesis in adenovirus-infected cells. *Virology*, **168**, 112–18.

Ohlmann, T. M., Rau, S. J. M. & Pain, V. M. (1995). Proteolytic cleavage of initiation factor eIF-4y in the reticulocyte lysate inhibits translation of capped mRNAs but enhances that of uncapped mRNAs. *Nucleic Acids Research*, **23**, 334–40.

Patel, R. C. & Sen, G. C. (1992). Identification of the double-stranded RNA-binding domain of the human interferon-inducible protein kinase. *Journal of Biological Chemistry*, **267**, 7671–6.

Peabody, D. S., Subramani, S. & Berg, P. (1986). Effect of upstream reading frames on translational efficiency in simian virus 40 recombinants. *Molecular and Cellular Biology*, **6**, 2704–11.

Pelletier, J. & Sonenberg, N. (1988). Internal initiation of translation of eukaryotic mRNA directed by a sequence derived from poliovirus RNA. *Nature*, **334**, 320–5.

Penner, M., Morad, I., Snyder, L. & Kaufmann, G. (1995). Phage T4-coded Stp: double-edged effector of coupled DNA and tRNA-restrition systems. *Journal of Molecular Biology*, **249**, 857–68.

Rajan, P., Swaminathan, S., Zhu, J., Cole, C. N., Barber, G., Tevethia, M. J. & Thimmapaya, B. (1995). A novel translation regulation function of the simian virus 40 large-T antigen gene. *Journal of Virology*, **69**, 785–95.

Ravetch, J. V., Model, P. & Robertson, H. D. (1977). Isolation and characterization of phi- x 174 ribosome binding sites. *Nature*, **265**, 698–702.

Roy, S., Katze, M. G., Parkin, N. T., Edery, I., Hovanessian, A. G. & Sonenberg, N. (1990). Control of the interferon-induced 68-kilodalton protein kinase by the HIV-1 *tat* gene product. *Science*, **247**, 1216–19.

Samuel, C. E. (1988). Mechanisms of the antiviral action of interferons. *Progress in Nucleic Acid Research and Molecular Biology*, **35**, 27–72.

Schmidt, M. R., Gravel, K. A. & Woodland, R. T. (1995). Progression of a vesicular stomatitis virus infection in primary lymphocytes is restricted at multiple levels during B cell activation. *Journal of Immunology*, 2533–44.

Schneider, R. J. (1996). Adenovirus and vaccinia virus translational control. In

Hershey, J. W. B., Mathews, M. B. & Sonenberg, N., eds. *Translational Control*, p. 575. Cold Spring Harbor Laboratory Press, Cold Spring Harbor, NY.

Shterman, N., Elroy-Stein, O., Morad, I., Amitsur, M. & Kaufmann, G. (1995). Cleavage of the HIV replication primer $tRNA^{Lys,3}$ in human cells expessing bacterial anticodon nuclease. *Nucleic Acids Research*, **23**, 1744–9.

Schwartz, S., Felber, B. K. & Pavlakis, G. N. (1992). Mechanism of translation of monocistronic and multicistronic human immunodeficiency virus type 1 mRNAs. *Molecular and Cellular Biology*, **12**, 207–19.

SenGupta, D. N., Berkhout, B., Gatignol, A., Zhou, A. & Silverman, R. H. (1990). Direct evidence for translational regulation by leader RNA and Tat protein of human immunodeficiency virus type 1. *Proceedings of the National Academy of Sciences, USA*, **87**, 7492–6.

Sonenberg, N. (1990). Measures and countermeasures in the modulation of initiation factor activities by viruses. *The New Biologist*, **2**, 402–9.

Sonenberg, N. (1996). mRNA 5′ cap-binding protein eIF4E and control of cell growth. In Hershey, J. W. B., Mathews, M. B. & Sonenberg, N., eds. *Translational Control*, p. 245. Cold Spring Harbor Laboratory Press, Cold Spring Harbor, NY.

Staeheli, P. (1990). Interferon-induced proteins and the antiviral state. *Advances in Virus Research*, **38**, 147–200.

Toczyski, D. P., Matera, A. G., Ward, D. C. & Steitz, J. A. (1994). The Epstein–Barr virus (EBV) small RNA EBER1 binds and relocalizes ribosomal protein L22 in EBV-infected human B lymphocytes. *Proceedings of the National Academy of Sciences, USA*, **91**, 3463–7.

Trachsel, H. (1996). Binding of initiator methionyl-tRNA to ribosomes. In Hershey, J. W. B., Mathews, M. B. & Sonenberg, N., eds. *Translational Control*, p. 113. Cold Spring Harbor Laboratory Press, Cold Spring Harbor, NY.

van Duin, J. (1988). Single-stranded RNA bacteriophages. In Calendar, R., ed. *The Bacteriophages*, p. 117. Plenum Press, New York and London.

Weiss, R. B., Huang, W. M. & Dunn, D. M. (1990). A nascent peptide is required for ribosomal bypass of the coding gap in bacteriophage T4 gene 60. *Cell*, **62**, 117–26.

Wimmer, E., Chang, A. Y., Clark, J. M., Jr & Reichmann, M. E. (1968). Sequence studies of satellite tobacco necrosis virus RNA. Isolation and characterization of a 5′-terminal trinucleotide. *Journal of Molecular Biology*, **38**, 59–73.

Witherell, G. W., Gott, J. M. & Uhlenbeck, O. C. (1991). Specific interaction between RNA phage coat proteins and RNA. *Progress in Nucleic Acid Research and Molecular Biology*, **40**, 185–220.

Wulczyn, F. G. & Kahmann, R. (1991). Translational stimulation: RNA sequence and structure requirements for binding of com protein. *Cell*, **65**, 259–69.

Yueh, A. & Schneider, R. J. (1996). Selective translation initiation by ribosome jumping in adenovirus-infected and heat-shocked cells. *Genes and Development*, **10**, 1557–67.

Zhang, Y. & Schneider, R. J. (1994). Adenovirus inhibition of cell translation facilitates release of virus particles and enhances degradation of the cytokeratin network. *Journal of Virology*, **68**, 2544–55.

Zhou, A., Hassel, B. A. & Silverman, R. H. (1993). Expression cloning of 2–5A-dependent RNAase: a uniquely regulated mediator of interferon action. *Cell*, **72**, 753–65.

SHIGELLA-INDUCED CYTOSKELETAL REORGANIZATION DURING HOST CELL INVASION

G. TRAN VAN NHIEU, T. ADAMS, C. DEHIO, R. MÉNARD, A. SKOUDY, J. MOUNIER, R. HELLIO, P. GOUNON, AND P. SANSONETTI

Pathogénie Microbienne Moléculaire, Institut Pasteur, 28, rue du Dr Roux, 75724 Paris Cedex 15, France

INTRODUCTION

Many pathogens can enter epithelial cells, a property that is considered to be essential for host tissue colonization by the microorganism and the establishment of disease. Different strategies of cellular entry have been elaborated by enteroinvasive pathogens: entry into cells can be mediated by interactions between a ligand exposed on the bacterial surface and a cellular receptor: this is the case for *Yersinia* entry mediated by interaction between the bacterial invasin and multiple β_1 integrins (Isberg, 1991), or *Listeria* entry, mediated by the internalin-cadherin interaction (Mengaud *et al.*, 1996). For these pathogens, entry is reminiscent of the 'zippering' occurring during phagocytosis of particles by macrophages (Griffin *et al.*, 1975): internalization is correlated to a strong adhesion of bacteria to the cell surface, cellular pseudopods extend in close contact with the bacterial body and merge to engulf the bacterium in a tight phagosome. For *Salmonella* and *Shigella*, as will be discussed in this chapter, entry into epithelial cells involves many bacterial genes and an important reorganization of the cell cytoskeleton at the site of bacterial entry.

Shigella flexneri is a Gram-negative bacterium responsible for diarrhoeal diseases in humans. Severe symptoms of shigellosis are due to the destruction of the colonic epithelium linked to an intense inflammatory reaction. Animal studies indicate that entry of *Shigella* in the colonic mucosal epithelium occurs primarily via M cells (Wassef, Keren & Mailloux, 1989; Sansonetti, *et al.*, 1991), which are specialized structures devoid of brush borders contacting intestinal lymph nodes. After ingestion by M cells, and phagocytosis by resident macrophages of the underlying lymph nodes, it is thought that *Shigella* induces the release of cytokines by killing the macrophage in an apoptotic process (Zychlinsky, Prévost & Sansonetti, 1992; Zychlinsky *et al.*, 1994). The inflammatory reaction resulting from the influx of PMNs and monocytes to the site of the infection, leads to the destabilization of the epithelium and favours a massive infection of the enterocytes by bacteria present in the intestinal tract lumen. The host inflammatory reaction plays a key role in the severity of the disease: for example, treatment with an IL1-receptor antagonist suppresses the shigellosis symptoms in a rabbit ileal loop model (Sansonetti *et al.*, 1995). Therefore, the strategy used by this pathogen appears to be based on a keen manipulation of the host defence system by specific virulence determinants.

Among these virulence factors, the abilities of *Shigella* to invade and spread from cell-to-cell have been particularly studied using cultured cells. In cultured cells, *Shigella*-induced phagocytosis, much like that in *Salmonella* (Galan *et al.*, 1992; Francis *et al.*, 1993), presents many similarities with growth-factor activation. This interaction leads to internalization of the bacteria by a macropinocytic-like process. After internalization by epithelial cells, *Shigella* lyses the phagocytic vacuole and multiplies freely within the cell cytosol. During this phase *Shigella* moves within the cell cytosol using the cell cytoskeleton and following different modes: *Shigella* can move rapidly and in a seemingly random fashion by polymerizing actin at one end of the bacterial body in a process leading to the formation of 'actin comet tails' (Ics phenotype) (Bernardini *et al.*, 1989). Alternatively, *Shigella* appears to be able to slide along actin cables in a process reminiscent of organelle movement (Olm phenotype) (Vasselon *et al.*, 1991). In polarized epithelial cells, both motions are probably required for cell-to-cell spread (Vasselon *et al.*, 1992). This spreading ability is characterized by the formation of bacteria-containing protrusions, which push into adjacent cells. After lysis of protrusion- and recipient-cell membrane, *Shigella* can reinitiate the infection cycle, and spread from cell-to-cell within the epithelial tissue protected from humoral defence mechanisms.

In this chapter, we will discuss the central role of the cell cytoskeleton during *Shigella* invasion of epithelial cells, in view of progress accomplished in the last few years. If this role is prominent in the mechanisms of intracellular motillity, it appears that *Shigella* -induced entry within cells is also an example of a complex manipulation of cellular components regulating cytoskeletal organization.

SHIGELLA DETERMINANTS OF CELLULAR INVASION

Bacterial genetic studies have allowed the identification of loci involved in *Shigella* entry into cultured epithelial cells (for review, see Parsot & Sansonetti, 1996). It was first shown that a 30 kb region of the *Shigella* 220 kDa virulence plasmid was sufficient to confer the invasive ability to an *E. coli* K-12 strain (Maurelli *et al.*, 1985). Mutagenic analysis of this region allowed the delineation of genes that are critical for the entry process (Sasakawa *et al.*, 1986; Ménard, Sansonetti & Parsot, 1993): unlike invasion systems of other intracellular pathogens for which a single locus is involved in the entry process, up to 30 *Shigella* genes are required for cellular entry. These genes are organized on two divergently transcribed regions: (i) the genes part of the *mxi-spa* locus, required for the formation of a specific secretion/translocation machinery (Allaoui, Sansonetti & Parsot, 1992; Parsot *et al.*, 1995); (ii) the *ipa* genes, considered as the bacterial effectors of entry (Ménard *et al.*, 1996), and whose secretion depends on the *mxi-spa* locus (Ménard *et al.*, 1994*a*). It has recently been shown that latex particles coated with Ipa proteins are internalized by cultured HeLa cells, implicating these proteins as effectors of bacterial uptake. Entry of these coated particles, however, was accompanied by a reorganization of the cytoskeleton at the site of particle interaction with the membrane that was much less important than cytoskeletal reorganization induced by *Shigella* during entry (Ménard *et al.*, 1996). There may be several explanations for these differences: for example, immobilization of the Ipa protein complex on the surface of the particle by specific antibodies could interfere with the mode of action of bacterial invasins, as it remains controversial whether the Ipa proteins localize on the surface of *Shigella*. Even if any are localized on the bacterial surface, however, it is clear that the Ipa proteins do not promote *Shigella* adherence to epithelial cells. Although the Ipa protein complex has been shown to bind β_1 integrins (Watarai, Funato & Sasakawa, 1996), the mode of action of *Shigella* invasins therefore appears different from particle uptake based on a high affinity cell surface receptor-bacterial ligand interaction. In fact, secretion of Ipas by *Shigella* is rapidly triggered upon host cell contact (Ménard *et al.*, 1994*b*; Watarai *et al.*, 1995), and this rapid and massive secretion could result in a locally high concentration of Ipas in the cell cytosol, a prerequisite for actin polymerization at the site of bacterial entry. Consistent with this, the *mxi-spa* translocation apparatus presents a significant homology with the *Yersinia* apparatus used for translocation of the Yop proteins (Rosqvist *et al.*, 1995). In the case of *Yersinia*, it has been shown that Yop proteins are translocated into the macrophage cytosol upon cell contact (Rosqvist *et al.*, 1994), where they can exert their antiphagocytic effects. As IpaB can also translocate via the *Yersinia* translocation system (Rosqvist *et al.*, 1995), it is conceivable that Ipa proteins are injected into the cell cytosol during entry where they can activate membrane rufflings at the site of bacterial interaction. The identification of cellular targets of the invasins responsible for cytoskeletal reorganization has been an area of intense investigation in the past years: if the entry process remains, by and large, to be characterized at the molecular level, there is now evidence that *Shigella* entry into cells is a multi-step process, utilizing signals mediated by distinct invasins. Also, and as will be discussed in following sections, the focal adhesion protein vinculin is directly targeted by the *Shigella* invasion apparatus.

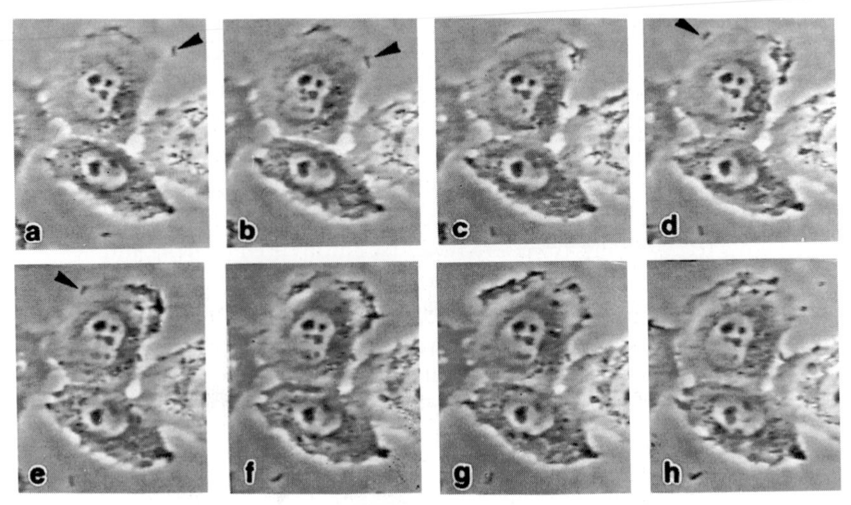

Fig. 1. Cinematographic microscopy of HeLa cells challenged with wild-type *Shigella flexneri*. Semi-confluent HeLa cells grown on tissue culture dishes were challenged with wild-type *Shigella* in MEM medium at 37 °C, and filmed using a 3CCD videocamera (Sony) connected to a time-lapse videorecorder. Panels a–h represent pictures from the same field taken at 2 min intervals. Occasionally, a bacterium moving in the medium associates with a cellular edge (Panels a and b, arrowheads), and within a few minutes, membrane folding occurs at the site of interaction, which extends to involve a wide area of the cell periphery (Panels c–f). Membrane rufflings progressively decrease, as the cell reestablishes contact with the plastic surface (Panels g and h). The arrowheads in Panels d and e show a second bacterium associating with the cell, and responsible for the rufflings observed in the corresponding cellular area in Panels f–h.

SHIGELLA AND *SALMONELLA* INDUCED 'RUFFLINGS' OF HOST CELL MEMBRANES

Shigella invasion of cultured HeLa cells is characterized by the formation of cellular extensions at the site of bacterial interaction with the cell membrane, which are very similar to structures induced by the other enteropathogen *Salmonella* (Francis *et al.*, 1993). It is likely that both *Shigella* and *Salmonella* utilize a similar strategy to induce their phagocytosis by epithelial cells, as *Salmonella* proteins have been identified that share (30 to 57%) homology to the *Shigella* Ipa proteins (Hermant *et al.*, 1995; Kaniga *et al.*, 1995). In contrast to *Salmonella* which shows cell binding activity (Altmeyer *et al.*, 1993), *Shigella* does not adhere to cultured cells, and this difference might account for significant interaction of *Salmonella* with enterocytes in the course of the infection, not observed in the early stages of *Shigella* infection. Figure 1 shows a cinematographic study of HeLa cells internalizing *Shigella*: at early time points, very few bacteria are seen associated with cells (Fig. 1, panel a). Occasionally, a bacterium contacting the cell membrane (Fig. 1, panels a and b, arrows) induces a deformation of the cellular membrane localized at the area of bacterial interaction detected by phase-contrast

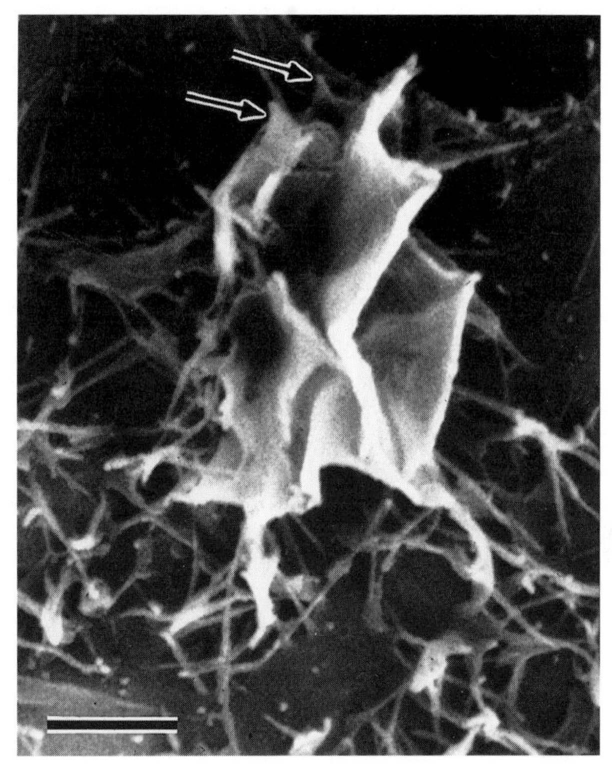

Fig. 2. Scanning electron microscopy of *Shigella*-induced membrane folds at the site of bacterial entry. HeLa cells were challenged with wild-type *Shigella* expressing the AfaE adhesin (Labigne-Roussel *et al.*, 1984) for 12 min at 37 °C. Samples were fixed with paraformaldehyde and processed for SEM analysis. Membrane leaflets raise from the cell dorsal surface at the site of bacterial interaction. The arrows points at microspike-like structures present at the base of the entry structure, and which appear to precede membrane leaflet formation (Adam *et al.*, 1995). Note the presence of membrane structures present on the dorsal cell surface at the vicinity of the bacterial entry site, associated with a local concentration of actin filaments. Bar: 2 μm.

microscopy, and occurring within minutes after cell contact (Fig. 1, panels c–f). These cellular extensions can reach up to tens of microns in length, extend around the bacterial body and allow its internalization by the formation of large vacuoles which are not restricted to the bacterial body and also lead to the endocytosis of large amounts of extracellular fluid. After internalization is completed, rufflings often persist for a few minutes but eventually disappear as the cell re-establishes adhesion to the substrate (Fig. 1, panels g and h). Although these cellular extensions are almost exclusively formed at the cell leading edges, with which *Shigella* interacts preferentially, they can also form at the dorsal cell surface. This is so in particular if one increases *Shigella* binding to cells by expression of the AfaE adhesin (Labigne-Roussel *et al.*, 1984). Figure 2 shows a scanning electron micrograph of an *Shigella*-

induced entry structure: typically, the *Shigella*-induced cellular extensions are organized as membrane leaflets that extend from the site of bacterial interaction. Microspike-like structures are often seen at the early stages of the entry process, and are also visible at the base of this entry focus (Fig. 2, arrows). Clearly, actin reorganization is critical for these processes, as cell treatment with the F-actin inhibitor cytochalasin inhibits *Shigella* entry.

ACTIN-BUNDLING PROTEINS

Fluorescence labelling of F-actin shows that *Shigella* entry implicates an important polymerization of actin in the cellular extensions raising around the bacterial body, suggesting that actin polymerization represents the driving force responsible for the membrane deformation observed upon *Shigella* contact (Fig. 3A; see Fig. 6). A denser coat of polymerized actin is also visible in the immediate vicinity of the nascent bacterial phagosome (Fig. 3A, arrow), which also co-localizes with an increased labelling for focal adhesion components, as will be discussed further. Consistent with the dynamic role of actin, electron microscopy analysis of actin decoration with myosin S1 headpiece indicates that cellular extensions induced by *Shigella* during entry contain tightly bundled actin filaments, organized in parallel orientation with their barbed ends orientated towards the tip of the cell extension (Adam *et al.*, 1995). Actin nucleation foci are also found associated with the membrane juxtaposing the bacterium, as well as at the tip of nascent cellular extensions (Fig. 3B; Adam *et al.*, 1995). Interestingly, Ipa proteins have been shown to associate with the cell membrane contacting *Shigella* after secretion, suggesting that they participate in the nucleation of actin filaments (Mounier *et al.*, submitted).

The actin-bundling protein plastin is directly implicated in the organization of the *Shigella* entry structure (Adam *et al.*, 1995). Plastin shows a strong recruitment in cellular extensions induced by *Shigella* at the site of entry, and introduction of a transdominant negative form of plastin deleted for its aminoterminal actin-binding site in HeLa cells, results in about 60% inhibition of *Shigella* entry. These findings indicate that *Shigella* induced cellular extensions involve mechanisms similar to microvillus formation on the apical surface of epithelial cells, which also depend on the expression of the plastin T-isoform. It is believed that plastin, along with other cytoskeletal components, is part of the scaffold required for stabilization of *Shigella*-induced cellular extensions, rather than playing an active role in the induction of these cellular processes. Given the diversity of cytoskeletal proteins recruited at the site of *Shigella* entry, direct targets of *Shigella* invasins most probably consist of upstream regulatory components of cytoskeletal organization. To better characterize the *Shigella*-induced cell response, recruitment of cytoskeletal and focal adhesion components have been analyzed by immuno-localization.

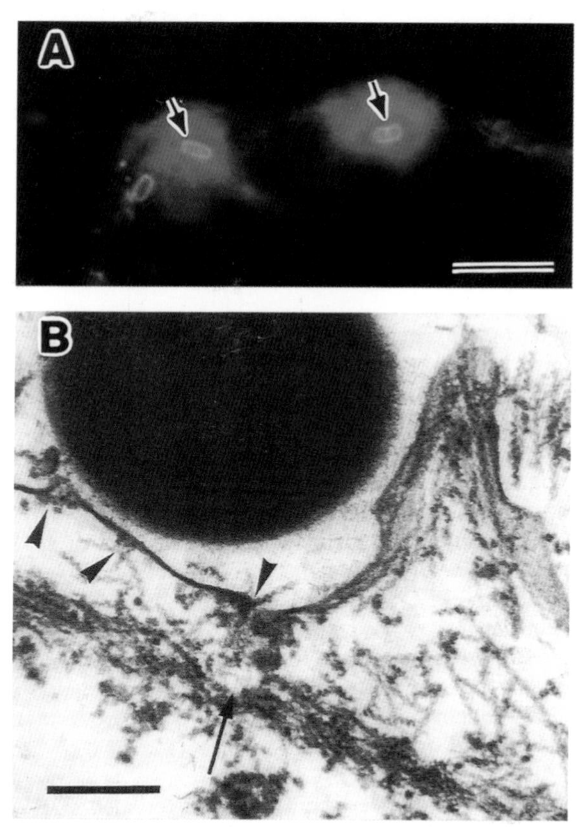

Fig. 3. Panel A: F-actin staining of cells challenged with wild-type *Shigella*. F-actin was stained with Bodipy-phalloidin. The bacteria are stained with an anti-LPS antibody and an anti-rabbit antibody conjugated to Texas-Red. The bacterial entry structure is characterized by actin polymerization in a cell area involving several square microns, associated with membrane folds observed at the site of bacterial interaction. A denser coat of polymerized actin is often visible at the immediate vicinity of the bacterium in the process of being internalized (arrows). Bar: 10 μm. Panel B. Myosin S1 headpiece decoration of actin filaments at the site of *Shigella* entry. Bundles of actin filaments in parallel orientation are visible with their barbed ends pointing towards the tip of the cellular extension raising along the bacterial body. Clusters of polymerized actin are also visible in association with the cell membrane juxtaposing the bacterium (arrowheads), and which may correspond to foci of actin nucleation (Adam *et al.*, 1995). Bar: 1 μm.

RECRUITMENT OF FOCAL ADHESION COMPONENTS

Although the interpretation of such experiments has to take into account the possibility of a passive recruitment of actin-binding proteins due to the important polymerization of actin at the site of *Shigella* entry, it is possible to establish a heriarchy in the levels of recruitment of different cytoskeletal components, by comparing their fluorescence intensity levels in the entry structure to that of the cellular background. These studies point at focal

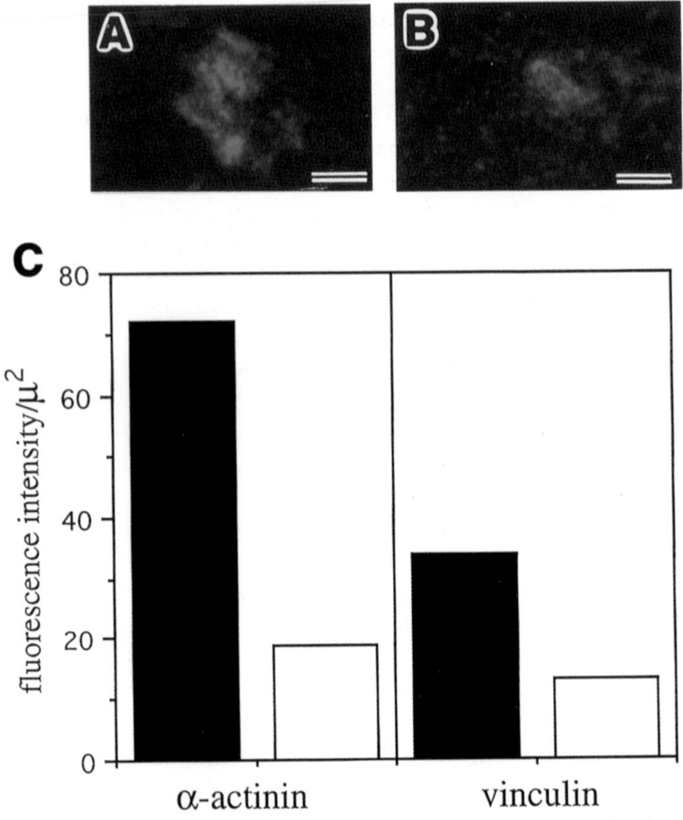

Fig. 4. Recruitment of α-actinin and vinculin at the site of *Shigella* entry. Confocal microscopy analysis of *Shigella* entry structures fluorescently labelled with anti-α-actinin monoclonal antibody (Panel A) or anti-vinculin monoclonal antibody (Panel B). α-actinin is massively recruited in *Shigella* entry structures (Panel A), with a staining that superimposes that of F-actin. Vinculin is recruited preferentially in the close vicinity of the bacterial body, and forms a dense coat that superimposes with the coat of F-actin (Panel B). Panels A and B, bar: 2 μm. Quantification of the fluorescence intensity by scanning confocal microscopy indicates that α-actinin is four- to five-fold more concentrated in areas of the *Shigella* entry structure compared to the average cellular staining (Panel C, α-actinin, compare solid and empty bars), whereas vinculin shows a two- to three-fold recruitment (Panel C, vinculin, compare solid and empty bars).

adhesion components as important players of *Shigella* entry. The actin-bundling protein α-actinin, which binds the cytoplasmic domain of β1 integrins, is markedly recruited to *Shigella*-induced cellular extensions (Fig. 4A; see Fig. 6). Also, albeit to a lower extent, vinculin (Fig. 4B; see Fig. 6) and talin, which also associate with the cytoplasmic domain of β1 integrins, are specifically recruited to the site of entry. In this latter case, however, recruitment is much more pronounced at the levels of the membrane interacting with the bacterium, with a localization coinciding with that of the F-actin coat. These findings appear to be consistent with the interaction that was described between the Ipa protein complex and $\beta 1$ integrins (Watarai *et al.*, 1996). Hence, *Shigella* would be capable of inducing the formation of an adhesion complex at the level of the phagosomal membrane by sequestering $\beta 1$ integrins at the site of bacterial-cell interaction via the Ipa proteins. Recruitment of focal adhesion components would then occur after signalling of the $\beta 1$ integrins, and by association with the integrin $\beta 1$ cytoplasmic domain. This model, however, does not reflect satisfactorily the complexity of cellular processes occurring during *Shigella* entry into epithelial cells. For example, it is not clear how $\beta 1$ integrin activation by itself can induce the formation of cellular extensions observed at the site of *Shigella* entry. Also, tyrosine phosphokinase inhibitors, which inhibit the formation of focal adhesions, have no effect on *Shigella* entry (Dehio, unpublished observations). An explanation for these results lies in the possibility that Ipa proteins are translocated into the cell cytosol, thus bypassing classical schemes of receptor-mediated activation. In fact, there is now evidence that the *Shigella* IpaA protein translocates into the cell cytosol where it directly binds vinculin, this interaction being responsible for the recruitment of focal adhesion proteins α-actinin and talin at the phagosomal membrane (Tran Van Nhieu *et al.*, submitted). Multiple signals are therefore transmitted by *Shigella* during entry: an IpaA-vinculin-dependent signal responsible for the formation of a pseudo-focal adhesion at the phagosomal membrane, and signals that remain to be identified responsible for the cytoskeletal reorganization leading to ruffling at the site of entry of *Shigella*. This bacterial-induced focal adhesion appears to be required for efficient internalization, probably by stabilizing the bacterium within the ruffles at the site of entry.

ROLE OF SIGNALLING MOLECULES

pp60c-src activation

Although tyrosine kinase inhibitors have little effect on *Shigella* entry, tyrosine phosphorylation could still implicate important signals during this process. For example, it is conceivable that the bacterial effectors responsible for entry are present in such a local concentration at the site of entry that specific inhibition can not be achieved. Immunofluorescence labelling with anti-phosphotyrosine antibody shows that tyrosyl-phosphorylated proteins are specifically recruited in *Shigella*-induced entry structures (Dehio *et al.*, 1995). The actin-binding protein cortactin probably contributes largely to the pool of tyrosyl-phosphorylated proteins recruited at the site of entry, as cortactin is the major protein that is specifically tyrosyl-phosphorylated upon *Shigella* entry (Fig. 5; Dehio *et al.*, 1995). Interestingly, cortactin has been shown to be a substrate for the $pp60^{c\text{-}src}$ tyrosine kinase, suggesting that $pp60^{c\text{-}src}$ is activated during *Shigella* entry. Consistent with this, overexpres-

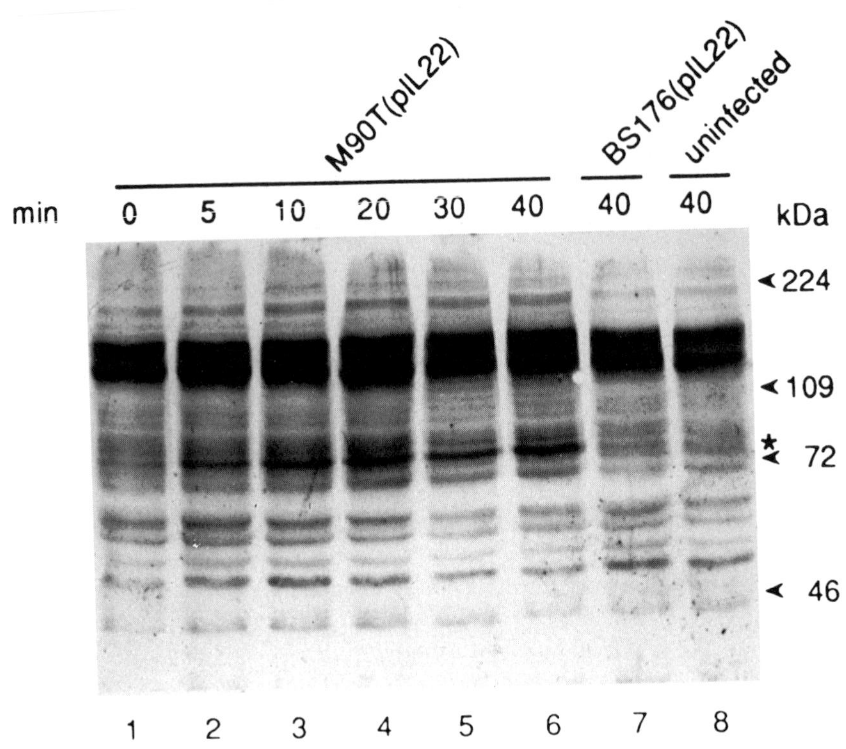

Fig. 5. Tyrosine phosphorylation of cortactin upon *Shigella* entry. HeLa cells were incubated at 37 °C with *Shigella* strains expressing the AfaI adhesin. At time points indicated above each lane, samples were subjected to Western blot analysis using an anti-phosphotyrosine monoclonal antibody. M90T(pil22): wild-type *Shigella* expressing the AfaI adhesin. BS176(pil22): isogenic non-invasive strain of *Shigella* cured from the 220 Kb virulence plasmid. The 72 KDa species that is predominantly tyrosyl-phosphorylated upon *Shigella* entry has been identified as the pp60$^{c\text{-}src}$ substrate cortactin (Dehio, Prévost & Sansonetti, 1995).

sion of pp60c-src in HeLa cells results in increased bacterial internalization by up to an order of magnitude (Dehio *et al.*, 1995). Also, pp60$^{c\text{-}src}$ is specifically recruited to *Shigella* entry structures (Fig. 6A, plane 3, arrowheads; see Fig. 6) as well as at the level of the phagosomal membrane early after internalization (Fig. 6A, plane 1 and Fig. 6B, arrowheads; see Fig. 6). The role of pp60$^{c\text{-}src}$ in *Shigella* entry, however, remains to be firmly established because attempts to inhibit pp60$^{c\text{-}src}$ by means of transdominant negative expression in HeLa cells have been inconclusive (Dehio, unpublished observations). It is possible that pp60$^{c\text{-}src}$ local activation is responsible for the formation of ruffles at the site of *Shigella* entry, as overexpression of this protein also leads to increased ruffling activity at the cell periphery by a yet to be characterized mechanism. Also, pp60$^{c\text{-}src}$ activation could participate in the formation of a pseudo-focal adhesion at the phagosomal

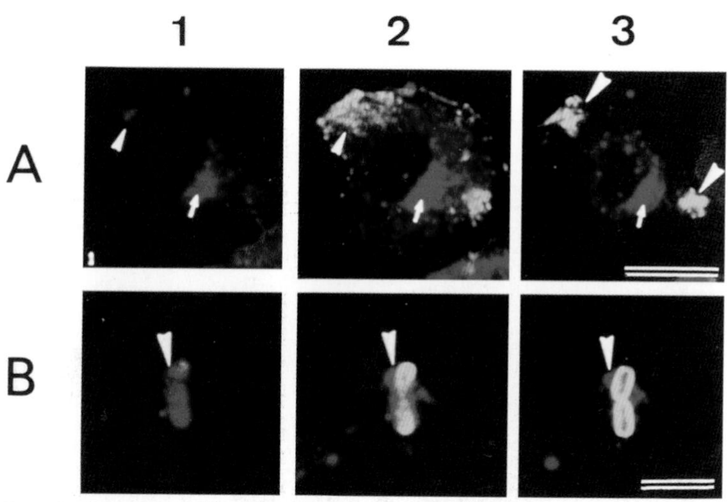

Fig. 6. Recruitment of pp60$^{c\text{-}src}$ to *Shigella*-containing phagosome in HeLa cells overexpressing pp60$^{c\text{-}src}$. Different optical sections from the same field spaced by 1 μm are shown from the basal (plane 1), to the apical side of the cell (plane 3). Panel A: Cells are labelled with an anti-pp60$^{c\text{-}src}$ monoclonal antibody and an anti-mouse IgG antibody linked to Texas red. F-actin is stained with NBD-phallicidin (green). pp60src shows a juxtanuclear localization (arrows), and is recruited to the site of *Shigella* entry (planes 2 and 3, arrowheads). Bar: 20 μm. Panel B: Bacteria are labelled with an anti-LPS antibody and a secondary antibody linked to FITC (green). Cells are labelled with an anti-pp60$^{c\text{-}src}$ monoclonal antibody and an anti-mouse IgG antibody linked to Texas red. Co-localization of pp60$^{c\text{-}src}$ with the bacterial phagosome is visualized as yellow fluorescence (planes 2 and 3). Note the association of pp60$^{c\text{-}src}$ with the cell membrane underlying the bacteria in plane 1. Bar: 5 μm.

membrane. Indeed, Src kinases have been implicated in assembly of focal adhesions (Kaplan *et al.*, 1995): although their role remains to be clearly defined, it is generally admitted that pp60c-src could regulate focal adhesion formation by directly associating with focal adhesion src-substrates, such as the non-receptor tyrosine kinase paxillin (Schaller & Parsons, 1993; Clark & Brugge, 1995). Further analysis of phosphorylation of focal adhesion components might prove useful for the identification of signalling pathways implicated in *Shigella* entry.

Implication of Rho in Shigella *entry*

Small G-proteins from the Rac/Rho family have been implicated in rufflings induced in response to growth factor activation (Ridley *et al.*, 1992; Nishiyama *et al.*, 1994), a route of activation which has been proposed for *Salmonella*-induced internalization by epithelial cells (Galan *et al.*, 1992). Attempts to inhibit *Salmonella* uptake by a dominant-negative form of Rac or after microinjection of the ADP-ribosyltransferase exoenzyme C3 specific for Rho, however, proved unsuccessful (Jones *et al.*, 1993). For *Shigella*,

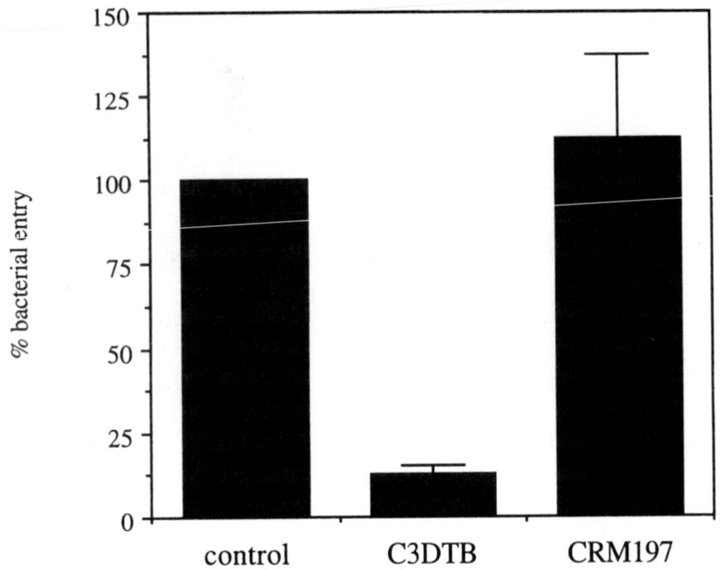

Fig. 7. C3 inhibition of *Shigella* entry into HeLa cells. HeLa cells grown to semi-confluency were challenged with wild-type *Shigella* in the presence or absence of recombinant exoenzyme C3, which specifically ADP-ribosylate Rho. The percentage of internalized bacteria was determined by the gentamicin protection procedure. control: untreated HeLa cells. C3DTB: cells treated with recombinant C3 exoenzyme, consisting of C3 fused in its carboxyterminus to a non-toxic mutant of the diphtheria toxin to allow cell penetration. CRM197: cells treated with a non-toxic form of the diphtheria toxin, used as a negative control. C3 treatment results in about a 90% inhibition of *Shigella* entry in HeLa cells (Adam *et al.*, 1996).

pretreatment of cells with a C3 derivative that was fused to the B subunit of the diphtheria toxin to allow delivery into the cell cytosol, results in a strong inhibition of bacterial internalization (Fig. 7; Adam *et al.*, 1996). This result points at a key role of Rho in the cytoskeletal reorganization occurring during *Shigella* entry. Interestingly, cell treatment with the C3 exoenzyme abolishes the formation of cellular extensions at the immediate vicinity of the bacterium, although foci of actin nucleation still form underneath the aborted phagosomal membrane (Adam *et al.*, 1996). These results implicate a role for Rho in the elongation of actin filaments required for the formation of cellular extensions, rather than in the actual nucleation of such filaments. It is tempting to speculate that initiation of the pseudo-focal adhesion structure corresponding to the F-actin coat at the bacterial phagosomal membrane also does not depend on Rho activity, as C3 treatment does not appear to prevent the formation of the actin coat surrounding *Shigella* in the process of internalization. At least two signalling events can be distinguished during *Shigella* entry into epithelial cells: (i) one leading to cytoskeletal reorganization via Rho activation, and (ii) a Rho-independent event leading to the formation of a pseudo-focal adhesion at the level of the nascent phagosomal membrane which could be mediated by the interaction between the *Shigella* protein IpaA and vinculin. Obviously, numerous important question marks remain: in particular, it is not known what is the bacterial effector responsible for activation of Rho, and how this activation occurs. Other inhibitor studies suggest that *Shigella* activation of Rho does not use classical receptor-mediated outside-in signalling: for example, PDGF or EGF activation of Rac, as well as HGF activation of Rho which lead to membrane rufflings involve PI3K activation following receptor binding and prior to small G-protein activation (for review, see Craig & Johnson, 1996). Treatment of HeLa cells with wortmannin, a potent inhibitor of PI3K which inhibits membrane rufflings induced by growth-factor activation, has little effect on *Shigella* entry into cells (Skoudy, unpublished observations). These results suggest that *Shigella* effectors either intercept the signalling pathway downstream of PI3K activation, or that these effectors utilize a, yet to be defined, activation mode of Rho.

CONCLUSION

Bacterial effectors responsible for entry may affect several cellular components to modulate *Shigella*-induced cytoskeletal reorganization: evidence for a two-step level of subversion of the cellular machinery has already been discussed. Up to 15 different *Shigella* proteins are secreted via the *mxi-spa* translocation apparatus (Parsot *et al.*, 1995). Although some of these products might not be directly relevant for bacterial entry, it is possible that many of them play a role in fine-tuning the cell response and modulate it at different levels of regulatory steps of cytoskeletal organization.

ACKNOWLEDGEMENTS

We wish to thank Dr Wandy Beatty for review of the manuscript, and people from the Laboratoire de Pathogénie Microbienne Moléculaire for their help and comments. Figure 5 is from Dehio *et al.*, *EMBO J.* (1995). 14: 2471–82. Figure 7 is from Adam *et al.*, *EMBO J.* (1996). 15: 3315–21. This work was supported by Grant 94092 from the DRET (Direction des Recherches et Techniques).

REFERENCES

Adam, T., Arpin, M., Prévost, M.C., Gounon, P. & Sansonetti, P.J. (1995). Cytoskeletal rearrangements and the functional role of T-plastin during entry of *Shigella flexneri* into HeLa cells. *Journal of Cell Biology*, **129**, 367–81.

Adam, T., Giry, M., Boquet, P. & Sansonetti, P.J. (1996). Rho-dependent membrane folding causes *Shigella* entry into epithelial cells. *EMBO Journal*, **15**, 3315–21.

Allaoui, A., Sansonetti, P.J. & Parsot, C. (1992). MxiD, an outer membrane protein

necessary for the secretion of the *S. flexneri* Ipa invasins. *Molecular Microbiology*, **7**, 59–68.
Altmeyer, R.M., McNern, J.K., Bossio, J.C., Rosenshine, I., Finlay, B.B. & Galan, J.E. (1993). Cloning and molecular characterization of a gene involved in Salmonella adherence and invasion of cultured epithelial cells. *Molecular Microbiology*, **7**, 89–98.
Bernardini, M.L., Mounier, J., d'Hauteville, H., Coquis-Rondon, M. & Sansonetti, P.J. (1989). Identification of *icsA*, a plasmid locus of *Shigella flexneri* that governs bacterial intra and intercellular spread through interaction with F-actin. *Proceedings of the National Academy of Sciences, USA*, **86**, 3867–71.
Clark, E.A. & Brugge, J.S. (1995). Integrins and signal transduction pathways: the road taken. *Science*, **268**, 233–9.
Craig, S.W. & Johnson, R.P. (1996). Assembly of focal adhesions: progress, paradigms, and portents. *Current Opinion in Cell Biology*, **8**, 74–85.
Dehio, C., Prévost, M.C. & Sansonetti, P.J. (1995). Invasion of epithelial cells by *Shigella flexneri* induces tyrosine phosphorylation of cortactin by a $pp60^{c\text{-}src}$ mediated signalling pathway. *EMBO Journal*, **14**, 2471–82.
Francis, C.L., Ryan, T.A., Jones, B.D., Smith, S.J. & Falkow, S. (1993). Ruffles induced by Salmonella and other stimuli direct macropinocytosis of bacteria. *Nature*, **364**, 639–42.
Galan, J.E., Pace, J. & Hayman, M.J. (1992). Involvement of the epidermal growth factor receptor in the invasion of cultured mammalian cells by *Salmonella typhimurium*. *Nature*, **357**, 588–9.
Griffin, F.M., Griffin, J.A., Leider, J.E. & Silverstein, S.C. (1975). Studies on the mechanism of phagocytosis. I. Requirements for the circumferential attachment of particle-bound ligands to specific receptors on the macrophage plasma membrane. *Journal of Experimental Medicine*, **142**, 1263–82.
Hermant, D., Ménard, R., Arricau, N., Parsot, C. & Popoff, M.Y. (1995). Functional conservation of the *Salmonella* and *Shigella* effectors of entry into epithelial cells. *Molecular Microbiology*, 781–9.
Isberg, R.R. (1991). Discrimination between intracellular uptake and surface adhesion of bacterial pathogens. *Science*, **252**, 934–8.
Jones, B.D., Paterson, H.F., Hall, A. & Falkow, S. (1993). *Salmonella typhimurium* induces membrane ruffling by a growth-factor receptor independent mechanism. *Proceedings of the National Academy of Sciences, USA*, **90**, 10390–4.
Kaniga, K., Tucker, S., Trollinger, J.E. & Galan, J.E. (1995). Homologues of the *Shigella* IpaB and IpaC invasins are required for *Salmonella typhimurium* entry into cultured epithelial cells. *Journal of Bacteriology*, **177**, 7078–85.
Kaplan, K.B., Swedlow, J.R., Morgan, D.O. & Varmus, H.E. (1995). c-Src enhances the spreading of src-/- fibroblasts on fibronectin by a kinase independent mechanism. *Genes and Development*, **9**, 1505–17.
Labigne-Roussel, A.F., Lark, L., Schoolnik, G. & Falkow, S. (1984). Cloning and expression of an afimbrial adhesin (AFA) responsible for P blood group-independent mannose-resistant hemagglutination from a pyelone-phritic *E. coli* strain. *Infection and Immunity*, **46**, 251–9.
Maurelli, A.T., Baudry, B., d'Hauteville, H., Hale, T.L. & Sansonetti, P.J. (1985). Cloning of plasmid DNA sequences involved in invasion of HeLa cells by *Shigella flexneri*. *Infection and Immunity*, **49**, 164–71.
Ménard, R., Sansonetti, P.J. & Parsot, C. (1993). Non polar mutagenesis of the *ipa* gene defines IpaB, IpaC and IpaD as effectors of *Shigella flexneri* entry into epithelial cells. *Journal of Bacteriology*, **175**, 5899–906.
Ménard, R., Sansonetti, P.J., Parsot, C. & Vasselon, T. (1994*a*). Extracellular

association and cytoplasmic partitioning of the IpaB and IpaC invasins of *Shigella flexneri*. *Cell*, **79**, 515–25.

Ménard, R., Sansonetti, P.J. & Parsot, C. (1994*b*). The secretion of the *Shigella flexneri* Ipa invasins is induced by the epithelial cell and controlled by IpaB and IpaD. *EMBO Journal*, **13**, 5293–302.

Ménard, R., Prévost, M.C., Gounon, P., Sansonetti, P.J. & Dehio, C. (1996). The secreted Ipa complex of *Shigella flexneri* promotes entry into mammalian cells. *Proceedings of the National Academy of Sciences, USA*, **93**, 1254–8.

Mengaud, J., Ohayon, H., Gounon, P., Mège, R.-M. & Cossart, P. (1996). E-cadherin is the receptor for internalin, a surface protein required for entry of *L. monocytogenes* into epithelial cells. *Cell*, **84**, 923–33.

Nishiyama, T., Sasaki, T., Takaishi, K., Masaki, K., Yaku, H., Araki, K. *et al.* (1994). Rac p21 is involved in insulin-induced membrane ruffling and Rho p21 is involved in hepatocyte growth factor- and 12-O-tetradecanoylphorbol-13-acetate (TPA)-induced membrane ruffling in KB cells. *Molecular and Cellular Biology*, **14**, 2447–56.

Parsot, C., Ménard, R., Gounon, P. & Sansonetti, P.J. (1995). Enhanced secretion through the *Shigella flexneri* Mxi-Spa translocon leads to assembly of extracellular proteins into macromolecular structures. *Molecular Microbiology*, **16**, 291–300.

Parsot, C. & Sansonetti, P.J. (1996). In Miller, V.L., eds. *Bacterial Invasiveness*, pp. 25–42. Springer Verlag.

Ridley, A., Paterson, H.F., Johnston, C.L., Diekmann, D. & Hall, A. (1992). The small GTP-binding protein rac regulates growth factor-induced membrane ruffling. *Cell*, **70**, 401–10.

Rosqvist, R., Magnusson, K.E. & WolWatz, H. (1994). Target cell contact triggers expression and polarized transfer of *Yersinia* YopE cytotoxin into mammalian cells. *EMBO Journal*, **13**, 964–72.

Rosqvist, R., Hakansson, S., Forsberg, A. & Wolf-Watz, H. (1995). Functional conservation of the secretion and translocation machinery for virulence proteins of *Yersiniae*, *Salmonellae* and *Shigellae*. *EMBO Journal*, **14**, 4187–95.

Sansonetti, P.J., Arondel, J., Cavaillon, J.-M. & Huerre, M. (1995). Role of Interleukin-1 in the pathogenesis of experimental shigellosis. *Journal of Clinical Investigations*, **96**, 884–92.

Sansonetti, P.J., Arondel, J., Fontaine, A., d'Hauteville, H. & Bernardini, M.L. (1991). *ompB* (osmo-regulation) and *icsA* (cell-to-cell spread) mutants of *Shigella flexneri*: vaccine candidates and probes to study the pathogenesis of shigellosis. *Vaccine*, **9**, 416–22.

Sasakawa, C., Makino, S., Kamata, K. & Yoshikawa, M. (1986). Isolation, characterization and mapping of Tn*5* insertions into the 140-Megadalton invasion plasmid defective in the mouse Sereny test in *Shigella flexneri*. *Infection and Immunity*, **54**, 32–6.

Schaller, M.D. & Parsons, J.T. (1993). Focal adhesion kinase: an integrin-linked protein tyrosine kinase. *Trends in Cell Biology*, **3**, 258–62.

Vasselon, T., Mounier, J., Hellio, R. & Sansonetti, P.J. (1992). Movement along actin filaments of the perijunctional area and De Novo polymerization of cellular actin are required for *Shigella flexneri* colonization of epithelial Caco-2 cell monolayers. *Infection and Immunity*, **60**, 1031–40.

Vasselon, T., Mounier, J., Prévost, M.C., Hellio, R. & Sansonetti, P.J. (1991). A stress fiber-based movement of *Shigella flexneri* within cells. *Infection and Immunity*, **59**, 1723–32.

Wassef, J.S., Keren, D.F. & Mailloux, J.L. (1989). Role of M cells in initial antigen uptake and in ulcer formation in the rabbit intestinal loop model of shigellosis. *Infection and Immunity*, **57**, 858–63.

Watarai, M., Tobe, T., Yoshikawa, M. & Sasakawa, C. (1995). Contact of *Shigella* with host cells triggers release of Ipa invasins and is an essential function of invasiveness. *EMBO Journal*, **14**, 2461–70.

Watarai, M., Funato, S. & Sasakawa, C. (1996). Interaction of Ipa proteins of Shigella flexneri with alpha5-beta1 integrin promotes entry of the bacteria into mammalian cells. *Journal of Experimental Medicine*, **183**, 991–9.

Zychlinsky, A., Fitting, C., Cavaillon, J.-M. & Sansonetti, P.J. (1994). Interleukin 1 is released by murine macrophages during apoptosis induced by *Shigella flexneri*. *Journal of Clinical Investigations*, **94**, 1328–32.

Zychlinsky, A., Prévost, M.-C. & Sansonetti, P.J. (1992). *Shigella flexneri* induces apoptosis in infected macrophages. *Nature*, **358**, 167–9.

TRANSPORT THROUGH PLASMODESMATA AND NUCLEAR PORES: CELL-TO-CELL MOVEMENT OF PLANT VIRUSES AND NUCLEAR IMPORT OF *AGROBACTERIUM* T-DNA

R. LARTEY, S. GHOSHROY, J. SHENG AND V. CITOVSKY

Department of Biochemistry and Cell Biology, State University of New York, Stony Brook, NY 11794-5215, USA

PLASMODESMAL TRANSPORT OF PLANT VIRUSES

Structure of Plasmodesmata

Plasmodesmata are cytoplasmic bridges across the walls separating plant cells. Bernhardi suggested the existence of such channels, as well as the possibility of the intercellular communication which they would permit, almost two centuries ago (Bernhardi, 1805). Three-quarters of a century passed before Tangl first observed plasmodesmata (Tangl, 1879), and it was not until the turn of the century that Strasburger gave them their present name (Strasburger, 1901). None the less, the isolation of plasmodesmata and the elucidation of their constituent proteins and regulatory mechanisms have remained elusive to this day. In recent years, plasmodesmata have been implicated not only in transport of small molecules such as water, ions, and photoassimilates, but in intercellular traffic of proteins, nucleic acids, and protein-nucleic acid complexes. Interestingly, this ability to transport large molecules and multimolecular complexes has been described for only one other biological channel, the nuclear pore.

The current structural model of a simple plasmodesma derives from the electron microscopy experiments that used high-pressure freezing to preserve fine details of tobacco leaf plasmodesmata (Ding, Turgeon & Parthasarathy, 1992*b*). According to this study, the endoplasmic reticulum passes through the plasmodesma, and is both surrounded by and filled with regularly-spaced globular particles approximately 3 nm in diameter; this membrane-protein complex is referred to as the desmotubule or appressed endoplasmic reticulum (Fig. 1). The particles associated with the outer leaflet of the appressed endoplasmic reticulum appear to be connected to those on the inner leaflet by most likely proteinaceous filaments. Further, globular particles are embedded in the inner leaflet of the plasmalemma, greatly restricting the open space (Ding *et al.*, 1992*b*). In addition to the endoplasmic reticulum, plasmodesmata may associate

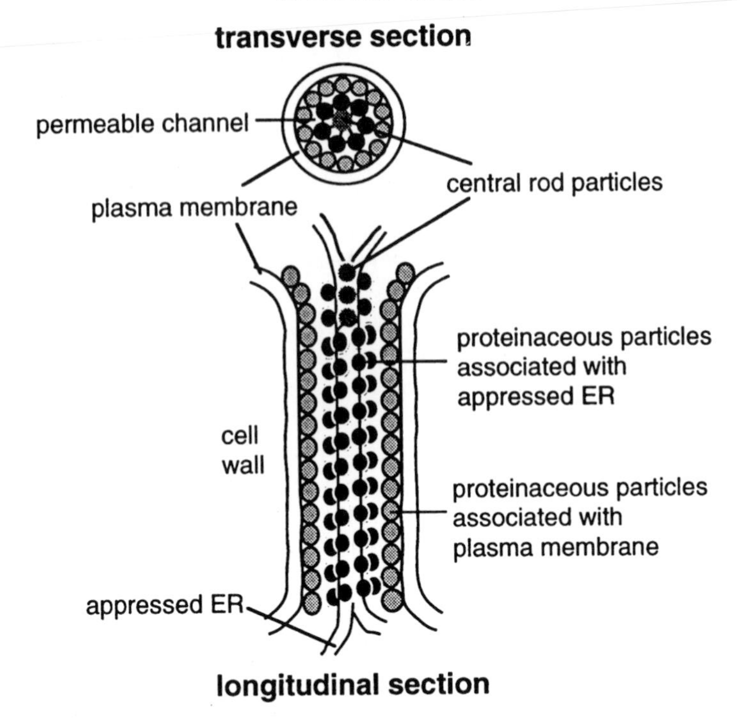

Fig. 1. Structure of a simple plasmodesma. See text for details. (Adapted from Ding, Turgeon & Parthasarathy, 1992*b*; Citovsky, 1993; Citovsky & Zambryski, 1993.)

with cytoskeletal elements. For example, recent results suggested that myosin, which interacts with actin and produces force through hydrolysis of ATP, colocalizes with plasmodesmata (Radford & White, 1996). Similarly, the presence of actin in plasmodesmata has been proposed (White *et al.*, 1994).

Plasmodesmata follow a complex developmental pathway. There are two types of plasmodesmata, which may be morphologically defined as simple and branched, or developmentally defined as primary and secondary (Ding *et al.*, 1992*a*; Lucas & Gilbertson, 1994). Primary or simple plasmodesmata are found predominantly in young tissue and consist of a simple, single channel, as pictured in Fig. 1 (Ding *et al.*, 1992*a*). The creation of primary plasmodesmata is a function of cell plate formation during cytokinesis (Lucas *et al.*, 1990; Lucas, Ding & van der Schoot, 1993; Franceschi, Ding & Lucas, 1994). The ER is positioned across, and perpendicular to, the cell plate (Porter & Machado, 1960; Hepler, 1982). It then becomes appressed by the developing cell plate and, together with the plasmalemma, provides cytoplasmic continuity between cells. The primary plasmodesma is essentially an incomplete separation between two daughter cells. Secondary or branched plasmodesmata are found in older tissues and show a higher degree of variability, often with many single channels leading into a larger central cavity (Ding *et al.*, 1992*a*). As the number of plasmodesmata in a given area is constant in both older and younger tissue, it is believed that secondary plasmodesmata are derived, in most cases, from pre-existing primary plasmodesmata (Ding *et al.*, 1993; Lucas & Gilbertson, 1994). Functionally, the secondary plasmodesmata differ in several ways from simple ones, most notably in response to viral infection (Ding *et al.*, 1992*a*) (see below).

Transport through plasmodesmata appears to be very complex. Ordinarily, only small microchannels are available for passive transport between the globular particles. This unassisted transport through plasmodesmata, studied using microinjected dyes, appears to be limited to molecules up to 1.5–2.0 nm in diameter, equivalent to a molecular mass of 0.75–1.0 kDa (Terry & Robards, 1987; Wolf *et al.*, 1989). Several factors act to decrease plasmodesmal permeability. These include divalent cations (Ca^{2+}, Mn^{2+}, and Sr^{2+}), phorbol esters, aromatic amino acids, and phosphoinositides (Erwee & Goodwin, 1984; Baron-Eppel *et al.*, 1988; Tucker, 1988; Lucas *et al.*, 1993). Only one known endogenous plant protein, KN1, encoded by the maize *knotted-1* homeobox gene, is known to increase plasmodesmal permeability (Lucas *et al.*, 1995, see also below). However, many plant viruses have evolved to increase the plasmodesmal size exclusion limit during local and systemic spread of infection. The mechanism by which this increase in plasmodesmal permeability occurs is unknown. For example, it is possible that a conformational change in the aforementioned filaments enlarges the permeable space of plasmodesmal channels by pulling the globular particles into the appressed ER (Fig. 2) (Citovsky, 1993).

Our present knowledge of plasmodesmal structure and composition is purely descriptive and has been derived mainly from electron microscopy studies. No functional plasmodesmal proteins have been definitely identified. Initial reports on cloning plasmodesmata genes or purifying plasmodesmata-associated proteins (Meiners & Schindler, 1989; Meiners, Xu & Schindler, 1991) either have been disputed or require further substantiation (Mushegian & Koonin, 1993). This is in contrast to the animal counterparts of plasmodesmata, gap junctions, which have been long since purified, characterized, and their encoding genes cloned (Kumar & Gilula, 1996). The main reason for such difference is technical. Unlike gap junctions, plasmodesmata are firmly embedded in the plant cell wall and, thus, recalcitrant to purification. One way to circumvent this difficulty is to use plant viruses, which specifically interact with plasmodesmata during infection, as a natural molecular tool to identify and characterize the plasmodesmal transport pathway and its protein components.

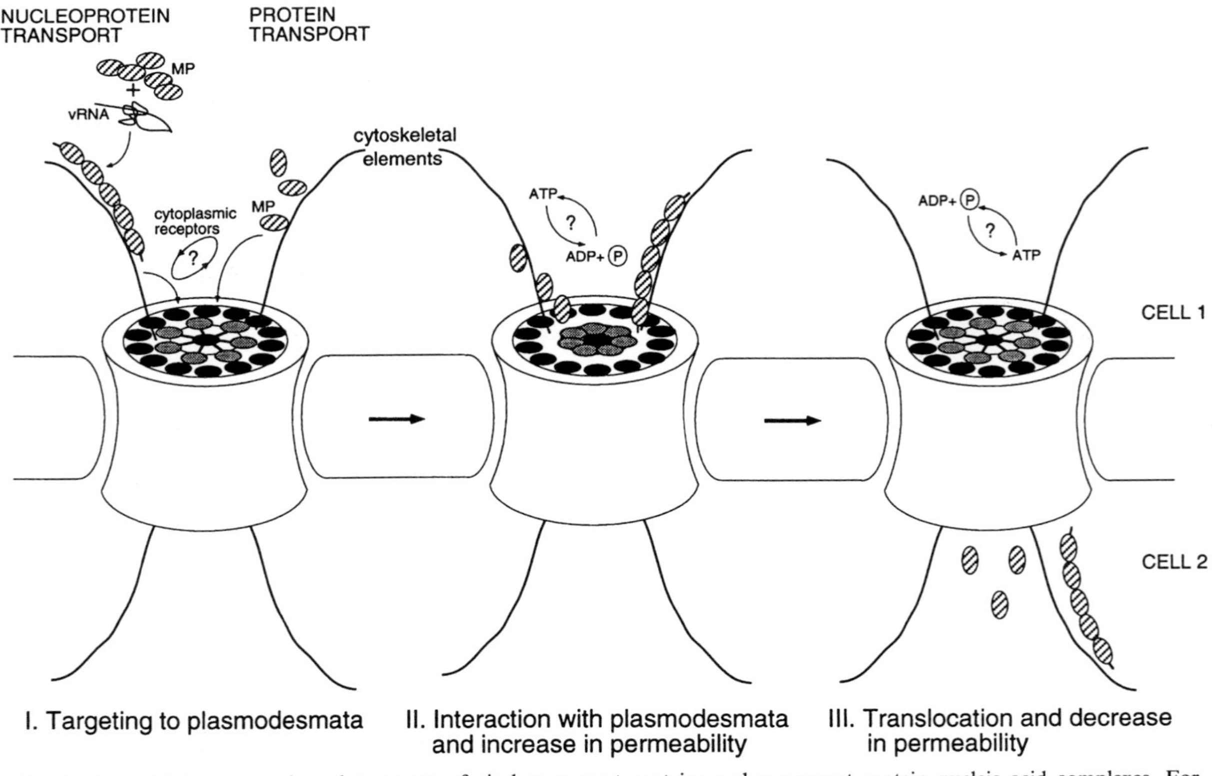

Fig. 2. A model for plasmodesmal transport of viral movement proteins and movement protein–nucleic acid complexes. For explanation, see text.

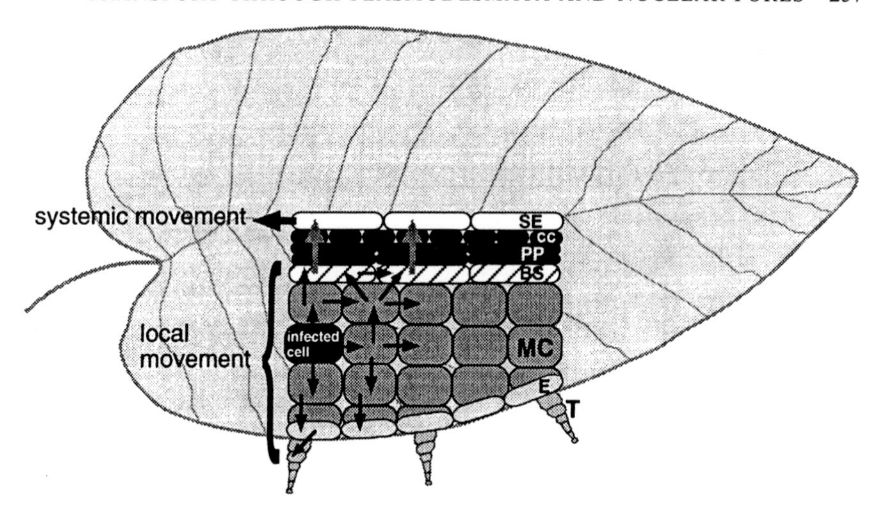

Fig. 3. Cellular routes for local and systemic movement of plant viruses. T, trichome; E, epidermal cells; MC, mesophyll cells; BS, bundle sheath cells; PP, phloem parenchyma cells; CC, companion cells; and SE, sieve elements. Arrows indicate viral movement. Viral spread between trichome, epidermis, and mesophyll cells represents local, cell-to-cell movement. Plasmodesmata between bundle sheath and phloem parenchyma cells are thought to mediate transition from local to systemic movement which then proceeds through the sieve elements to other plant organs (Ding *et al.*, 1992*a*; Lucas & Gilbertson, 1994).

Viral cell-to-cell movement proteins

Katherine Esau first postulated that viruses moved throughout the plant via plasmodesmata (Esau, 1948). Since then, viral spread through plant intercellular connection has been shown to occur in two major steps: local and systemic (Fig. 3). Following initial infection, usually by mechanical or insect-mediated inoculation, many plant viruses spread from cell to cell through plasmodesmata until they reach the vascular system; the viruses are then transported systemically through the vasculature. Presumably, viral spread through the vascular tissue is a passive process, occurring with the flow of photoassimilates (for review see Leisner & Howell, 1993); in contrast, the cell-to-cell movement is an active function, requiring specific interaction between the virus and plasmodesmata. This interaction is mediated by virus-encoded nonstructural movement proteins which act to increase plasmodesmal permeability and transport viral nucleic acids through the enlarged plasmodesmal channels (for review see Lucas *et al.*, 1990; Citovsky, 1993; Lucas & Gilbertson, 1994).

Movement protein–nucleic acid complexes

The best studied cell-to-cell movement protein is the 30 kDa protein (P30) of tobacco mosaic virus (TMV) (Deom, Shaw & Beachy, 1987). To date, P30 has been suggested to possess three biological activities. It is thought to bind TMV RNA, forming an extended P30-RNA complex that can penetrate the plasmodesmal channel (Citovsky *et al.*, 1990; Citovsky *et al.*, 1992*b*). It also may interact with the cytoskeletal elements to facilitate transport of the P30-TMV RNA complexes from the cell cytoplasm to plasmodesmata (Heinlein, Epel & Beachy, 1995; McLean, Zupan & Zambryski, 1995). Finally, P30 functions to increase the size exclusion limit of plasmodesmata (Wolf *et al.*, 1989). Based on these activities, a model for the P30-mediated intercellular transport of TMV RNA has been proposed (Fig. 2) (Citovsky, 1993; Citovsky & Zambryski, 1995; Zambryski, 1995). In the initially infected cell, P30 is produced by transcription of the invading genomic TMV RNA. This protein then associates with a certain portion of the viral RNA molecules, sequestering them from replication and mediating their transport into the neighboring uninfected host cells. *In vitro* studies showed that P30 binds both RNA and single stranded (ss)DNA, but not double stranded (ds)DNA, cooperatively and without sequence specificity (Citovsky *et al.*, 1990; Citovsky *et al.*, 1992*b*). This sequence non-specific binding explains the observations that coinfection with TMV can complement cell-to-cell movement of plant viruses which normally do not spread through plasmodesmata (for review see Carr & Kim, 1983; Atabekov & Taliansky, 1990). Electron microscopic observations revealed that P30 binding unfolds the nucleic acid molecule creating an extended protein-RNA complex of 2.0–2.5 nm in diameter (Citovsky *et al.*, 1992*b*) (Fig. 2). Because the free folded TMV RNA has been estimated to be 10 nm in diameter (Gibbs, 1976), association with P30 likely shapes it in a thinner transferable form capable of transport through plasmodesmal channels.

Subsequent to the demonstration that P30 of TMV binds single-stranded nucleic acids (Citovsky *et al.*, 1990; Citovsky *et al.*, 1992*b*), cell-to-cell movement proteins of many other plant viruses were also found to exhibit similar activity. For example, P1 of cauliflower mosaic virus (CaMV) was shown to associate with both ssDNA and RNA into long and thin complexes which closely resemble P30-ssDNA and P30-RNA complexes (Citovsky, Knorr & Zambryski, 1991; Thomas & Maule, 1995). Furthermore, P1 binding affinity to RNA was higher than that towards ssDNA, suggesting that P1-CaMV RNA complexes may be involved in the cell-to-cell spread of this virus (Citovsky *et al.*, 1991). This TMV-like mechanism for cell-to-cell movement may complement the better-characterized spread of CaMV in the form of a whole viral particle through virus-modified plasmodesmata (Kitajima & Lauritis, 1969). In addition to TMV and CaMV, the movement proteins of alfalfa mosaic virus (AMV), red clover necrotic mosaic dianthovirus (RCNMV) and several other viruses have been shown to bind single-stranded nucleic acids (Osman, Hayes & Buck, 1992; Schoumacher *et al.*, 1992; Pascal *et al.*, 1994). Similar to P30, these movement proteins bound nucleic acids cooperatively (Citovsky *et al.*, 1991; Osman *et al.*, 1992;

Schoumacher *et al.*, 1992); however, the movement protein of RCNMV did not appear to significantly extend the bound RNA molecules (Fujiwara *et al.*, 1993). Thus, transport through plasmodesmata via movement protein–nucleic acid intermediates may represent a common mechanism for cell-to-cell spread of many plant viruses.

Although practically all movement proteins function to transport viral nucleic acids, the degree of transport selectivity varies. For example, the P30 protein of TMV can bind and, by implication, transport any single-stranded nucleic acid (Citovsky *et al.*, 1990, 1992*b*). In contrast, the RCNMV movement protein is capable of trafficking ssRNA but not ssDNA or dsDNA (Fujiwara *et al.*, 1993), whereas the bean dwarf mosaic geminivirus (BDMV) BL1 movement protein facilitates transport of dsDNA but not ssDNA or ssRNA molecules (Noueiry, Lucas & Gilbertson, 1994). In the latter case, however, the transport of dsDNA seems incompatible with biochemical and genetic evidence obtained with another bipartite geminivirus, squash leaf curl virus (SqLCV), for which the transported form has been proposed to be its genomic ssDNA (Pascal *et al.*, 1994). Because BL1 only weakly binds nucleic acids (Pascal *et al.*, 1994), it interacts with the second geminivirus movement protein, BR1 (Sanderfoot & Lazarowitz, 1995), which directly associates with the transported nucleic acid molecule (Pascal *et al.*, 1994). Thus, BR1 most likely binds the viral nucleic acid and transports it out of the host cell nucleus where geminiviruses replicate (Goodman, 1981). BR1-ssDNA complexes (or BR1-dsDNA complexes, in the case of BDMV) associate with BL1 which then mediates the plasmodesmal transport (see also below).

Movement protein–cytoskeleton interaction

Because TMV RNA translation and therefore, the production of P30 occurs in the host cell cytoplasm (Palikaitis & Zaitlin, 1986), P30-TMV RNA cell-to-cell transport complexes most likely are also formed in the cytoplasmic compartment. How, then, do these complexes arrive at plasmodesmata prior to the cell-to-cell movement? Recent data suggest that P30 interacts with microtubuli and, to a lesser extent, with actin microfilaments (Heinlein *et al.*, 1995; McLean *et al.*, 1995). This interaction was inferred from colocalization of the wild-type P30 transiently expressed in tobacco protoplasts with tubulin as well as with actin filaments (McLean *et al.*, 1995). P30 association with actin and tubulin was also demonstrated using *in vitro* binding assays (McLean *et al.*, 1995). Furthermore, P30 tagged by translational fusion with the jellyfish green fluorescent protein (GFP) formed a filamentous network following transient expression in plant protoplasts. Interestingly, these filamentous arrays of GFP-P30 were best detected at 18 to 20 hours following expression; after 48 to 72 hours, most GFP-P30 appeared as aggregates along the periphery of the protoplasts (McLean *et al.*, 1995). Protoplasts from transgenic tobacco expressing P30, which represent a steady state versus a transient expression system, contained P30 predominantly near the plasma membrane and the residual cell walls, where, in intact tissue, the plasmodesmata would be found (McLean *et al.*, 1995). Finally, GFP-P30 was introduced into the TMV genome and the resulting modified virus retained infectivity (Heinlein *et al.*, 1995). Fluorescent GFP expressed in tobacco protoplasts and leaf tissue following infection formed an intracellular network which coaligned with cellular microtubuli (however, no association of P30 with F-actin was detected in these experiments) (Heinlein *et al.*, 1995). Although movement proteins of other plant viruses have not yet been tested for interaction with the cytoskeleton, it is likely that such an interaction will be found for many viral species.

Taken together, these observations suggest that P30 may interact with the cytoskeletal elements in the host cell cytoplasm and use them as tracks to migrate to the cell periphery and, ultimately, to plasmodesmata (Fig. 2). Because P30 is most likely associated with the viral RNA during infection, it is the P30-TMV RNA complex that may interact with the host cell microtubuli and microfilaments during transport to plasmodesmata (Fig. 2). Alternatively, interaction with the cytoskeleton may anchor P30 or P30-TMV RNA complexes in the cytoplasm, gradually releasing them in response to an, as yet unidentified, signal for plasmodesmal transport to occur. In this case, association with microtubuli and microfilaments would function as a regulatory mechanism for plasmodesmal transport rather than the targeting apparatus. Similar regulation by cytoplasmic anchoring has been described for nuclear import of the NF-κB transcription factor (for review see Dingwall, 1991) (see also below).

Increase in plasmodesmal permeability

Once the P30-TMV RNA cell-to-cell transport complex reaches the plasmodesmal channel, it must transverse it to enter the neighbouring host cell. Although the estimated 2.0–2.5 nm diameter of this nucleoprotein complex (Citovsky *et al.*, 1992*b*) is relatively small, it still is incompatible with the 1.5 nm diameter of the intact plasmodesmal channel (Wolf *et al.*, 1989). To allow movement therefore, P30 induces an increase in plasmodesmal permeability (Fig. 2). Originally, the ability of P30 to elevate plasmodesmal size exclusion limit was detected by injection of fluorescently labeled dextrans into leaf mesophyll of transgenic tobaccos expressing P30 (Wolf *et al.*, 1989). Unlike the wild-type tobacco mesophyll plasmodesmata which can traffic only dextrans of up to 0.75–1.0 kDa, the P30 transgenic plants exhibited a plasmodesmal size exclusion limit of almost 10 kDa (Wolf *et al.*, 1989). Because the transgenic plant tissue represents a steady state of P30 accumulation and activity, it remained unclear whether the increase in plasmodesmal size exclusion limit was due to activation of an endogenous plasmodesmal transport pathway or induction of a specific host response to this viral protein. Direct microinjection of purified P30 into the wild-type tobacco mesophyll resulted in a relatively fast (3–5 minutes) size exclusion limit increase up to 20 kDa, indicating that P30 functions via an existing plasmodesmal transport machinery (Waigmann *et al.*, 1994). Importantly, the increased size exclusion limit of 10–20 kDa (Wolf *et al.*, 1989; Waigmann *et al.*, 1994) corresponds to a 5–9 nm diameter of the dilated channel, potentially allowing unrestricted traffic of 2.0–2.5 nm-wide P30-TMV RNA complexes (Citovsky *et al.*, 1992*b*).

P30 microinjection experiments provided another important clue to its function. It was noted that large fluorescent dextrans moved not only into the cells adjacent to the microinjected cell, but traveled as far as 20 to 50 cells away from the site of injection (Waigmann *et al.*, 1994). The observations indicated that P30 itself must have moved through plasmodesmata to induce the increase in size exclusion limit in the distant mesophyll cells, providing the first evidence that these plasmodesmata can traffic protein molecules. An alternative and unlikely possibility that P30 induced an intracellular signaling pathway without its own cell-to-cell movement was ruled out later by immunolocalization experiments which showed that microinjected P30 indeed moves between plant cells (Waigmann & Zambryski, 1995).

Similarly to single-stranded nucleic acid binding, the increase in plasmodesmal size exclusion limit is most likely a property of many viral movement proteins. Presently, the cell-to-cell movement proteins of RCNMV, AMV, cucumber mosaic virus (CMV), tobacco rattle virus (TRV), potato virus X, and the BL1 movement protein of BDMV have been shown to enable transport of large fluorescent dextrans between plant cells (Derrick, Barker & Oparka, 1992; Fujiwara *et al.*, 1993; Poirson *et al.*, 1993; Noueiry *et al.*, 1994; Vaquero *et al.*, 1994; Angell, Davies & Baulcombe, 1996). RCNMV, CMV and BDMV movement proteins were also shown to move from cell to cell themselves (Fujiwara *et al.*, 1993; Noueiry *et al.*, 1994; Ding *et al.*, 1995; Waigmann & Zambryski, 1995). Collectively, these experiments have established the use of large fluorescent dextrans as an assay for interaction between plasmodesmata and viral movement proteins. Recent observations, however, questioned the biological relevance of this approach. P30 was coinjected with fluorescent dextrans into the tobacco trichomes which are arranged in a linear file of cells and, consequently, allow better visualization of intercellular transport (Waigmann & Zambryski, 1995). Surprisingly, no size exclusion limit increase could be detected although trichome cells support viral infection and cell-to-cell movement. However, when P30 was fused translationally to a reporter β-glucuronidase (GUS) enzyme, the microinjected GUS-P30 protein efficiently moved between trichome cells (Waigmann & Zambryski, 1995). This result strongly suggests that P30 is a *cis*-acting mediator of plasmodesmal transport. In other words, P30 has to be physically associated with the transported molecule such as viral RNA or the reporter GUS enzyme.

This idea is consistent with most known examples of protein transport and targeting. For example, the increase in the nuclear pore size exclusion limit during nuclear import does not allow passage of protein molecules which are not directly associated with the nuclear localization signal (NLS) sequence (Goldfarb *et al.*, 1986). Cell-to-cell transport of large fluorescent dextrans coinjected with P30 into the leaf mesophyll, then, may be only an 'afterglow' of the P30 biological activity when mesophyll (but not trichome) cell plasmodesmata remain open following microinjection or overexpression of the movement protein. The true plasmodesmal transport, however, likely requires direct interaction between the movement protein and the transported molecule.

Possible mechanisms for plasmodesmal transport

The mechanism by which macromolecules are transported through plasmodesmata is completely unknown. However, it may be possible to predict the key features of this process by analogy to nuclear transport which is the only other known example of traffic of large proteins and protein-nucleic acid complexes through a membrane pore. To make such a prediction, we first summarize the mechanism of the nuclear import process.

Nuclear localization signal (NLS)

Molecular transport across the nuclear envelope involves a great diversity of proteins and nucleic acids. This transport is bi-directional and occurs exclusively through the nuclear pore complex (NPC). While relatively small molecules (up to 60 kDa) (for review see Goldfarb & Michaud, 1991; Nigg, Baeuerle & Luhrmann, 1991; Akey, 1992) diffuse through the NPC, transport of larger molecules occurs by an active mechanism mediated by specific nuclear localization signal (NLS) sequences contained in the transported molecule (for review see Garcia-Bustos, Heitman & Hall, 1991); interestingly, transport of some small endogenous nuclear proteins such as the H1 histone (21 kDa) also occurs by an active process (Breeuwer & Goldfarb, 1990).

With few exceptions (e.g. influenza virus nucleoprotein NLS, yeast Gal4 protein NLS), all NLSs can be classified in two general groups: (i) the single basic motif exemplified by the SV40 large T antigen NLS (PKKKRKV), and (ii) the bipartite motif consisting of two basic domains separated by a variable number (but not less than four) of spacer amino acids and exemplified by the nucleoplasmin NLS (KR-X_{10}-KKKL) (Robbins *et al.*, 1991). The first domain of a bipartite NLS usually consists of two adjacent basic residues while the second domain contains three out of five basic amino acids (reviewed in Dingwall & Laskey, 1991). To date, most NLSs found in plant proteins belong to the bipartite type (Raikhel, 1992).

NLS receptors

NLS receptors, which have been implicated in protein nuclear import in several organisms, belong to a multiprotein family which includes animal karyopherin α (importin 60) and yeast Kap60 (formerly Srp1) (Gorlich *et al.*, 1994; Powers & Forbes, 1994 and references therein: Enenkel, Blobel & Rexach, 1995; Rexach & Blobel, 1995; Weis, Mattaj & Lamond, 1995). They are thought to recognize and bind the NLS sequence of the transported protein molecule and direct it to the nuclear pore. Once the receptor–NLS bearing protein complex is at the nuclear pore, animal karyopherin β (importin 90) or yeast Kap95 proteins mediate its binding to the NPC proteins, nucleoporins. This binding is followed by translocation into the nucleus which requires the GTPase Ran (or its yeast homolog Gsp1) (Melchior *et al.*, 1993; Moore & Blobel, 1993; Corbett *et al.*, 1995) and the Ran-interacting protein p10 (Nehrbass & Blobel, 1996). During translocation, karyopherin α is thought to enter the nucleus together with the bound NLS-containing protein while karyopherin β remains at the nuclear pore orifice (Gorlich *et al.*, 1995).

Thus, nuclear import involves two discrete steps: (i) binding of NLSs to the cytoplasmic receptors which direct the transported molecule to the NPC, and (ii) transport through the nuclear pore (Newmeyer & Forbes, 1988; Richardson *et al.*, 1988). While interaction with NLS receptors and NPC binding are energy-independent, the actual translocation requires metabolic energy in the form of ATP and GTP (Newmayer *et al.*, 1986; Melchior *et al.*, 1993).

Regulation of nuclear import

Many nuclear proteins function only at a specific developmental stage or in response to certain external stimuli such as hormones. Nuclear entry of these proteins, therefore, is tightly regulated. Signal masking and cytoplasmic anchoring represent the two major mechanisms for regulation of nuclear import (Dingwall, 1991). Regulation by NLS masking is exemplified by the nuclear import of the glucocorticoid receptor. In the absence of hormone, this protein is bound in the cytoplasm to the HSP90 chaperone protein, masking the NLS (Picard & Yamamoto, 1987; Cadepond *et al.*, 1991). When activated by the binding of a steroid hormone, the receptor molecule is released from HSP90 and its NLS is revealed, directing the receptor-hormone complex into the cell nucleus to activate gene expression (Picard & Yamamoto, 1987; Ham & Parker, 1989).

Nuclear import of the NF-κB transcription factor is a paradigm of regulation by cytoplasmic anchors. NF-κB, which usually consists of 50 kDa and 65 kDa subunits (Gilmore, 1990), is present in the cytoplasm of non-activated cells as a complex with an inhibitory protein I-κB (Baeuerle & Baltimore, 1988). I-κB contains ankyrin repeats (Hatada *et al.*, 1992) originally found in the erythrocyte membrane anchoring protein ankyrin

(Lux, John & Bennett, 1990). These repeats are thought to promote association of I-κB, and its cognate NF-κB, with the cellular matrix (Whiteside & Goodbourn, 1993). Stimulation of cells with a variety of agents (e.g. phorbol esters, interleukin 1) activates cellular kinases which phosphorylate I-κB, resulting in the release of NF-κB and its migration into the cell nucleus (Shirakawa & Urizel, 1989). Thus, the selectivity of the nuclear pore is itself constant and the regulation of transport involves conversion of the transported molecule from a non-transferable form to a transferable form (Dingwall, 1991).

Implications for the mechanism of plasmodesmal transport

Similar to nuclear import, transport through plasmodesmata most likely consists of two major steps: (i) recognition of the transported molecule in the cell cytoplasm and its targeting to the plasmodesmal channel, and (ii) translocation (Fig. 2). Proteins and protein-nucleic acids complexes destined for cell-to-cell transport may be recognized by their putative targeting sequence, the plasmodesmata localization signal (PLS). The carboxyl terminal part of the 100 amino acid-long P30 domain required for interaction with plasmodesmata (Waigmann *et al.*, 1994) may carry such a signal. PLSs may also mediate transport of nucleic acids associated with the PLS-containing protein, such as P30-TMV RNA complexes (Fig. 2). Similar transport of nucleic acids via a protein import pathway has been proposed for the nuclear import of *Agrobacterium* T-DNA associated with the bacterial VirD2 and VirE2 proteins (Citovsky *et al.*, 1992*c*; Howard *et al.*, 1992; Citovsky & Zambryski, 1993; Citovsky, Warnick & Zambryski, 1994; Guralnick, Thomsen & Citovsky, 1996; Zupan, Citovsky & Zambryski, 1996) as well as for the influenza virus genomic RNA-NP nucleoprotein complexes (O'Neill *et al.*, 1995).

The putative PLS potentially interacts with specific cytoplasmic receptors (Fig. 2). These, as yet unidentified, receptor proteins may function to transport the PLS-containing protein to the plasmodesmal annulus. Alternatively, the transported protein may be guided to plasmodesmata simply by association with cytoskeletal tracks (see above). In this case, the question of specific targeting, i.e. whether there are microfilament and microtubuli arrangements that lead only to plasmodesmata, remains to be resolved.

Once at the plasmodesmal channel, the transported protein or protein-PLS receptor complexes must increase plasmodesmal permeability to allow translocation. By analogy to nuclear import, GTPase and/or ATPase activities may be involved. Indeed, an ATPase activity has been localized to plasmodesmata (Zheng *et al.*, 1985; Didehvar & Baker, 1986), but further studies are necessary to ascertain the molecular nature of this enzyme. It is also possible that interaction of the transported protein with cytoskeleton directly 'relaxes' the plasmodesmal annulus increasing its size exclusion limit for translocation. The actual mechanism of plasmodesmal gating will be elucidated only with purification and characterization of the protein components of this channel.

Interaction of viral movement proteins with plasmodesmata may interfere with normal intercellular communication and, thus, be detrimental to the host plant. It is likely therefore, that a mechanism exists to regulate the activity of P30 and, possibly, cellular proteins capable of plasmodesmal transport (Lucas *et al.*, 1995). Irreversible deposition of P30 in the central cavity of secondary plasmodesmata has been proposed to represent one such mechanism (Citovsky *et al.*, 1993). Electron microscopy studies have localized P30 to the secondary but not primary plasmodesmata (Ding *et al.*, 1992*a*). Conversely, microinjected P30 increases plasmodesmal permeability in young tobacco leaves devoid of secondary plasmodesmata (Waigmann *et al.*, 1994). Thus, secondary plasmodesmata may represent sites of deposition of inactivated P30. It is also possible that P30 inactivation is mediated by a phosphorylation reaction (Fig. 2). A plant cell wall-associated protein kinase has been shown to specifically phosphorylate P30 at its carboxyl terminal serine and threonine residues (Citovsky *et al.*, 1993). The P30 kinase activity was developmentally regulated (Citovsky *et al.*, 1993), correlating with the formation of secondary plasmodesmata within the tobacco leaf (Ding *et al.*, 1992*a*). The possibility that the P30 kinase is a functional component of secondary plasmodesmata has not yet been tested.

By analogy to nuclear import, another possibility for plasmodesmal regulation is cytoplasmic anchoring. P30 interaction with the cytoskeleton may serve such a function by immobilizing P30 in the cell cytoplasm. This interaction may also mask the putative PLS sequence on the transported protein. Regardless of its molecular basis, the regulation pattern of plasmodesmal transport most likely determines the communication domains thought to exist in a plant organism.

NUCLEAR IMPORT OF *AGROBACTERIUM* T-DNA

Agrobacterium–plant cell interaction is the only known natural example of DNA transport between kingdoms. In this process, DNA is transported from wild-type *Agrobacterium* into the plant cell nucleus. Expression of this transferred DNA (T-DNA) results in neoplastic growths on the host plant. The wild-type T-DNA carries genes involved in synthesis of plant growth hormones and production of amino acid derivatives, opines, a major carbon/nitrogen source for *Agrobacterium*. Agrobacteria are usually classified based on the type of opines specified by the bacterial T-DNA, the most common strains being octopine- and nopaline-specific (Hooykaas & Beijersbergen, 1994).

Three genetic components of *Agrobacterium* are required for plant cell transformation (Fig. 4). The first component is the T-DNA which is actually transported from the bacterium to the plant cell. The T-DNA is a discrete segment of DNA located on the 200 kbp Ti (tumour-inducing) plasmid of *Agrobacterium*; it is delineated by two 25 bp imperfect direct repeats, known as the T-DNA borders. The second component is a 35 kbp virulence (vir) region located on the Ti plasmid and composed of seven major loci (virA, virB, virC, virD, virE, virG, and virH). The protein products of these genes, termed virulence (Vir) proteins, respond to the wounded plant environment, generate a copy of the T-DNA and mediate its transfer into the host cell. The third component is the chromosomal virulence (chv) genes, located on the *Agrobacterium* chromosome. Chv genes are involved in bacterial chemotaxis towards, and attachment to, the wounded plant cell (for review see Citovsky *et al.*, 1992*a*; Zambryski, 1992).

Because the T-DNA element is defined by its borders, the coding region of the wild-type T-DNA can be replaced by any DNA sequence without affecting its transfer from *Agrobacterium* to the plant. Thus, *Agrobacterium* is often used to produce transgenic plants expressing genes of interest. In addition to this technical application, *Agrobacterium* represents a fascinating model system to study the wide variety of biological processes which ultimately result in genetic transformation of the host plant cell.

T-DNA element

Induction of vir gene expression eventually produces a T-DNA copy capable of genetically transforming plant cells. Different types of Ti plasmids carry T-DNA elements of different composition. For example, the T-DNA in the nopaline Ti plasmid is a contiguous stretch of about 22 kbp whereas the octopine-specific T-DNA is composed of three adjacent T-DNAs: left (13 kbp), central (1.5 kbp), and right (7.8 kbp). The borders of a T-DNA element are defined as conserved 25 bp sequences which delimit the transferred segments. Genetic studies using deletion mutants have shown that the right border is absolutely required for *Agrobacterium* pathogenicity, whereas deletions of the left border have little effect. Furthermore, inversion of the right border leads to reduced virulence and transfer of nearly the entire Ti plasmid instead of the T-DNA region (for review see Citovsky *et al.*, 1992*a*; Zambryski, 1992). These results suggest that transfer of the T-DNA is polar from right to left, determined by the orientation of the T-DNA border repeats.

Production of the transferable T-strand

Vir-induced *Agrobacterium* cells generate a linear single-stranded copy of the T-DNA region, designated the T-strand (Stachel, Timmerman &

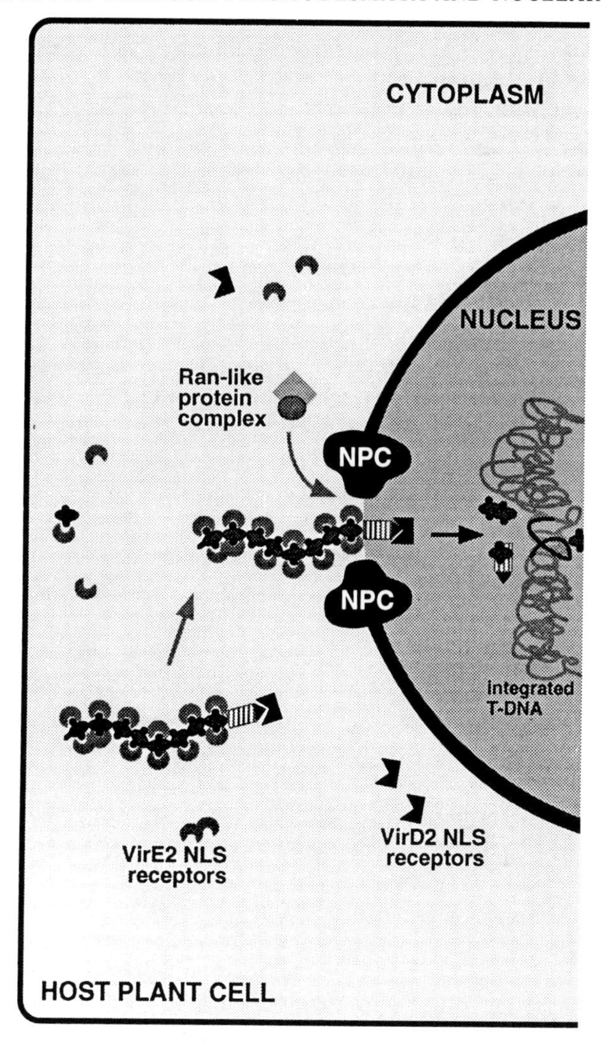

Fig. 4. A model for nuclear import of *Agrobacterium* T-complex. NPC, nuclear pore complex. For detailed description, see text.

Zambryski, 1986). The T-strand is found at about one copy per induced *Agrobacterium* cell, and is homologous to the bottom strand of the T-DNA (Stachel, Timmerman & Zambryski, 1986). Presumably, T-strand production occurs in a 5′ to 3′ direction, initiating at the right T-DNA border and terminating at the left border (for review see Citovsky *et al.*, 1992*a*). VirD1 and virD2 proteins function as an endonuclease responsible for site- and strand-specific nicks between the 3rd and 4th base pair of the bottom strand of the T-DNA borders (Fig. 4) (Wang *et al.*, 1987).

Following cleavage, VirD2 covalently attaches to the 5′ end of the T-strand at the right border nick and to the 5′ end of the remaining bottom strand of the Ti plasmid at the left border nick (Herrera-Estrella *et al.*, 1988). The excised T-strand is removed and the resulting single-stranded gap is repaired most likely by replacement DNA strand synthesis. Potentially, this reaction removes the VirD2 molecule attached to the 5′ end of the left border, restoring the circular DNA molecule of the Ti plasmid. *In vitro* studies demonstrated that purified VirD1 and VirD2, indeed, act as a site-specific endonuclease complex on a supercoiled plasmid containing a 25 bp border repeat (Scheiffele, Pansegrau & Lanka, 1995). Recent data indicate that VirD2 may also participate in ligating the left border nick (Pansegrau *et al.*, 1993). Finally, another virulence protein, VirC1, can enhance T-strand production from the octopine Ti plasmid when VirD1 and VirD2 are limiting (De Vos & Zambryski, 1989). Because so few plant-induced *Agrobacterium* proteins are necessary for T-strand production, bacterial housekeeping enzymes of DNA repair and metabolism (e.g. helicases) may also be involved in this process.

Although specific accumulation of T-strands in vir-induced *Agrobacterium* strongly suggested that these molecules are destined for transfer into the recipient cell (Stachel *et al.*, 1986), the direct proof for this idea has been obtained only recently. First, a sensitive PCR assay showed that single-stranded but not double-stranded T-DNA is present in the host plant cells shortly after the onset of *Agrobacterium* infection (Yusibov *et al.*, 1994). Independently, an assay based on the different extrachromosomal recombination properties of double-stranded and single-stranded DNA indicated that the T-DNA derivatives enter the plant cell nucleus in a single-stranded form (Tinland, Hohn & Puchta, 1994).

Structural model of T-DNA transfer intermediate, the T-complex

Presently, the T-strand is thought to be transferred out of the bacterium and into the plant cell as a protein–nucleic acid complex. This T-DNA transport intermediate, designated the T-complex (Howard & Citovsky, 1990), is composed of at least three components: T-strand which carries the genetic information, and its cognate VirD2 and VirE2 proteins which protect the T-strand, shape it in a transferable (thin and unfolded) form, and supply the specific targeting signals (see below). As mentioned previously, each T-strand is covalently attached to one molecule of VirD2 (Herrera-Estrella *et al.*, 1988); although both VirD1 and VirD2 are involved in border-specific nicking (De Vos & Zambryski, 1989), there is no experimental evidence that VirD1 remains bound to the generated T-strand. Also associated with the T-strand is VirE2, a ssDNA binding protein (SSB) (Citovsky, De Vos & Zambryski, 1988; Citovsky, Wong & Zambryski, 1989). Binding of VirE2 to ssDNA *in vitro* is strong and cooperative, leading to formation of very stable unfolded VirE2-ssDNA complexes, which are largely inaccessible to external nucleolytic activity (Citovsky *et al.*, 1989). Based on electron microscopy data and *in vitro* VirE2-ssDNA binding kinetics (Citovsky *et al.*, 1989), the nopaline-specific T-complex would be 3600 nm long, 2 nm wide, contain approximately 600 molecules of VirE2 and one molecule of VirD2, and have a predicted molecular mass of 50×10^3 kDa.

The structural and functional model of the T-complex implies that both VirD2 and VirE2 proteins are transported into the recipient plant cell together with the T-strand. However, coinoculation of the same plant with *Agrobacterium* carrying a T-DNA but lacking VirE2 and a strain producing VirE2 but lacking T-DNA resulted in a productive infection by these separately non-pathogenic bacteria (Otten *et al.*, 1984). This intercellular complementation suggests that (i) VirE2 and T-strands can be transported independently from *Agrobacterium* into plant cell, and (ii) VirE2 functions primarily inside the plant cell. The latter conclusion is supported by the observation that VirE2 expressed in a transgenic plant restores infectivity of a VirE2-deficient *Agrobacterium* (Citovsky *et al.*, 1992*c*, see also below).

Recent data indicate that a VirE1-deficient mutant having normal amounts of VirE2 and T-strands is not infectious, presumably, because VirE1 may be required for export of VirE2-coated T-strands (Sundberg *et al.*, 1996). However, when this *Agrobacterium* mutant was coinoculated with a strain producing both VirE1 and VirE2 but lacking T-DNA, the VirE1 mutant became infectious, indicating functional export of T-strands from this strain (Sundberg *et al.*, 1996). These results led to the suggestion that, within *Agrobacterium*, T-strands may not be complexed with VirE2 (although they would still carry a covalently attached VirD2 molecule) and that VirE1 assists export of VirE2 but not of T-strands (Sundberg *et al.*, 1996). Alternative explanations, however, cannot be excluded. For example, VirE1 may mediate its own export (and, possibly, that of VirE2) into the coinoculated VirE1-deficient bacteria rather than into the host plant cell. In this case, the presence of the exported VirE1 in the same cell with VirE2 and T-strands will allow the export of the entire T-complexes. Generally, it is unlikely that such a strong SSB as VirE2 (Citovsky *et al.*, 1989) will not complex with the T-strand already inside the wild-type *Agrobacterium*, especially if both VirE2 and the T-strand are transported into the plant cell through the same channel (as proposed by Binns, Beaupre & Dale, 1995). Formation of T-complexes is also supported by the observation that, in vir-induced *Agrobacterium* extracts, T-strands and VirE2 are coimmunoprecipitated by anti-VirE2 antibodies (Christie *et al.*, 1988). To determine conclusively whether T-strands and VirE2 exit *Agrobacterium* separately or as complex will require development of a more direct and quantitative export assay.

Nuclear import of the T-complex

VirD2 and VirE2 likely mediate nuclear import of the T-complex

Because pathogenic microorganisms often use existing cellular machinery for their own needs, *Agrobacterium* likely employs an endogenous cellular pathway for nuclear transport of the invading T-complex. Consequently, several recent studies have used *Agrobacterium* T-complex as a model system to examine the general process of nuclear import of nucleic acids in plant cells (Fig. 4). The estimated size of the T-complex (50×10^3 kDa) far exceeds the size-exclusion limit of the nuclear pore (60 kDa), precluding passive diffusion into the nucleus and requiring active transport. Since the T-strand presumably does not itself carry targeting signals, T-complex nuclear import is most likely mediated by VirD2 and VirE2 proteins. Indeed, VirD2 was shown to accumulate in the plant cell nucleus using both VirD2 translational fusion to a reporter enzyme (β-glucuronidase, GUS) (Howard *et al.*, 1992) and direct immunolocalization of VirD2 (Tinland *et al.*, 1992).

VirD2 was found to contain a functional NLS in the carboxyl terminus of the protein conforming to the bipartite consensus motif (Howard *et al.*, 1992). The biological relevance of this NLS was confirmed by the observations that *Agrobacterium* T-DNA expression and tumourigenicity were reduced in NLS deletion mutants of VirD2 (Shurvinton, Hodges & Ream, 1992; Narasimhulu *et al.*, 1996). Collectively, these results suggest that the VirD2 protein, attached to the 5′ end of the T-strand, acts to direct the T-complex to the host cell nucleus.

The very large predicted size of the T-complex and the residual tumourigenicity of *Agrobacterium* mutants with deleted VirD2 NLS suggest that VirD2 is not the sole mediator of nuclear uptake of the T-complex. Subsequently, nuclear localization of VirE2, the major structural component of the T-complex, was demonstrated using VirE2-GUS fusions. These experiments identified two functional NLSs within the central region of VirE2 molecule. Although each VirE2 NLS is independently active, maximal VirE2 nuclear import required the presence of both signals (Citovsky *et al.*, 1992*c*).

That VirE2 is involved in nuclear uptake of the T-complex is further supported by the observation that plants transgenic for VirE2 complement the virulence of an *Agrobacterium* strain with an inactivated virE locus (Citovsky *et al.*, 1992*c*). In control experiments, plants expressing an unrelated SSB did not complement virulence, indicating that VirE2 does not simply bind and protect the transported T-strand from cellular nucleases. In the second set of experiments, transgenic plants expressing mutant VirE2 that was unable to bind ssDNA but retained both NLSs developed a significant resistance to the wild-type *Agrobacterium* infection, suggesting that the mutant VirE2 may compete with the incoming T-complex for the host cell nuclear import machinery (Citovsky *et al.*, 1994).

The role of VirE2 in T-complex nuclear uptake was inferred from its proposed association with the transported DNA (Citovsky *et al.*, 1992*c*). To obtain such evidence directly, protein-ssDNA complexes, constructed *in vitro* from purified VirE2 and fluorescently labeled ssDNA, were microinjected into stamen hair cells of the flowering plant *Tradescantia virginiana* and their intracellular localization determined by epifluorescence microscopy (Zupan *et al.*, 1996). Whereas microinjected fluorescent ssDNA alone remained cytoplasmic, fluorescent ssDNA-VirE2 complexes efficiently accumulated in the cell nucleus. In control experiments, double-stranded DNA which is not bound by VirE2 did not enter the nucleus when coinjected with this protein, indicating that VirE2-mediated nuclear import of ssDNA depends on the formation of a nucleoprotein complex (Zupan *et al.*, 1996). Wheat germ agglutinin (WGA) and a non-hydrolysable GTP analogue (GMP-PMP), known as specific inhibitors of protein nuclear import in animal systems (for review see Goldfarb & Michaud, 1991; Goldfarb, 1994), blocked nuclear accumulation of the fluorescent VirE2-ssDNA complexes (Zupan *et al.*, 1996).

Taken together these observations indicate the direct and active role of VirE2 in nuclear uptake of ssDNA. They also demonstrate that VirE2 binding to ssDNA does not mask its NLS sequences which partially overlap the ssDNA binding domain. This model, however, does not support the recent hypothesis that VirE2 function may be limited to protection of the T-strand. This conclusion was based on the observation that, in tobacco, deletion of the entire VirE2 did not reduce tumourigenicity of the VirD2 NLS deficient strain below the already low basal level of 4–7% as compared to the wild-type *Agrobacterium* (Rossi, Hohn & Tinland, 1996). Unlike microinjection of fluorescent VirE2-ssDNA complexes, tumourigenicity measurements do not directly assay nuclear import and, thus, are difficult to interpret in these terms. Furthermore, these results contradict other reports that tobacco infection by *Agrobacterium* with deleted VirD2 NLS still results in 20–30% tumour formation (Shurvinton, Hodges & Ream, 1992) as well as T-DNA transcription (Narasimhulu *et al.*, 1996). Infection of other plant hosts (e.g. *Arabidopsis* roots and potato tubers) by this mutant showed even smaller dependency on the VirD2 NLS (tumour formation and T-DNA transcription of 60% or greater), indicating that VirD2 NLS is not absolutely required for nuclear uptake of T-strands (Shurvinton *et al.*, 1992; Narasimhulu *et al.*, 1996).

VirD2 and VirE2 NLSs represent two functional types of plant nuclear targeting signals

In the case of the nopaline Ti-plasmid, the T-complex is predicted to have a VirE2:VirD2 molar ratio of 600 and a VirE2 NLS:VirD2 NLS ratio of 1,200. Why, then, has *Agrobacterium* evolved to specify the VirD2 NLS at all? One possibility is that VirD2 is imported into the plant cell nucleus by a pathway which is different from that of VirE2 and that both mechanisms are required for the optimal nuclear import of the T-complex. To test this idea, the function of VirD2 and VirE2 NLSs was assayed in a heterologous animal system which may lack one of the potential plant nuclear import pathways, thereby allowing discrimination between the VirD2 and VirE2 NLS activities (Guralnick *et al.*, 1996). In these experiments, VirD2 and VirE2 were fluorescently labeled and microinjected into *Xenopus* oocytes and *Drosophila* embryos. All microinjected VirD2 accumulated in cell nuclei as determined by confocal microscopy. This nuclear uptake was specifically inhibited by an excess of free VirD2 NLS peptide, probably due to competition for the nuclear import machinery (Guilizia *et al.*, 1994). VirD2 nuclear import was also inhibited by coinjection of GTPγS, a non-hydrolysable GTP analogue known to block the GTPase Ran which is absolutely essential for the transport of proteins through the nuclear pore (for review see Goldfarb, 1994). These results indicated that VirD2 is actively imported into the nuclei of animal cells and that this import is mediated by the VirD2 NLS which therefore represents a signal sequence evolutionarily conserved between plants and animals.

In contrast to VirD2, microinjected fluorescently labeled VirE2 remained cytoplasmic in both *Drosophila* and *Xenopus*, suggesting that the VirE2 NLSs are not recognized in an animal system and, therefore, are plant-specific. Amino acid sequence analysis revealed that in both VirE2 NLSs one uncharged amino acid residue interrupted the normally adjacent basic residues of the first basic signal domain of the bipartite NLS. When this intervening residue was switched with the adjacent basic amino acid, changing the first domain of the bipartite signal from KLR to KRL (first VirE2 NLS) or from KTK to KKT (second VirE2 NLS), the resulting mutant proteins accumulated in the animal cell nuclei. Furthermore, the second VirE2 NLS mutant (designated VirE2s20), which retained the full ssDNA binding activity, was shown to mediate import of fluorescently-labeled ssDNA into the nuclei of *Xenopus* oocytes (Guralnick *et al.*, 1996). These observations lend additional support to the potential function of VirE2 as a nuclear transport protein for *Agrobacterium* T-strands.

Implications for nuclear import of T-DNA and nucleic acids in general

That VirE2 NLSs function in plant but not animal cells suggests differences between plant and animal nuclear import machinery. This idea is consistent with the observations that the yeast Matα2 NLS functions in plants (Hicks & Raikhel, 1995) but not in mammalian cells (Chelsky, Ralph & Jonak, 1989; Lanford *et al.*, 1990). What is the molecular basis for this difference? NLS recognition most likely occurs by interaction with cellular receptors, usually belonging to the karyopherin α protein family (for review see Powers & Forbes, 1994; Gorlich & Mattaj, 1996). Thus, it is possible that plant cells have a subset of NLS receptors which recognize the VirE2 NLSs and which are absent in animal cells. Other plant cell NLS receptors may recognize the conserved bipartite-type NLS of VirD2, sharing this recognition with animal NLS binding proteins (Fig. 4). Alternatively, the same plant NLS receptors may recognize both the VirD2 and VirE2 NLSs but with different affinities whereas animal receptors are more stringent, interacting only with the consensus NLS sequences. In both scenarios, the functional variations in the NLS sequence may reflect cellular regulation of nuclear import of proteins and/or protein-nucleic acid complexes. For example, the nuclear import of *Agrobacterium* T-complex may occur in a polar and linear fashion, potentially important for the subsequent integration of the T-strand into the plant cell genome (for review see Zambryski, 1992; Citovsky & Zambryski, 1993; Citovsky & Zambryski, 1995). The T-complex model (Howard, Citovsky & Zambryski, 1990; Citovsky & Zambryski, 1993; Citovsky & Zambryski, 1995) suggests that the 5′ end of the T-strand is associated with the VirD2 molecule whereas the 3′ end most likely has a VirE2 molecule attached in its proximity; the functional variation between the NLS signals of these proteins may specify the ends of the T-strand and determine the polarity of its transport and integration.

As suggested by experiments with VirE2, formation of complexes with a specialized transport protein(s) may be necessary for DNA nuclear import in many eukaryotic organisms. This model for protein-mediated nuclear import of nucleic acids is supported by the recent observation that influenza virus nucleoprotein transports the viral genomic RNA into the cell nucleus in an *in vitro* system (O'Neill *et al.*, 1995). Polarity may represent another common feature of nucleic acid transport through the nuclear pore (Citovsky & Zambryski, 1993). For example, nuclear export of a 75S premessenger ribonucleoprotein particle in *Chironomus tentans* initiates exclusively at the 5′ end of the RNA (Mehlin, Daneholt & Skoglund, 1992).

ACKNOWLEDGEMENTS

Our research is supported by grants from National Institutes of Health (Grant R01-GM50224), US Department of Agriculture (Grant 94–02564), and US-Israel Binational Research and Development Fund (BARD) (Grant US-2247–93) to V.C. R.L. is a recipient of a NIH Supplement award (Grant GM46134).

REFERENCES

Akey, C. W. (1992). The nuclear pore complex. *Current Opinions in Structural Biology*, **2**, 258–63.

Angell, S. M., Davies, C. & Baulcombe, D. C. (1996). Cell-to-cell movement of potato virus X is associated with a change in the size exclusion limit of plasmodesmata in trichome cells of *Nicotiana clevelandii*. *Virology*, **216**, 197–201.

Atabekov, J. G. & Taliansky, M. E. (1990). Expression of a plant virus-coded transport function by different viral genomes. *Advances in Virus Research*, **38**, 201–48.

Baeuerle, P. A. & Baltimore, D. (1988). I-κB: a specific inhibitor of the NF-KB transcription factor. *Science*, **242**, 540–6.

Baron-Eppel, O., Hernandes, D., Jiang, L.-W., Meiners, S. & Schindler, M. (1988). Dynamic continuity of cytoplasmic and membrane compartments between plant cells. *Journal of Cell Biology*, **106**, 715–21.

Bernhardi, J. J. (1805). *Beobachtungen über Pflanzengefässe und eine neue Artderselben*: Erfurt.

Binns, A. N., Beaupre, C. E. & Dale, E. M. (1995). Inhibition of VirB-mediated transfer of diverse substrates from *Agrobacterium tumefaciens* by the InQ plasmid RSF1010. *Journal of Bacteriology*, **177**, 4890–9.

Breeuwer, M. & Goldfarb, D. G. (1990). Facilitated nuclear transport of histone H1 and other small nucleophilic proteins. *Cell*, **60**, 999–1008.

Cadepond, F., Schweizer-Groyer, G., Segard-Maurel, I., Jibard, N., Hollenberg, S. M., Giguere, V., Evans, R. M. & Baulieu, E. E. (1991). Heat shock protein 90 is a critical factor in maintaining glucocorticosteroid receptor in a nonfunctional state. *Journal of Biological Chemistry*, **266**, 5834–41.

Carr, R. J. & Kim, K. S. (1983). Evidence that bean golded mosaic virus invaded non-phloem tissue in double infections with tobacco mosaic virus. *Journal of General Virology*, **64**, 2489–92.

Chelsky, D., Ralph, R. & Jonak, G. (1989). Sequence requirements for synthetic peptide-mediated translocation to the nucleus. *Molecular and Cellular Biology*, **9**, 2487–92.

Christie, P. J., Ward, J. E., Winans, S. C. & Nester, E. W. (1988). The *Agrobacterium tumefaciens virE2* gene product is a single-stranded-DNA-binding protein that associates with T-DNA. *Journal of Bacteriology*, **170**, 2659–67.

Citovsky, V. (1993). Probing plasmodesmal transport with plant viruses. *Plant Physiology*, **102**, 1071–6.

Citovsky, V. & Zambryski, P. (1993). Transport of nucleic acids through membrane channels: snaking through small holes. *Annuual Review of Microbiology*, **47**, 167–97.

Citovsky, V. & Zambryski, P. (1995). Transport of protein-nucleic acid complexes within and between plant cells. *Membrane Protein Transport*, **1**, 39–57.

Citovsky, V., De Vos, G. & Zambryski, P. (1988). Single-stranded DNA binding protein encoded by the *virE* locus of *Agrobacterium tumefaciens*. *Science*, **240**, 501–4.

Citovsky, V., Wong, M. L. & Zambryski, P. (1989). Cooperative interaction of *Agrobacterium* VirE2 protein with single stranded DNA: implications for the T-DNA transfer process. *Proceedings of the National Academy of Sciences, USA*, **86**, 1193–7.

Citovsky, V., Knorr, D., Schuster, G. & Zambryski, P. (1990). The P30 movement protein of tobacco mosaic virus is a single strand nucleic acid binding protein. *Cell*, **60**, 637–47.

Citovsky, V., Knorr, D. & Zambryski, P. (1991). Gene I, a potential movement locus of CaMV, encodes an RNA binding protein. *Proceedings of the National Academy of Sciences, USA*, **88**, 2476–80.

Citovsky, V., McLean, G., Greene, E., Howard, E., Kuldau, G., Thorstenson, Y., Zupan, J. & Zambryski, P. (1992*a*). *Agrobacterium*–plant cell interaction: induction of *vir* genes and T-DNA transfer. In Verma, D. P. S., ed. *Molecular Signals in Plant–Microbe Communications*, pp. 169–98. Boca Raton, FL: CRC Press, Inc.

Citovsky, V., Wong, M. L., Shaw, A., Prasad, B. V. V. & Zambryski, P. (1992*b*).

Visualization and characterization of tobacco mosaic virus movement protein binding to single-stranded nucleic acids. *The Plant Cell*, **4**, 397–411.

Citovsky, V., Zupan, J., Warnick, D. & Zambryski, P. (1992*c*). Nuclear localization of *Agrobacterium* VirE2 protein in plant cells. *Science*, **256**, 1803–5.

Citovsky, V., McLean, B. G., Zupan, J. & Zambryski, P. (1993). Phosphorylation of tobacco mosaic virus cell-to-cell movement protein by a developmentally-regulated plant cell wall-associated protein kinase. *Genes and Development*, **7**, 904–10.

Citovsky, V., Warnick, D. & Zambryski, P. (1994). Nuclear import of *Agrobacterium* VirD2 and VirE2 proteins in maize and tobacco. *Proceedings of the National Academy of Sciences, USA*, **91**, 3210–14.

Corbett, A. H., Koepp, D. M., Schlenstedt, G., Lee, M. S., Hopper, A. K. & Silver, P. A. (1995). Rna1p, a Ran/TC4 GTPase activating protein, is required for nuclear import. *Journal of Cell Biology*, **130**, 1017–26.

De Vos, G. & Zambryski, P. (1989). Expression of *Agrobacterium* nopaline specific VirD1, VirD2, and VirC1 proteins and their requirement for T-strand production in *E. coli. Molecular Plant–Microbe Interactions*, **2**, 43–52.

Deom, C. M., Shaw, M. J. & Beachy, R. N. (1987). The 30-kilodalton gene product of tobacco mosaic virus potentiates virus movement. *Science*, **327**, 389–94.

Derrick, P. M., Barker, H. & Oparka, K. J. (1992). Increase in plasmodesmatal permeability during cell-to-cell spread of tobacco rattle tobravirus from individually inoculated cells. *The Plant Cell*, **4**, 1405–12.

Didehvar, F. & Baker, D. A. (1986). Localization of ATPase in sink tissue of *Ricinus. Annals of Botany*, **40**, 823–8.

Dingwall, C. (1991). Transport across the nuclear envelope: enigmas and explanations. *BioEssays*, **13**, 213–18.

Dingwall, C. & Laskey, R. A. (1991). Nuclear targeting sequences a consensus? *Trends in Biochemical Sciences*, **16**, 478–81.

Ding, B., Haudenshield, J. S., Hull, R. J., Wolf, S., Beachy, R. N. & Lucas, W. J. (1992*a*). Secondary plasmodesmata are specific sites of localization of the tobacco mosaic virus movement protein in transgenic tobacco plants. *The Plant Cell*, **4**, 915–28.

Ding, B., Turgeon, R. & Parthasarathy, M. V. (1992*b*). Substructure of freeze-substituted plasmodesmata. *Protoplasma*, **169**, 28–41.

Ding, B., Haudenshield, J. S., Willmitzer, L. & W. J. Lucas, W. J. (1993). Correlation between arrested secondary plasmodesmal development and onset of accelerated leaf senescence in yeast acid invertase transgenic tobacco plants. *The Plant Journal*, **4**, 179–90.

Ding, B., Li, Q., Nguyen, L., Palukaitis, P. & Lucas, W. J. (1995). Cucumber mosaic virus 3a protein potentiates cell-to-cell trafficking of CMV RNA in tobacco plants. *Virology*, **207**, 345–53.

Enenkel, C., Blobel, G. & Rexach, M. (1995). Identification of a yeast karyopherin heterodimer that targets import substrate to mammalian nuclear pore complexes. *Journal of Biological Chemistry*, **270**, 16499–502.

Erwee, M. G. & Goodwin, P. B. (1984). Characterization of the *Egeria densa* leaf symplast: response to plasmolysis, deplasmolysis and to aromatic amino acids. *Protoplasma*, **122**, 162–8.

Esau, K. (1948). Some anatomical aspects of plant virus disease problems. II. *Botanical Reviews*, **14**, 413–49.

Franceschi, V. R., Ding, B. & Lucas, W. J. (1994). Mechanism of plasmodesmata formation in characean algae in relation to evolution of intercellular communication in higher plants. *Planta*, **192**, 347–58.

Fujiwara, T., Giesman-Cookmeyer, D., Ding, B., Lommel, S. A. & Lucas, W. J.

(1993). Cell-to-cell trafficking of macromolecules through plasmodesmata potentiated by the red clover necrotic virus movement protein. *The Plant Cell*, **5**, 1783–94.

Garcia-Bustos, J., Heitman, J. & Hall, M. N. (1991). Nuclear protein localization. *Biochimica et Biophysica Acta*, **1071**, 83–101.

Gibbs, A. J. (1976). Viruses and plasmodesmata. In Gunning, B. E. S. & Robards, A. W., eds. *Intercellular Communication in Plants: Studies on Plasmodesmata*, pp. 149–164. Berlin: Springer-Verlag.

Gilmore, T. D. (1990). NF-KB, KBF1, *dorsal* and related matters. *Cell*, **62**, 841–3.

Goldfarb, D. (1994). GTPase cycle for nuclear transport. *Current Biology*, **4**, 57–60.

Goldfarb, D. & Michaud, N. (1991). Pathways for the nuclear transport of proteins and RNAs. *Trends in Cell Biology*, **1**, 20–4.

Goldfarb, D. S., Gariepy, J., Schoolnik, G. & Kornberg, R. D. (1986). Synthetic peptides as nuclear localization signals. *Nature*, **322**, 641–4.

Goodman, R. M. (1981). Geminiviruses. In Kurstak, E., ed. *Handbook of Plant Virus Infection and Comparative Diagnosis*, pp. 879–910. New York: Elsevier/North Holland Biomedical Press.

Gorlich, D. & Mattaj, I. W. (1996). Nucleocytoplasmic transport. *Science*, **271**, 1513–18.

Gorlich, D., Prehn, S., Laskey, R. A. & Hartmamn, E. (1994). Isolation of a protein that is essential for the first step of nuclear import. *Cell*, **79**, 767–78.

Gorlich, D., Vogel, F., Mills, A. D., Hartmann, E. & Laskey, R. A. (1995). Distinct functions for the two importin subunits in nuclear protein import. *Nature*, **377**, 246–8.

Guilizia, J., Dempsey, M., Sharova, N., Burkinsky, M., Goldfarb, D. S. & Stevenson, M. (1994). Reduced nuclear import of HIV-1 preintegration complexes in the presence of a prototypic nuclear targeting signal. *Journal of Virology*, **68**, 2021–5.

Guralnick, B., Thomsen, G. & Citovsky, V. (1996). Transport of DNA into the nuclei of *Xenopus* oocytes by a modified VirE2 protein of *Agrobacterium*. *The Plant Cell*, **8**, 363–73.

Ham, J. & Parker, M. G. (1989). Regulation of gene expression by nuclear hormone receptors. *Current Opinions in Cell Biology*, **1**, 503–11.

Hatada, E. N., Nieters, A., Wulczyn, G., Naumann, M., Meyer, R., Nucifora, G., McKeithan, T. W. & Scheidereit, C. (1992). The ankyrin repeat domains of the NF-κB precursor p105 and the protooncogene *bcl-3* act as specific inhibitors of NF-κB DNA binding. *Proceedings of the National Academy of Sciences, USA*, **89**, 2489–93.

Heinlein, M., Epel, B. L. & Beachy, R. N. (1995). Interaction of tobamovirus movement proteins with the plant cytoskeleton. *Science*, **270**, 1983–5.

Hepler, P. K. (1982). Endoplasmic reticulum in the formation of the cell plate and plasmodesmata. *Protoplasma*, **111**, 121–33.

Herrera-Estrella, A., Chen, Z., Van Montagu, M. & Wang, K. (1988). VirD proteins of *Agrobacterium tumefaciens* are required for the formation of a covalent DNA protein complex at the 5' terminus of T-strand molecules. *EMBO Journal*, **7**, 4055–62.

Hicks, G. R. & Raikhel, N. R. (1995). Nuclear localization signal binding proteins in higher plant nuclei. *Proceedings of the National Academy of Sciences, USA*, **92**, 734–8.

Hooykaas, P. J. J. & Beijersbergen, A. G. M. (1994). The virulence system of *Agrobacterium tumefaciens*. *Annuual Review of Phytopathology*, **32**, 157–79.

Howard, E. A. & Citovsky, V. (1990). The emerging structure of the *Agrobacterium* T-DNA transfer complex. *Bioessays*, **12**, 103–8.

Howard, E. A., Citovsky, V. & Zambryski, P. (1990). The T-complex of *Agrobacter-*

ium tumefaciens. *UCLA Symposia in Molecular and Cell Biology News Series*, **129**, 1–11.

Howard, E., Zupan, J., Citovsky, V. & Zambryski, P. (1992). The VirD2 protein of *A. tumefaciens* contains a C-terminal bipartite nuclear localization signal: implications for nuclear uptake of DNA in plant cells. *Cell*, **68**, 109–18.

Kitajima, E. W. & Lauritis, J. A. (1969). Plant virions in plasmodesmata. *Virology*, **37**, 681–5.

Kumar, N. M. & Gilula, N. B. (1996). The gap junction communication channel. *Cell*, **84**, 381–8.

Lanford, R. E., Feldherr, C. M., White, R. G., Dunham, R. G. & Kanda, P. (1990). Comparison of diverse transport signals in synthetic peptide-induced nuclear transport. *Experimental Cell Research*, **186**, 32–8.

Leisner, S. M. & Howell, S. H. (1993). Long-distance movement of viruses in plants. *Trends in Microbiology*, **1**, 314–17.

Lucas, W. J., Ding, B. & van der Schoot, C. (1993). Plasmodesmata and the supracellular nature of plants. *The New Phytologist*, **125**, 435–76.

Lucas, W. J. & Gilbertson, R. L. (1994). Plasmodesmata in relation to viral movement within leaf tissues. *Annual Review of Phytopathology*, **32**, 387–411.

Lucas, W. J., Wolf, S., Deom, C. M., Kishmore, G. M. & Beachy, R. N. (1990). Plasmodesmata–virus interaction. In Robards, A. W., Jongsma, H., Lucas, W. J., Pitts, J. & Spray, D., eds. *Parallels in Cell-to-Cell Junctions in Plant and Animals*, pp. 261–72. Berlin: Springer-Verlag.

Lucas, W. L., Bouche-Pillon, S., Jackson, D. P., Nguyen, L., Baker, L., Ding, B. & Hake, S. (1995). Selective trafficking of KNOTTED1 homeodomain protein and its mRNA through plasmodesmata. *Science*, **270**, 1980–3.

Lux, S. E., John, K. M. & Bennett, Y. (1990). Analysis of cDNA for human erythrocyte ankyrin indicates a repeated structure with homology to tissue-differentiation and cell-cycle control proteins. *Nature*, **344**, 36–42.

McLean, B. G., Zupan, J. & Zambryski, P. (1995). Tobacco mosaic virus movement protein associates with the cytoskeleton in tobacco cells. *The Plant Cell*, **7**, 2101–14.

Mehlin, H., Daneholt, B. & Skoglund, U. (1992). Translocation of a specific premessenger ribonucleoprotein particle through the nuclear pore studied with electron microscope tomography. *Cell*, **69**, 605–13.

Meiners, S. & Schindler, M. (1989). Characterization of a connexin homologue in cultured soybean cells and diverse plant organs. *Planta*, **179**, 148–55.

Meiners, S., Xu, A. & Schindler, M. (1991). Gap junction protein homologue from *Arabidopsis thaliana*: evidence for connexins in plants. *Proceedings of the National Academy of Sciences, USA*, **88**, 4119–22.

Melchior, F., Paschal, B., Evans, J. & Gerace, L. (1993). Inhibition of nuclear import by the nohydrolyzable analogues of GTP and identification of the small GTPase Ran/TC4 as an essential transport factor. *Journal of Cell Biology*, **123**, 1649–59.

Moore, M. S. & Blobel, G. (1993). The GTP-binding protein Ran/TC4 is required for protein import into the nucleus. *Nature*, **365**, 661–3.

Mushegian, A. R. & Koonin, E. V. (1993). The proposed plant connexin is a protein kinase-like protein. *The Plant Cell*, **5**, 998–9.

Narasimhulu, S. B., Deng, X.-B., Sarria, R. & Gelvin, S. B. (1996). Early transcription of *Agrobacterium* T-DNA genes in tobacco and maize. *The Plant Cell*, **8**, 873–86.

Nehrbass, U. & Blobel, G. (1996). Role of the nuclear transport factor p10 in nuclear import. *Science*, **272**, 120–2.

Newmeyer, D. D. & Forbes, D. J. (1988). Nuclear import can be separated into distinct steps *in vitro*: nuclear pore binding and translocation. *Cell*, **52**, 641–53.

Newmayer, D. D., Lucocq, J. M., Burglin, T. R. & De Robertis, E. M. (1986). Assembly *in vitro* of nuclei active in nuclear protein transport: ATP is required for nucleoplasmin accumulation. *EMBO Journal*, **5**, 501–10.

Nigg, E. A., Baeuerle, P. A. & Luhrmann, R. (1991). Nuclear import–export: in search of signals and mechanisms. *Cell*, **66**, 15–22.

Noueiry, A. O., Lucas, W. L. & Gilbertson, R. L. (1994). Two proteins of a plant DNA virus coordinate nuclear and plasmodesmal transport. *Cell*, **76**, 925–32.

O'Neill, R. E., Jaskunas, R., Blobel, G., Palese, P. & Moroianu, J. (1995). Nuclear import of influenza virus RNA can be mediated by viral nucleoprotein and transport factors required for protein import. *Journal of Biological Chemistry*, **270**, 22701–4.

Osman, T. A. M., Hayes, R. J. & Buck, K. W. (1992). Cooperative binding of the red clover necrotic mosaic virus movement protein to single-stranded nucleic acids. *Journal of General Virology*, **73**, 223–7.

Otten, L., DeGreve, H., Leemans, J., Hain, R., Hooykass, P. & Schell, J. (1984). Restoration of virulence of *vir* region mutants of *A. tumefaciens* strain B6S3 by coinfection with normal and mutant *Agrobacterium* strains. *Molecular and General Genetics*, **195**, 159–63.

Palikaitis, P. & Zaitlin, M. (1986). Tobacco mosaic virus: infectivity and replication. In Van Regenmortel, M. H. V. & Fraenkel-Conrat, H., eds. *The Rod-Shaped Viruses*, pp. 105–31. New York: Plenum Press.

Pansegrau, W., Schoumacher, F., Hohn, B. & Lanka, E. (1993). Site-specific cleavage and joining of single-stranded DNA by Vird2 protein of *Agrobacterium tumefaciens* Ti plasmids: analogy to bacterial conjugation. *Proceedings of the National Academy of Sciences, USA*, **90**, 11538–42.

Pascal, E., Sanderfoot, A. A., Ward, B. M., Medville, R., Turgeon, R. & Lazarowitz, S. G. (1994). The geminivirus BR1 movement protein binds single-stranded DNA and localizes to the cell nucleus. *The Plant Cell*, **6**, 995–1006.

Picard, D. & Yamamoto, K. (1987). Two signals mediate hormone-dependent nuclear localization of the glucocorticoid receptor. *EMBO Journal*, **6**, 3333–40.

Poirson, A., Turner, A. P., Giovane, C., Berna, A., Roberts, K. & Godefroy-Colburn, T. (1993). Effect of the alfalfa mosaic virus movement protein expressed in transgenic plants on the permeability of plasmodesmata. *Journal of General Virology*, **74**, 2459–61.

Porter, K. R. & Machado, R. D. (1960). Studies on the endoplasmic reticulum. IV. Its form and distribution during mitosis in cells of onion root tip. *Journal of Biophysical and Biochemical Cytology*, **7**, 167–80.

Powers, M. A. & Forbes, D. J. (1994). Cytosolic factors in nuclear import: what's importin? *Cell*, **79**, 931–4.

Radford, J. & White, R. G. (1996). Preliminary localization of myosin to plasmodesmata. *Third International Workshop on Basic and Applied Research in Plasmodesmal Biology*, Zichron-Yakov, Israel, pp. 37–38.

Raikhel, N. V. (1992). Nuclear targeting in plants. *Plant Physiology*, **100**, 1627–32.

Rexach, M. & Blobel, G. (1995). Protein import into nuclei: association and dissociation reactions involving transport substrate, transport factors, and nucleoporins. *Cell*, **83**, 683–92.

Richardson, W. D., Mills, A. D., Dilworth, S. M., Laskey, R. A. & Dingwall, C. (1988). Nuclear pore migration involves two steps: rapid binding at the nuclear envelope followed by slower translocation through nuclear pores. *Cell*, **52**, 655–64.

Robbins, J., Dilworth, S. M., Laskey, R. A. & Dingwall, C. (1991). Two interdependent basic domains in nucleoplasmin nuclear targeting sequence: identification of a class of bipartite nuclear targeting sequence. *Cell*, **64**, 615–23.

Rossi, L., Hohn, B. & Tinland, B. (1996). Integration of complete transferred DNA units is dependent on the activity of virulence E2 protein of *Agrobacterium tumefaciens*. *Proceedings of the National Academy of Sciences, USA*, **93**, 126–30.

Sanderfoot, A. A. & Lazarowitz, S. G. (1995). Cooperation in viral movement: the geminivirus BL1 movement protein interacts with BR1 and redirects it from the nucleus to the cell periphery. *The Plant Cell*, **7**, 1185–94.

Scheiffele, P., Pansegrau, W. & Lanka, E. (1995). Initiation of *Agrobacterium tumefaciens* T-DNA processing. *Journal of Biological Chemistry*, **270**, 1269–76.

Schoumacher, F., Erny, C., Berna, A., Godefroy-Colburn, T. & Stussi-Garaud, C. (1992). Nucleic acid binding properties of the alfalfa mosaic virus movement protein produced in yeast. *Virology*, **188**, 896–9.

Shirakawa, F. & Urizel, S. B. (1989). *In vitro* activation and nuclear translocation of NF-κB catalysed by cyclic AMP-dependent protein kinase and protein kinase C. *Molecular and Cellular Biology*, **9**, 2424–30.

Shurvinton, C. E., Hodges, L. & Ream, W. (1992). A nuclear localization signal and the C-terminal omega sequence in the *Agrobacterium tumefaciens* VirD2 endonuclease are important for tumor formation. *Proceedings of the National Academy of Sciences, USA*, **89**, 11837–41.

Stachel, S. E., Timmerman, B. & Zambryski, P. (1986). Generation of single-stranded T-DNA molecules during the initial stages of T-DNA transfer for *Agrobacterium tumefaciens* to plant cells. *Nature*, **322**, 706–12.

Strasburger, E. (1901). Ueber Plasmaverbindungen pflanzlicher Zellen. *Jahrbuch Wissenschaft Botanische*, **36**, 493–601.

Sundberg, C., Meek, L., Carrol, K., Das, A. & Ream, W. (1996). VirE1 protein mediates export of single-stranded DNA binding protein VirE2 from *Agrobacterium tumefaciens* into plant cells. *Journal of Bacteriology*, **178**, 1207–12.

Tangl, E. (1879). Ueber offene Communicationen zwischen den Zellen des Endosperms einiger Samen. *Jahrbuch Wissenschaft Botanische*, **12**, 170–90.

Terry, B. R. & Robards, A. W. (1987). Hydrodynamic radius alone governs the mobility of molecules through plasmodesmata. *Planta*, **171**, 145–57.

Thomas, C. L. & Maule, A. J. (1995). Identification of the cauliflower mosaic virus movement protein RNA binding domain. *Virology*, **206**, 1145–9.

Tinland, B., Koukolikova-Nicola, Z., Hall, M. N. & Hohn, B. (1992). The T-DNA-linked VirD2 protein contains two distinct nuclear localization signals. *Proceedings of the National Academy of Sciences, USA*, **89**, 7442–6.

Tinland, B., Hohn, B. & Puchta, H. (1994). *Agrobacterium tumefaciens* transfers single-stranded transferred DNA (T-DNA) into the plant cell nucleus. *Proceedings of the National Academy of Sciences, USA*, **91**, 8000–4.

Tucker, E. B. (1988). Inositol biphosphate and inositol triphosphate inhibit cell-to-cell passage of carboxyfluorescein in staminal hairs of *Setcreasea purpurea*. *Planta*, **174**, 358–63.

Vaquero, C., Turner, A. P., Demangeat, G., Sanz, A., Serra, M. T., Roberts, K. & Garcia-Luque, I. (1994). The 3a protein from cucumber mosaic virus increases the gating capacity of plasmodesmata in transgenic tobacco plants. *Journal of General Virology*, **75**, 3193–7.

Waigmann, E. & Zambryski, P. (1995). Tobacco mosaic virus movement protein-mediated transport between trichome cells. *The Plant Cell*, **7**, 2069–79.

Waigmann, E., Lucas, W., Citovsky, V. & Zambryski, P. (1994). Direct functional assay for tobacco mosaic virus cell-to-cell movement protein and identification of a

domain involved in increasing plasmodesmal permeability. *Proceedings of the National Academy of Sciences, USA*, **91**, 1433–7.
Wang, K., Stachel, S. E., Timmerman, B., Van Montagu, M. & Zambryski, P. (1987). Site-specific nick occurs within the 25 bp transfer promoting border sequence following induction of vir gene expression in *Agrobacterium tumefaciens. Science*, **235**, 587–91.
Weis, K., Mattaj, I. W. & Lamond, A. I. (1995). Identification of hSRP1PF025as a functional receptor for nuclear localization sequences. *Science*, **268**, 1049–53.
White, R. G., Badelt, K., Overall, R. L. & Vesk, M. (1994). Actin associated with plasmodesmata. *Protoplasma*, **180**, 169–84.
Whiteside, S. T. & Goodbourn, S. (1993). Signal transduction and nuclear targeting: regulation of transcription factor activity by subcellular localization. *Journal of Cell Science*, **104**, 949–55.
Wolf, S., Deom, C. M., Beachy, R. N. & Lucas, W. J. (1989). Movement protein of tobacco mosaic virus modifies plasmodesmatal size exclusion limit. *Science*, **246**, 377–9.
Yusibov, V. M., Steck, T. R., Gupta, V. & Gelvin, S. B. (1994). Association of single-stranded transferred DNA from *Agrobacterium tumefaciens* with tobacco cells. *Proceedings of the National Academy of Sciences, USA*, **91**, 2994–8.
Zambryski, P. (1992). Chronicles from the *Agrobacterium*–plant cell DNA transfer story. *Annual Review of Plant Physiology and Plant Molecular Biology*, **43**, 465–90.
Zambryski, P. (1995). Plasmodesmata: plant channels for molecules on the move. *Science*, **270**, 1943–4.
Zheng, G-C., Nie, X-V., Wang, Y-X., L-C., J., Sun, L-H. & Sun, D-L. (1985). Cytochemical localization of ATPase activity during cytomixis in pollen mother cells of David lily-*Lilium davidii* var. Willmottiae and its relation to the intercellular migrating chromatin substance. *Acta Academia Sinica*, **27**, 26–32.
Zupan, J., Citovsky, V. & Zambryski, P. (1996). *Agrobacterium* VirE2 protein mediates nuclear uptake of ssDNA in plant cells. *Proceedings of the National Academy of Sciences, USA*, **93**, 2392–7.

VIRAL CELL RECOGNITION AND ENTRY

MICHAEL G. ROSSMANN

Department of Biological Sciences, Purdue University, West Lafayette, Indiana 47907-1392, USA

INTRODUCTION

Unlike plant viruses, most animal, insect and bacterial viruses attach to specific cellular receptors that, in part, determine host range and tissue tropism. Viruses have adapted themselves to utilize a wide variety of cell-surface molecules as their receptors, including proteins, carbohydrates and glycolipids (Table 1). Some viruses recognize very specific molecules, e.g. a large group of rhinoviruses recognize intercellular adhesion molecule-1 (ICAM-1), while other viruses recognize widely distributed chemical groups, e.g. influenza viruses recognize sialic acid moieties. The tissue distribution of the receptor will, in part, determine the tropism of the virus and, hence, the symptoms of the infection. Similarly, species differences between receptor molecules can limit host range. For instance, only humans and apes have been shown to be susceptible to rhinovirus infections, a property correlated to the inability of human rhinoviruses to bind to the receptor ICAM-1 molecule in other species.

Receptor binding is only the first, albeit essential, step in the infection process. The virus, or the virus genome alone, then has to enter the cell, a process which requires translocation of the viral genome or a subviral particle across the membrane into the cytoplasm, and, in some cases, into the nucleus. Because delivery of the viral genome into the cell involves major rearrangements of the capsid structure, entry must be a tightly regulated process, which is triggered by the cell. The mechanism of entry can be, in the case of enveloped viruses, by fusion of the viral envelope with the limiting cellular membrane. This process has been well characterized in several viruses (Semliki Forest virus (SFV), influenza virus, Sendai virus) where fusion is induced by specific viral envelope proteins, activated by conformational changes induced by the low pH environment of endosomes. The mechanism by which protein-encapsidated viruses like picornaviruses (Rueckert, 1990) enter the cytoplasm has not been well elucidated, but must differ significantly in detail from the membrane-fusion strategy demonstrated by enveloped viruses in that RNA must be translocated through the membrane.

Table 1. *Some known receptors for animal viruses*

Receptor	Virus	Reference
1. Sialic acid	Reoviruses	Paul, Choi & Lee, 1989; Choi, Paul & Lee, 1990
	Influenza virus	Weis *et al.*, 1988
	Polyoma virus	Fried, Cahan & Paulson, 1981
2. Immunoglobulin family:		
(i) CD4	Human immunodeficiency viruses	Dalgleish *et al.*, 1984
(ii) Poliovirus receptor	Polioviruses	Mendelsohn *et al.*, 1989
(iii) ICAM-1	Major serotype group of HRV	Greve *et al.*, 1989; Staunton *et al.*, 1989; Tomassini *et al.*, 1989
3. Integrins	Foot-and-mouth disease virus	Acharya *et al.*, 1989; Mason *et al.*, 1993
4. Complement receptor type 2 (a B lymphocyte surface glycoprotein)	Epstein-Barr viruses	Moore *et al.*, 1987; Tanner *et al.*, 1987
5. Amino acid permeases	Ecotropic murine leukemia virus	J. W. Kim *et al.*, 1991; H. Wang *et al.*, 1991
6. Carcinoembryonic antigen	Mouse hepatitis virus (a coronavirus)	Williams *et al.*, 1991
7. Erythrocyte P antigen	Human B19 parvovirus	Brown *et al.*, 1993
8. LDL receptor	Minor serotype group of HRV (?)	Hofer *et al.*, 1994

RHINOVIRUSES

Picornavirus receptors

Although there are extensive similarities of sequence, structure and physical properties among picornaviruses that show these viruses have evolved from a common ancestor (Palmenberg, 1989; Rossmann *et al.*, 1985; Rueckert, 1990), they nevertheless recognize a variety of receptors (Table 2). Possibly the primordial virus had the ability to bind weakly to a large number of different molecules. With time, different viruses evolved that became progressively more efficient and specialized toward recognizing one particular molecule as a way of infecting specific cells. Indeed, the grouping of viruses might suggest such a scenario. Thus, all polioviruses appear to recognize the same receptor and most coxsackie A viruses recognize their own receptor, whereas coxsackie B viruses recognize yet another receptor. Therefore, it is surprising that rhinovirus serotypes can be divided into three groups that recognize different receptors (Abraham & Colonno, 1984; Uncapher, De Witt & Colonno, 1991). Furthermore, the receptor for the major group of rhinoviruses, ICAM-1, belongs to the immunoglobulin superfamily (Greve *et al.*, 1989; Staunton *et al.*, 1989), whereas the receptor for the minor group has been reported to be the low density lipoprotein (LDL) receptor (Hofer *et al.*, 1994).

The canyon hypothesis

It was hypothesized (Rossmann *et al.*, 1985) that the canyon (one around each five-fold vertex; Fig. 1) in HRV was the site of receptor attachment, largely inaccessible to the broad antigen-binding region seen on antibodies. Thus, residues in the lining of the canyon, which should be resistant to accepting mutations that might inhibit receptor attachment, would avoid presenting an unchanging target to neutralizing antibodies. Indeed, the neutralizing immunogenic sites that had been mapped by escape mutations were not in the canyon, but on the most exposed and variable parts of the virion both in HRV (Rossmann *et al.*, 1985; Sherry *et al.*, 1986; Sherry & Rueckert, 1985) and in poliovirus (Hogle, Chow & Filman, 1985; Page *et al.*, 1988). The 'canyon hypothesis' suggests that one strategy for viruses to escape the host's immune surveillance is to protect the receptor attachment site in a surface depression. Similar depressions related to host cell attachment have also been found on the surface of the hemagglutinin spike of influenza virus (Weis *et al.*, 1988; Wilson, Skehel & Wiley, 1981) and may be the case for human immunodeficiency virus (Matthews *et al.*, 1987).

A number of lines of evidence emerged to support the canyon hypothesis. First, a comparison of the variability of surface-exposed residues between a number of picornaviruses indicated that amino acid residues lining the canyon are significantly more conserved than other surface-exposed residues

Table 2. *Receptor families for picornaviruses based on virus competition for cell receptors*

Virus	Receptor molecule	Receptor family	Reference
Human rhinovirus major group: 78 serotypes, including 3, 5, 9, 12, 14, 15, 22, 32, 36, 39, 41, 51, 58, 59, 60, 66, 67, 89	ICAM-1	Ig (5 Ig domains)	Abraham & Colonno, 1984; Greve *et al.*, 1989; Staunton *et al.*, 1989
Human rhinovirus minor group: 11 serotypes, including 1A, 2, 44, 49	Low density lipoprotein (LDL) receptor	LDLR	Abraham & Colonno, 1984; Hofer *et al.*, 1994
Polioviruses	Poliovirus receptor (PVR)	Ig (3 Ig domains)	Mendelsohn *et al.*, 1989
Coxsackievirus A13, 18, 21	ICAM-1	Ig (5 Ig domains)	Colonno, Callahan & Long, 1986; Roivainen *et al.*, 1991
Coxsackievirus A2, 5, 13, 15, 18	?	?	Colonno *et al.*, 1986; Roivainen *et al.*, 1991; Schultz & Crowell, 1983
Coxsackievirus B3 and adenovirus 2	?	?	Lonberg-Holm, Crowell & Philipson, 1976
Echovirus 1	VLA-2	Integrin	Bergelson *et al.*, 1992
Echovirus 6	?	?	Crowell, 1966
Foot-and-mouth disease viruses, types A_{12}119, O_{1B}, C_{3Res}, SAT_{1-3}	RGD integrin	Integrin	Sekiguchi, Franke & Baxt, 1982; Mason *et al.*, 1993
Mengo virus	?	Glycophorin (?)	Burness & Pardoe, 1981, 1983

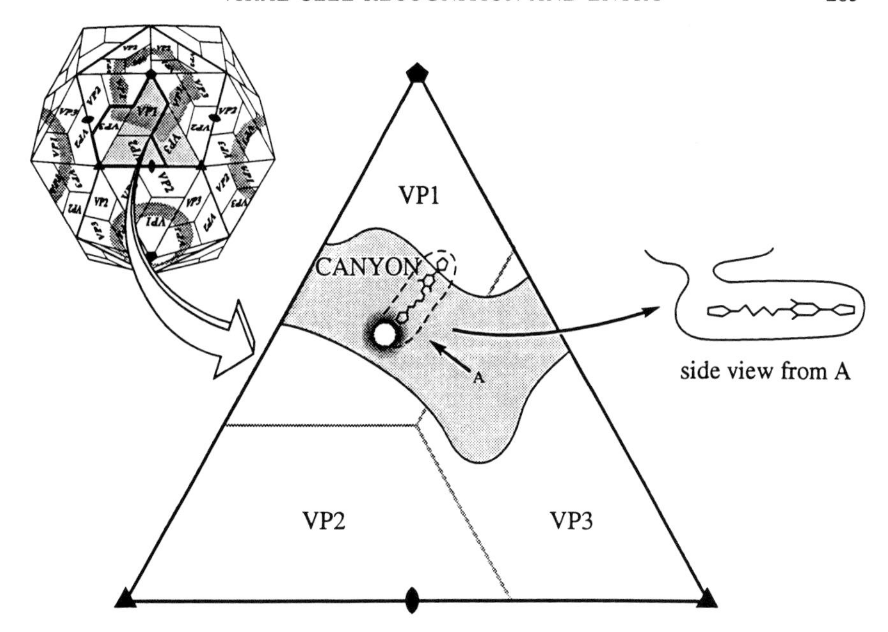

Fig. 1. *Top left*: Diagrammatic view of picornavirus showing VP1, VP2 and VP3 and the deep cleft or 'canyon' running around each five-fold vertex. The 6S protomeric assembly unit (which differs from the geometric definition of the asymmetric unit) is shown in heavy outline on the icosahedron. *Centre*: Enlargement of one icosahedral asymmetric unit showing the outline of the canyon and the entrance to the WIN pocket. The terms 'north' (top) and 'south' rims of the canyon refer to this standard orientation. (Reprinted with permission from Oliveira *et al.*, 1993. Copyright by Current Biology Ltd.)

(Chapman & Rossmann, 1993*a*; Rossmann & Palmenberg, 1988). Secondly, the hypothesis rationalized the contrast between many vertebrate virus structures and plant viruses, e.g. tomato bushy stunt virus (Harrison *et al.*, 1978), southern bean mosaic virus (Abad-Zapatero *et al.*, 1980), satellite tobacco necrosis virus (Liljas *et al.*, 1982) and cowpea mosaic virus (Stauffacher *et al.*, 1987) or insect viruses, e.g. black beetle virus (Hosur *et al.*, 1987). Namely, animal viruses tend to have surface depressions (a notable exception is FMDV; Acharya *et al.* (1989)), but viruses whose hosts do not have immune systems tend to have smooth surfaces or protrusions on their surfaces. Thirdly, site-directed mutagenesis of HRV14 indicated that modification of several amino acid residues located in the base of the canyon has an impact upon virus–receptor affinity (Colonno *et al.*, 1988). Specifically, mutants with substitutions at residues 1273[1], 1223, 1103 and 1220 exhibited an alteration in virus–receptor affinity. Fourthly, certain capsid-binding 'WIN' antiviral compounds block the binding of some of the major receptor rhinoviruses, including HRV14 (Pevear *et al.*, 1989). These compounds bind

[1]Residues are numbered sequentially for each of VP1, VP2, VP3 and VP4, but start at 1001, 2001, 3001 and 4001, respectively.

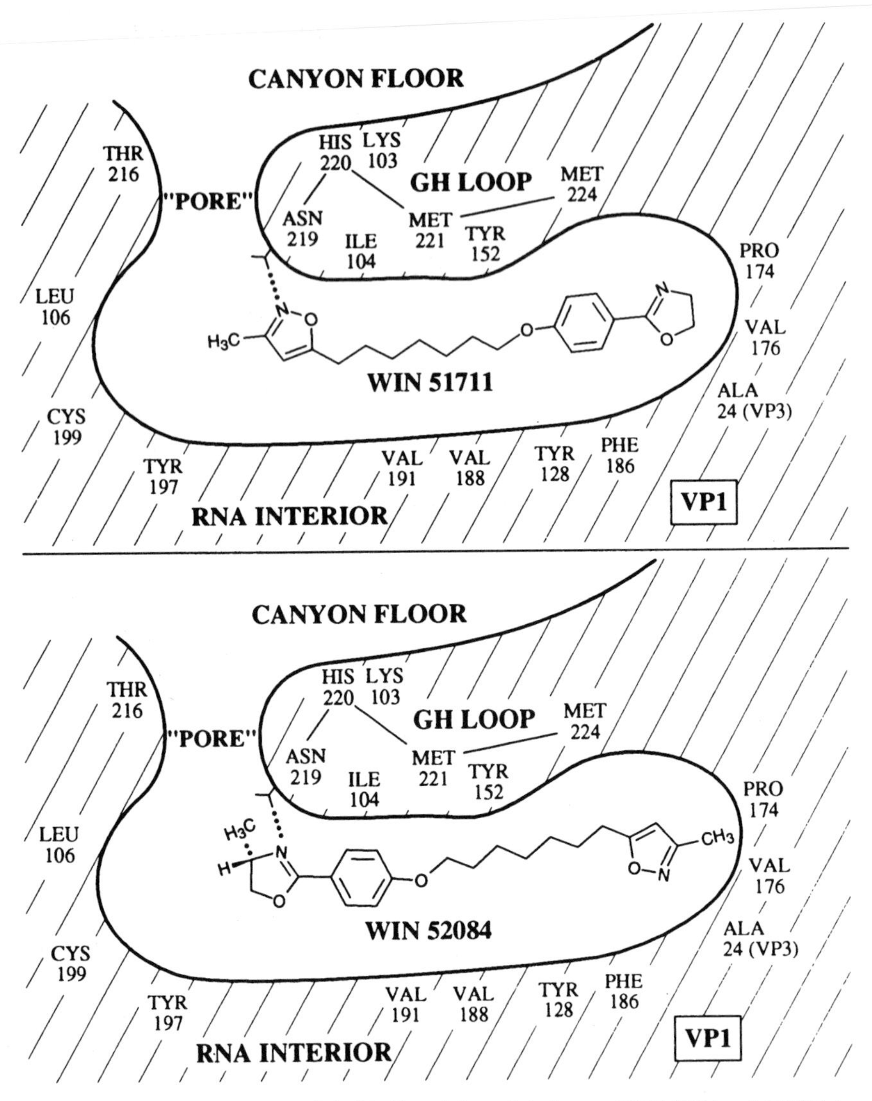

Fig. 2. Schematic representation of the binding of the antiviral agents WIN 51711 and 52084 into a pocket underneath the canyon in HRV14. This causes enlargement of the pocket and conformational changes in the floor of the canyon, inhibiting attachment of the virus to HeLa cells in some cases, and also increasing the stability of the virus in all cases. (Reprinted with permission from Dutko *et al.*, 1989. Copyright by Springer-Verlag, New York.)

to many picornaviruses in a hydrophobic pocket located under the canyon floor (Fig. 1) and, in most cases, block virus from uncoating (Badger *et al.*, 1988; Kim *et al.*, 1993; Smith *et al.*, 1986). Upon binding to HRV14, a conformational change occurs in the roof of the pocket, which is also the floor of the canyon (Fig. 2). Several amino acid residues are displaced by as

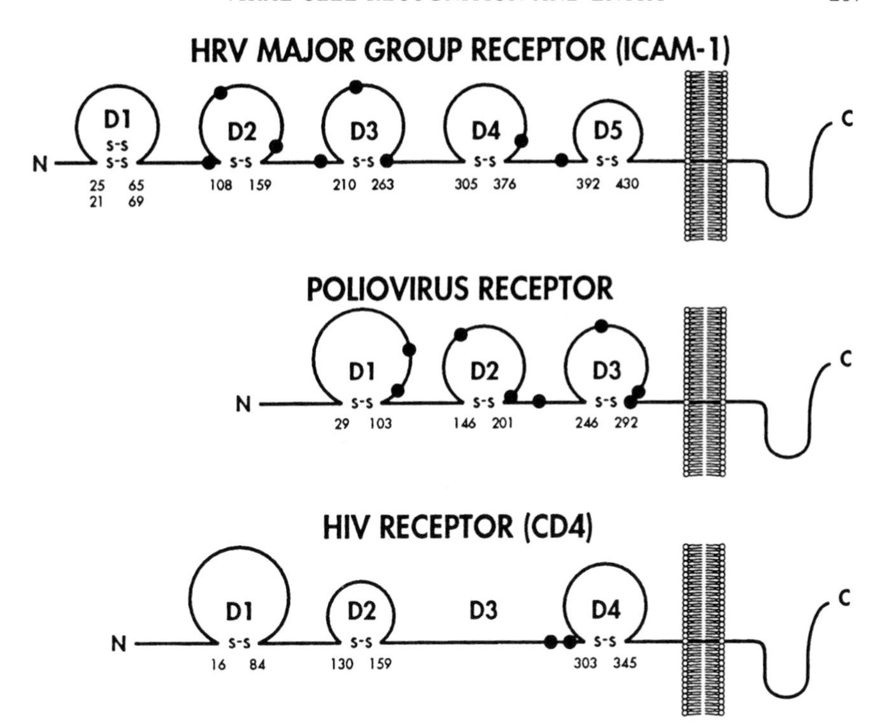

Fig. 3. Schematic diagram of viral receptors. The relative size and distribution of immunoglobulin-like domains are shown. The black circles show the position of potential glycosylation sites. Numbers indicate the amino acid positions of Cys residues involved in predicted disulfide (S–S) bridges. (Reprinted with permission from Colonno, 1992. Copyright by Academic Press Limited.)

much as 4 Å in their C_α positions. These findings suggested that the conformational changes at the base of the canyon prevent viral attachment to cells. Although the observations for rhinovirus were consistent with the canyon being the receptor binding site, they did not provide conclusive proof nor did they identify a complete footprint of the receptor on the virus surface.

Binding of ICAM-1, the major group rhinovirus receptor, to virus surface

There are at least 78 serotypes (Tomassini, Maxson & Colonno, 1989) that bind to ICAM-1, the major group rhinovirus receptor (Greve *et al.*, 1989; Staunton *et al.*, 1989). The ICAM-1 molecule has five immunoglobulin-like domains (D1–D5, numbered sequentially from the amino end), a transmembrane portion and a small cytoplasmic domain (Simmons, Makgoba & Seed, 1988; Staunton *et al.*, 1988). Domains D2, D3 and D4 are glycosylated (Fig. 3). Unlike immunoglobulins, ICAM-1 appears to be monomeric (Staunton *et al.*, 1989). Mutational analysis of ICAM-1 has shown that domain D1 contains the primary binding site for rhinoviruses as well as the binding site for its natural ligand, lymphocyte function-associated antigen-1 (LFA-1) (Lineberger *et al.*, 1990; McClelland *et al.*, 1991; Staunton *et al.*, 1988, 1990). Other surface antigens within the immunoglobulin superfamily that are used by viruses as receptors include CD4 for human immunodeficiency virus type 1 (Dalgleish *et al.*, 1984; Klatzmann *et al.*, 1984; Maddon *et al.*, 1986; Robey & Axel, 1990), the poliovirus receptor (Mendelsohn, Wimmer & Racaniello, 1989) and the mouse coronavirus receptor (Williams, Jiang & Holmes, 1991). In ICAM-1, in the poliovirus receptor (Freistadt & Racaniello, 1991; Koike, Ise & Nomoto, 1991) and in CD4 (Arthos *et al.*, 1989), the primary receptor-virus binding site is domain D1. The structures of the two amino-terminal domains of CD4 have been determined to atomic resolution (Brady *et al.*, 1993; Ryu *et al.*, 1990; Wang *et al.*, 1990). Truncated proteins corresponding to the two amino-terminal domains of ICAM-1 (D1D2 consisting of 185 amino acids) as well as the intact extracellular portion of ICAM-1 (D1–D5 consisting of 453 amino acids) have been expressed in CHO cells (Greve *et al.*, 1991). The desialated form of D1D2 has been crystallized (Kolatkar *et al.*, 1992).

The structure of the complex of D1D2 with HRV16 (Olson *et al.*, 1993) and with HRV14 (P. R. Kolatkar, N. H. Olson, C. Music, J. M. Greve, T. S. Baker and M. G. Rossmann, unpublished results) and of D1D5 with HRV16 (Kolatkar *et al.*, unpublished results), has been determined using cryo-electron microscopy and image reconstruction procedures (Fig. 4). The position of the ICAM-1 molecule relative to the icosahedral symmetry axes of the virus is unambiguous (Kolatkar *et al.*, unpublished results) and shows the receptor binding into the canyon (Fig. 5). Each D1D2 molecule has an approximate dumbbell shape, consistent with the presence of a two-domain structure. A difference map between the EM density and the 20 Å resolution HRV16 or HRV14 densities confirmed that the D1D2 molecule binds to the central portion of the canyon roughly as predicted by Giranda, Chapman & Rossmann (1990). There are some small differences in orientation of D1D2 when complexed to HRV16 or HRV14 that may relate to the change in length of the VP1 BC loop forming the north rim of the canyon (Kolatkar *et al.*, unpublished results). The D1D2 ICAM fragment is oriented roughly perpendicular to the viral surface and extends to a radius of about 205 Å. Its total length is about 75 Å.

Extensive structural similarity between D1D2 of ICAM-1 and CD4 was shown by means of a cross-rotation function between the known structure of D1D2 for CD4 (Ryu *et al.*, 1990; Wang *et al.*, 1990) and the crystal diffraction data for ICAM-1 D1D2 (P. R. Kolatkar, J. M. Greve & M. G. Rossmann, unpublished results). Thus, it seemed reasonable to use the known structures of CD4 for fitting the reconstructed density map (Fig. 4), although there was slightly too little density for domain D1 and too much density for D2. A better assessment of the fit of domain D1 to the density was

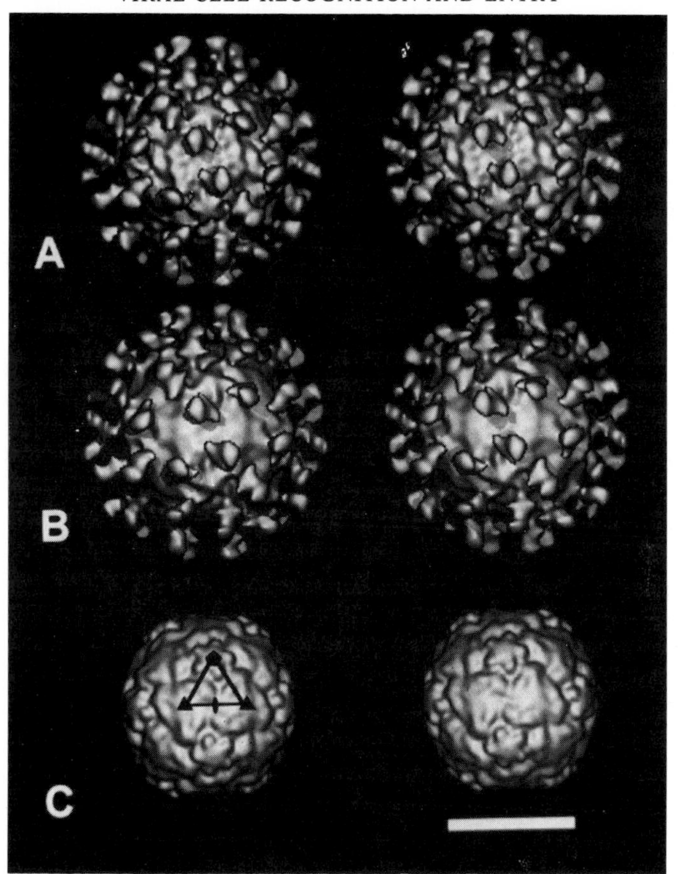

Fig. 4. Stereo views of cryo EM image reconstructions of A, HRV16 (green)-D1D2 (orange) and B, HRV14 (blue)-D1D2 (orange) complex, viewed along an icosahedral two-fold axis in approximately the same orientation as in Fig. 1. Both A and B show 60 D1D2 molecules bound to symmetry-equivalent positions in the canyons on the virion surface. C, Shaded-surface view of HRV14 (blue), computed from the known atomic structure (Rossmann *et al.*, 1985), truncated to 20 Å resolution. (Reprinted with permission from Rossmann, 1994. Copyright by The Protein Society.)

obtained by taking the predicted D1 structure of ICAM-1, including all side chains, and superimposing it onto the fitted C_α backbone of CD4. One major difference is that, although domain D1 of CD4 resembles a variable immunoglobulin-like domain with two extra β-strands, the ICAM-1 sequence is shorter and more like a constant C1 domain (Giranda *et al.*, 1990); however, Berendt *et al.* (1992) suggest that the topology might be like a constant C2 domain in which strand C is not part of either sheet region. This gives domain D1 of ICAM-1 a sleeker appearance, consistent with the observed difference density. The extra density in D2 (in the region furthest away from the virus) compared with domain D2 of CD4 is probably due to the four associated carbohydrate groups located in this region.

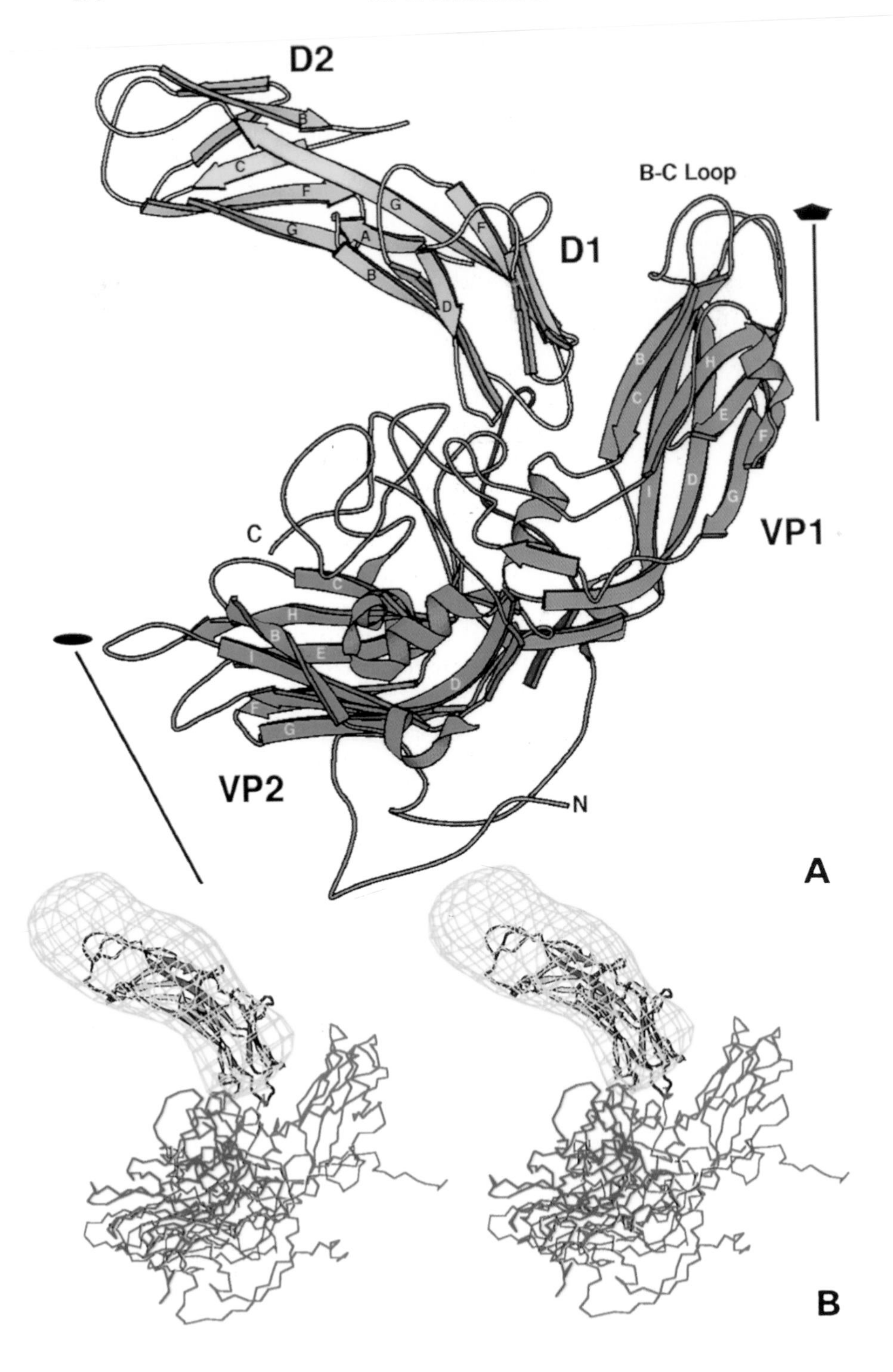
D2
B
C
F
G
G
A
B
D
F
D1
B-C Loop
B
H
C
E
F
D
I
G
VP1
C
C
H
B
E
I
D
F
G
VP2
N
A
B

The footprint of ICAM-1 onto the HRV14 structure correlates very well with Colonno's mutational studies of residues in the canyon that alter affinity of the virus to HeLa cell membranes (Colonno *et al.*, 1988). All the residues are part of the canyon floor and lie centrally within the footprint of the D1D2 molecule binding site. Similarly, there is excellent agreement between the ICAM-1 footprint and residues on the virus surface whose conformation is changed by antiviral agents (Heinz *et al.*, 1989; Pevear *et al.*, 1989; Smith *et al.*, 1986).

Immunoglobulin-like domains consist of seven β-strands (βA to βG) arranged into two β-sheets that form a β-sandwich (Fig. 5). The sequence of the first domain of ICAM-1 (D1) has two unusual features for an immunoglobulin-like domain: it is relatively short, being 88 residues instead of the more typical size of approximately 100 residues; and, instead of the typical two cysteine residues, located in the βB strand and the βF strand, there are four cysteines (Fig. 3). The βB and the βF cysteines usually participate in an intrachain disulfide bond across the β-sandwich in most members of the immunoglobulin supergene family. However, the additional two cysteine residues in ICAM-1 D1 have an $i + 4$ spacing relative to Cys21 and Cys65, which in a β-strand would place them in proper register for forming a second disulfide bond between the βB and βF strands.

The parts of the predicted ICAM-1 structure (based on Giranda *et al.*, 1990) that contact HRV14 or HRV16 are the amino-terminal four residues and loops BC (residues 24–26), DE (residues 45–49) and FG (residues 71–72). This is roughly in correspondence with the 'malarial' binding side of ICAM-1, rather than the LFA-1 binding region (Berendt *et al.*, 1992). This part of ICAM-1 has been associated with adherence to erythrocytes infected with the malarial parasite *Plasmodium falciparum*. Staunton *et al.* (1990), McClelland *et al.* (1991) and Register *et al.* (1991) have examined the effects of a number of site-directed mutations and mouse-human substitutions in domain D1 of ICAM-1 on rhinovirus binding (mouse ICAM-1 does not bind to rhinoviruses). There is correspondence to the four regions of ICAM-1 seen to be in contact with rhinovirus and four of the seven regions implicated in virus binding by site-directed mutagenesis, but there are also inconsistencies between the mutational and structural data. These should be resolved when the crystal structure of ICAM-1 (Kolatkar *et al.*, 1992), or better still of the complex, has been determined.

Virus entry and uncoating

Rhinovirus and poliovirus 149S infectious virions undergo several progressive transformations (Everaert, Vrijsen & Boeyé, 1989; Lonberg-Holm &

Fig. 5. A, Structure of HRV16 VP1 (blue), VP2 (green) and part of VP3 (red) complexed with D1D2 of ICAM-1 (orange) modelled from the known structure of CD4 and B, the difference map between those shown in Figs. 4A and 4B. (Reprinted with permission from Rossmann, 1994. Copyright by The Protein Society.)

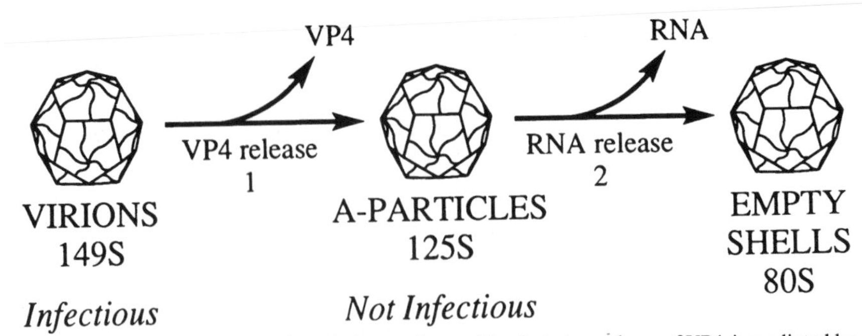

Fig. 6. Two steps in the uncoating of picornaviruses. The first step, release of VP4, is mediated by interaction with viral receptor, by decreasing pH or by heating to 52 °C. The second step, release of RNA, may be caused by acidification of the membrane-bound particle. (Reprinted with permission from Giranda *et al.*, 1992. Copyright by the National Academy of Sciences.)

Korant, 1972) when bound to cells (Fig. 6) that can be followed by sedimentation through sucrose gradients. The 149S virions are initially converted to 135–125S particles which have lost VP4 but retain RNA (altered or 'A'-particles). Subsequently, the RNA is released with the formation of 80S empty capsids as well as small capsid fragments.

The A-particles have a number of properties that suggest a role in virus entry. They have been shown to be hydrophobic and able to bind to liposomes (Hoover-Litty & Greve, 1993; Korant *et al.*, 1975). It has also been shown that the formation of poliovirus A-particles is associated with externalization of the N-terminus of VP1 and that removal of approximately 30 residues from the N-terminus of VP1 by proteolysis abolishes the ability of poliovirus to bind to liposomes (Fricks & Hogle, 1990). The sequence of the amino-terminal 23 residues of VP1 suggests that it could form an amphipathic α-helix and, thus, could promote interactions with lipid bilayers.

A-like particles can be generated under certain conditions *in vitro* (Hoover-Litty & Greve, 1993; Koike *et al.*, 1992; Yafal *et al.*, 1993). HRV14 incubated at pH 5–6, the pH likely to be found in endosomes, is converted to 135S A-particles. HRV14 incubated with soluble ICAM-1 is converted, through a virus–receptor complex intermediate, to 80S empty capsids, suggesting that receptor binding can destabilize the virion (Hoover-Litty & Greve, 1993).

Because the conformational changes required for uncoating that occur on acidification are probably similar to those that occur on viral interaction with receptor, a structural determination of these changes could be useful. It has been possible to study the initial changes that occur in wild-type HRV14 crystals upon lowering the pH by using a very high intensity synchrotron X-ray source (Giranda *et al.*, 1992), permitting the rapid recording of the diffraction pattern before the crystals completely disintegrated. It was found that an ion binding site on the icosahedral five-fold axes, the interior of the virus shell near the five-fold axes (including the amino-terminal residue of VP3), much of the ordered part of VP4 and the

GH loop of VP1 all became disordered. Furthermore, the magnitude of the disorder increased as the time of acid exposure increased. An expansion of the β-cylinder (even beyond the first residue) and cation release, therefore, may be among the first events permitting eventual escape of VP4s, possibly along the five-fold axial channels. There are parallels to this process in the externalization of VP1 through the five-fold axial channels of canine parvovirus (Tsao *et al.*, 1991) and the ejection of single-stranded DNA through the five-fold ion channel of ϕX174 (McKenna, Ilag & Rossmann, 1994; McKenna *et al.*, 1992*b*). An alternative proposal made by Fricks and Hogle (1990) based on mutational analyses and a comparison with properties of tomato bushy stunt virus (Robinson & Harrison, 1982) suggests that the first step in uncoating and the externalization of VP1 is a weakening of the contacts between protomeric units (Fig. 1).

Inhibition of uncoating and the pocket factor

Capsid-binding antiviral agents such as the 'WIN' compounds bind into a hydrophobic pocket in VP1 below the canyon floor. Not only do they inhibit attachment in HRV14 and other major group rhinoviruses, but they also stabilize major and minor group rhinoviruses *in vitro* to acidification (Gruenberger *et al.*, 1991) and heat (Fox, Otto & McKinlay, 1986). HRV14 differs from other picornaviruses in that its pocket is empty in the native structure. For example, there is electron density in the homologous pockets of poliovirus Mahoney 1, poliovirus Sabin 3 and in a chimera of poliovirus 2 (Filman *et al.*, 1989; Hogle *et al.*, 1985; Yeates *et al.*, 1991). This density has been interpreted as a sphingosine or palmitate-like molecule because of the hydrophobic nature of the pocket and the polar environment at one end of the pocket. Similarly, the somewhat smaller electron density in the pocket of HRV1A (K. H. Kim *et al.*, 1993; S. Kim *et al.*, 1989) and HRV16 (Oliveira *et al.*, 1993) has been tentatively interpreted as a fatty acid, eight or more carbon atoms long. A rather longer 'pocket factor' is found in this pocket for coxsackievirus B3 (CVB3) (Muckelbauer *et al.*, 1995). Although it is possible that the pocket factor might be a small impurity picked up in the extraction procedure with detergent or during crystallization with polyethylene glycol, these conditions differ greatly among the known structures. Smith *et al.* (1986) imply, whereas Filman *et al.* (1989) and Flore *et al.* (1990) propose, that the pocket factor might be cellular in origin and might regulate viral assembly and uncoating.

Binding of WIN compounds to HRV14 causes major conformational changes in the pocket and, hence, also in the canyon floor (the receptor attachment site). These changes were correlated to inhibition of attachment in the presence of the antiviral compounds (Heinz *et al.*, 1989; Pevear *et al.*, 1989). In contrast, in HRV1A (a minor receptor group virus) and polioviruses, where the WIN compounds merely displace the pocket factor without a correspondingly large conformational change, there is inhibition of uncoating but not of attachment. Preliminary results suggested that rhinoviruses of the minor receptor group exhibited no inhibition of attachment, whereas those of the major receptor group behaved like HRV14, for which attachment is inhibited. Thus, it was a surprise to find 'pocket factor' electron density in HRV16, causing the shape of the pocket to resemble that of the 'WIN filled' form of HRV14 (K. H. Kim *et al.*, 1993; S. Kim *et al.*, 1989).

In HRV16 and CVB3, the height for the density of the pocket factor is comparable to that of amino acid side chains, indicating that most pockets are fully occupied. However, in HRV16, the height decreases beyond the sixth carbon atom, suggesting that the density might represent a mixture of fatty acids 6, 8 or 10 carbon atoms long.

In HRV1A and HRV16, the more active antiviral compounds tend to have an aliphatic chain less than or equal to five carbon atoms long (Mallamo *et al.*, 1992), correlating with the available space within the binding pocket (Diana *et al.*, 1990, 1992; K. H. Kim *et al.*, 1993). In HRV14, the most active antiviral agents tend to be longer, with seven-carbon aliphatic chains. For example, WIN 56291 has an aliphatic chain of only three carbons (compare Fig. 2) and is equally active against HRV16 and HRV1A but less active against HRV14. Thus, for each serotype, there is an optimal drug size that displays the greatest activity and binding affinity (Diana *et al.*, 1990, 1992) and best fills the volume of the pocket. It follows that the smaller pocket factors, which can be easily displaced by WIN compounds in HRV16 and HRV1A (K. H. Kim *et al.*, 1993; Oliveira *et al.*, 1993), bind with less affinity than the antiviral compounds. Nevertheless, the pocket factors seen in the electron densities remain in the pocket even after extensive dialysis of the virus sample. The WIN compounds have a binding constant comparable to their minimal inhibitory concentrations of $\sim 10^{-8}$ M (Fox *et al.*, 1986, 1991).

The role of the pocket factor

When the antiviral binding pocket in HRV14 is filled with WIN compounds or fragments of WIN compounds that do not inhibit infectivity, there is an increase in the thermal stability of the virus (Bibler-Muckelbauer *et al.*, 1994; Heinz, Shepard & Rueckert, 1990), presumably as a consequence of placing a hydrophobic molecule into an internal hydrophobic cavity (Eriksson *et al.*, 1992*a*,*b*). Similarly, drug-dependent mutants of poliovirus require WIN compounds to maintain their stability (Mosser & Rueckert, 1993). The pocket factor may, therefore, be required to stabilize the virus in transit from one cell to another. However, the delivery of the infectious RNA into the cytoplasm must require a destabilizing step that might be affected by expulsion of the pocket factor during receptor-mediated uncoating.

Because ICAM-1 binds to HRV14 and to HRV16 (Fig. 7), the shape of the canyon for HRV16 should be similar to that in HRV14 when ICAM-1 binding occurs. As soluble ICAM-1 binds to purified HRV14, which does not contain any pocket factor, presumably the pocket is empty when ICAM-1 binds to HRV16. However, the structure of HRV16 shows the presence of a pocket factor in the purified virus (Oliveira *et al.*, 1993). Hence, it must be assumed that the pocket factor is displaced before the receptor can seat itself into the canyon. In essence, there are two competing equilibria: the binding of ICAM-1 and the binding of the pocket factor to the virus. Although the sites of binding of ICAM-1 and of the pocket factor are not the same, they are in close proximity and interfere with each other. The floor of the canyon is also the roof of the pocket for the pocket factor or WIN compounds. When ICAM-1 binds, the floor is depressed downwards, which is possible only when there is nothing in the pocket. Conversely, when there is a compound in the pocket, its roof is raised upward. The displacement of the pocket factor *per se* does not cause the virus to fall apart. For instance, when HRV14 is crystallized, it does not contain a pocket factor, and the complex of HRV16 with ICAM-1 is reasonably stable. Nevertheless, the absence of pocket factor increases the potential for disruption by lowered pH or by formation of the receptor–virus complex.

Presumably, the destabilization of the virus on cell attachment is made possible by the displacement of a sufficient number of pocket factors when the receptor competes for the overlapping binding site. Progressive recruitment of receptors is then sufficient to trigger release of the VP4s. The terminal myristate moieties of VP4 and the exposure of the amino terminus of VP1 will permit entry through the cell membrane, possibly by creating a channel along the five-fold axes of the virus (Giranda *et al.*, 1992).

A class of HRV14 drug-resistant (compensation) mutants can be selected by growing the virus in the presence of antiviral WIN compounds. Such mutants occur at a frequency of about one per 10^4 virions. They have been shown to be mostly single mutations (Heinz *et al.*, 1990; Shepard *et al.*, 1993) and six of the seven characterized to date are situated near the walls and floor of the canyon. WIN compounds bind into the pocket of these mutant viruses and deform the canyon floor in a similar manner to their effect on wild-type viruses (M. A. Oliveira, I. Minor, R. R. Rueckert and M. G. Rossmann, unpublished data). In some of these mutants, the affinity of ICAM-1 for the virus is enhanced (R. R. Rueckert, private communication; M. P. Fox, D. C. Pevear and F. J. Dutko, unpublished data). Thus, it is reasonable to conclude that ICAM-1 binds better to these mutant viruses than the WIN compounds (Fig. 7B).

In the case of poliovirus or HRV1A (a minor group rhinovirus), only uncoating is inhibited by WIN compounds, and not attachment. If the pocket factor needs to be absent for the virus to uncoat, binding of receptor to these viruses should lead to displacement of the pocket factor, just as is the

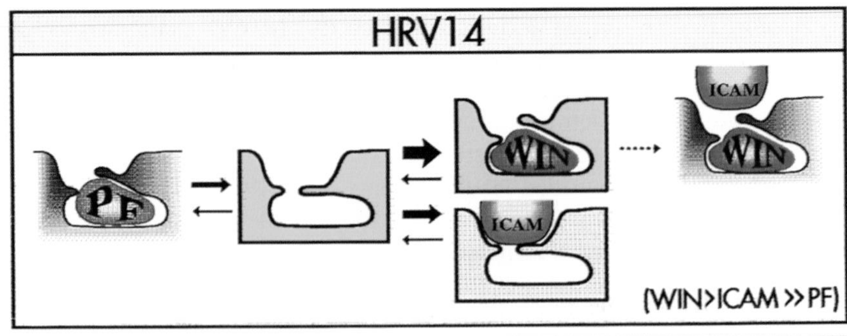

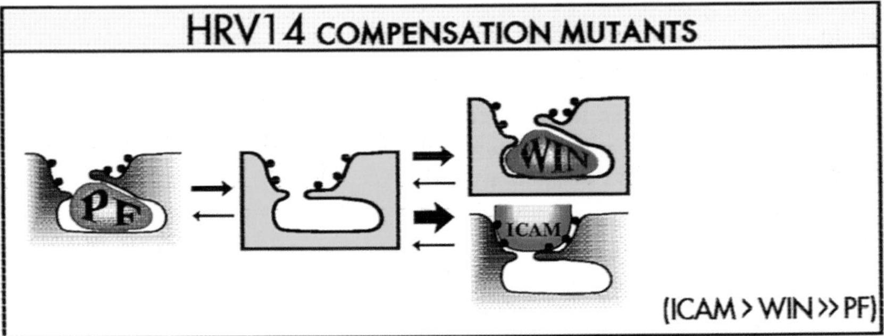

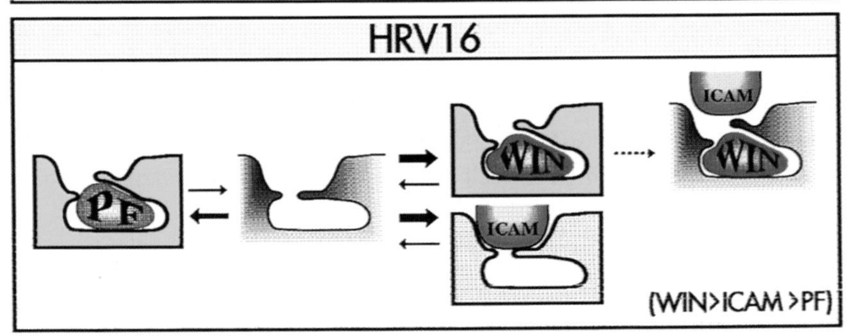

Fig. 7. Conditions for inhibition of viral attachment by WIN compounds. Crystallographically and electron microscopically determined structures are in yellow and pink, respectively, while hypothetical structures are in grey. A, In wild-type HRV14, the pocket factor binds weakly and is not observed in crystallographic studies. When WIN compounds bind into the pocket, they deform the roof of the pocket which is also the floor of the canyon. This inhibits the attachment of the virus to the ICAM-1 receptor and, hence, presumably the binding affinity of WIN is greater than that of ICAM-1. When ICAM-1 recognizes the canyon floor, the putative pocket factor must be displaced by ICAM-1 and, hence, the binding affinity of ICAM-1 is greater than that of pocket factor. B, Drug-resistant compensation mutants of HRV14 cluster around the canyon walls and floor (•) and increase the affinity of ICAM-1 for the virus. Although WIN compounds can bind to the virus, they do not inhibit infectivity. Thus, the binding affinity of the mutant virus to ICAM-1 is greater than that of WIN. C, Wild-type HRV16 contains a pocket factor. This can be replaced by WIN compounds which inhibit attachment. Hence, in this case the affinity of HRV16 for WIN is greater than that of ICAM-1 which is greater than that of pocket factor. (Reprinted with permission from Rossmann, 1994. Copyright by The Protein Society.)

case for the major group rhinoviruses. Similarly, the WIN compounds must also be displaced by the receptor as there is no inhibition of attachment, thus requiring the remaining WIN compounds to stabilize the virus sufficiently to inhibit uncoating.

PARVOVIRUSES

Parvovirus receptors

Parvovirus B19 is the only member of *Parvoviridae* pathogenic to man (Brown, 1995; Brown, Young & Liu, 1994). Acute infection causes a childhood measles-like rash condition, erythrema infectosium, also known as fifth disease, and acute or chronic arthritis in adults. In persons with underlying haemolysis, acute B19 infection results in transient aplastic crises, due to cessation in red blood cell production, that is caused by tropism of B19 to erythroid progenitor cells. In immunocompromised individuals, such as AIDS patients and patients undergoing immunosuppressive drug therapy, persistent B19 infection causes chronic severe anaemia due to erythroid marrow failure. B19 infection of pregnant women may cause hydrops fetalis (congestive heart failure) of the foetus and spontaneous abortions due to the inability of the foetus to mount an adequate immune response.

The initiation of viral replication for parvoviruses requires interactions with specific cell surface receptors. Brown, Anderson and Young (1993) have shown that one of the blood group P antigens, the glycolipid globoside ($\sim$1400 daltons), is the receptor for B19. Erythrocytes that contain globoside haemagglutinate in the presence of B19, *in vitro* (Brown *et al.*, 1993; Brown & Cohen, 1992). Some parvoviruses, such as CPV (Basak, Turner & Parr, 1994) and minute virus of mouse (Crawford, 1966), use carbohydrates for cellular attachment. Furthermore, CPV and FPV bind to glycolipids that have a structure similar to P antigens (Y. Suzuki, personal communication). Globoside is also found in endothelial cells, which may be the site of B19 replication resulting in fifth disease (Rouger, Gane & Salmon, 1987). The presence of globoside in foetal cardiac myocytes may lead to B19 infection and myocarditis (Morey *et al.*, 1992; Naides & Weiner, 1989).

Human B19 virus complexed with its receptor

The reconstructed cryo-EM density map of the B19 VP2 capsid shows depressions at both the icosahedral two-fold and three-fold axes and canyon-like regions about each of the five-fold axes (Fig. 8c and d) consistent with the earlier 8 Å resolution X-ray map (Agbandje *et al.*, 1994). Unlike CPV (Fig. 8a and b), B19 (Fig. 8c and d) lacks a channel at the five-fold icosahedral axes. The reconstructed map of the B19: globoside complex (Fig. 8e and f) showed

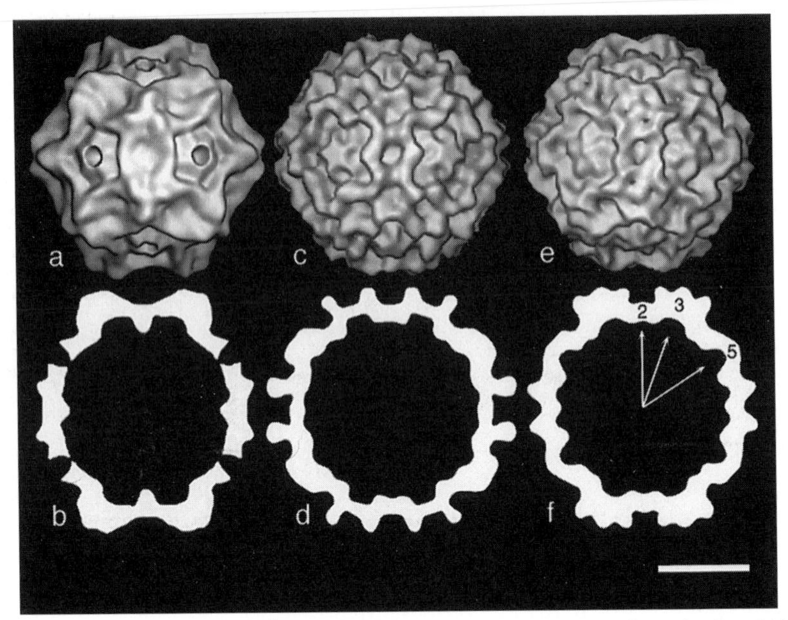

Fig. 8. The structure of CPV (21 Å resolution) derived from X-ray crystallography (a and b) is compared with image reconstructions of B19 (c and d) and the B19:globoside complex with its receptor (e and f). Surface views are shown in a, c and e, and central sections are shown in b, d and f. Small differences can also be seen at the five-fold axes. Scale bar = 100 Å. (Reprinted with permission from Chipman *et al.*, 1996. Copyright by The National Academy of Sciences.)

extra density on the three-fold depression when compared with the B19 VP2 capsid alone. A difference map between the complex and B19 VP2 capsid (Fig. 9) showed that the receptor molecules bind into the depression at the icosahedral three-fold axes. The globoside density extends approximately 19 Å out from the surface of the depression. Small differences can also be seen around the five-fold axes (compare Fig. 8d and f). These may indicate conformational changes as a result of globoside binding, but, more probably, represent noise, as a consequence of the fewer number of particles used in the image reconstruction of the complex. The height of the globoside was 4.5σ and the height of the features about the five-fold axes 2σ above background, where σ was the root-mean-square deviation from zero in the difference map.

The volume of the roughly spherical globoside difference density attributed to globoside is approximately 3600 Å^3, as determined at a contour level just above the noise of the map. Assuming a density of about 1.1 gm/cm^3 for a glycolipid (Matthews, 1968), this volume would be occupied by three molecules each having a mass of 655 Da. The interaction of the receptor with the capsid mainly involves the sugar portion of the glycolipid, as has been observed for other viruses which use glycolipids or glycoproteins as receptor molecules (Sauter *et al.*, 1992; Stehle *et al.*, 1994; Weis *et al.*, 1988). The mass

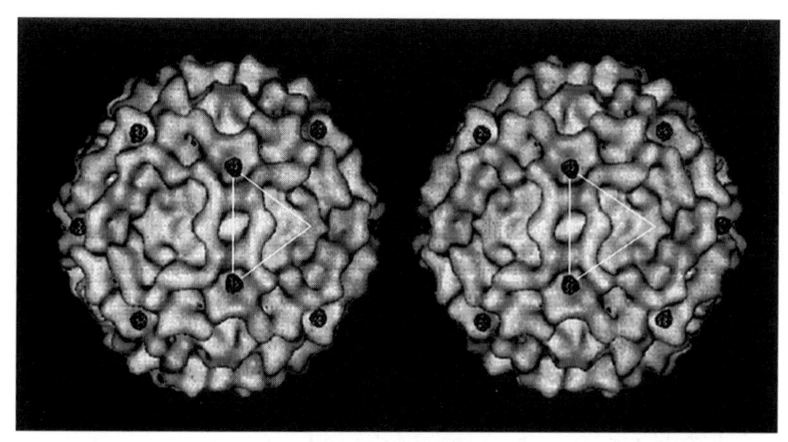

Fig. 9. A stereoscopic view of the difference map between B19 complexed with the globoside receptor and B19 on its own superimposed onto the B19 cryo-EM reconstructed surface image. Globoside molecules bind into the surface depressions at the three-fold vertices. The triangle outlines one icosahedral asymmetric unit. (Reprinted with permission from Chipman *et al.*, 1996. Copyright by The National Academy of Sciences.)

of the carbohydrate component of globoside is 690 Da, which gives a reasonable fit to the 655 Da expected to fill the difference electron density. Thus, the lipid component of globoside (Fig. 10) is likely to be disordered and, therefore, not visible in the density map.

The segment having the sequence Glc(β1-3)Glc(β1-3)Glc(β1-3)Glc(β1-3), as observed in the naturally occurring triple helical fibre structure of 1-3-β-D glucan (Deslandes, Marchessault & Sarko, 1980), was used as the tetrasaccharide model to fit into the difference map. This structure has a rise per residue of 2.94 Å. Every sugar residue has the OH group at the C_2 position facing the three-fold axis, forming a hydrogen bond with a three-fold-related helical counterpart. The (β1-3) linkage between the first and second sugar residues of globoside is the same as in the glucan segment. The only possible adjustments were a rotation about and a small translation along the three-fold axis which were needed to make a hydrophobic contact with Tyr401 of B19 VP2. Similar hydrophobic contacts between sugar and aromatic residues are common features of protein–carbohydrate interactions (Kreusch & Schulz, 1994; Schirmer *et al.*, 1995). The resulting fit of this trimeric tetrasaccharide is fully consistent with the difference electron density map (Fig. 11a and b).

The probable structure of the globoside carbohydrate (Fig. 10) was predicted from the allowed dihedral angles for each linkage and the most stable chair conformation of the monosaccharides (Sathyanarayana & Rao, 1971, 1972). The principal difference between the carbohydrate moiety of globoside and that of the glucan structure arises at the (α1-4) linkage between the second and third sugar residues, giving the globoside tetrasaccharide a

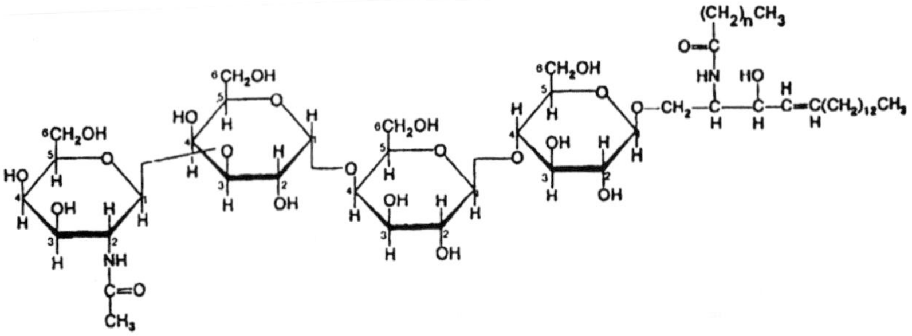

Fig. 10. Globoside, (Gal-N-Ac)(β1-3)Gal(αl-4)Gal(β1-4)Glc-Cer, where Gal-N-Ac is N-Acetylgalactosamine, Gal is galactose, Glc is glucose, Cer is ceramide and n = 16, 18, 22 or 24. (Reprinted with permission from Chipman *et al.*, 1996. Copyright by The National Academy of Sciences.)

kink at this junction. Its first two sugars were superimposed onto the first two residues of the glucan triple helix. Only a small adjustment was required to avoid steric interference around the three-fold axis. In the model, the O_6 atom of the third sugar residue faces the three-fold axis, where it will make a hydrogen bond with a three-fold-related molecule. This fit (Fig. 11b and d) leaves the ceramide extending outwards, presumably able to anchor into the cellular membrane. The trimeric lipid arrangement is likely to be incorporated into the usual hexagonal packing of lipid bilayers (Fig. 12).

An attempt to model the B19 structure based on that of CPV and FPV used the 8 Å resolution B19 map as a constraint. The atomic structure of CPV was modified in two ways; (a) to be consistent with the aligned B19 sequence (Chapman & Rossmann, 1993*b*) by substituting individual residues and (b) the CPV insertions that are absent in B19 were removed according to a modified alignment scheme (Fig. 13). The footprint of the trimeric receptor on the viral surface is shown in Fig. 14. The depression at the icosahedral three-fold axes has a hydrophobic centre (including Tyr401) that is surrounded by polar residues within the asymmetric unit of the globoside footprint.

The relationship between the receptor attachment site within a surface depression at the three-fold axes of B19 surrounded by neutralization epitopes is analogous to depressions found on the surfaces of rhinoviruses (Rossmann *et al.*, 1985), polioviruses (Racaniello, 1992) and influenza viruses (Weis *et al.*, 1988). This arrangement ensures that naturally occurring mutants at the base of the depression can escape host immune surveillance, but recognize and bind to the receptor. The binding of globoside in the icosahedral three-fold depression (but not in the canyon-like depression as is the case for rhinoviruses) has identified the region of the B19 capsid involved in host recognition and binding. It remains to be determined how attachment initiates cell entry and viral transport to the nucleus.

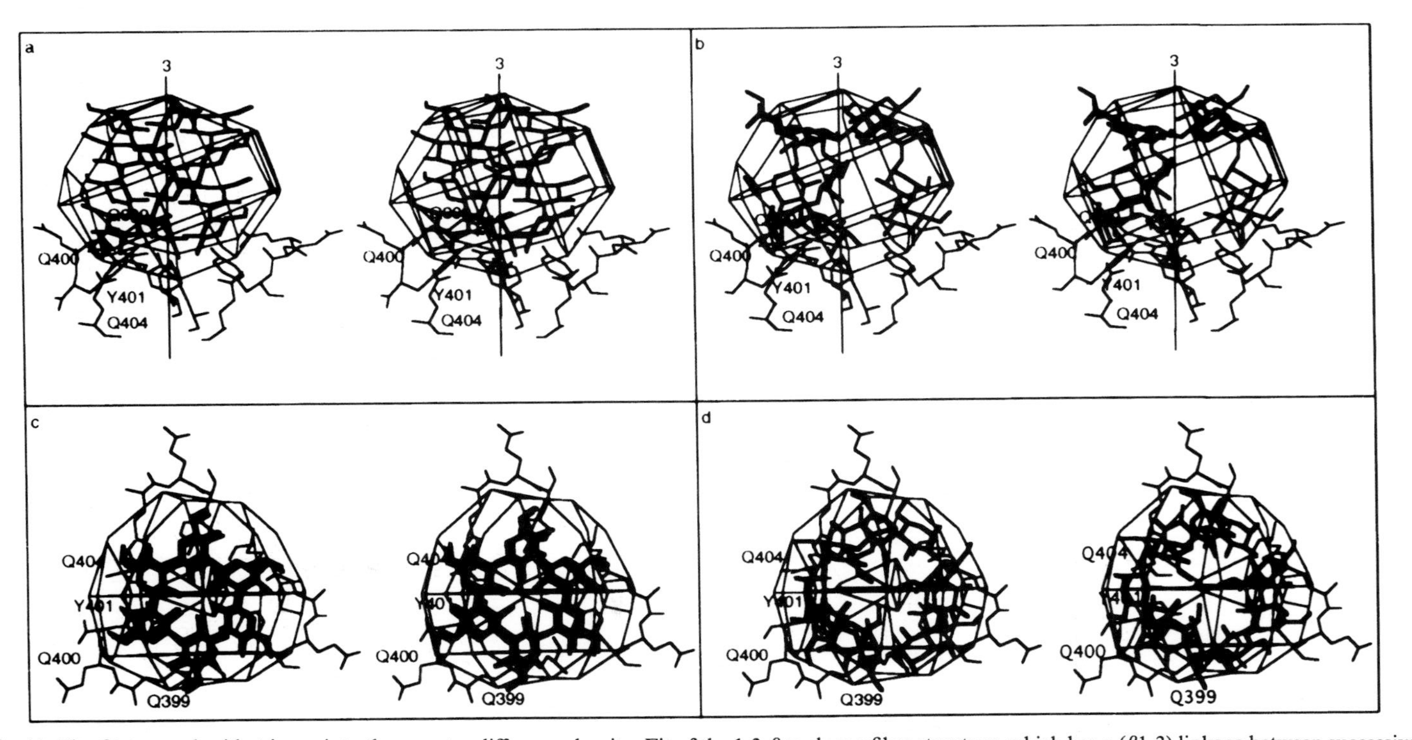

Fig. 11. Fit of tetrasaccharide trimers into the receptor difference density. Fit of the 1-3-β-D glucan fibre structure, which has a (β1-3) linkage between successive glucose moieties (a and c) and of the preferred structure of the globoside carbohydrate moiety positioned as suggested by the glucan fibre structure (b and d). a and b are side views (perpendicular to the icosahedral three-fold axis); c and d are views down the three-fold axis. (Reprinted with permission from Chipman *et al.*, 1996. Copyright by The National Academy of Sciences.)

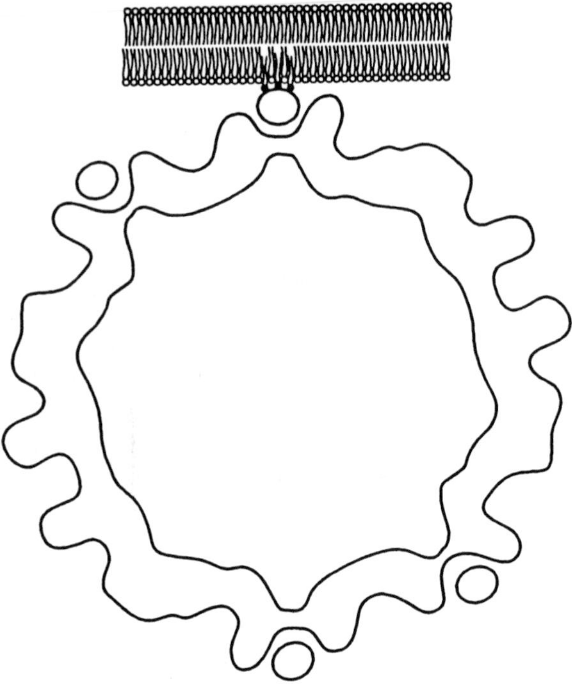

Fig. 12. Schematic representation of the docking of B19 to a cellular receptor membrane glycolipid. (Reprinted with permission from Chipman *et al.*, 1996. Copyright by The National Academy of Sciences.)

```
                            <--------loop 2-------->
   Canine      208  WRYYFQWDRTL..IPSHTGTSGT..PTNIYHGTD 237
Human_B19      194  YAYL.......TV..GDVNT...QGISGDSKKLA 205

                        <---------------loop 4--------------->
   Canine      407  GRYPEGDWIQNINFNLPVTNDNVLLPTDPIGGKTGINYT 445
Human_B19      384  LQGLNMHTY........FPNKGTQQYTD.........YT 402
```

Fig. 13. Modified sequence alignment (Chapman & Rossmann, 1993b) of CPV to B19 in the loop 2 and 4 regions. The large insertions in loop 2 and loop 4 form a small spike on each three-fold axis in CPV. Conversely, the deletions in B19 leave a slight depression on the three-fold axes. (Reprinted with permission from Chipman *et al.*, 1996. Copyright by The National Academy of Sciences.)

ACKNOWLEDGEMENTS

I thank Cheryl Towell for help in the preparation of this manuscript. The work was supported by National Institutes of Health grants (AI11219 and AI33468) and a grant from the Sterling-Winthrop Pharmaceuticals Research Division to M.G.R. and a Lucille P. Markey Foundation Award for the expansion of structural studies at Purdue University.

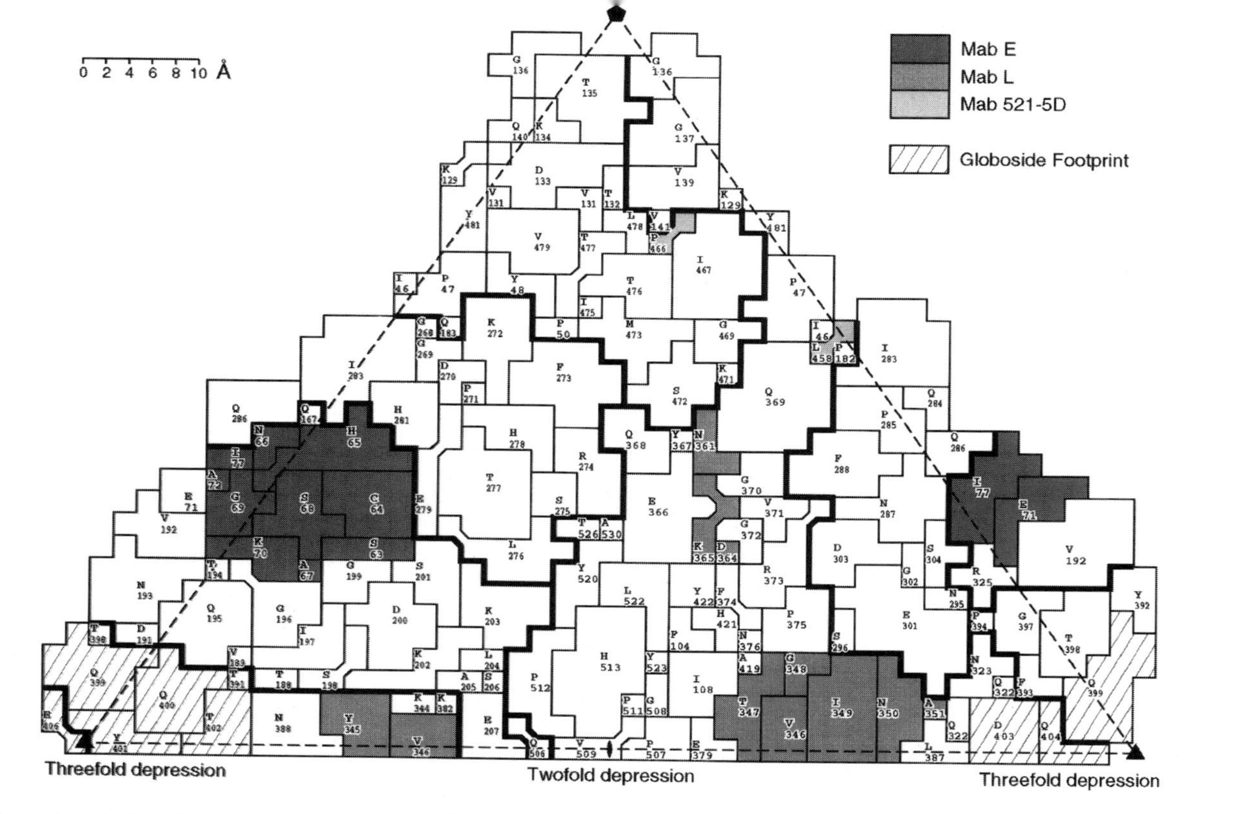

Fig. 14. A 'roadmap' of one icosahedral asymmetric unit showing the surface residues of the B19 virus based on a modified sequence alignment (Fig. 13) relative to CPV. The footprint of the globoside receptor on the viral surface (striped region) and the site of escape mutations for the neutralizing antibodies MAb E, L and 521-5D (shaded regions) are shown. Thick lines denote boundaries between symmetry-related subunits. (Reprinted with permission from Chipman *et al.*, 1996. Copyright by The National Academy of Sciences.)

REFERENCES

Abad-Zapatero, C., Abdel-Meguid, S. S., Johnson, J. E., Leslie, A. G. W., Rayment, I., Rossmann, M. G., Suck, D. & Tsukihara, T. (1980). Structure of southern bean mosaic virus at 2.8 Å resolution. *Nature (London)*, **286**, 33–9.

Abraham, G. & Colonno, R. J. (1984). Many rhinovirus serotypes share the same cellular receptor. *Journal of Virology*, **51**, 340–5.

Acharya, R., Fry, E., Stuart, D., Fox, G., Rowlands, D. & Brown, F. (1989). The three-dimensional structure of foot-and-mouth disease virus at 2.9 Å resolution. *Nature (London)*, **337**, 709–16.

Agbandje, M., Kajigaya, S., McKenna, R., Young, N. S. & Rossmann, M. G. (1994). The structure of human parvovirus B19 at 8 Å resolution. *Virology*, **203**, 106–15.

Arthos, J., Dean, K. C., Chaikin, M. A., Fornwald, J. A., Sathe, G., Sattentau, Q. J., Clapham, P. R., Weiss, R. A., McDougal, J. S., Pietropaolo, C., Axel, R., Truneh, A., Maddon, P. J. & Sweet, R. W. (1989). Identification of the residues in human CD4 critical for the binding of HIV. *Cell*, **57**, 469–81.

Badger, J., Minor, I., Kremer, M. J., Oliveira, M. A., Smith, T. J., Griffith, J. P., Guerin, D. M. A., Krishnaswamy, S., Luo, M., Rossmann, M. G., McKinlay, M. A., Diana, G. D., Dutko, F. J., Fancher, M., Rueckert, R. R. & Heinz, B. A. (1988). Structural analysis of a series of antiviral agents complexed with human rhinovirus 14. *Proceedings of the National Academy of Sciences, USA*, **85**, 3304–8.

Basak, S., Turner, H. & Parr, S. (1994). Identification of a 40- to 42-kDa attachment polypeptide for canine parvovirus in A72 cells. *Virology*, **205**, 7–16.

Berendt, A. R., McDowall, A., Craig, A. G., Bates, P. A., Sternberg, M. J. E., Marsh, K., Newbold, C. I. & Hogg, N. (1992). The binding site on ICAM-1 for *Plasmodium falciparum*-infected erythrocytes overlaps, but is distinct from, the LFA-1 binding site. *Cell*, **68**, 71–81.

Bergelson, J. M., Shepley, M. P., Chan, B. M. C., Hemler, M. E. & Finberg, R. W. (1992). Identification of the integrin VLA-2 as a receptor for echovirus 1. *Science*, **255**, 1718–20.

Bibler-Muckelbauer, J. K., Kremer, M. J., Rossmann, M. G., Diana, G. D., Dutko, F. J., Pevear, D. C. & McKinlay, M. A. (1994). Human rhinovirus 14 complexed with fragments of active antiviral compounds. *Virology*, **202**, 360–9.

Brady, R. L., Dodson, E. J., Dodson, G. G., Lange, G., Davis, S. J., Williams, A. F. & Barclay, A. N. (1993). Crystal structure of domains 3 and 4 of rat CD4: relation to the NH_2-terminal domains. *Science*, **260**, 979–83.

Brown, K. E. (1995). The identification of the parvovirus B19 receptor. M Sc Thesis, Cambridge University.

Brown, K. E., Anderson, S. M. & Young, N. S. (1993). Erythrocyte P antigen: cellular receptor for B19 parvovirus. *Science*, **262**, 114–17.

Brown, K. E. & Cohen, B. J. (1992). Haemagglutination by parvovirus B19. *Journal of General Virology*, **73**, 2147–9.

Brown, K. E., Young, N. S. & Liu, J. M. (1994). Molecular, cellular and clinical aspects of parvovirus B19 infection. *Critical Reviews in Oncology and Hematology*, **16**, 1–31.

Burness, A. T. H. & Pardoe, I. U. (1981). Effect of enzymes on the attachment of influenza and encephalomyocarditis viruses to erythrocytes. *Journal of General Virology*, **55**, 275–88.

Burness, A. T. H. & Pardoe, I. U. (1983). A sialoglycopeptide from human erythrocytes with receptor-like properties for encephalomyocarditis and influenza viruses. *Journal of General Virology*, **64**, 1137–48.

Chapman, M. S. & Rossmann, M. G. (1993*a*). Comparison of surface properties of picornaviruses: strategies for hiding the receptor site from immune surveillance. *Virology*, **195**, 745–56.

Chapman, M. S. & Rossmann, M. G. (1993*b*). Structure, sequence and function correlations among parvoviruses. *Virology*, **194**, 491–508.

Chipman, P. R., Agbandje-McKenna, M., Kajigaya, S., Brown, K. E., Young, N. S., Baker, T. S. & Rossmann, M. G. (1996). Cryo-electron microscopy studies of empty capsids of human parvovirus B19, complexed with its cellular receptor. *Proceedings of the National Academy of Sciences, USA*, **93**, 7502–6.

Choi, A. H. C., Paul, R. W. & Lee, P. W. K. (1991). Reovirus binds to multiple plasma membrane proteins of mouse L fibroblasts. *Virology*, **178**, 316–20.

Colonno, R. J. (1992). Molecular interactions between human rhinoviruses and their cellular receptors. *Seminars in Virology*, **3**, 101–7.

Colonno, R. J., Callahan, P. L. & Long, W. J. (1986). Isolation of a monoclonal antibody that blocks attachment of the major group of human rhinoviruses. *Journal of Virology*, **57**, 7–12.

Colonno, R. J., Condra, J. H., Mizutani, S., Callahan, P. L., Davies, M. E. & Murcko, M. A. (1988). Evidence for the direct involvement of the rhinovirus canyon in receptor binding. *Proceedings of the National Academy of Sciences, USA*, **85**, 5449–53.

Crawford, L. V. (1966). A minute virus of mice. *Virology*, **29**, 605–12.

Crowell, R. L. (1966). Specific cell-surface alteration by enteroviruses as reflected by viral-attachment interference. *Journal of Bacteriology*, **91**, 198–204.

Dalgleish, A. G., Beverley, P. C. L., Clapham, P. R., Crawford, D. H., Greaves, M. F. & Weiss, R. A. (1984). The CD4 (T4) antigen is an essential component of the receptor for the AIDS retrovirus. *Nature (London)*, **312**, 763–7.

Deslandes, Y., Marchessault, R. H. & Sarko, A. (1980). Triple-helical structure of (1-3)-β-d glucan. *Macromolecules*, **13**, 1466–71.

Diana, G. D., Treasurywala, A. M., Bailey, T. R., Oglesby, R. C., Pevear, D. C. & Dutko, F. J. (1990). A model for compounds active against human rhinovirus-14 based on X-ray crystallography data. *Journal of Medicinal Chemistry*, **33**, 1306–11.

Diana, G. D., Kowalczyk, P., Treasurywala, A. M., Oglesby, R. C., Pevear, D. C. & Dutko, F. J. (1992). CoMFA analysis of the interactions of antipicornavirus compounds in the binding pocket of human rhinovirus-14. *Journal of Medicinal Chemistry*, **35**, 1002–6.

Eriksson, A. E., Baase, W. A., Wozniak, J. A. & Matthews, B. A. (1992*a*). A cavity-containing mutant of T4 lysozyme is stabilized by buried benzene. *Nature (London)*, **355**, 371–3.

Eriksson, A. E., Baase, W. A., Zhang, X.-J., Heinz, D. W., Blaber, M., Baldwin, E. P. & Matthews, B. W. (1992*b*). Response of a protein structure to cavity-creating mutations and its relation to the hydrophobic effect. *Science*, **255**, 178–83.

Everaert, L., Vrijsen, R. & Boeyé, A. (1989). Eclipse products of poliovirus after cold-synchronized infection of HeLa cells. *Virology*, **171**, 76–82.

Filman, D. J., Syed, R., Chow, M., Macadam, A. J., Minor, P. D. & Hogle, J. M. (1989). Structural factors that control conformational transitions and serotype specificity in type 3 poliovirus. *The EMBO Journal*, **8**, 1567–79.

Flore, O., Fricks, C. E., Filman, D. J. & Hogle, J. M. (1990). Conformational changes in poliovirus assembly and cell entry. *Seminars in Virology*, **1**, 429–38.

Fox, M. P., Otto, M. J. & McKinlay, M. A. (1986). The prevention of rhinovirus and poliovirus uncoating by WIN 51711: a new antiviral drug. *Antimicrobioal Agents Chemotherapy*, **30**, 110–16.

Fox, M. P., McKinlay, M. A., Diana, G. D. & Dutko, F. J. (1991). Binding affinities of structurally related human rhinovirus capsid-binding compounds are correlated to their activities against human rhinovirus type 14. *Antimicrobial Agents Chemotherapy*, **35**, 1040–7.

Freistadt, M. S. & Racaniello, V. R. (1991). Mutational analysis of the cellular receptor for poliovirus. *Journal of Virology*, **65**, 3873–6.

Fricks, C. E. & Hogle, J. M. (1990). Cell-induced conformational change in poliovirus: externalization of the amino terminus of VP1 is responsible for liposome binding. *Journal of Virology*, **64**, 1934–45.

Fried, H., Cahan, L. D. & Paulson, J. C. (1981). Polyoma virus recognizes specific sialyloligosaccharide receptors on host cells. *Virology*, **109**, 188–92.

Giranda, V. L., Chapman, M. S. & Rossmann, M. G. (1990). Modeling of the human intercellular adhesion molecule-1, the human rhinovirus major group receptor. *Proteins: Structure, Function, and Genetics*, **7**, 227–33.

Giranda, V. L., Heinz, B. A., Oliveira, M. A., Minor, I., Kim, K. H., Kolatkar, P. R., Rossmann, M. G. & Rueckert, R. R. (1992). Acid-induced structural changes in human rhinovirus 14: possible role in uncoating. *Proceedings of the National Academy of Sciences, USA*, **89**, 10213–17.

Greve, J. M., Davis, G., Meyer, A. M., Forte, C. P., Yost, S. C., Marlor, C. W., Kamarck, M. E. & McClelland, A. (1989). The major human rhinovirus receptor is ICAM-1. *Cell*, **56**, 839–47.

Greve, J. M., Forte, C. P., Marlor, C. W., Meyer, A. M., Hoover-Litty, H., Wunderlich, D. & McClelland, A. (1991). Mechanisms of receptor-mediated rhinovirus neutralization defined by two soluble forms of ICAM-1. *Journal of Virology*, **65**, 6015–23.

Gruenberger, M., Pevear, D., Diana, G. D., Kuechler, E. & Blaas, D. (1991). Stabilization of human rhinovirus serotype 2 against pH-induced conformational change by antiviral compounds. *Journal of General Virology*, **72**, 431–3.

Harrison, S. C., Olson, A. J., Schutt, C. E., Winkler, F. K. & Bricogne, G. (1978). Tomato bushy stunt virus at 2.9 Å resolution. *Nature (London)*, **276**, 368–73.

Heinz, B. A., Rueckert, R. R., Shepard, D. A., Dutko, F. J., McKinlay, M. A., Fancher, M., Rossmann, M. G., Badger, J. & Smith, T. J. (1989). Genetic and molecular analyses of spontaneous mutants of human rhinovirus 14 that are resistant to an antiviral compound. *Journal of Virology*, **63**, 2476–85.

Heinz, B. A., Shepard, D. A. & Rueckert, R. R. (1990). Escape mutant analysis of a drug-binding site can be used to map functions in the rhinovirus capsid. In Laver, W. G. & Air, G. M., eds. *Use of X-ray Crystallography in the Design of Antiviral Agents*, pp. 173–86. San Diego: Academic Press.

Hofer, F., Gruenberger, M., Kowalski, H., Machat, H., Huettinger, M., Kuechler, E. & Blaas, D. (1994). Members of the low density lipoprotein receptor family mediate cell entry of a minor-group common cold virus. *Proceedings of the National Academy of Sciences, USA*, **91**, 1839–42.

Hogle, J. M., Chow, M. & Filman, D. J. (1985). Three-dimensional structure of poliovirus at 2.9 Å resolution. *Science*, **229**, 1358–65.

Hoover-Litty, H. & Greve, J. M. (1993). Formation of rhinovirus-soluble ICAM-1 complexes and conformational changes in the virion. *Journal of Virology*, **67**, 390–7.

Hosur, M. V., Schmidt, T., Tucker, R. C., Johnson, J. E., Gallagher, T. M., Selling, B. H. & Rueckert, R. R. (1987). Structure of an insect virus at 3.0 Å resolution. *Proteins: Structure, Function, and Genetics*, **2**, 167–76.

Kim, J. W., Closs, E. I., Albritton, L. M. & Cunningham, J. M. (1991). Transport of cationic amino acids by the mouse ecotropic retrovirus receptor. *Nature (London)*, **352**, 725–8.

Kim, K. H., Willingmann, P., Gong, Z. X., Kremer, M. J., Chapman, M. S., Minor, I., Oliveira, M. A., Rossmann, M. G., Andries, K., Diana, G. D., Dutko, F. J., McKinlay, M. A. & Pevear, D. C. (1993). A comparison of the anti-rhinoviral drug binding pocket in HRV14 and HRV1A. *Journal of Molecular Biology*, **230**, 206–25.

Kim, S., Smith, T. J., Chapman, M. S., Rossmann, M. G., Pevear, D. C., Dutko, F. J., Felock, P. J., Diana, G. D. & McKinlay, M. A. (1989). The crystal structure of human rhinovirus serotype 1A (HRV1A). *Journal of Molecular Biology*, **210**, 91–111.

Klatzmann, D., Champagne, E., Chamaret, S., Gruest, J., Guetard, D., Hercend, T., Gluckman, J. C. & Montagnier, L. (1984). T-lymphocyte T4 molecule behaves as the receptor for human retrovirus LAV. *Nature (London)*, **312**, 767–8.

Koike, S., Ise, I. & Nomoto, A. (1991). Functional domains of the poliovirus receptor. *Proceedings of the National Academy of Sciences, USA*, **88**, 4104–8.

Koike, S., Ise, I., Sato, Y., Mitsui, K., Horie, H., Umeyama, H. & Nomoto, A. (1992). Early events of poliovirus infection. *Seminars in Virology*, **3**, 109–15.

Kolatkar, P. R., Oliveira, M. A., Rossmann, M. G., Robbins, A. H., Katti, S. K., Hoover-Litty, H., Forte, C., Greve, J. M., McClelland, A. & Olson, N. H. (1992). Preliminary X-ray crystallographic analysis of intercellular adhesion molecule-1. *Journal of Molecular Biology*, **225**, 1127–30.

Korant, B. D., Lonberg-Holm, K., Yin, F. H. & Noble-Harvey, J. (1975). Fractionation of biologically active and inactive populations of human rhinovirus type 2. *Virology*, **63**, 384–94.

Kreusch, A. & Schulz, G. E. (1994). Refined structure of the porin from *Rhodopseudomonas blastica*. Comparison with the porin from *Rhodobacter capsulatus*. *Journal of Molecular Biology*, **243**, 891–905.

Liljas, L., Unge, T., Jones, T. A., Fridborg, K., Lövgren, S., Skoglund, U. & Strandberg, B. (1982). Structure of satellite tobacco necrosis virus at 3.0 Å resolution. *Journal of Molecular Biology*, **159**, 93–108.

Lineberger, D. W., Graham, D. J., Tomassini, J. E. & Colonno, R. J. (1990). Antibodies that block rhinovirus attachment map to domain 1 of the major group receptor. *Journal of Virology*, **64**, 2582–7.

Lonberg-Holm, K. & Korant, B. D. (1972). Early interaction of rhinoviruses with host cells. *Journal of Virology*, **9**, 29–40.

Lonberg-Holm, K., Crowell, R. L. & Philipson, L. (1976). Unrelated animal viruses share receptors. *Nature (London)*, **259**, 679–81.

McClelland, A., deBear, J., Yost, S. C., Meyer, A. M., Marlor, C. W. & Greve, J. M. (1991). Identification of monoclonal antibody epitopes and critical residues for rhinovirus binding in domain 1 of ICAM-1. *Proceedings of the National Academy of Sciences, USA*, **88**, 7993–7.

McKenna, R., Ilag, L. L. & Rossmann, M. G. (1994). Analysis of the single-stranded DNA bacteriophage ϕX174, refined at a resolution of 3.0 Å. *Journal of Molecular Biology*, **237**, 517–43.

McKenna, R., Xia, D., Willingmann, P., Ilag, L. L., Krishnaswamy, S., Rossmann, M. G., Olson, N. H., Baker, T. S. & Incardona, N. L. (1992*a*). Atomic structure of single-stranged DNA bacteriophage ϕX174 and its functional implications. *Nature (London)*, **355**, 137–43.

McKenna, R., Xia, D., Willingmann, P., Ilag, L. L. & Rossmann, M. G. (1992*b*). Structure determination of the bacteriophage ϕX174. *Acta Crystallographica*, **B48**, 499–511.

Maddon, P. J., Dalgleish, A. G., McDougal, J. S., Clapham, P. R., Weiss, R. A. & Axel, R. (1986). The T4 gene encodes the AIDS virus receptor and is expressed in the immune system and the brain. *Cell*, **47**, 333–48.

Mallamo, J. P., Diana, G. D., Pevear, D. C., Dutko, F. J., Chapman, M. S., Kim, K. H., Minor, I., Oliveira, M. & Rossmann, M. G. (1992). Conformationally restricted analogues of disoxaril: a comparison of the activity against human rhinovirus types 14 and 1A. *Journal of Medicinal Chemistry*, **35**, 4690–5.

Mason, P. W., Baxt, B., Brown, F., Harber, J., Murdin, A. & Wimmer, E. (1993). Antibody-complexed foot-and-mouth disease virus, but not poliovirus, can infect normally insusceptible cells via the Fc receptor. *Virology*, **192**, 568–77.

Matthews, B. W. (1968). Solvent content of protein crystals. *Journal of Molecular Biology*, **33**, 491–7.

Matthews, T. J., Weinhold, K. J., Lyerly, H. K., Langlois, A. J., Wigzell, H. & Bolognesi, D. P. (1987). Interaction between the human T-cell lymphotropic virus type III_B envelope glycoprotein gp120 and the surface antigen CD4: role of carbohydrate in binding and cell fusion. *Proceedings of the National Academy of Sciences, USA*, **84**, 5424–8.

Mendelsohn, C. L., Wimmer, E. & Racaniello, V. R. (1989). Cellular receptors for poliovirus: molecular cloning, nucleotide sequence, and expression of a new member of the immunoglobulin superfamily. *Cell*, **56**, 855–65.

Moore, M. D., Cooper, N. R., Tack, B. F. & Nemerow, G. R. (1987). Molecular cloning of the cDNA encoding the Epstein-Barr virus/C3d receptor (complement receptor type 2) of human B lymphocytes. *Proceedings of the National Academy of Sciences, USA*, **84**, 9194–8.

Morey, A. L., Patou, G., Myint, S. & Fleming, K. A. (1992). *In vitro* culture for the detection of infectious human parvovirus B19 and B19-specific antibodies using foetal haematopoietic precursor cells. *Journal of General Virology*, **73**, 3313–17.

Mosser, A. G. & Rueckert, R. R. (1993). WIN 51711-dependent mutants of poliovirus type 3: evidence that virions decay after release from cells unless drug is present. *Journal of Virology*, **67**, 1246–54.

Muckelbauer, J. K., Kremer, M., Minor, I., Tong, L., Zlotnick, A., Johnson, J. E. & Rossmann, M. G. (1995). The structure determination of coxsackievirus B3 at 3.5 Å resolution. *Acta Crystallographica*, **D51**, 653–67.

Naides, S. J. & Weiner, C. P. (1989). Antenatal diagnosis and palliative treatment of non-immune *Hydrops fetalis* secondary to fetal parvovirus B19 infection. *Prenatal Diagnosis*, **9**, 105–14.

Oliveira, M. A., Zhao, R., Lee, W. M., Kremer, M. J., Minor, I., Rueckert, R. R., Diana, G. D., Pevear, D. C., Dutko, F. J., McKinlay, M. A. & Rossmann, M. G. (1993). The structure of human rhinovirus 16. *Structure*, **1**, 51–68.

Olson, N. H., Kolatkar, P. R., Oliveira, M. A., Cheng, R. H., Greve, J. M., McClelland, A., Baker, T. S. & Rossmann, M. G. (1993). Structure of a human rhinovirus complexed with its receptor molecule. *Proceedings of the National Academy of Sciences, USA*, **90**, 507–11.

Page, G. S., Mosser, A. G., Hogle, J. M., Filman, D. J., Rueckert, R. R. & Chow, M. (1988). Three-dimensional structure of poliovirus serotype 1 neutralizing determinants. *Journal of Virology*, **62**, 1781–94.

Palmenberg, A. C. (1989). Sequence alignments of picornaviral capsid proteins. In Semler, B. L. & Ehrenfeld, E., eds. *Molecular Aspects of Picornavirus Infection and Detection*, pp. 211–41. Washington, DC: American Society for Microbiology.

Paul, R. W., Choi, A. H. C. & Lee, P. W. K. (1989). The α-anomeric form of sialic acid is the minimal receptor determinant recognized by reovirus. *Virology*, **172**, 382–5.

Pevear, D. C., Fancher, M. J., Felock, P. J., Rossmann, M. G., Miller, M. S., Diana, G., Treasurywala, A. M., McKinlay, M. A. & Dutko, F. J. (1989). Conformational change in the floor of the human rhinovirus canyon blocks adsorption to HeLa cell receptors. *Journal of Virology*, **63**, 2002–7.

Racaniello, V. R. (1992). Interaction of poliovirus with its cell receptor. *Seminars in Virology*, **3**, 473–81.

Register, R. B., Uncapher, C. R., Naylor, A. M., Lineberger, D. W. & Colonno, R. J. (1991). Human-murine chimeras of ICAM-1 identify amino acid residues critical for rhinovirus and antibody binding. *Journal of Virology*, **65**, 6589–96.

Robey, E. & Axel, R. (1990). CD4: collaborator in immune recognition and HIV infection. *Cell*, **60**, 697–700.

Robinson, I. K. & Harrison, S. C. (1982). Structure of the expanded state of tomato bushy stunt virus. *Nature (London)*, **297**, 563–8.

Roivainen, M., Hyypiä, T., Piirainen, L., Kalkkinen, N., Stanway, G. & Hovi, T. (1991). RGD-dependent entry of coxsackievirus A9 into host cells and its bypass after cleavage of VP1 protein by intestinal proteases. *Journal of Virology*, **65**, 4735–40.

Rossmann, M. G. (1994). Viral cell recognition and entry. *Protein Science*, **3**, 1712–25.

Rossmann, M. G. & Palmenberg, A. C. (1988). Conservation of the putative receptor attachment site in picornaviruses. *Virology*, **164**, 373–82.

Rossmann, M. G., Arnold, E., Erickson, J. W., Frankenberger, E. A., Griffith, J. P., Hecht, H. J., Johnson, J. E., Kamer, G., Luo, M., Mosser, A. G., Rueckert, R. R., Sherry, B. & Vriend, G. (1985). Structure of a human common cold virus and functional relationship to other picornaviruses. *Nature (London)*, **317**, 145–53.

Rouger, P., Gane, P. & Salmon, C. (1987). Tissue distribution of H, Lewis and P antigens as shown by a panel of 18 monoclonal antibodies. *Revue Française de Transfusion et Immuno-Hematologie*, **30**, 699–708.

Rueckert, R. R. (1990). Picornaviridae and their replication. In Fields, B. N. & Knipe, D. M., eds. *Virology*, pp. 507–548, New York: Raven Press.

Ryu, S.-E., Kwong, P. D., Truneh, A., Porter, T. G., Arthos, J., Rosenberg, M., Dai, X., Xuong, N.-h., Axel, R., Sweet, R. W. & Hendrickson, W. A. (1990). Crystal structure of an HIV-binding recombinant fragment of human CD4. *Nature (London)*, **348**, 419–25.

Sathyanarayana, B. K. & Rao, V. S. R. (1971). Conformational studies on β-glucans. *Biopolymers*, **10**, 1605–15.

Sathyanarayana, B. K. & Rao, V. S. R. (1972). Conformational studies on α-glucans. *Biopolymers*, **11**, 1379–94.

Sauter, N. K., Glick, G. D., Crowther, R. L., Park, S.-J., Eisen, M. B., Skehel, J. J., Knowles, J. R. & Wiley, D. C. (1992). Crystallographic detection of a second ligand binding site in influenza virus hemagglutinin. *Proceedings of the National Academy of Sciences, USA*, **89**, 324–8.

Schirmer, T., Keller, T. A., Wang, Y. & Rosenbusch, J. P. (1995). Structural bais for sugar translocation through maltoporin channels at 3.1 Å resolution. *Science*, **267**, 512–14.

Schultz, M. & Crowell, R. L. (1983). Eclipse of coxsackievirus infectivity: the restrictive event for a non-fusing myogenic cell line. *Journal of General Virology*, **64**, 1725–34.

Sekiguchi, K., Franke, A. J. & Baxt, B. (1982). Competition for cellular receptor sites among selected aphthoviruses. *Archives of Virology*, **74**, 53–64.

Shepard, D. A., Heinz, B. A. & Rueckert, R. R. (1993). WIN compounds inhibit both attachment and eclipse of human rhinovirus 14. *Journal of Virology*, **67**, 2245–54.

Sherry, B. & Rueckert, R. (1985). Evidence for at least two dominant neutralization antigens on human rhinovirus 14. *Journal of Virology*, **53**, 137–43.

Sherry, B., Mosser, A. G., Colonno, R. J. & Rueckert, R. R. (1986). Use of monoclonal antibodies to identify four neutralization immunogens on a common cold picornavirus, human rhinovirus 14. *Journal of Virology*, **57**, 246–57.

Simmons, D., Makgoba, M. W. & Seed, B. (1988). ICAM, an adhesion ligand of LFA-1, is homologous to the neural cell adhesion molecule NCAM. *Nature (London)*, **331**, 624–7.

Smith, T. J., Kremer, M. J., Luo, M., Vriend, G., Arnold, E., Kamer, G., Rossmann, M. G., McKinlay, M. A., Diana, G. D. & Otto, M. J. (1986). The site of attachment

in human rhinovirus 14 for antiviral agents that inhibit uncoating. *Science*, **233**, 1286–93.
Stauffacher, C. V., Usha, R., Harrington, M., Schmidt, T., Hosur, M. V. & Johnson, J. E. (1987). The structure of cowpea mosaic virus at 3.5 Å resolution. In Moras, D., Drenth, J., Strandberg, B., Suck, D. & Wilson, K., eds. *Crystallography in Molecular Biology*, pp. 293–308. New York & London: Plenum Press.
Staunton, D. E., Marlin, S. D., Stratowa, C., Dustin, M. L. & Springer, T. A. (1988). Primary structure of ICAM-1 demonstrates interaction between members of the immunoglobulin and integrin supergene families. *Cell*, **52**, 925–33.
Staunton, D. E., Merluzzi, V. J., Rothlein, R., Barton, R., Marling, S. D. & Springer, T. A. (1989). A cell adhesion molecule, ICAM-1, is the major surface receptor for rhinoviruses. *Cell*, **56**, 849–53.
Staunton, D. E., Dustin, M. L., Erickson, H. P. & Springer, T. A. (1990). The arrangement of the immunoglobulin-like domains of the ICAM-1 and the binding site of LFA-1 and rhinovirus. *Cell*, **61**, 243–54.
Stehle, T., Yan, Y., Benjamin, T. L. & Harrison, S. C. (1994). Structure of murine polyomavirus complexed with an oligosaccharide receptor fragment. *Nature (London)*, **369**, 160–3.
Tanner, J., Weis, J., Fearon, D., Whang, Y. & Kieff, E. (1987). Epstein–Barr virus gp350/220 binding to the B lymphocyte C3d receptor mediates adsorption, capping, and endocytosis. *Cell*, **50**, 203–13.
Tomassini, J. E., Maxson, T. R. & Colonno, R. J. (1989). Biochemical characterization of a glycoprotein required for rhinovirus attachment. *Journal of Biological Chemistry*, **264**, 1656–62.
Tsao, J., Chapman, M. S., Agbandje, M., Keller, W., Smith, K., Wu, H., Luo, M., Smith, T. J., Rossmann, M. G., Compans, R. W. & Parrish, C. R. (1991). The three-dimensional structure of canine parvovirus and its functional implications. *Science*, **251**, 1456–64.
Uncapher, C. R., DeWitt, C. M. & Colonno, R. J. (1991). The major and minor group receptor families contain all but one human rhinovirus serotype. *Virology*, **180**, 814–17.
Wang, H., Kavanaugh, M. P., North, R. A. & Kabat, D. (1991). Cell-surface receptor for ecotropic murine retroviruses is a basic amino-acid transporter. *Nature (London)*, **352**, 729–31.
Wang, J., Yan, Y., Garrett, T. P. J., Liu, J., Rodgers, D. W., Garlick, R. L., Tarr, G. E., Hussain, Y., Reinherz, E. L. & Harrison, S. C. (1990). Atomic structure of a fragment of human CD4 containing two immunoglobulin-like domains. *Nature (London)*, **348** 411–18.
Weis, W., Brown, J. H., Cusack, S., Paulson, J. C., Skehel, J. J. & Wiley, D. C. (1988). Structure of the influenza virus haemagglutinin complexed with its receptor, sialic acid. *Nature (London)*, **333**, 426–31.
Williams, R. K., Jiang, G. S. & Holmes, K. V. (1991). Receptor for mouse hepatitis virus is a member of the carcinoembryonic antigen family of glycoproteins. *Proceedings of the National Academy of Sciences, USA*, **88**, 5533–6.
Wilson, I. A., Skehel, J. J. & Wiley, D. C. (1981). Structure of the haemagglutinin membrane glycoprotein of influenza virus at 3 Å resolution. *Nature (London)*, **289**, 366–73.
Yafal, A. G., Kaplan, G., Racaniello, V. R. & Hogle, J. M. (1993). Characterization of poliovirus conformational alteration mediated by soluble cell receptors. *Virology*, **197**, 501–5.
Yeates, T. O., Jacobson, D. H., Martin, A., Wychowski, C., Girard, M., Filman, D. J. & Hogle, J. M. (1991). Three-dimensional structure of a mouse-adapted type 2/type 1 poliovirus chimera. *EMBO Journal*, **10**, 2331–41.

ENTEROPATHOGENIC *E. COLI* AND *SALMONELLA* SPECIES EXPLOITATION OF HOST CELLS

B. B. FINLAY

Biotechnology Laboratory and the Departments of Biochemistry & Molecular Biology, and Microbiology & Immunology, University of British Columbia, Vancouver, BC, Canada V6T-1Z3

INTRODUCTION

Many bacterial pathogens interact with surfaces on the body which can result in disease. Enteropathogenic *E. coli* (EPEC) and *Salmonella* species interact intimately with intestinal epithelial cells, causing diarrhoea and associated diseases. EPEC bind to epithelial cells using sophisticated mechanisms that exploit existing epithelial signal transduction pathways and host cytoskeletal components. Unlike EPEC, *Salmonella* species actually enter into epithelial cells (invade) and function as intracellular parasites. Comparision of the virulence mechanisms used by these two pathogens and their interactions with epithelial cells illustrates several principles that are used by bacterial pathogens that adhere to and invade mammalian host cells.

EPEC-MEDIATED DISEASE

Escherichia coli is an extremely versatile pathogen. In addition to being a member of the normal intestinal flora, *E. coli* also causes bladder infections, meningitis, and diarrhoea. Diarrhoeagenic *E. coli* contain at least five types of *E. coli*, which cause various symptoms ranging from cholera-like to extreme colitis (Hart, Batt & Saunders, 1993). Each type of diarrhoeagenic *E. coli* possesses a particular set of virulence factors, including specific adhesins, invasins, and/or toxins, which are responsible for causing a specific type of diarrhoea. One of these groups, enteropathogenic *E. coli* (EPEC), is a predominant cause of infant diarrhoea worldwide. In addition to isolated outbreaks in daycare centres and nurseries in developed countries, EPEC poses a major endemic health threat to young children (< 6 months) in developing countries, where it has a high mortality rate (Levine & Edelman, 1984). Worldwide, EPEC is the leading cause of bacterial mediated diarrhoea in children, and it has been estimated to kill up to one million children per year worldwide. EPEC disease is characterized by watery diarrhoea of varying severity, while vomiting and fever often accompany fluid loss.

Despite the significance of EPEC-mediated disease, little is known about how this pathogen actually causes disease. Unlike other *E. coli* diarrhoeas such as enterotoxigenic *E. coli*, EPEC diarrhoea is not mediated by a toxin. Instead, EPEC binds to intestinal surfaces of the small bowel. A characteristic histological lesion, called the attaching and effacing (A/E) lesion, occurs (Knutton *et al.*, 1989). A/E lesions are marked by dissolution of the intestinal brush border surface and loss of epithelial microvilli (effacement) at the sites of bacterial attachment. Once bound, EPEC reside upon a cup-like projection or pedestal on which adherent bacteria reside. It has been shown that, underlying this pedestal, in the epithelial cell are several cytoskeletal components, including actin, alpha-actinin, ezrin, talin, and myosin light chain (Finlay *et al.*, 1992; Knutton *et al.*, 1989). Formation of the A/E lesion appears to be responsible for fluid secretion and diarrhoea, although mechanistically this remains to be proven. It has been suggested that disruption of the brush border and microvilli may be responsible for diarrhoea. Although EPEC can enter (invade) tissue culture cells (Donnenberg, Donohue-Rolfe & Keusch, 1990), it does not normally cause invasive disease, and rarely penetrates the intestinal barrier.

EPEC belongs to a group of pathogenic organisms that form A/E lesions, including enterohemorrhagic *E. coli* (EHEC), several EPEC-like animal pathogens that cause disease in rabbits (RDEC), dogs, pigs (PEPEC), etc, some isolates of *Citrobacter freundii, Hafnia alvei*, and probably *Helicobacter pylori*. These organisms all cause cytoskeletal rearrangement and pedestal formation on relevant host epithelial cells. EHEC, which causes enteric colitis ('hamburger disease') can also cause haemolytic uremic syndrome in approximately 10% of cases. It appears that EHEC possess all of the EPEC virulence factors needed for A/E lesion formation, but has an additional shiga-like toxin, which contributes to its increased pathogenesis.

MECHANISMS OF EPEC PATHOGENICITY

Initial adherence

Recently, significant progress has been made in defining the bacterial and host factors involved in formation of attaching and effacing lesions. Initial bacterial adherence is dependent on the presence of a 55–70 MDa plasmid that is common to EPEC strains. This process is mediated by a plasmid-encoded bundle forming pilus (BFP), and possibly other factors (Giron, Ho & Schoolnik, 1993). Mutants in EPEC that are defective in initial adherence produce fewer A/E lesions on epithelial cells, but these lesions are indistinguishable from those caused by parental EPEC.

Signal transduction

When EPEC interact with cultured epithelial cells, several signal transduction pathways are activated in the epithelial cells, including the release of the eukaryotic secondary messengers, IP_3, and intracellular calcium (Dytoc, Fedorko & Sherman, 1994; Foubister, Rosenshine & Finlay, 1994*a*). EPEC binding to cultured epithelial cells also causes tyrosine phosphorylation of a host 90 kDa membrane protein, Hp90, which is not normally phosphorylated in uninfected cultured cells (Rosenshine *et al.*, 1992). Addition of tyrosine kinase inhibitors inhibits the phosphorylation of Hp90 and EPEC uptake into epithelial cells. It appears that Hp90 phosphorylation precedes IP_3 fluxes and cytoskeletal rearrangements (Foubister, Rosenshine & Finlay, 1994*a*).

All of the EPEC genes known to be involved in A/E formation (except the plasmid-encoded regulator, *per*) are found within a unique contiguous region in the EPEC chromosome (McDaniel *et al.*, 1995). Several bacterial loci have been identified that are involved in activating epithelial signal transduction. Strains containing mutations in *espB* (*E. coli* *s*ecreted *p*rotein B, formerly *eaeB*), a gene found downstream of the intimin gene *eaeA* (see below), do not stimulate signal transduction or cytoskeletal rearrangement (Foubister *et al.*, 1994*b*). Strains cured of the EPEC virulence plasmid are still capable of activating epithelial signal transduction pathways and organizing the underlying cytoskeletal structure, although their efficiency at these events is significantly decreased, presumably due to the loss of the plasmid-encoded bundle-forming pilus and a plasmid-encoded positive regulator. In addition to *espB*, Tn*phoA* mutants belonging to Class 4 (*cfm* mutants) are also unable to stimulate signal transduction (Rosenshine *et al.*, 1992). Another locus upstream of *espB*, called *espA* is also needed to stimulate epithelial signals (Kenny *et al.*, 1996).

When EPEC is grown in tissue culture media, five bacterial proteins (110, 40, 39, 37, and 25 kDa) are secreted into the supernatant medium (Kenny & Finlay, 1995). Amino terminal sequencing identified the 37 kDa as EspB (a protein needed to trigger signal transduction; Foubister *et al.*, 1994*b*), and the 25 kDa protein matched the predicted product of *espA*, EspA. The 39 kDa protein is homologous to glyceraldehyde-3-phosphate dehydrogenase (GAPDH), with 14 of 16 amino acids at its amino terminus being identical to GAPDH. *cfm* mutants were unable to secrete any of these proteins except the 110 kDa protein (Kenny & Finlay, 1995). We have recently cloned and sequenced the gene encoding the 110 kDa protein (Stein *et al.*, 1996*a*). This protein is homologous to a haemagglutinin found in an avian pathogen *E. coli* and uses an IgA protease secretion mechanism. However, by constructing internal gene deletions, we have found that the 110 kDa protein is not needed for signal transduction, and is not secreted by several A/E-causing organisms such as RDEC, *Citrobacter freundii*, and *Hafnia alvei*. It also appears that the 39 and 40 kDa proteins are not encoded within the 35 kb locus of the enterocyte effacement (LEE) region, based on DNA hybridization studies and DNA sequence analysis. This region encodes all the factors necessary for A/E formation when placed in HB101 (J. Kaper, personal communication), and thus these two proteins are probably not needed for A/E lesion formation. Characterization of the *cfm* insertions led to the identification of a 'type III' secretion system in EPEC, encoded by the *sep* genes (Jarvis *et al.*, 1995). Such secretion systems are used by several bacterial pathogens to export virulence factors out of the bacteria and into contact with mammalian cells. Examples include the invasion systems of *Shigella* and *Salmonella* species, the *Yersinia* secreted proteins (YOPs), including a tyrosine phosphatase which enters phagocytic cells, and harpins from plant pathogens (Van Gijsegem, Genin & Boucher, 1993). Collectively, this information indicates that EPEC secretes at least two molecules (EspB and EspA) that are critical for activating signal transduction and cytoskeletal rearrangement in epithelial cells, and EPEC has a specialized secretion system for exporting these molecules.

Intimate adherence

Intimin is the product of a bacterial chromosomal locus, *eaeA*, and is a 94 kDa EPEC outer membrane protein that is needed for intimate adherence (Jerse *et al.*, 1990). *eaeA* mutants form immature A/E lesions and do not organize phosphotyrosine proteins and cytoskeletal components beneath adherent bacteria, although epithelial signal transduction is still activated (Rosenshine *et al.*, 1992). Intimin appears to participate in reorganization of the underlying host cytoskeleton after other bacterial factors stimulate epithelial signal transduction (Rosenshine *et al.*, 1992). Although cloned intimin in non-pathogenic *E.coli* does not mediate adherence (Jerse *et al.*, 1990), if an *eaeA*-defective strain of EPEC is added to epithelial cells (to stimulate signal transduction) prior to addition of the cloned intimin expressed in *E.coli* HB101, bacteria expressing the cloned intimin now adhere to epithelial cells (Rosenshine *et al.*, 1996). If EPEC mutants that are defective for stimulating signal transduction (such as *espB* or *cfm*) are used to preinfect monolayers, organisms containing the cloned intimin do not adhere. This indicates that EPEC-induced signal transduction pathways are needed prior to successful intimin-mediated adherence to epithelial cells.

Further support for this hypothesis comes from studies done using a purified fusion peptide consisting of 280 amino acids of the carboxyl terminus of intimin fused to the maltose binding protein (MBP) (Frankel *et al.*, 1994, Rosenshine *et al.*, 1996). It was found that this fusion peptide (MBP/Int) adheres to epithelial cells *only* when the monolayer is previously infected with EPEC (or the *eaeA* mutant) to preinduce signals (Rosenshine *et al.*, 1996). If these signals are blocked with a tyrosine kinase inhibitor (which blocks signalling), binding of the fusion peptide to epithelial cells is also blocked. Not surprisingly, peptide binding is not affected by treatment of the monolayer with cytochalasin D, indicating that actin rearrangement is not needed for intimin-mediated binding.

Hp90 is an epithelial membrane-localized protein and it localizes beneath adherent organisms at the tip of extended pseudopods (Rosenshine *et al.*, 1996). This protein interacts with intimin, and is an intimin receptor. If EPEC is added to epithelial cells and then removed by detergent extraction, Hp90 remains associated with the bacteria; however, if an *eaeA* mutant is used, Hp90 is not isolated with the bacteria (Rosenshine *et al.*, 1996). Hp90 is also coimmunoprecipitated with the intimin–maltose binding protein fusion peptide, but only if prior signals are induced in the epithelial cells by EPEC strains. Intimin fused to MBP also binds *directly* and specifically to Hp90 when HeLa cell membrane extracts from *eaeA* -preinduced cells are separated by PAGE, renatured, and hybridized with MBP/Int (Rosenshine *et al.*, 1996). MBP/Int does not bind to membrane extracts from uninduced monolayers, or those infected with signaling mutants such as *espB, cfm* or *sep*. Collectively, this evidence indicates that MBP/Int binds directly to Hp90, but only if Hp90 is tyrosine-phosphorylated prior to intimin-mediated binding.

Cytoskeletal rearrangement and pedestal formation

As described above, the attaching/effacing (A/E) lesion (or pedestal) formed by EPEC upon association with epithelial cells is associated with the assembly of highly organized cytoskeletal structures in the epithelial cells immediately beneath adherent bacteria. Although this pedestal usually raises the bacterium slightly above the epithelial cell surface, EPEC can trigger extended pseudopod formation, with projections extending up to 10 microns above the epithelial cell surface with the bacteria located extracellularly at the tip of these extensions (Rosenshine *et al.*, 1996). The stalk of these extended pseudopods contains polymerized actin, while Hp90 is localized only at the tip of these structures beneath EPEC. Extended pedestals are not seen when strains containing mutations in *eaeA, espB,* or *cfm* are used, reinforcing the linkage between signal transduction events and cytoskeletal rearrangement.

It appears the product of the *eaeA* gene, intimin, is critical for organizing cytoskeletal rearrangements. *eaeA* mutants trigger signals in epithelial cells and cause generalized actin accumulation near adherent organisms, but are unable to organize the cytoskeleton into defined structures that lead to pedestal and extended pseudopod formation. Additionally, they do not invade epithelial cells, even if complemented with signal transduction-defective EPEC mutants. Further support for the role of intimin comes from experiments performed with the cloned EPEC intimin expressed in non-pathogenic *E. coli* HB101. If EPEC containing a defective *eaeA* gene are first added to epithelial cells followed by strains containing cloned intimin, only HB101 harbouring intimin, but not the *eaeA* mutant, organize the cytoskeleton into pedestals and extended pseudopods (Rosenshine *et al.*, 1996). This indicates that intimin molecules direct the final condensation and organization of the host cytoskeleton from their outer membrane location on the adherent *E. coli*. Hp90 also probably plays a significant role in organizing the host cytoskeleton, in addition to mediating intimin binding. It may even serve as a bridge, linking intimin in the bacterial outer membrane to the host cytoskeleton on the other side of the epithelial cell membrane.

Production of diarrhoea?

Despite our increasing knowledge of the bacterial factors and host molecules that mediate EPEC interactions with epithelial cells, the actual molecular mechanisms that cause diarrhoea remain undefined. EPEC strains lacking intimin are significantly decreased in their ability to cause diarrhoea in human volunteers (Donnenberg *et al.*, 1993). It is probable that one or more of the events associated with formation of attaching and effacing lesions also causes diarrhoea. However, signal transduction mutants such as *espB* or *espA* have not been tested for virulence in humans or relevant animal models. In addition to the morphological rearrangements that occur on the apical surface of epithelial cells, EPEC also causes a large decrease in transepithelial resistance in polarized Caco-2 epithelial cell monolayers (Canil *et al.*, 1993). This disruption does not appear to be due to alterations in tight junctions, but instead affects a transcellular pathway. Mutants defective for signal transduction and *eaeA* mutants do not cause this loss in transepithelial resistance, indicating that this process is linked to these events. It is possible that such transepithelial disruptions occur *in vivo*, which would lead to ionic imbalances and possibly diarrhoea.

By using whole cell patch clamping technology, we have recently found that EPEC causes a significant depolarization of individual HeLa cells (Stein *et al.*, 1996*b*). Although *eaeA* mutants still caused depolarization, both *espB* and *cfm* mutants did not depolarize cells, indicating that EPEC secreted proteins that affect epithelial signaling are needed for these events. Occurrence of such a process in the gut would reduce the electrochemical gradient available for sodium ion absorption from the gut lumen, thereby contributing to ionic imbalance, fluid loss, and diarrhoea.

SALMONELLA AS MODEL INTRACELLULAR PATHOGENS

There are several advantages to using *Salmonella* species as models to study the biology of intracellular parasitism. *Salmonella* species are close relatives of *E. coli*, are not obligate intracellular parasites, and are easy to grow. Much is known about *S. typhimurium* biology (when grown extracellularly). *S. typhimurium* is easily manipulated using molecular genetic techniques such as transduction, transposon mutagenesis, and electroporation/transformation, and molecular techniques have recently been developed for *S. typhi*, including transduction and a genomic map. Finally, *S. typhimurium* infection of mice is an excellent model of human typhoid fever.

Salmonellosis

Diseases caused by *Salmonella* species (salmonellosis) pose major health problems throughout the world. In Canada, it has been estimated that each case of reported non-typhoidal salmonellosis costs approximately $1350 (only 10% of cases are reported). In addition, there are 16×10^6 cases of typhoid fever worldwide/year, resulting in 600 000 deaths, mainly in Asia. Salmonellosis is a serious disease in immunosuppressed patients, and with more immunocompromised individuals (such as AIDS and bone marrow transplant patients) appearing in the hospital setting, the number of non-typhoidal *Salmonella* infections is rising. In developing nations, diarrhoeal diseases are a major cause of illness and death among infants and young children. *Salmonella* infections comprise a large proportion of these infections. Compounding these problems is the increasing resistance to antibiotics. Many *S. typhi* isolates worldwide are multi-drug resistant. On a more optimistic note, significant progress has been made in developing vaccines to *S. typhi*.

Salmonella bacteria are considered facultative intracellular parasites. Most *Salmonella* infections arise from oral ingestion of tainted food or water. It is in the small intestine that *Salmonella* species begin to penetrate the intestinal mucosa. These bacteria cause the intestinal microvilli to disappear, and the bacteria enter into a membrane-bound inclusion of the epithelial cell (Takeuchi, 1967). The intracellular bacteria can then pass through these cells to the opposite surface (Finlay, Gumbiner & Falkow, 1988). Cells of the reticuloendothelial system then ingest the *Salmonella*, but, at least with macrophages, do not kill the bacteria. The infected macrophages then migrate to the intestinal lymph nodes, where the *Salmonella* replicate further before escaping these cells and entering the blood to disseminate throughout the body.

There are three types of diseases caused by *Salmonella* species in humans, and all occur by variations of the infection route described above. *S. typhi* is the causative agent of typhoid fever. Gastroenteritis, commonly known as 'food poisoning', is caused by many *Salmonella* species including *S. typhimurium* and *S. enteritidis*. The bacteria penetrate through a few layers of the intestinal mucosa, but are usually cleared from the host before they cause serious infections. In immunodeficient individuals the bacteria proceed to deeper tissue, and often cause systemic infections such as septicaemia (infection of the blood). *S. choleraesuis* is the prototype of 'invasive' *Salmonella*. These bacteria proceed through the intestinal epithelium, enter the blood, and disseminate throughout the body. These infections are difficult to treat, as the organisms are often within cells in bone and meninges, and have an untreated fatality rate three-fold greater than typhoid fever.

Salmonella invasion

In vitro tissue culture monolayers have been used to examine bacterial interactions with eukaryotic cells, including bacterial invasion (or entry) (Tang *et al.*, 1993). These systems have the advantage of consistency and are easier to use than primary isolates of mammalian cells. *Salmonella* uptake into eukaryotic cells appears to follow similar mechanisms with all cell types, although most of the work has concentrated on epithelial cell lines. The mechanisms used by different *Salmonella* species appear the same for the various species. Following initial contact of a bacterium with the mammalian cell surface, several events occur. Several signal transduction cascades are activated, including IP_3 (Ruschkowski, Rosenshine & Finlay, 1992) and Ca^{2+} fluxes (Pace, Hayman & Galan, 1993). Chelating intracellular Ca^{2+} blocked *S. typhimurium* invasion (Ruschkowski *et al.*, 1992). There is also a marked rearrangement in actin-containing cytoskeleton, with the actin and related molcules polymerizing beneath the invading organisms (Finlay, Rushkowski & Dedhar, 1991), and polymerized actin is required for *Salmonella* invasion (Finlay & Falkow, 1988). Membrane ruffling (Finlay & Falkow, 1990, Francis *et al.*, 1993) and macropinocytosis (Garcia-del Portillo & Finlay, 1994) are also stimulated in the immediate vicinity of the invading organisms. Host cell surface molecules are also capped to the areas around invading bacteria (Garcia-del Portillo *et al.*, 1994). *Salmonella* appear to be engulfed by the actin driven membrane ruffles, and are internalized within a membrane bound vacuole. The entire process occurs within minutes, and the cell surface returns to its previous state following bacterial invasion.

Although cultured cells enhance the study of bacterial invasion, one cannot address later intracellular events such as penetration through the cell and exit to the opposite host cell surface. Polarized epithelial monolayers of MDCK and Caco-2 epithelial cells grown on filters with 3 μm pores to have been used to facilitate these studies (Finlay *et al.*, 1988, Finlay & Falkow, 1990). *Salmonella* species cause the characteristic membrane ruffling on the apical surface of these cells, and penetrate these barriers within a few hours. *Salmonella* disrupts the epithelial tight junctions, causing a loss in transepithelial electrical resistance and cell polarity.

Several bacterial loci have been identified that are involved in *Salmonella* invasion (for review see Finlay, 1994, and references therein). These genes are clustered at one location (58–60 min) on the bacterial chromosome. They encode several factors, including regulatory proteins. Like EPEC, one operon encodes a 'type III' secretory apparatus that appears to be involved in mediating export of proteins out of the bacteria. These surface/secreted proteins form a bacterial surface structure, and are probably involved in triggering host signal transduction pathways. Mutations in bacterial invasion genes decrease *Salmonella* virulence. The regulation of *Salmonella* invasion factors is complex and influenced by several parameters, including anaerobiosis, growth state, and DNA supercoiling.

Intracellular vacuole targeting

There is general agreement that *Salmonella* species reside within a membrane bound vacuole within both phagocytic and non-phagocytic cells. More recently, the trafficking of this vacuole has been examined. In cultured epithelial cells, *S. typhimurium* triggers capping of several host cell surface proteins (Garcia-del Portillo *et al.*, 1994). However, only one of several surface proteins, MHC class I, is present at early post-infection times in vacuoles containing this bacterium. Other host cell surface markers are internalized into the host cell, but do not colocalize with bacteria. This finding implies that there is a sorting mechanism at the host cell surface that excludes some cytoskeletal-associated proteins from entering the bacteria-containing vacuole or fast recycling to plasma membranes. No apparent sorting of host surface markers was observed in vacuoles containing bacteria which invade using a *Yersinia* invasin-mediated pathway.

Once the *Salmonella* vacuole is formed, its composition changes. 90 minutes after bacterial entry, some lysosomal markers begin to appear in the bacteria-containing vacuoles, including lysosomal glycoproteins (lgps) and lysosomal acid phosphatase (Garcia-del Portillo & Finlay, 1995). However, both of these markers are targeted to lysosomes via a mannose-6-phosphate receptor (M6PR) independent pathway. Markers that rely upon M6PR targeting, such as cathepsin D and M6PR itself, do not localize with *S. typhimurium* vacuoles. Additionally, fluid phase markers do not fuse with the bacteria-containing vacuoles when added prior to or after bacterial invasion. Collectively these data suggest that *S. typhimurium* enters initially into a specialized vacuole which then develops into a specialized compartment that is isolated from normal endocytic traffic. *S. typhi* takes a similar route within epithelial cells.

Several investigators have begun to characterize the intracellular targeting and environment of *S. typhimurium* inside macrophages. One report indicated that *S. typhimurium* resides within phagosomes that have fused with lysosomes (Carrol *et al.*, 1979). Other workers concluded that *S. typhimurium* inhibited phagosome–lysosome fusion within several types of mouse derived macrophages (Buchmeier & Heffron, 1991; Ishibashi & Aria, 1990). Perhaps these discrepancies are due to incomplete inhibition of phagosome lysosome fusion, or only part of the intracellular population is capable of blocking this event. It was also suggested that viable intracellular bacteria are needed for this inhibition (Carrol *et al.*, 1979). Phagosomes containing *S. typhimurium* are acidified slowly, and it takes 4–5 hours before the pH drops below 5.0 (Alpuche-Aranda *et al.*, 1994). In contrast, vacuoles containing killed organisms are rapidly acidified (pH < 4.5 within 1 hour). This data would suggest that viable organisms are either needed for inhibition of acidification, or need to invade via a bacterial mediated pathway which delivers the organism to an intracellular location which is acidified slower than with the phagocytic pathway. These workers also found that fluid phase markers fused with internalized bacteria, which is different from that seen with epithelial cells. They concluded that *S. typhimurium* resides within a lysosome, yet is capable of blocking endosome acidification. It is also possible that the organisms reside within an intracellular environment that contains some of the lysosomal markers, yet this environment is not a normal phagolysosome.

The intracellular environment

We have begun to characterize the intracellular environment that *S. typhimurium* resides within inside epithelial cells by using bacterial reporter genes (Garcia del Portillo *et al.*, 1992). Measurement of β-galactosidase activity of various *lacZ* fusions using a fluorescent substrate led to the conclusion that the concentrations of free Fe^{2+} and Mg^{2+} in the vacuole of epithelial cells are low, that the vacuole has a mild-acidic pH, and that lysine and oxygen are present within the intracellular environment. This work demonstrates the utility of using bacterial gene fusions to measure genes that are expressed intracellularly, and gives a glimpse of the intracellular environment that *Salmonella* resides within inside epithelial cells.

Intracellular survival

The past few years have seen the identification of several bacterial factors that enhance *S. typhimurium* survival within macrophages (Fields *et al.*, 1986). These survival factors are macrophage specific, since these mutants survive equally well within non-phagocytic cells. The PhoP/PhoQ system is a two-component regulatory system that activates at least five bacterial products (*pag*) and represses others (*prg*) (for review see Miller, 1991). One of the phenotypes that PhoP/PhoQ regulates is the capacity to survive bactericidal cationic peptides which are thought to be involved in killing intracellular bacteria (Fields, Groisman & Heffron, 1989; Miller, Kukral &

Mekalanos, 1989). A cytolysin has recently been identified in *Salmonella* species that is required for survival within macrophages and virulence, although its function remains obscure (Libby *et al.*, 1994). Other bacterial products may also contribute to intracellular survival within phagocytic cells.

Intracellular replication and filamentous lysosome formation

There are conflicting data on the growth of *Salmonella* within macrophages. Mutant strains that are unable to grow within epithelial cells are unaffected within macrophages. However, it has recently been proposed that two populations of *S. typhimurium* exist within macrophages: one which is static, and the other which is rapidly growing (Abshire & Neidhardt, 1993). The existence of these two pools may perhaps explain the conflicting data regarding lysosome fusion and intracellular growth within phagocytic cells.

Salmonella species have the capacity to multiply within vacuoles in non-phagocytic cells, after an initial lag of approximately 4 hours (Finlay & Falkow, 1988). The lag period seen in non-phagocytic cells that precedes initiation of bacterial replication indicates that processes occur prior to bacterial replication, and that specific bacterial genes may be required for replication in this unique niche. Non-virulent *E. coli* (and EPEC) do not replicate within vacuoles in epithelial cells.

Mutants in *S. typhimurium* that are defective for intracellular replication in epithelial cells have been identified (Leung & Finlay, 1991). Both auxotrophic and prototrophic mutants were found, and the auxotrophic mutants could be complemented by addition of the appropriate nutrients to the tissue culture medium. The three prototrophic mutants identified were highly attenuated for virulence in mice, yet persisted within livers and spleens for at least 3 weeks. The identification of these mutants suggests that there are bacterial genes specific for intracellular replication and these participate in virulence.

A potential role for these bacterial genes has been identified (Garcia-del Portillo *et al.*, 1993). As discussed above, *S. typhimurium* is localized to vacuoles that contain lysosomal glycoproteins (lgps) in epithelial cells. However, 4–6 hours after invasion, intracellular *Salmonella* induce the formation of stable filamentous structures that contain lgps that are connected to the vacuoles containing bacteria. The kinetics of formation of these lgp-rich filamentous structures parallel the rate of intracellular replication, including the initial lag period. Filament formation requires viable intracellular bacteria, since addition of antibiotics blocks formation of these novel structures. Endosome acidification inhibitors or microtubule-disrupting agents also inhibit tubule formation. These unique structures are never observed in uninfected cells, nor those infected with *Yersinia* species, although all *Salmonella* species tested trigger their formation. The prototrophic *Salmonella* mutants that are unable to multiply inside epithelial cells are also completely defective for triggering formation of these filamentous structures.

Molecular characterization of one locus that mediates tubule formation (Stein *et al.*, 1996*c*) indicates that it is a unique *Salmonella* sequence that is inserted between housekeeping genes involved in spermidine uptake, and has no homology to other sequences within the gene banks. Thus it appears that *Salmonella* have specific loci that are responsible for triggering filament formation from within the host cell. How the bacteria direct the formation of these novel host processes is unknown. One potential function for these structures is that they provide access to nutrients for the intracellular bacteria, possibly by intersecting with endocytic or exocytic vesicular transport routes.

CONCLUSIONS

It is becoming apparent that many successful bacterial pathogens are capable of exploiting host cell processes and functions. Both EPEC and *Salmonella* species exploit several such pathways in host epithelial cells. By activating host cell signal transduction and rearranging the host cytoskeleton, bacterial adherence (EPEC) and bacterial invasion (*Salmonella*) is facilitated. Additionally, once inside a host cell, *Salmonella* species direct the targeting of the bacterial-containing vacuole such that it avoids the normal endocytic pathway. Additional host structures are affected that appear to be linked to intracellular growth. Collectively, these results suggest intimate interactions with epithelial cells play a key role in pathogenesis for these organisms and ultimately production of disease.

ACKNOWLEDGEMENTS

Work in my laboratory is supported by the Medical Research Council of Canada, the Canadian Bacterial Disease Centre of Excellence, a Howard Hughes International Research Scholar award and NSERC.

REFERENCES

Abshire, K. Z. & Neidhardt, F. C. (1993). Growth rate paradox of *Salmonella typhimurium* within host macrophages. *Journal of Bacteriology*, **175**, 3744–8.

Alpuche-Aranda, C. M., Racoonsin, E. L., Swanson, J. A. & Miller, S. I. (1994). *Salmnonella* stimulate macophage macropinocytosis and persist within spacious phagosomes. *Journal of Experimental Medicine*, **179**, 601–8.

Buchmeier, N. A. & Heffron, F. (1991). Inhibition of macrophage phagosome-lysosome fusion by *Salmonella typhimurium*. *Infection and Immunity*, **59**, 2232–8.

Canil, C., Rosenshine, I., Ruschkowski, S., Donnenberg, M. S., Kaper, J. B. & Finlay, B. B. (1993). Enteropathogenic *Escherichia coli* decreases the transepithelial electrical resistance of polarized epithelial monolayers. *Infection and Immunity*, **61**, 2755–62.

Carrol, M. E., Jackett, P. S., Aber, V. R. & Lowrie, D. B. (1979). Phagolysosome formation, cyclic adenosine 3′:5′-monophosphate and the fate of *Salmonella typhimurium* within mouse peritoneal macrophages. *Journal of General Microbiology*, **110**, 421–9.

Donnenberg, M. S., Donohue-Rolfe, A. & Keusch, G. T. (1990). A comparison of HEp-2 cell invasion by enteropathogenic and enteroinvasive *Escherichia coli*. *FEMS Microbiology Letters*, **57**, 83–6.

Donnenberg, M. S., Tacket, C. O., James, S. P., Losonsky, G., Nataro, J. P., Wasserman, S. S., Kaper, J. B. & Levine, M. M. (1993). Role of the *eaeA* gene in experimental enteropathogenic *Escherichia coli* infection. *Journal of Clinical Investigation*, **92**, 1412–7.

Dytoc, M., Fedorko, L. & Sherman, P. M. (1994). Signal transduction in human epithelial cells infected with attaching and effacing *Escherichia coli in vitro*. *Gastroenterology*, **106**, 1150–61.

Fields, P. I., Groisman, E. A. & Heffron, F. (1989). A *Salmonella* locus that controls resistance to microbicidal proteins from phagocytic cells. *Science*, **243**, 1059–62.

Fields, P. I., Swanson, R. V., Haidaris, C. G. & Heffron, F. (1986). Mutants of *Salmonella typhimurium* that cannot survive within the macrophage are avirulent. *Proceedings of the National Academy of Sciences, USA*, **83**, 5189–93.

Finlay, B. B. (1994). Molecular and cellular mechanisms of *Salmonella* pathogenesis. *Current Topics in Microbiology and Immunology*, **192**, 163–85.

Finlay, B. B. & Falkow, S. (1988). Comparison of the invasion strategies used by *Salmonella cholerae-suis, Shigella flexneri* and *Yersinia enterocolitica* to enter cultured animal cells: endosome acidification is not required for bacterial invasion or intracellular replication. *Biochimie*, **70**, 1089–99.

Finlay, B. B. & Falkow, S. (1990). *Salmonella* interactions with polarized human intestinal Caco-2 epithelial cells. *Journal of Infectious Diseases*, **162**, 1096–106.

Finlay, B. B., Gumbiner, B. & Falkow, S. (1988). Penetration of *Salmonella* through a polarized Madin-Darby canine kidney epithelial cell monolayer. *Journal of Cell Biology*, **107**, 221–30.

Finlay, B. B., Rosenshine, I., Donnenberg, M. S. & Kaper, J. B. (1992). Cytoskeletal composition of attaching and effacing lesions associated with enteropathogenic *Escherichia coli* adherence to HeLa cells. *Infection and Immunity*, **60**, 2541–3.

Finlay, B. B., Ruschkowski, S. & Dedhar, S. (1991). Cytoskeletal rearrangements accompanying *Salmonella* entry into epithelial cells. *Journal of Cell Science*, **99**, 283–96.

Foubister, V., Rosenshine, I. & Finlay, B. B. (1994*a*). A diarrheal pathogen, enteropathogenic *Escherichia coli* (EPEC), triggers a flux of inositol phosphates in infected epithelial cells. *Journal of Experimental Medicine*, **179**, 993–8.

Foubister, V., Rosenshine, I., Donnenberg, M. S. & Finlay, B. B. (1994*b*). The *eaeB* gene of enteropathogenic *Escherichia coli* is necessary for signal transduction in epithelial cells. *Infection and Immunity*, **62**, 3038–40.

Francis, C. L., Ryan, T. A., Jones, B. D., Smith, S. J. & Falkow, S. (1993). Ruffles induced by *Salmonella* and other stimuli direct macropinocytosis of bacteria. *Nature*, **364**, 639–42.

Frankel, G., Candy, D. C., Everest, P. & Dougan, G. (1994). Characterization of the C-terminal domains of intimin-like proteins of enteropathogenic and enterohemorrhagic *Escherichia coli, Citrobacter freundii,* and *Hafnia alvei*. *Infection and Immunity*, **62**, 1835–42.

Garcia del Portillo, F., Foster, J. W., Maguire, M. E. & Finlay, B. B. (1992).

Characterization of the micro-environment of *Salmonella typhimurium*-containing vacuoles within MDCK epithelial cells. *Molecular Microbiology*, **6**, 3289–97.

Garcia-del Portillo, F. & Finlay, B. B. (1994). *Salmonella* invasion of nonphagocytic cells induces formation of macropinosomes in the host cell. *Infection and Immunity*, **62**, 4641–5.

Garcia-del Portillo, F. & Finlay, B. B. (1995). Targeting of *Salmonella typhimurium* to vesicles containing lysosomal membrane glycoproteins bypasses compartments with mannose 6-phosphate receptors. *Journal of Cell Biology*, **129**, 81–97.

Garcia-del Portillo, F., Pucciarelli, M. G., Jefferies, W. A. & Finlay, B. B. (1994). *Salmonella typhimurium* induces selective aggregation and internalization of host cell surface proteins during invasion of epithelial cells. *Journal of Cell Science*, **107**, 2005–20.

Garcia-del Portillo, F., Zwick, M. B., Leung, K. Y. & Finlay, B. B. (1993). *Salmonella* induces the formation of filamentous structures containing lysosomal membrane glycoproteins in epithelial cells. *Proceedings of the National Academy of Sciences, USA*, **90**, 10544–8.

Giron, J. A., Ho, A. S. & Schoolnik, G. K. (1993). Characterization of fimbriae produced by enteropathogenic *Escherichia coli*. *Journal of Bacteriology*, **175**, 7391–403.

Hart, C. A., Batt, R. M. & Saunders, J. R. (1993). Diarrhoea caused by *Escherichia coli*. *Annals in Tropical Paedatrics*, **13**, 121–31.

Ishibashi, Y. & Arai, T. (1990). Specific inhibition of phagosome-lysosome fusion in murine macrophages mediated by *Salmonella typhimurium* infection. *FEMS Microbiology and Immunology*, **2**, 35–43.

Jarvis, K. G., Giron, J. A., Jerse, A. E., McDaniel, T. K. & Kaper, J. B. (1995). Enteropathogenic *Escherichia coli* contains a specialized secretion system necessary for the export of proteins involved in attaching and effacing lesion formation. *Proceedings of the National Academy of Sciences, USA*, **92**, 7996–8000.

Jerse, A. E., Yu, J., Tall, B. D. & Kaper, J. B. (1990). A genetic locus of enteropathogenic *Escherichia coli* necessary for the production of attaching and effacing lesions on tissue culture cells. *Proceedings of the National Academy of Sciences, USA*, **87**, 7839–43.

Kenny, B. & Finlay, B. B. (1995). Protein secretion by enteropathogenic *Escherichia coli* is essential for transducing signals to epithelial cells. *Proceedings of the National Academy of Sciences, USA*, **92**, 7991–5.

Kenny, B., Lai, L., Finlay, B. B. & Donnenberg, M. S. (1996). EspA, a protein secreted by enteropathogenic *Escherichia coli*, is required to induce signals in epithelial cells. *Molecular Microbiology*, **20**, 313–23.

Knutton, S., Baldwin, T., Williams, P. H. & McNeish, A. S. (1989). Actin accumulation at sites of bacterial adhesion to tissue culture cells: basis of a new diagnostic test for enteropathogenic and enterohemorrhagic *Escherichia coli*. *Infection and Immunity*, **57**, 1290–8.

Leung, K. Y. & Finlay, B. B. (1991). Intracellular replication is essential for the virulence of *Salmonella typhimurium*. *Proceedings of the National Academy of Sciences, USA*, **88**, 11470–4.

Levine, M. M. & Edelman, R. (1984). Enteropathogenic *Escherichia coli* of classic serotypes associated with infant diarrhea: epidemiology and pathogenesis. *Epidemiology Reviews*, **6**, 31–51.

Libby, S. J., Goebel, W., Ludwig, A., Buchmeier, N., Bowe, F., Fang, F. C., Guiney, D. G., Songer, J. G. & Heffron, F. (1994). A cytolysin encoded by *Salmonella* is required for survival within macrophages. *Proceedings of the National Academy of Sciences, USA*, **91**, 489–93.

McDaniel, T. K., Jarvis, K. G., Donnenberg, M. S. & Kaper, J. B. (1995). A genetic locus of enterocyte effacement conserved among diverse enterobacterial pathogens. *Proceedings of the National Academy of Sciences, USA*, **92**, 1664–8.

Miller, S. I. (1991). PhoP/PhoQ: macrophage-specific modulators of *Salmonella* virulence? *Molecular Microbiology*, **5**, 2073–8.

Miller, S. I., Kukral, A. M. & Mekalanos, J. J. (1989). A two-component regulatory system (*phoP phoQ*) controls *Salmonella typhimurium* virulence. *Proceedings of the National Academy of Sciences, USA*, **86**, 5054–8.

Pace, J., Hayman, M. J. & Galan, J. E. (1993). Signal transduction and invasion of epithelial cells by *S. typhimurium*. *Cell*, **72**, 505–14.

Rosenshine, I., Donnenberg, M. S., Kaper, J. B. & Finlay, B. B. (1992). Signal transduction between enteropathogenic *Escherichia coli* (EPEC) and epithelial cells: EPEC induces tyrosine phosphorylation of host cell proteins to initiate cytoskeletal rearrangement and bacterial uptake. *EMBO Journal*, **11**, 3551–60.

Rosenshine, I., Ruschkowski, S., Stein, M., Reinscheid, D. & Finlay, B. B. (1996). A pathogenic bacterium triggers epithelial signals to form a functional bacterial receptor that mediates actin pseudopod formation. *EMBO Journal*, **15**, in press.

Ruschkowski, S., Rosenshine, I. & Finlay, B. B. (1992). *Salmonella typhimurium* induces an inositol phosphate flux in infected epithelial cells. *FEMS Microbiology Letters*, **95**, 121–6.

Stein, M., Kenny, B., Stein, M. A. & Finlay, B. B. (1996*a*). Characterization of EspC, a 110-kilodalton protein secreted by enteropathogenic *Escherichia coli* which is homologous to members of the immunoglobulin A protease-like family of secreted proteins. *Journal of Bacteriology*, **178**, 6546–54.

Stein, M. A., Mathers, D. A., Yen, H., Bainbridge, K. G., & Finlay, B. B. (1996*b*). Enteropathogenic *Escherichia coli* markedly decreases the resting membrane potential of Caco-2 and HeLa human epithelial cell. *Infection and Immunity*, **64**, 4820–5.

Stein, M., Garcia-del Portillo, F., Leung, K. Y. & Finlay, B. B. (1996*c*). Identification of a *Salmonella* virulence gene required for formation of filamentous structures containing lysosomal membrane glycoproteins within epithelial cells. *Molecular Microbiology*, **20**, 151–64.

Takeuchi, A. (1967). Electron microscope studies of experimental Salmonella infection. I. Penetration into the intestinal epithelium by *Salmonella typhimurium*. *American Journal of Pathology*, **50**, 109–36.

Tang, P., Foubister, V., Pucciarellia, M. G. & Finlay, B. B. (1993). Methods to study bacterial invasion. *Journal of Microbiological Methods*, **18**, 227–40.

Van Gijsegem, F., Genin, S. & Boucher, C. (1993). Conservation of secretion pathways for pathogenicity determinants of plant and animal bacteria. *Trends in Microbiology*, **1**, 175–80.

HOST–VIRAL PROTEIN–PROTEIN INTERACTIONS IN INFLUENZA VIRUS REPLICATION

P. PALESE, P. WANG, T. WOLFF AND R. E. O'NEILL

Department of Microbiology, Mount Sinai School of Medicine, 1 Gustave L. Levy Place, New York, NY 10029, USA

INTRODUCTION

Viruses are among the most simple pathogens, and thus require the host cell machinery for replication and survival. How viruses interact with, and use, their host cell is, however, very different for each virus group and involves a perplexing array of replication strategies. DNA-containing viruses have genomes ranging from several thousand base pairs to almost 400 thousand base pairs, and consequently they can code for anywhere from a few viral proteins to more than 300. Similarly, there is a considerable range of genome sizes for RNA-containing viruses (several thousand nucleotides to more than 30 000). This diversity indicates that viruses differ significantly in how they affect the host cell and in compartments/components of the cell which they co-opt to support their replication. In order to explore these principles of virology, much attention in the last two decades was aimed at studying the molecular biology of the different viruses. The genomes of many important human viruses have been cloned and sequenced, and structure/function studies have been performed on viral genes and gene products. Recent advances in cell biology now allow a shift from this reductionist approach of merely studying the virus itself (or a single viral gene/gene product) to an approach which explores the molecular interactions of viruses with their host cells.

The purpose of this chapter is to summarize attempts to identify specific interactions between influenza viral proteins and host proteins. Specifically, using the yeast two-hybrid system, cellular proteins were identified which interact with the viral nucleoprotein (NP) or with the non-structural protein (NS1). Two of these cellular proteins, NPI-1 and NPI-3, have been studied in detail. They were found to facilitate the nuclear import of the NP and of the influenza viral RNA, which is transcribed and replicated in the nucleus of the infected cell. These studies are expected to increase our fundamental knowledge of how influenza viruses replicate. Also, by elucidating virus–cell interactions at the molecular level, it is hoped that a clearer understanding of influenza virus virulence will be gained.

Table 1. *RNA segments and proteins of influenza A virus*

Segment	Protein	AA length	Function/activity
1	PB2	759	Cap binding, endonuclease, component of RNA polymerase
2	PB1	757	Catalytic subunit of RNA polymerase
3	PA	716	Proteolytic activity, component of RNA polymerase
4	HA	~560	Attachment to sialic acid, acid pH-activated membrane fusion
5	NP	498	Structural component of ribonucleoprotein (RNP) complex, RNA-binding activity, nuclear/cytoplasmic transport of vRNA
6	NA	~450	Neuraminidase (sialidase) activity, release of virus
7	M1	252	Structural protein underlying the lipid bilayer
	M2	96	Ion channel
8	NS1	124–237	Effects on RNA transport, splicing and translation
	NS2	121	Unknown

REPLICATION STRATEGY OF INFLUENZA VIRUSES

Influenza viruses are among the best studied viruses and detailed information has been obtained concerning the genome (Luo & Palese, 1992), and the biochemical structure of the virus (Lamb & Krug, 1996). The viruses are spherical and lipid-containing, with a diameter of approximately 120 nm; they contain a segmented genome of eight negative-strand RNAs (Table 1). Each of the eight RNA segments of type A and type B influenza viruses codes for a different protein, with the two smallest RNAs each transcribing an additional (spliced) mRNA which encodes a second protein. Type C influenza viruses lack a neuraminidase (NA) gene and thus have only seven RNA segments. The total genomic length of influenza A, B and C viruses is approximately 13 600, 14 600 and 12 000 nt, respectively (Hayden & Palese, 1996).

Influenza viruses start the replication cycle (Fig. 1) by attaching to sialic acid-containing receptors on cell surfaces via their hemagglutinins (HA). The particles are then internalized into endosomes which have a low pH. The acid pH-activated ion channels made up of the viral M2 protein play an important role during this initial step by facilitating the uncoating of the virus. The RNA of the incoming virus particle remains associated with viral protein, and enters the nucleus as ribonucleoprotein (RNP) through the nuclear pore complex.

The transcription and replication of the influenza virus genome takes place in the nucleus (Braam, Ulmanen & Krug, 1983; Krug *et al.*, 1989), and the incoming viral RNP gives rise to three different species of virus-specific RNA: (a) an mRNA which derives its first 9–15 nucleotides from host mRNA, is capped at the 5′ end, and carries a polyA tail lacking the 3′ 15–16 nucleotides of the template RNA; (b) full-length complementary copy

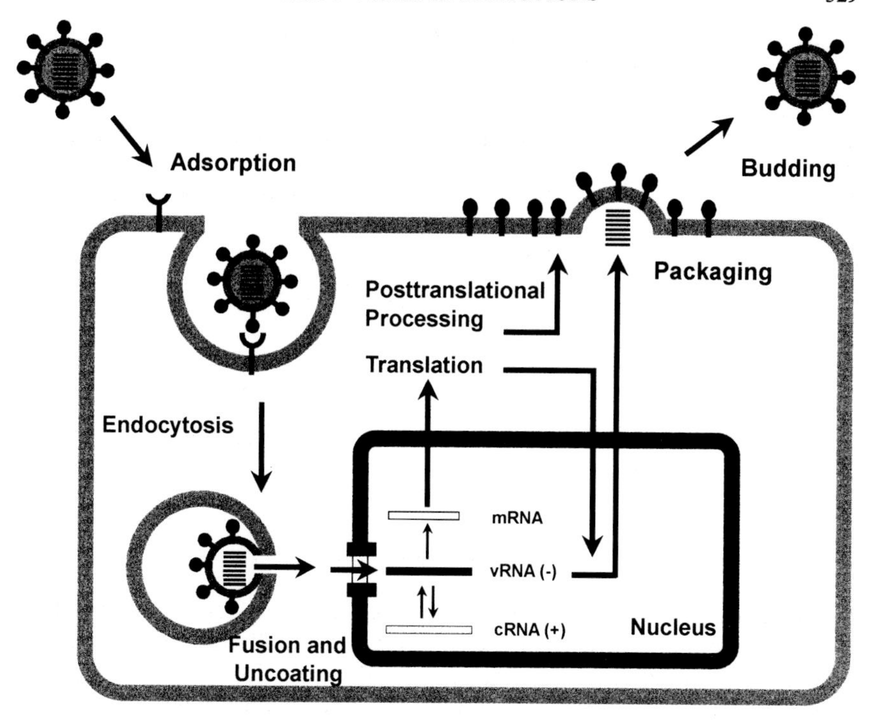

Fig. 1. Schematic diagram of influenza virus replication in an infected cell. For description of the steps involved, see text. Virus and cell compartments are not drawn to scale.

(cRNA) of the template RNA, and (c) additional copies of the virion (v)RNA amplified from the cRNA.

The minimal promoter sequences of the vRNA and cRNA molecules appear to be their highly conserved 3′ terminal sequences (Luo & Palese, 1992). In addition, the 5′ sequences of the templates are part of the transcription/replication process (Cianci, Tiley & Krystal, 1995; Hagen, Chung & Krystal, 1994). Genetic and *in vitro* experiments have suggested that both the vRNA- and the cRNA-RNP templates have pan handle or fork structures (Hsu *et al.*, 1987; Fodor, Pritlove & Brownlee, 1995). These structures, involving both the 3′ and 5′ ends, exist during the active transcription/replication of viral RNAs. Most recently it has been found that the non-conserved nucleotides (outside the minimal promoter sequences) also play a role in viral RNA replication (Zheng, Palese & García-Sastre, 1996).

What regulates the switch from transcription (mRNA synthesis) to replication (amplification of vRNA) is not fully understood, but it appears that the concentration of free nucleoprotein (NP) determines whether mRNA or cRNA is produced (Shapiro & Krug, 1988). We also know that mRNA synthesis is regulated. Although the different vRNA molecules appear to be equimolar in the virus and in infected cells, the mRNAs of the haemagglutinin (HA), membrane protein (M), nucleoprotein (NP) and non-structural protein (NS) genes are much more abundant than those of the three polymerase (P) genes or the NA gene. Also, mRNA synthesis is down-regulated altogether late in the replication cycle, while vRNA synthesis continues unabated. A possible regulator of viral RNA synthesis is the NS1 protein, which inhibits nuclear export of spliced and cellular mRNAs and inhibits pre-mRNA splicing (Lamb & Krug, 1996).

Assembly and packaging of RNAs into infectious viruses involves several compartments in the cell: the nucleus, the cytoplasm and the cytoplasmic membrane. The viral P proteins and the NP have specific nuclear localization signals so that they can travel into the nucleus where they associate with viral RNAs to form ribonucleoprotein complexes (RNPs). The export of the RNPs from the nucleus likewise requires a unique mechanism. Following their export into the cytoplasm, the RNPs assemble at the cytoplasmic membrane under patches of the viral glycoproteins HA and NA. Budding of virus particles occurs from the cytoplasmic membrane, while the NA clears the virus and the cell membrane of sialic acids to prevent virus–virus aggregation and virus–cell surface retention, respectively (Palese *et al.*, 1974). the viral NA thus facilitates the release of virus particles into the extracellular space. The role of the NS2 protein (encoded by the spliced NS gene-specific mRNA) in the replication cycle is unknown and the phenotype of the only known temperature-sensitive NS2 mutant (Odagiri *et al.*, 1994) is not informative.

GENETIC ENGINEERING OF INFLUENZA VIRUSES

The genome of negative sense RNA viruses is not infectious and thus transfection of cDNA-derived influenza virus RNAs alone does not result in the formation of infectious virus particles. However, a system was developed which allows the *in vitro* reconstitution of biologically active RNP from synthetic RNAs and purified viral proteins. Transfection of synthetic RNPs into helper virus-infected cells allows the rescue of transfectant viruses which contain one or more genetically engineered segments (Luytjes *et al.*, 1989; Enami & Palese, 1991). This powerful system permits the introduction of specific mutations into the genomes of infectious influenza viruses, so that the *cis*-acting signals involved in RNA transcription, replication and packaging can now be studied. In addition, this methodology allows the structure/function analysis of individual viral proteins involved in virus replication. A similar approach may also lead to the construction of improved vaccine strains and to the development of influenza viruses for use as viral vectors (García-Sastre & Palese, 1993, 1995; Palese, 1995; Palese *et al.*, 1996).

HOST–VIRUS INTERACTIONS

As discussed, genetic engineering methods have become available to study site-specific mutations in the context of an infectious influenza virus, and thus effects on virulence caused by mutations in the genome can be studied. Furthermore, X-ray diffraction analyses of the viral glycoproteins HA and NA were completed (Weis *et al.*, 1988; Bullough *et al.*, 1994; Colman, 1994), which permitted greater insight into how the virus interacts with the sialic acid receptor on the cell surface, what the mechanism of acid pH-induced fusion is, and how the NA interacts with cellular substrates. Exciting as these advances have been, they tell only one side of the story because they focus mostly on the virus and do not address the intracellular host machinery needed for virus replication. It was thus decided to expand the approach by turning to the study of intracellular proteins which interact with viral proteins and by investigating the biological consequences of these interactions.

CELLULAR PROTEINS INTERACTING WITH VIRAL NP

Nucleoprotein interactor 1 (NPI-1)

The approach to identify cellular proteins which interact with specific influenza A virus proteins was based on the hypothesis that cellular factors modulate viral gene expression and thus affect host specificity and virulence. The NP was selected first because it is part of the polymerase complex (Huang, Palese & Krystal, 1991), and it had previously been shown to contribute to the determination of host range (Shimizu *et al.*, 1983; Scholtissek *et al.*, 1985). Initial attempts using monoclonal and polyclonal anti-NP antibodies to co-immunoprecipitate cellular proteins were unsuccessful. It was then decided to utilize the yeast two-hybrid system as described by Fields & Song (1989) and Gyuris *et al.* (1993).

This powerful methodology uses a transcriptional ‘read-out’ in yeast cells, which is based on the interaction of two fusion proteins. The first is a hybrid consisting of a DNA binding protein and a ‘bait’ protein (in the study described here, the viral NP). The second fusion protein consists of a transcriptional activator protein and the ‘catch’ or ‘prey’ protein (Fields & Sternglanz, 1994; Luban & Goff, 1995). In a study conducted in this laboratory (O’Neill & Palese, 1995), a cDNA library of ‘preys’ from HeLa cells was screened, and several cDNA clones were identified that coded for polypeptides which interacted with the viral NP used as the ‘bait’ (Table 2). A promising lead was the clone coding for NPI-1 (nucleoprotein interactor 1). The protein appeared to be a homologue of the yeast SRP1 protein (Yano *et al.*, 1992) which can suppress the defect in a temperature sensitive RNA polymerase I mutant. It was speculated that the mammalian homologue of

Table 2. *Influenza A virus nucleoprotein interactor proteins*

Interactor protein	Identity of protein/function
NPI-1	Karyopherin alpha 1 (Moroianu *et al.*, 1995*a*) hSRP1 (Cortes *et al.*, 1994; Weis *et al.*, 1995) Docking protein involved in import of viral protein and RNA (O'Neill *et al.*, 1995)
NPI-2	AU-rich element RNA binding factor (AUF1) (Zhang *et al.*, 1993); hnRNP C1-like protein (Lahiri & Thomas, 1986)
NPI-3	Karyopherin alpha 2 (Moroianu *et al.*, 1995*c*); Rch1 (Cuomo *et al.*, 1994)
NPI-4	–
NPI-5	BAT1 (DEAD box protein) (Peelman *et al.*, 1995); Putative helicase; possible function in viral RNA replication
NPI-6	–

the SRP1 protein binds to a component of the viral RNA polymerase because of a functional interaction similar to that observed for the yeast SRP1 gene product and the yeast RNA polymerase. However, the original hope that the NPI-1 protein would be a transcription factor or a protein which modulates the influenza virus RNA polymerase did not materialize. In collaboration with Dr Blobel's laboratory, it was shown that the NPI-1 is a nuclear localization signal receptor that is involved in the import of the viral NP from the cytoplasm into the nucleus. These results were obtained using an *in vitro* nuclear import assay involving cells whose plasma membrane was permeabilized with digitonin. It was possible to demonstrate that four cytoplasmic factors – NPI-1 (karyopherin alpha), karyopherin beta, Ran and p10 – are required for the transport of the NP into the nucleus (O'Neill *et al.*, 1995; Moroianu *et al.*, 1995*a*) (Fig 2).

MAPPING OF THE NPI-1 INTERACTING DOMAIN ON INFLUENZA A VIRUS NP

Interaction between the NPI-1 and the viral NP was first demonstrated in the yeast two-hybrid genetic assay system. This assay was then used to further define the specific domain within the NP which interacts with the NPI-1 by making fusion proteins containing defined regions of the NP molecule and testing them. These experiments delimited the NPI-1 interaction domain to the first 20 (N-terminal) amino acids of NP. This same fragment (20 amino acids) – when fused to the measles virus P protein and expressed from a transfected plasmid – locates to the nucleus. The measles virus P protein is normally a cytoplasmic protein, but when it possesses the NP-derived extension (20 amino acids) it is pulled into the nucleus. This experiment demonstrates a correlation between the presence of a nuclear localization signal (NLS) and the interaction with the nucleus import factor NPI-1.

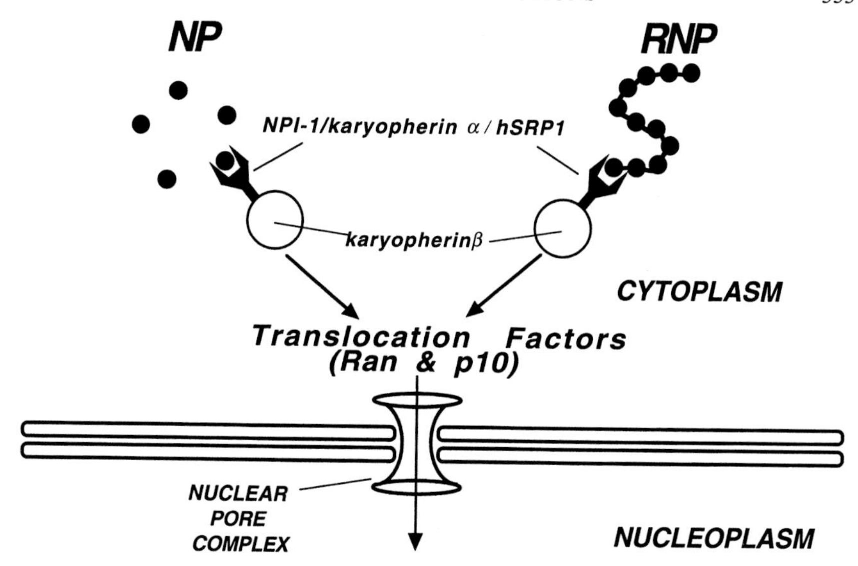

Fig. 2. Schematic diagram of components involved in the nuclear import of influenza viral NP and RNA. Free NP (filled circles) or NP associated with RNA (RNP) binds to the cellular NPI-1 (also known as karyopherin α or hSRP1). The NPI-1 (a docking factor) binds to karyopherin β to form a complex which is translocated through the nuclear pore with the help of ran and p10. Thus, four soluble factors (NPI-1, karyopherin β, ran and p10) appear to be minimally required for nuclear import of the NP. In *in vitro* experiments utilizing the digitonin-permeabilized cell assay, the NPI-1 can be replaced by NPI-3.

A fine mapping of the NPI-1 interacting domain was done by alanine scanning of the first 20 amino acids and assaying the products in the yeast two-hybrid system. As can be seen, the NPI-1 binding site has the following motif: ^{3}SxGTKRSYxxM^{13} (Fig. 3). It covers amino acids 3 to 13. Several of the alanine mutants have also been checked for NLS activity. These experiments suggest that the NPI-1 binding site is associated with an NLS site on the NP. Most NLSs resemble the archetypal SV40 large T-antigen NLS (PKKKRK) or the bipartite NLS of nucleoplasmin (KRXXXXXXXXXXKKKK) in which the two basic amino acid clusters are separated by ten (non-basic) amino acids (Fig. 4). The first 20 N-terminal amino acids of the influenza A/PR/8/34 viral NP do not contain either type of conventional NLS (nor, for that matter, does the remainder of the NP molecule) (Fig. 4).

Earlier studies had suggested (Davey, Dimmock & Colman, 1985) that a nuclear accumulation signal was associated with amino acids 327–345 of the influenza viral NP. The proposed karyophilic signal (AAFEDLRLS) was found to be important for the nuclear accumulation of the NP as well as of NP-globin fusion proteins in *Xenopus* oocytes. However, a fusion protein containing amino acids 255–363 (encompassing the putative NLS) of the NP

		NPI-1
w.t.	MASQGTKRSYEQMETDGERQ	+
A3	MA**A**QGTKRSYEQMETDGERQ	-
A4	MAS**A**GTKRSYEQMETDGERQ	+
A5	MASQ**A**TKRSYEQMETDGERQ	-
A6	MASQG**A**KRSYEQMETDGERQ	-
A7	MASQGT**A**RSYEQMETDGERQ	-
A8	MASQGTK**A**SYEQMETDGERQ	-
A9	MASQGTKR**A**YEQMETDGERQ	-
A10	MASQGTKRS**A**EQMETDGERQ	-
A11	MASQGTKRSY**A**QMETDGERQ	+
A12	MASQGTKRSYE**A**METDGERQ	+
A13	MASQGTKRSYEQ**A**ETDGERQ	-
A14	MASQGTKRSYEQM**A**TDGERQ	+
A15	MASQGTKRSYEQME**A**DGERQ	+
A16	MASQGTKRSYEQMET**A**GERQ	+
A17	MASQGTKRSYEQMETD**A**ERQ	+
A18	MASQGTKRSYEQMETDG**A**RQ	+
A19	MASQGTKRSYEQMETDGE**A**Q	+
A20	MASQGTKRSYEQMETDGER**A**	+

NPI-1 binding motif ^{3}**SxGTKRSYxxM**^{13}

Fig. 3. Mutational analysis of the NPI-1 binding site on the influenza virus NP. Deletion mutants of the NP were shown by the yeast two hybrid system to contain the NPI-1 binding site in the N-terminal 20 amino acids. Alanine scanning and analysis of the mutants in the yeast two-hybrid system revealed the SxGTKRSYxxM motif as the NPI-1 binding site. Cellular localization studies employing fusion proteins (containing the N-terminal 20 amino acids of the NP or different mutants) revealed the presence of a nuclear localization signal (NLS) in this domain of the NP.

Influenza A virus NP (aa 1–20)	MA**SQGTKRSY**EQ**M**ETDGERQ
SV40 antigen	**PKKKRKV**
Neucleoplasmin	**KR**PAATLLAGQA**KKKK**

Fig. 4. Non-conventional nuclear localization signal of the influenza virus NP.

and the measles virus P protein was cytoplasmic in mammalian cells in our hands. This finding suggests that the earlier data concerning the nucleophilic signal of the NP were most likely an artefact of the *Xenopus* system.

Requirements for the nuclear import of influenza viral RNA

Studies on the nuclear import of influenza virus components were extended to the viral RNA. Viral RNA must enter the nucleus of infected cells, where transcription, replication and splicing of the viral RNAs take place. It was demonstrated that naked viral RNA does not enter the nucleus in the digitonin-permeabilized cell assay system; however, in the presence of NP the RNA is transported into the nucleus with requirements indistinguishable from those for docking and entry of NP (O'Neill *et al.*, 1995). That is, the four cytoplasmic factors described above are also needed for the import of RNA. Furthermore, these data indicate that nuclear uptake of influenza virus RNA is not via a signal in the RNA. Instead, it must be facilitated by an NLS in a viral protein such as NP. These results concerning the influenza viruses are significant because they may be pertinent to other viruses as well as to normal cellular processes: NLS-containing proteins of other viruses (e.g. HIV, adenovirus and some herpesviruses) may have similar functions in mediating the nuclear import of their viral genomes. In addition, nuclear export of influenza viral RNAs/mRNAs may be mediated by signals on associated proteins, and not by signals on the RNAs themselves.

Other NP-interacting proteins

NPI-3, an NPI-1-related protein, was also identified as a result of the search using the yeast two-hybrid system (Table 2). Although this protein shows only a 46% sequence identity with NPI-1, it is able to replace NPI-1 in the *in vitro* nuclear import assay for influenza virus NP and for influenza virus RNA (O'Neill *et al.*, 1995). It is thus possible that there is redundancy of function in the interaction of NPI-1 and NPI-3 with the viral NP. However, since nothing is known about the cellular distribution or the temporal regulation of these cellular proteins, the precise role of these proteins (working alone or in concert) during influenza virus replication remains to be elucidated.

Another NP-binding protein, NPI-2, has previously been described and was found to bind to RNA and to be localized in the nucleus (Table 2). No further experiments to define its function for virus replication have been conducted. Similarly, the gene termed NPI-5 (Table 2) has previously been described. It codes for the BAT1 protein which is a putative DEAD-box RNA helicase. It can be speculated that such a protein, if indeed it has a helicase activity, may affect the transcription/replication of the viral RNP templates. Two additional clones, NPI-4 and NPI-6, have been identified (Table 2) which express cellular proteins (or fragments) that interact with the influenza A virus NP. At the present time, nothing is known about the biological role of these genes in uninfected cells or during virus replication.

CELLULAR PROTEINS THAT BIND TO THE INFLUENZA VIRUS NS1 PROTEIN

The yeast two-hybrid system was also used to identify cDNA clones that express proteins which interact with the NS1 protein of influenza A viruses. The NS1-I (NS1-Interacting) protein was found to interact with NS1 proteins in all *in vitro* and *in vivo* essays employed (Wolff, O'Neill & Palese, 1996). Also, it was found that the NS1-I protein bound not only to the NS1 proteins of influenza A viruses but also to those of influenza B viruses. The influenza B NS1 proteins share only 20% of their sequence with those of the A viruses, suggesting that there is a functional interaction between the cellular NS1-I and the NS1 proteins of influenza A and of influenza B viruses.

The functions of the NS1 protein appear to be pleiotropic. They include the inhibition of the transport of poly A-containing mRNAs from the nucleus into the cytoplasm, the inhibition of splicing, and the enhancement of translation of viral mRNAs (Enami *et al.*, 1994; De La Luna *et al.*, 1995; Lamb & Krug, 1996). The gene coding for NS1-I has been previously described and it was suggested that it codes for the precursor of a protein with 17-β-hydroxysteroid-dehydrogenase activity. It is not clear at this point how the known functions of the NS1 protein would be affected by the interaction with the cellular NS1-I protein (Wolff *et al.*, 1996).

As in the case of the NP-interacting cellular proteins, additional clones of NS1-interacting proteins have been identified. One of them – clone 59-1 (in the form of a GST fusion protein) – interferes with the splicing of a ^{32}P-labeled pre-mRNA *in vitro* (Wolff, unpublished observations). The block in splicing appears to occur at a step after spliceosome assembly, which parallels the splicing-inhibitory effect of the NS1 protein (Fortes, Beloso & Ortin, 1994; Lu, Qian & Krug, 1994). One interpretation of the results is that the endogenous 59-1 protein has a function in splicing and that exogenous GST-59-1 protein titrates out other factors essential for splicing, resulting in the formation of stalled spliceosomes. During viral infection, the NS1 protein binds to the 59-1 protein, which might result in an overall decrease in mRNA splicing.

CONCLUSIONS

Knowledge of the influenza virus genome structure and the functions and characteristics of viral proteins has not been matched by an equally detailed understanding of how the virus interacts with the host cell. New techniques, such as the yeast two-hybrid system, now allow us to identify specific cellular proteins which function during virus replication. Such studies should enhance our understanding of how the viral components fare in the cell and advance our knowledge of the fundamental processes of host–virus interactions. Clearly, the yeast two-hybrid system is but one of the techniques which will give us a window to explore how the virus co-opts components of the cell for its own replication. Other methodologies will include the use of transgenics in tissue culture as well as in the animal model, and efficient *in vitro* systems to study transport of cellular and viral components within the cell. Powerful tetracycline-regulated gene expression systems may also play an important role in studying the effects of cellular (trans-dominant mutant) proteins on virus replication (Shockett & Schatz, 1966). Considering the past success in exploring the structure of influenza viruses and the functions of their components, it is likely that the application of these new techniques will provide a new dimension in our understanding of the pathways of infection and replication of the virus.

REFERENCES

Braam, J., Ulmanen, I. & Krug, R. M. (1983). Molecular model of a eukaryotic transcription complex: Functions and movements of influenza P proteins during capped RNA-primer transcription. *Cell*, **34**, 609–18.

Bullough, P. A., Hughson, F. M., Skehel, J. J. & Wiley, D. C. (1994). Structure of influenza hemagglutinin at the pH of membrane fusion. *Nature*, **371**, 37–43.

Cianci, C., Tiley, L. & Krystal, M. (1995). Differential activation of the influenza virus polymerase via template RNA binding. *Journal of Virology*, **69**, 3995–9.

Colman, P. M. (1994). Influenza virus neuraminidase: structure, antibodies and inhibitors. *Protein-Science*, **3**, 1687–96.

Cortes, P., Ye, Z. & Baltimore, D. (1994). RAG-1 interacts with the repeated amino acid motif of the human homologue of the yeast protein SRP-1. *Proceedings of the National Academy of Sciences, USA*, **91**, 7633–7.

Cuomo, C. A., Kirch, S. A., Gyuris, J., Brent, R. & Oettinger, M. A. (1994). Rch1, a protein that specifically interacts with the RAG-1 recombination-activating protein. *Proceedings of the National Academy of Sciences, USA*, **91**, 6156–60.

Davey, J., Dimmock, N. J. & Colman, A. (1985). Identification of the sequence responsible for the nuclear accumulation of the influenza virus nucleoprotein in *Xenopus* oocytes. Cell, **40**, 667–75.

De La Luna, S., Fortes, P., Beloso, A. & Ortin, J. (1995). Influenza virus NS1 protein enhances the rate of translation initiation of viral mRNAs. *Journal of Virology*, **69**, 2427–33.

Enami, K., Sato, T. A., Nakada, S. & Enami, M. (1994). Influenza virus NS1 protein stimulates translation of the M1 protein. *Journal of Virology*, **68**, 1432–7.

Enami, M. & Palese, P. (1991). High-efficiency formation of influenza virus transfectants. *Journal of Virology*, **65**, 2711–13.

Fields, S. & Song, O.-K. (1989). A novel genetic system to detect protein–protein interactions. *Nature*, **340**, 245–6.

Fields, S. & Sternglanz, R. (1994). The two-hybrid system: an assay for protein–protein interactions. *Trends in Genetics*, **10**, 286–92.

Fodor, E., Pritlove, D. C. & Brownlee, G. G. (1995). Characterization of the RNA-fork model of virion RNA in the initiation of transcription of influenza A virus. *Journal of Virology*, **69**, 4012–19.

Fortes, P., Beloso, A. & Ortin, J. (1994). Influenza virus NS1 protein inhibits pre-mRNA splicing and blocks mRNA nucleocytoplasmic transport. *EMBO Journal*, **13**, 704–12.

García-Sastre, A. & Palese, P. (1993). Genetic manipulation of negative-strand RNA virus genomes. *Annual Reviews in Microbiology*, **47**, 765–90.

García-Sastre, A. & Palese, P. (1995). Influenza virus vectors. *Biologicals*, **23**, 171–8.

Gyuris, J., Golemis, E., Chertkov, H. & Brent, R. (1993). Cdi1, a human G1 and S phase protein phosphatase that associates with cdk2. *Cell*, **75**, 791–803.

Hagen, M., Chung, T. D. Y. & Krystal, M. (1994). Recombinant influenza virus polymerase requires both 5′ and 3′ ends for activation. *Journal of Virology*, **68**, 1509.

Hayden, F. G. & Palese, P. (1996). Influenza virus. In Richman, D. *et al.*, eds. *Clinical Virology*, Chapter 39. New York: Churchill-Livingstone.

Hsu, M.-T., Parvin, J. D., Gupta, S., Krystal, M. & Palese, P. (1987). The genomic RNAs of influenza viruses are held in a circular conformation in virions and in infected cells by a terminal panhandle. *Proceedings of the National Academy of Sciences, USA*, **84**, 8140–4.

Huang, T., Palese, P. & Krystal, M. (1991). Determination of influenza virus proteins required for genome replication. *Journal of Virology*, **64**: 5669–73.

Krug, R. M., Alonso-Caplen, F. V., Julkunen, I. & Katze, M. G. (1989). Expression and replication of the influenza virus genome. In Krug, R. M., ed. *The Influenza Viruses*, p. 89. New York: Plenum.

Lahiri, D. K. & Thomas, J. O. (1986). A cDNA clone of the hnRNP C protein and its homology with the single-stranded DNA binding protein UP2. *Nucleic Acids Research*, **14**, 4077–94.

Lamb, R. A. & Krug, R. M. (1996). Orthomyxoviridae: the viruses and their replication. In Fields, B. N., Knipe, D. M., Howley, P. M. *et al.*, eds. *Fields Virology, 3rd edition*, pp. 1353–95. Philadelphia: Lippincott-Raven.

Lu, Y., Qian, X.-Y. & Krug, R. M. (1994). The influenza virus NS1 protein: a novel inhibitor of pre-mRNA splicing. *Genes Development*, **8**, 1817–28.

Luban, J. & Goff, S. P. (1995). The yeast two-hybrid system for studying protein–protein interactions. *Current Opinion in Biotechnology*, **6**, 59–64.

Luo, G. & Palese, P. (1992). Genetic analysis of influenza virus. *Current Opinion in Genetic Development*, **2**, 77–81.

Luytjes, W., Krystal, M., Enami, M., Parvin, J. D. & Palese, P. (1989). Amplification, expression and packaging of a foreign gene by influenza virus. *Cell*, **59**, 1107–13.

Moroianu, J., Blobel, G. & Moore, M. S. (1995*a*). Previously identified protein of uncertain function is karyopherin alpha and together with karyopherin beta docks import substrate at nuclear pore complexes. *Proceedings of the National Academy of Sciences, USA*, **92**, 2008–11.

Moroianu, J., Hijikata, M., Blobel, G. & Radu, A. (1995*b*). Mammalian karyopherin alpha 1 beta and alpha 2 beta heterodimers: alpha 1 or alpha 2 subunit binds nuclear localization signal and beta subunit interacts with peptide repeat-containing nucleoporins. *Proceedings of the National Academy of Sciences, USA*, **92**, 6532–6.

Odagiri, T., Tominaga, K., Tobita, K. & Ohta, S. (1994). An amino acid change in the nonstructural NS2 protein of an influenza A virus mutant is responsible for the generation of defective interfering (DI) particles by amplifying DI RNAs and suppressing complementary RNA synthesis. *Journal of General Virology*, **75**, 43–53.

O'Neill, R. E. & Palese, P. (1995). NPI-1, the human homolog of SRP-1, interacts with influenza virus nucleoprotein. *Virology*, **206**, 116–25.

O'Neill, R. E., Jaskunas, S. R., Blobel, G., Palese, P. & Moroianu, J. (1995). Nuclear import of influenza virus RNA can be medicated by viral nucleoprotein and transport factors required for protein import. *Journal of Biological Chemistry*, **270**, 22701–4.

Palese, P., Tobita, K., Ueda, M. & Compans, R. W. (1974). Characterization of temperature sensitive influenza virus mutants defective in neuraminidase. *Virology*, **61**, 397–410.

Palese, P. (1995). Genetic engineering of infectious negative-strand RNA viruses. *Trends in Microbiology*, **3**, 123–5.

Palese, P., Zheng, H., Engelhardt, O. G. & García-Sastre, A. (1996). Negative-strand RNA viruses: genetic engineering and applications. *Proceedings of the National Academy of Sciences, USA* (in press).

Peelman, L. J., Chardon, P., Nunes, M., Renard, C., Geffrottin, C., Waiman, M., van Zeveren, A., Copieters, W., van de Weghe, Y., Bouquet, Y., Choy, W. W., Strominger, J. L. & Spies, T. (1995). The BAT1 gene in the MHC encodes an evolutionary putative nuclear RNA helicase of the DEAD family. *Genomics*, **26**, 210–18.

Pleschka, S., Jaskunas, S. R., Engelhardt, O. G., Zürcher, T., Palese, P. & García-Sastre, A. (1996). A plasmid-based reverse genetics system for influenza A virus. *Journal of Virology*, **70**, 4188–92.

Scholtissek, C., Burger, H., Kister, O., Shortridge, K. F. (1985). The nucleoprotein as a major factor in determining host specificity of influenza H3N2 viruses. *Virology*, **147**, 287–94.

Shapiro, G. I. & Krug, R. M. (1988). Influenza Virus replication *in vitro*: synthesis of viral template RNAs and virion RNAs in the absence of added primer. *Journal of Virology*, **62**, 2285–90.

Shimizu, K., Mullinix, M. G., Chanock, R. M., Murphy, B. R. (1983). Temperature-sensitive mutants of influenza A/Udorn/72 (H3N2) virus. III. Genetic analysis of temperature-dependent host range mutants. *Virology*, **147**, 35–44.

Shockett, P. E. & Schatz, D. G. (1996). Diverse strategies for tetracycline-regulated inducible gene expression. *Proceedings of the National Academy of Sciences, USA*, **93**, 5173–6.

Weis, K., Mattaj, I. W. & Lamond, A. I. (1995). Identification of hSRP alpha as a functional receptor for nuclear localization sequence. *Science*, **268**, 1049–53.

Weis, W., Brown, J. H., Cusack, S., Paulsen, J. C., Skehel, J. J. & Wiley, D. C. (1988). Structure of the influenza virus hemagglutinin complexed with its receptor, sialic acid. *Nature*, **333**, 426–31.

Wolff, T., O'Neill, R. E. & O'Neill, P. (1996). Interaction cloning of NSI-1, a human protein that binds to the nonstructural NS1 proteins of influenza A and B viruses. *Journal of Virology*, **70**, 5363–72.

Yano, R., Oakes, M., Yamaghishi, M., Dodd, J. A. & Nomura, M. (1992). Cloning and characterization of SRP1, a suppressor of temperature-sensitive RNA polymerase I mutations in *Saccharomyces cerevisiae. Molecular and Cellular Biology*, **12**, 5640–51.

Zhang, W., Wagner, B. J., Ehrenman, K., Schaefer, A. W., DeMaria, C. T., Crater, D., DeHaven, K., Long, L. & Brewer, G. (1993). Purification, characterization and cDNA cloning of an AU-rich element RNA-binding protein, AUF1. *Molecular and Cellular Biology*, **13**, 7652–65.

Zheng, H., Palese, P. & García-Sastre, A. (1996). Nonconserved nucleotides at the 3′ and 5′ ends of an influenza A virus RNA play an important role in viral RNA replication. *Virology*, **217**, 242–51.

TRANSMISSIBLE SPONGIFORM ENCEPHALOPATHIES AND THE FORMATION OF PROTEASE-RESISTANT PRION PROTEIN

B. CAUGHEY

Laboratory of Persistent Viral Diseases, Rocky Mountain Laboratories, National Institute of Allergy and Infectious Diseases, NIH, Hamilton, Montana 59840, USA

Considerable public interest in the transmissible spongiform encephalopathies (TSE) or prion protein (PrP) diseases has been kindled by the recent epidemic of bovine spongiform encephalopathy (BSE or mad cow disease) (Wells & Wilesmith, 1995) and the possibility that BSE might be transmissible to humans (Will *et al.*, 1996). TSEs are fatal neurodegenerative diseases that occur in a wide variety of mammals including ruminants, cats, mink, rodents and primates (for review, see Pocchiari, 1994; Weissmann, 1996). In humans, TSE diseases take the form of Creutzfeldt–Jakob disease (CJD), kuru, Gerstmann–Straussler–Scheinker syndrome (GSS) and fatal familial insomnia (FFI). So far, the TSEs occur only rarely in humans with overall annual incidence of approximately one per million population. However, scrapie is a widespread problem in sheep and BSE has seriously hurt the British cattle industry.

Host-to-host transmission of TSEs can occur by inoculation or ingestion of diseased tissues (Gibbs, Jr. & Gajdusek, 1973; Gibbs, Jr. *et al.*, 1980). In many cases TSEs can be passaged between different host species under either practical or experimental circumstances. The BSE epidemic provides an important example of interspecies transmission of TSEs through the food chain (Wells & Wilesmith, 1995). It is widely suspected that the BSE epidemic was initiated by a change in the commercial protein rendering process which allowed scrapie infectivity in sheep carcasses to be fed to cattle in the form of protein feed supplements. The problem was presumably amplified in cattle by the use of BSE infected cattle carcasses in the rendering process. BSE was then transmitted to a number of other species that were fed either infected tissues or protein supplements. The most troubling possibility is that the many humans who have been exposed to BSE infectivity through contaminated beef might be susceptible to BSE infection. The recent appearance of ten cases of a new variant of CJD in humans in the UK has provided the first hint that BSE might have been transmitted to humans, but the link to BSE has not been proven (Will *et al.*, 1996). This review will summarize some of the important basic issues that remain in the TSE field, with a particular focus on the involvement of prion protein (PrP).

THE INFECTIOUS AGENT OF THE TSE DISEASES

The precise nature of the infectious TSE agents remains a mystery despite decades of investigation and theorizing. Their infectious nature and ability to greatly amplify themselves in the host indicates that they are not merely environmental neurotoxins. Their apparent small size suggests that they might be viruses as opposed to other conventional microbial pathogens (Eklund, Hadlow & Kennedy, 1963). Because of their long incubation periods, ranging from months to decades, and their high resistance to inactivation by irradiation, heat and harsh chemical treatments, they have often been referred to as slow or unconventional viruses (for review see Rohwer, 1991). So far, however, no virus or unique nucleic acid genome has been ascribed to TSE infectious agent. This might be because the TSE virus is hard to find, as has been the case with numerous other viruses, or because no TSE virus exists. Available evidence suggests that if such a virus exists, it would have to be ubiquitous throughout many mammalian host species and have an unusually small genome that is highly protected (presumably by PrP-res) from treatments that degrade nucleic acids.

The major alternative to the virus theory that has been considered over the last three decades is the 'protein-only' or 'prion' hypothesis. As initially formulated (Griffith, 1967), the protein-only hypothesis holds that the TSE agent might be an abnormal, infectious form of a host protein that can somehow direct its own formation from its normal homolog in the host. Neither the viral nor protein-only models have been proven, so the most basic mystery in the TSE field remains unresolved.

THE IMPORTANCE OF PRP

Merz and colleagues identified a unique fibrillar structure in scrapie-infected, and not normal, brain tissue which were dubbed scrapie associated fibrils' (Merz *et al.*, 1981). Biochemical analysis of these fibrils and of other infectious TSE tissue preparations led to the discovery that they contained an abnormal form of a host protein which is now most widely known as prion protein (PrP) (Bolton, McKinley & Prusiner, 1982; Diringer *et al.*, 1983). The abnormal TSE-associated PrP forms insoluble aggregates (such as amyloid plaques) and is partially resistant to proteinase K (PK) (Diringer *et al.*, 1983; Prusiner *et al.*, 1983), which removes approximately 67 amino acid residues (6–7 kDa) from the N-terminus of each molecule (Oesch *et al.*, 1985; Hope *et al.*, 1986). One can refer to this abnormal form generically as protease-resistant PrP (PrP-res), but the forms associated with specific TSE diseases are often labelled with a superscript designating that disease, such as PrP^{Sc} for PrP-scrapie. In contrast, the normal, non-pathogenic form, PrP-sen or PrP^{C}, is fully sensitive by proteinase K and soluble in mild detergents (Meyer *et al.*,

1986; Rubenstein *et al.*, 1986). The close association of TSEs with PrP abnormalities has led to the frequent use of PrP diseases or prion diseases as alternative names for TSEs (Prusiner, 1991).

Besides the presence of PrP-res in TSE-infected tissues, tissue culture cells and enriched preparations of infectivity, there are many other indications that PrP plays a critical role in TSEs. Selective inhibition of PrP-res formation in scrapie-infected cells by Congo red treatment cures the cells of scrapie infection (Caughey, Ernst & Race, 1993). Elimination of the PrP gene altogether in mice makes them resistant to scrapie infection (Bueler *et al.*, 1993). In hosts that have PrP, the susceptibility to various TSE agent strains can be determined by their PrP genotype and how well it matches with the PrP genotype of the source species (Carlson *et al.*, 1986, 1988; Hunter *et al.*, 1987; Race *et al.*, 1990; Goldmann *et al.*, 1994; Belt *et al.*, 1995). In humans, an ever-increasing number of familial TSE diseases have been identified that are associated with rare mutations in the PrP gene (for review, see Pocchiari, 1994). The transfer of one such mutant PrP gene into mice can give them a neurodegenerative disease, but one that has very little PrP-res or infectivity associated with it (Hsiao *et al.*, 1990, 1994). These considerations are consistent with the proposal that PrP-res might be the self-inducing infectious protein modelled by Griffith (1967) and Prusiner (1982). Alternatively, PrP may act as a receptor or host-encoded component of an, as yet, unidentified viral agent (for review, see Rohwer, 1991). Regardless of the precise relationship between PrP-res and the infectious agent, PrP-res formation appears to be critical in pathogenesis.

PrP is a glycophosphatidylinositol (GPI)-linked sialoglycoprotein (Bolton, Meyer & Prusiner, 1985; Manuelidis, Valley & Manuelidis, 1985; Stahl *et al.*, 1987) expressed in a wide variety of cells in brain and many other tissues (Oesch *et al.*, 1985; Cho, 1986; Hope *et al.*, 1986; Meyer *et al.*, 1986; Robakis *et al.*, 1986; Rubenstein *et al.*, 1986; Caughey *et al.*, 1988; Mobley *et al.*, 1988; Bendheim *et al.*, 1992; Lieberburg, 1992; Manson *et al.*, 1992) and is developmentally regulated (Manson *et al.*, 1992). However, the normal physiological function of PrP is not known. In lymphocytes, PrP-sen appears to be involved in mitogen induced activation (Cashman *et al.*, 1990). Recent *in vitro* studies suggest that PrP-sen is involved in the regulation of intracellular Ca^{2+} concentrations in neuronal cells (Whatley *et al.*, 1995). Mice which lack PrP due to homozygous knockout of its gene develop into adults without gross abnormalities (Bueler *et al.*, 1992) but several differences between the PrP knockout mice and controls have been noted (Collinge *et al.*, 1994; Sakaguchi *et al.*, 1996; Tobler *et al.*, 1996). While these latter observations are intriguing, it is difficult to know at what point in development the PrP gene disruptions caused these abnormalities to occur, and, therefore, whether PrP itself is directly involved in the affected functions in adults.

Normal PrP biosynthesis and metabolism

PrP begins its metabolic cycle in the endoplasmic reticulum where the GPI anchor and high mannose glycans are attached (Caughey *et al.*, 1989). In the Golgi apparatus, the high mannose glycan moieties are converted to complex or hybrid glycans (Caughey *et al.*, 1989; Haraguchi *et al.*, 1989). Ultimately, most of the PrP is anchored to the cell surface by the phosphatidylinositol moiety (Stahl *et al.*, 1987) and, although PrP-sen can cycle between the plasma membrane and endocytic vesicles (Shyng *et al.*, 1995), most can be removed from intact cells with phospholipase or protease treatments (Stahl *et al.*, 1987; Caughey *et al.*, 1989, 1990). The half-life of PrP-sen is normally 3–6 h with a small proportion released into the medium (Caughey *et al.*, 1988, 1989; Borchelt *et al.*, 1990; Caughey & Raymond, 1991). Soluble forms of PrP have now been identified *in vivo* in human cerebrospinal fluid (Tagliavini *et al.*, 1992). N-terminally truncated forms of PrP-sen have been identified in both normal and CJD human brain tissue as well as in neuroblastoma cells (Chen *et al.*, 1995). One form, starting at residue 111 or 112, is abundant in normal tissues, and appears to be a major product of normal PrP metabolism. A similar PrP fragment (residues 121–231) has been found to form a soluble autonomous folding domain and the three-dimensional structure of this domain has been determined by NMR (Riek *et al.*, 1996).

PrP-res formation

In TSE-infected hosts, PrP-res usually accumulates in the central nervous system and lymphoreticular tissues (Bolton *et al.*, 1982; Diringer *et al.*, 1983; Rubenstein *et al.*, 1986, 1991; Shinagawa *et al.*, 1986; Race & Ernst, 1992). Both the normal and abnormal PrP isoforms are encoded by the same host gene (Basler *et al.*, 1986) and no TSE associated differences have been observed in either the PrP mRNA levels (Chesebro *et al.*, 1985; Oesch *et al.*, 1985) or the chemical structures of the PrP isoforms (Hope *et al.*, 1986; Stahl *et al.*, 1993). Pulse-chase metabolic labelling studies have shown that PrP-res is derived from mature PrP-sen posttranslationally (Borchelt *et al.*, 1990; Caughey & Raymond, 1991). These observations led to suggestions that the primary difference between PrP-res and PrP-sen might be conformational or due to interactions with cofactors (Hope *et al.*, 1986). Indeed, conformational analyses have provided evidence that PrP-res has a much higher beta sheet secondary structure composition than does PrP-sen (Caughey *et al.*, 1991*b*; Pan *et al.*, 1993; Safar *et al.*, 1993; Riek *et al.*, 1996).

Unlike normal PrP, PrP-res is resistant to phospholipase and protease treatments of intact scrapie-infected cells or their membranes (Caughey *et al.*, 1990; Stahl, Borchelt & Prusiner, 1990; Safar *et al.*, 1991) and shows no sign of turnover within the cells (Borchelt *et al.*, 1990; Caughey & Raymond,

1991). The low turnover of PrP-res can explain its accumulation *in vivo*. The formation of PrP-res occurs relatively slowly (Borchelt *et al.*, 1990) after its apparently normal (phospholipase- and protease-sensitive) PrP precursor reaches the cell surface (Caughey & Raymond, 1991). The conversion of PrP to the protease-resistant state likely occurs at the plasma membrane or along an endocytic pathway to the lysosomes (Caughey & Raymond, 1991; Caughey *et al.*, 1991; Borchelt, Taraboulos & Prusiner 1992). Once formed, PrP-res accumulates in secondary lysosomes (McKinley *et al.*, 1991), on the cell surface (Jeffrey *et al.*, 1992) or in extracellular spaces in the form of amorphous deposits, diffuse fibrils (Wiley *et al.*, 1987; Jeffrey *et al.*, 1994) or dense amyloid plaques (Bendheim *et al.*, 1984).

An important challenge that remains is to understand how, at the molecular level, the posttranslational conversion of PrP-sen to PrP-res occurs. Most protein-only models for the TSE agent predict that the putative infectious protein, such as PrP-res, directly interacts with the endogenous PrP-sen to induce the conversion of the latter to more PrP-res (Griffith, 1967; Bolton & Bendheim, 1988; Gadjusek, 1988; Prusiner, 1991; Jarrett & Lansbury, Jr. 1993). Passage of PrP-res from one host to another might then constitute 'infection' which could initiate pathogenic PrP-res formation in the new host. In the case of familial TSE diseases, mutations in the host PrP gene might facilitate spontaneous PrP-res formation without the need for infection with preformed PrP-res. Interestingly, PrP molecules containing mutations that are associated with humans familial TSE diseases exhibit altered metabolism and biochemical properties reminiscent of PrP-res when expressed in uninfected tissue culture cells (Lehmann & Harris, 1995, 1996; Petersen *et al.*, 1996). Even wild-type hamster PrP-sen can show a tendency to aggregate when expressed in the form of a dimer (Priola *et al.*, 1995).

Initial support for the importance of direct homologous PrP-sen PrP-res interactions in PrP-res formation came from a number of indirect lines of evidence. For example, in heterozygous humans with GSS (Indiana and Swedish kindreds) who express both the normal and mutant PrP alleles, only the mutant PrP is incorporated into amyloid fibrils (Tagliavini *et al.*, 1994). Studies in scrapie-infected transgenic mice and cell cultures have also shown that sequence compatibility between PrP-sen and PrP-res molecules is required for efficient PrP-res formation (Prusiner *et al.*, 1990; Scott *et al.*, 1992; Priola *et al.*, 1994). Conversely, expression of incompatible PrP-sen molecules differing by as little as one amino acid residue from the endogenous PrP can completely block PrP-res formation in initially scrapie-infected cells (Priola *et al.*, 1994; Priola & Chesebro, 1995). Analogous interference effects have been observed in the growth of synthetic amyloid fibrils by PrP peptide fragments when peptides of slightly different sequence are mixed together (Come & Lansbury, Jr. 1994).

Auto-induction of PrP-res formation in cell-free reactions

Direct evidence that PrP-res can induce its own formation has come from recent *in vitro* studies showing that PrP-sen can convert to the protease-resistant form in the presence of PrP-res isolated from scrapie-infected brain tissue (Kocisko *et al.*, 1994). The ability of PrP-res to induce this conversion of PrP-sen correlates with the presence of scrapie infectivity in guanidine denaturation studies (Caughey *et al.*, 1997). Striking specificity was observed in this cell-free conversion reaction which were consistent with, and might account for, biological aspects of TSE infections such as the existence of TSE strains and barriers in transmitting TSE strains between species (species barrier effects) (Bessen *et al.*, 1995; Kocisko *et al.*, 1995).

Species specificity in the conversion of PrP-res to PrP-sen

The molecular basis for species barrier effects was investigated by combining PrP-res and PrP-sen from different species in the cell-free conversion reaction (Kocisko *et al.*, 1995). Hamster PrP-res was unable to convert mouse PrP-sen to protease-resistant forms, whereas mouse PrP-res partially converted hamster PrP-sen. These results indicated that there was species specificity in the cell-free conversion reaction. Interestingly, the inability of hamster PrP-res to convert mouse PrP-sen to PK-resistant forms correlated with the inability of hamster scrapie strain 263K to infect mice. The contrasting ability of mouse PrP-res to convert hamster PrP-sen to PK-resistant forms correlated with the fact that the Chandler strain of mouse scrapie can be transmitted to hamsters, albeit with extended incubation periods. Thus, the species specificity of the cell-free conversion correlated with the corresponding scrapie species barriers observed *in vivo*. Conversion experiments performed with chimeric mouse/hamster PrP-sen precursors indicated that localized primary sequence differences between PrP-sen and PrP-res strongly affected the efficiency of the conversion reaction. These results provided strong support for the concept that the species specificity in the conversion of PrP-sen to protease-resistant forms contributes to the barriers to interspecies transmission and pathogenesis of TSE diseases *in vivo*.

Strain specificity in PrP-res formation

Various strains of TSE agents can be differentiated on the basis of species tropism, incubation period, clinical disease, neuropathological manifestations and PrP-res distribution in brain tissue (for review, see Bruce, 1996). Multiple TSE strains have been documented even within individual host species with invariant PrP genotype. This fact poses an interesting challenge for the protein-only hypothesis for the infectious agent; it requires that the 'inheritance' or propagation of the agent strain differences be mediated by stable variations in PrP-res structure rather than mutations in an agent-specific nucleic acid.

Strain differences in PrP-res structure have been described for the hyper (HY) and drowsy (DY) strains of hamster-adapted transmissible mink encephalopathy (TME) (Bessen & Marsh, 1992, 1994). Although HY and DY PrP-res are both derived from identical hamster PrP-sen molecules, they are cleaved at different N-terminal sites by proteinase K (Bessen & Marsh, 1994). Using the cell-free conversion reaction, we have demonstrated that HY and DY PrP-res molecules impart their strain-specific proteinase K resistance on hamster PrP-sen during formation of new PrP-res (Bessen *et al.*, 1995). In a sense, this represents the propagation of this strain specific biochemical phenotype of PrP-res by a nongenetic mechanism. It is likely that differences in polypeptide 3D structure account for the distinct proteinase K cleavages of HY and DY PrP-res. Although the role for strain-specific non-PrP cofactors has not been ruled out, these data are consistent with the concept that the self-propagation of PrP-res polymers with distinct 3D structures or conformations may be a molecular basis for scrapie strains.

Mechanism of PrP-res formation

The insolubility, protease resistance, birefringent staining with Congo red and high beta sheet content make at least some PrP-res deposits similar to the abnormal protein deposits, called amyloids, that occur in various other diseases (Prusiner *et al.*, 1983; Gadjusek, 1988). Amyloids can be composed of a number of different proteins, depending upon the disease (Glenner, 1980). Amyloid deposits are composed of linear fibrils that result from the polymerization of a usually soluble precursor protein or peptide. Preexisting amyloid fibril fragments can greatly accelerate the polymerization of amyloidogenic proteins much like seed crystals accelerating the crystallization of molecules from metastable supersaturated solutions (Gadjusek, 1988; Jarrett & Lansbury, Jr. 1993). Amyloid polymerization often involves an increase in the beta sheet content of the constituent protein. The similarities between PrP-res and other amyloids suggested that the mechanism of PrP-res formation is like that of other amyloids (Gadjusek, 1988; Jarrett & Lansbury, Jr. 1993). Additional support for this notion is the fact that small synthetic peptide fragments of the PrP sequence can form amyloid fibrils by a seeded polymerization mechanism (Gasset *et al.*, 1992; Come, Fraser & Lansbury, 1993; Forloni *et al.*, 1993; Goldfarb *et al.*, 1993).

More recent studies using the full PrP-res protein have provided evidence that only ordered aggregates of PrP-res, albeit widely variable in size, can induce the conversion of PrP-sen to the protease-resistant form (Caughey *et al.*, 1995). The polymerized state of PrP-res correlates with its proteinase K-resistance and its ability to renature to full proteinase K resistance after partial denaturation (Kocisko, Lansbury & Caughey, 1996). These observations are also consistent with, but do not yet prove, a seeded polymerization mechanism for PrP-res formation.

However, not all deposits of PrP-res *in vivo* give birefringent staining with Congo red or have readily visible amyloic fibril structures by electron microscopy (for example, Meyer *et al.*, 1986; Jeffrey *et al.*, 1992), leading some investigators to argue that polymerization of PrP is not required for PrP-res formation or neurotoxicity (Prusiner, 1991). One proposed mechanism derived from one of the Griffith models (Griffith, 1967), often called the heterodimer model, holds that a PrP-res monomer binds to a monomer of PrP-sen to form a heterodimer (Bolton & Bendheim, 1988; Prusiner, 1991). The heterodimer then spontaneously converts to a homodimer and then splits into two PrP-res monomers. The aggregation of PrP-res may then be a secondary phenomenon that results *in vivo* or *in vitro* only from partial proteolysis and/or extraction from the membrane environment (McKinley *et al.*, 1991). The fact that no proteinase K-resistant monomer of PrP has been identified runs counter to this argument. None the less, there have been reports of scrapie infectivity that cofractionates with monomeric forms of PrP (Brown *et al.*, 1990; Safar *et al.*, 1990), but these studies not been confirmed (Hope, 1994). Our own studies showing that ordered aggregates of PrP-res are active in converting PrP-sen to PrP-res demonstrate that there is at least no obligate requirement for a free PrP-res monomer, if one can exist, in the conversion mechanism (Caughey *et al.*, 1995). The apparent lack of visible fibrils in some tissue and membrane fractions containing PrP-res might be explained by a prevalence of short PrP-res polymers or the association of the PrP-res polymer with PrP-sen or other factors that obscure its underlying fibrillar ultrastructure and affect its birefringent staining with Congo red.

INHIBITION OF PRP-RES ACCUMULATION: A TSE THERAPY?

There is no effective treatment for TSE diseases once the clinical phase of diseases has begun. However, in animal or cell culture TSE models, some compounds, that is certain polyanions (Ehlers & Diringer, 1984; Farquhar & Dickinson, 1986; Kimberlin & Walker, 1986) and polyene antibiotics (Pocchiari, Schmittinger & Masullo, 1987; Demaimay *et al.*, 1994; McKenzie *et al.*, 1994), delayed the onset of clinical disease if given early enough after infection.

Two types of polyanions, sulfated glycans and a polyoxometalate, were initially tested because of their antiviral activities. It was later established that these polyanions, as well as the sulfonated dye Congo red, are potent inhibitors of PrP-res accumulation with IC_{50}s as low as ~100 pM (Caughey

& Race, 1992; Caughey & Raymond, 1993). The potency of various polyanionic compounds in inhibiting PrP-res accumulation correlated with their efficacies as anti-scrapie drugs. Congo red and sulfated glycans bind to both PrP-res (Prusiner *et al.*, 1983) and PrP-sen (Gabizon *et al.*, 1993; Caughey *et al.*, 1994; Shyng *et al.*, 1995) and sulfated glycosaminoglycans are components of PrP-res deposits *in vivo* (Snow *et al.*, 1989). These facts suggest that the mechanism of action of these compounds is to modulate an interaction between PrP and endogenous sulfated glycosaminoglycans that may be involved in PrP-res accumulation (Caughey *et al.*, 1994). One manifestation of sulfated glycan treatment that could be related to the prevention of PrP-res accumulation is a redistribution of PrP-sen within the cell (Shyng *et al.*, 1995).

The polyene antibiotics known to have anti-scrapie activity include the antifungal and antiviral drug amphotericin D and its analogues (Pocchiari *et al.*, 1987; Demaimay *et al.*, 1994; McKenzie *et al.*, 1994). The anti-scrapie mechanism of action of the amphotericins is not known, but it has been shown that these drugs delay the accumulation of PrP-res *in vivo*. A new derivative, MS8209, is less toxic and more broadly active than is amphotericin B itself, raising hopes that effective therapies might be based on such polyene antibiotics (Demaimay *et al.*, 1994).

The possibility that large numbers of people have been infected with BSE greatly increases the importance of developing effective therapies for TSE diseases. Although the polyanions and polyene antibiotics show promise as prophylactic drugs, there remains an urgent need for treatments that will be effective after clinical diagnoses of TSE disease have been made.

REFERENCES

Basler, K., Oesch, B., Scott, M., Westaway, D., Walchli, M., Groth, D. F., McKinley, M. P., Prusiner, S. B. & Weissmann, C. (1986). Scrapie and cellular PrP isoforms are encoded by the same chromosomal gene. *Cell*, **46**, 417–28.

Belt, P. B. G. M., Muileman, I. H., Schreuder, B. E. C., Ruijter, J. B., Gielkens, A. L. J. & Smits, M. A. (1995). Identification of five allelic variants of the sheep PrP gene and their association with natural scrapie. *Journal of General Virology*, **76**, 509–17.

Bendheim, P. E., Barry, R. A., DeArmond, S. J., Stites, D. P. & Prusiner, S. B. (1984). Antibodies to a scrapie prion protein. *Nature*, **310**, 418–21.

Bendheim, P. E., Brown, H. R., Rudelli, R. D., Scala, L. J., Goller, N. L., Wen, G. Y., Kascsak, R. J., Cashman, N. R. & Bolton, D. C. (1992). Nearly ubiquitous tissue distribution of the scrapie agent precursor protein. *Neurology*, **42**, 149–56.

Bessen, R. A. & Marsh, R. F. (1992). Identification of two biologically distinct strains of transmissible mink encephalopathy in hamsters. *Journal of General Virology*, **73**, 329–34.

Bessen, R. A. & Marsh, R. F. (1994). Distinct PrP properties suggest the molecular basis of strain variation in transmissible mink encephalopathy. *Journal of Virology*, **68**, 7859–68.

Bessen, R. A., Kocisko, D. A., Raymond, G. J., Nandan, S., Lansbury, P. T., Jr. &

Caughey, B. (1995). Nongenetic propagation of strain-specific phenotypes of scrapie prion protein. *Nature*, **375**, 698–700.

Bolton, D. C. and Bendheim, P. E. (1988). A modified host protein model of scrapie. In Bock, G. & Marsh, J., eds. *Novel Infectious Agents and the Central Nervous System*, pp. 164–81, Chichester: John Wiley & Sons.

Bolton, D. C., McKinley, M. P. & Prusiner, S. B. (1982). Identification of a protein that purifies with the scrapie prion. *Science*, **218**, 1309–11.

Bolton, D. C., Meyer, R. K. & Prusiner, S. B. (1985). Scrapie PrP 27–30 is a sialoglycoprotein. *Journal of Virology*, **53**, 596–606.

Borchelt, D. R., Scott, M., Taraboulos, A., Stahl, N. & Prusiner, S. B. (1990). Scrapie and cellular prion proteins differ in the kinetics of synthesis and topology in cultured cells. *Journal of Cell Biology*, **110**, 743–52.

Borchelt, D. R., Taraboulos, A. & Prusiner, S. B. (1992). Evidence for synthesis of scrapie prion protein in the endocytic pathway. *Journal of Biological Chemistry*, **267**, 16188–99.

Brown, P., Liberski, P., Wolff, A. & Gajdusek, D. C. (1990). Conservation of infectivity in purified fibrillary extracts of scrapie-infected hamster brain after sequential enzymatic digestion or polyacrylamide gel electrophoresis. *Proceedings of the National Academy of Sciences, USA*, **87**, 7240–4.

Bruce, M. E. (1996). Strain typing studies of scrapie and BSE. In Baker, H. & Ridley, R. M., eds. *Methods in Molecular Medicine: Prion Diseases*, pp. 223–36, Totowa: Humana Press.

Bueler, H., Fischer, M., Lang, Y., Bluethmann, H., Lipp, H.-P., DeArmond, S. J., Prusiner, S. B., Aguet, M. & Weissmann, C. (1992). Normal development and behavior of mice lacking the neuronal cell-surface PrP protein. *Nature*, **356**, 577–82.

Bueler, H., Aguzzi, A., Sailer, A., Greiner, R.-A., Autenried, P., Aguet, M. & Weissmann, C. (1993). Mice devoid of PrP are resistant to scrapie. *Cell*, **73**, 1339–47.

Carlson, G. A., Kingsbury, D. T., Goodman, P. A., Coleman, S., Marshall, S. T., DeArmond, S., Westaway, D. & Prusiner, S. B. (1986). Linkage of prion protein and scrapie incubation time genes. *Cell*, **46**, 503–11.

Carlson, G. A., Goodman, P. A., Lovett, M., Taylor, B. A., Marshall, S. T., Peterson-Torchia, M., Westaway, D. & Prusiner, S. B. (1988). Genetics and polymorphism of the mouse prion gene complex: control of scrapie incubation time. *Molecular and Cellular Biology*, **8**, 5528–40.

Cashman, N. R., Loertscher, R., Nalbantoglu, J., Shaw, I., Kascsak, R. J., Bolton, D. C. & Bendheim, P. E. (1990). Cellular isoform of the scrapie agent protein participates in lymphocyte activation. *Cell*, **61**, 185–92.

Caughey, B. & Raymond, G. J. (1991). The scrapie-associated form of PrP is made from a cell surface precursor that is both protease- and phospholipase-sensitive. *Journal of Biological Chemistry*, **266**, 18217–23.

Caughey, B. & Race, R. E. (1992). Potent inhibition of scrapie-associated PrP accumulation by Congo red. *Journal of Neurochemistry*, **59**, 768–71.

Caughey, B. & Raymond, G. J. (1993). Sulfated polyanion inhibition of scrapie-associated PrP accumulation in cultured cells. *Journal of Virology*, **67**, 643–50.

Caughey, B., Race, R. E. & Chesebro, B. (1988). Detection of prion protein mRNA in normal and scrapie-infected tissues and cell lines. *Journal of General Virology*, **69**, 711–16.

Caughey, B., Race, R. E., Vogel, M., Buchmeier, M. J. & Chesebro, B. (1988). *In vitro* expression in eukaryotic cells of the prion protein gene cloned from scrapie-infected mouse brain. *Proceedings of the National Academy of Sciences, USA*, **85**, 4657–61.

Caughey, B., Race, R. E., Ernst, D., Buchmeier, M. J. & Chesebro, B. (1989). Prion protein (PrP) biosynthesis in scrapie-infected and uninfected neuroblastoma cells. *Journal of Virology*, **63**, 175–81.

Caughey, B., Neary, K., Buller, R., Ernst, D., Perry, L., Chesebro, B. & Race, R. (1990). Normal and scrapie-associated forms of prion protein differ in their sensitivities to phospholipase and proteases in intact neuroblastoma cells. *Journal of Virology*, **64**, 1093–101.

Caughey, B., Raymond, G. J., Ernst, D. & Race, R. E. (1991*a*). N-terminal truncation of the scrapie-associated form of PrP by lysosomal protease(s): implications regarding the site of conversion of PrP to the protease-resistant state. *Journal of Virology*, **65**, 6597–603.

Caughey, B. W., Dong, A., Bhat, K. S., Ernst, D., Hayes, S. F. & Caughey, W. S. (1991*b*). Secondary structure analysis of the scrapie-associated protein PrP 27–30 in water by infrared spectroscopy. *Biochemistry*, **30**, 7672–80.

Caughey, B., Ernst, D. & Race, R. E. (1993). Congo red inhibition of scrapie agent replication. *Journal of Virology*, **67**, 6270–2.

Caughey, B., Brown, K., Raymond, G. J., Katzenstein, G. E. & Thresher, W. (1994). Binding of the protease-sensitive form of PrP (prion protein) to sulfated glycosaminoglycan and Congo red. *Journal of Virology*, **68**, 2135–41.

Caughey, B., Kocisko, D. A., Raymond, G. J. & Lansbury, P. T. (1995). Aggregates of scrapie associated prion protein induce the cell-free conversion of protease-sensitive prion protein to the protease-resistant state. *Chemistry and Biology*, **2**, 807–17.

Caughey, B., Raymond, G. J., Kocisko, D. A. & Lansbury, Jr., P. T. (1997). *Journal of Virology* (in press).

Chen, S. G., Teplow, D. B., Parchi, P., Teller, J. K., Gambetti, P. & Autilio-Gambetti, L. (1995). Truncated forms of the human prion protein in normal brain and in prion diseases. *Journal of Biological Chemistry*, **270**, 19173–80.

Chesebro, B., Race, R., Wehrly, K., Nishio, J., Bloom, M., Lechner, D., Bergstrom, S., Robbins, K., Mayer, L., Keith, J. M., Garon, C. & Haase, A. (1985). Identification of scrapie prion protein-specific mRNA in scrapie-infected and uninfected brain. *Nature*, **315**, 331–3.

Cho, H. J. (1986). Antibody to scrapie-associated fibril protein identifies a cellular antigen. *Journal of General Virology*, **67**, 243–53.

Collinge, J., Whittington, M. A., Sidle, K. C., Smith, C. J., Palmer, M. S., Clarke, A. & Jefferys, J. G. (1994). Prion protein is necessary for normal synaptic function. *Nature*, **370**, 295–7.

Come, J. H., Fraser, P. E. & Lansbury, P. T., Jr. (1993). A kinetic model for amyloid formation in the prion diseases: importance of seeding. *Proceedings of the National Academy of Sciences, USA*, **90**, 5959–63.

Come, J. H. & Lansbury, P. T., Jr. (1994). Predisposition of prion protein homozygotes to Creutzfeldt-Jakob disease can be explained by a nucleation-dependent polymerization mechanism. *Journal of the American Chemical Society*, **116**, 4109–10.

Demaimay, R., Adjou, K., Lasmezas, C., Lazarini, F., Cherifi, K., Seman, M., Deslys, J. P. & Dormont, D. (1994). Pharmacological studies of a new derivative of amphotericin B, MS-8209, in mouse and hamster scrapie. *Journal of General Virology*, **75**, 2499–503.

Diringer, H., Gelderblom, H., Hilmert, H., Ozel, M., Edelbluth, C. & Kimberlin, R. H. (1983). Scrapie infectivity, fibrils and low molecular weight protein. *Nature*, **306**, 476–8.

Dormont, D., Yeramian, P., Lambert, P. *et al.* (1986). *In vitro* and *in vivo* antiviral effects of HPA23. In Court, L. A., Dormont, D., Brown, P. & Kingsbury, D. T., eds. *Unconventional Virus Diseases of the Central Nervous System*, pp. 324–37, Paris: Masson.

Ehlers, B. & Diringer, H. (1984). Dextran sulphate 500 delays and prevents mouse scrapie by impairment of agent replication in spleen. *Journal of General Virology*, **65**, 1325–30.

Eklund, C. M., Hadlow, W. J. & Kennedy, R. C. (1963). Some properties of the scrapie agent and its behavior in mice. *Proceedings of the Society for Experimental Biology and Medicine*, **112**, 974–9.

Ernst, D. R. & Race, R. E. (1993). Comparative analysis of scrapie agent inactivation methods. *Journal of Virological Methods*, **41**, 193–202.

Farquhar, C. F. & Dickinson, A. G. (1986). Prolongation of scrapie incubation period by an injection of dextran sulphate 500 within the month before or after infection. *Journal of General Virology*, **67**, 463–73.

Forloni, G., Angeretti, N., Chiesa, R., Monzani, E., Salmona, M., Bugiani, O. & Tagliavini, F. (1993). Neurotoxicity of a prion protein fragment. *Nature*, **362**, 543–6.

Gabizon, R., Meiner, Z., Halimi, M. & Bensasson, S. A. (1993). Heparin-like molecules bind differentially to prion proteins and change their intracellular metabolic-fate. *Journal of Cellular Physiology*, **157**, 319–25.

Gadjusek, D. C. (1988). Transmissible and nontransmissible amyloidoses: Autocatalytic post-translational conversion of host precursor proteins to beta-pleated configurations. *Journal of Neuroimmunology*, **20**, 95–110.

Gasset, M., Baldwin, M. A., Lloyd, D. H., Gabriel, J., Holtzman, D. M., Cohen, F., Fletterick, R. & Prusiner, S. B. (1992). Predicted α-helical regions of the prion protein when synthesized as peptides form amyloid. *Proceedings of the National Academy of Sciences, USA*, **89**, 10940–44.

Gibbs, C. J., Jr. & Gajdusek, D. C. (1973). Experimental subacute spongiform virus encephalopathies in primates and other laboratory animals. *Science*, **182**, 67–8.

Gibbs, C. J., Jr., Amyx, H. L., Bacote, A., Masters, C. L. & Gajdusek, D. C. (1980). Oral transmission of Kuru, Creutzfeldt–Jakob disease, and scrapie to nonhuman primates. *Journal of Infectious Diseases*, **142**, 205–8.

Glenner, G. G. (1980). Amyloid deposits and amyloidosis: the beta-fibrillosa (second of two parts). *New England Journal of Medicine*, **302**, 1333–43.

Goldfarb, L. G., Brown, P., Haltia, M., Ghiso, J., Frangione, B. & Gajdusek, D. C. (1993). Synthetic peptides corresponding to different mutated regions of the amyloid gene in familial Creutzfeld-Jakob disease show enhanced in vitro formation of morphologically different amyloid fibrils. *Proceedings of the National Academy of Sciences, USA*, **90**, 4451–4.

Goldmann, W., Hunter, N., Smith, G., Foster, J. & Hope, J. (1994). PrP genotype and agent effects in scrapie: change in allelic interaction with different isolates of agent in sheep, a natural host of scrapie. *Journal of General Virology*, **75**, 989–95.

Griffith, J. S. (1967). Self-replication and scrapie. *Nature*, **215**, 1043–4.

Haraguchi, T., Fisher, S., Olofsson, S., Endo, T., Groth, D., Tarentino, A., Borchelt, D. R., Teplow, D., Hood, L., Burlingame, A., Lycke, E., Kobata, A. & Prusiner, S. B. (1989). Asparagine-linked glycosylation of the scrapie and cellular prion proteins. *Archives of Biochemistry and Biophysics*, **274**, 1–13.

Hope, J. (1994). The nature of the scrapie agent: the evolution of the virino. *Annals of the New York Academy of Sciences*, **724**, 282–9.

Hope, J., Morton, L. J. D., Farquhar, C. F., Multhaup, G., Beyreuther, K. & Kimberlin, R. H. (1986). The major polypeptide of scrapie-associated fibrils (SAF) has the same size, charge distribution and N-terminal protein sequence as predicted for the normal brain protein (PrP). *EMBO Journal*, **5**, 2591–7.

Hsiao, K. K., Scott, M., Foster, D., Groth, D. F., DeArmond, S. J. & Prusiner, S. B. (1990). Spontaneous neurodegeneration in transgenic mice with mutant prion protein. *Science*, **250**, 1587–90.

Hsiao, K. K., Groth, D., Scott, M., Serban, H., Rapp, D., Foster, D., Torchia, M., DeArmond, S. J. & Prusiner, S. B. (1994). Serial transmission in rodents of neurodegeneration from transgenic mice expressing mutant prion protein. *Proceedings of the National Academy of Sciences, USA*, **91**, 9126–30.

Hunter, N., Hope, J., McConnell, I. & Dickinson, A. G. (1987). Linkage of the scrapie-associated fibril protein (PrP) gene and sinc using congenic mice and restriction fragment length polymorphism analysis. *Journal of General Virology*, **68**, 2711–16.

Jarrett, J. T. & Lansbury, P. T., Jr. (1993). Seeding 'one-dimensional crystallization' of amyloid: a pathogenic mechanism in Alzheimer's disease and scrapie? *Cell*, **73**, 1055–8.

Jeffrey, M., Goodsir, C. M., Bruce, M. E., McBride, P. A., Scott, J. R. & Halliday, W. G. (1992). Infection specific prion protein (PrP) accumulates on neuronal plasmalemma in scrapie infected mice. *Neuroscience Letters*, **147**, 106–9.

Jeffrey, M., Goodsir, C. M., Bruce, M. E., McBride, P. A., Fowler, N. & Scott, J. R. (1994). Murine scrapie-infected neurons *in vivo* release excess prion protein into the extracellular space. *Neuroscience Letters*, **174**, 39–42.

Kimberlin, R. H. & Walker, C. A. (1986). Suppression of scrapie infection in mice by heteropolyanion 23, dextran sulfate, and some other polyanions. *Antimicrobial Agents and Chemotherapy*, **30**, 409–13.

Kocisko, D. A., Come, J. H., Priola, S. A., Chesebro, B., Raymond, G. J., Lansbury, P. T. & Caughey, B. (1994). Cell-free formation of protease-resistant prion protein. *Nature*, **370**, 471–4.

Kocisko, D. A., Priola, S. A., Raymond, G. J., Chesebro, B., Lansbury, P. T., Jr. & Caughey, B. (1995). Species specificity in the cell-free conversion of prion protein to protease-resistant forms: a model for the scrapie species barrier. *Proceedings of the National Academy of Sciences, USA*, **92**, 3923–27.

Kocisko, D. A., Lansbury, P. T. Jr. & Caughey, B. (1996). Partial unfolding and refolding of scrapie-associated prion protein: evidence for a critical 16-kDa C-terminal domain. *Biochemistry*, **35**, 13434–44.

Lehmann, S. & Harris, D. A. (1995). A mutant prion protein displays an aberrant membrane association when expressed in cultured cells. *Journal of Biological Chemistry*, **270**, 24589–97.

Lehmann, S. & Harris, D. A. (1996). Mutant and infectious prion proteins display common biochemical properties in cultured cells. *Journal of Biological Chemistry*, **271**, 1633–7.

Lieberburg, I. (1992). Developmental expression and regional distribution of the scrapie-associated protein mRNA in the rat central nervous system. *Brain Research*, **417**, 363–6.

Manson, J., West, J. D., Thomson, V., McBride, P., Kaufman, M. H. & Hope, J. (1992). The prion protein gene: a role in mouse embryogenesis? *Development*, **115**, 117–22.

Manuelidis, L., Valley, S. & Manuelidis, E. E. (1985). Specific proteins associated with Creutzfeldt-Jakob disease and scrapie share antigenic and carbohydrate determinants. *Proceedings of the National Academy of Sciences, USA*, **82**, 4263–7.

McKenzie, D., Kaczkowski, J., Marsh, R. & Aiken, J. M. (1994). Amphotericin B delays both scrapie agent replication and PrP-res accumulation early in infection. *Journal of Virology*, **68**, 7534–6.

McKinley, M. P., Meyer, R. K., Kenaga, L., Rahbar, F., Cotter, R., Servan, A. &

Prusiner, S. B. (1991*a*). Scrapie prion rod formation *in vitro* requires both detergent extraction and limited proteolysis. *Journal of Virology*, **65**, 1340–51.

McKinley, M. P., Taraboulos, A., Kenaga, L., Serban, D., Stieber, A., DeArmond, S. J., Prusiner, S. B. & Gonatas, N. (1991*b*). Ultrastructural localization of scrapie prion proteins in cytoplasmic vesicles of infected cultured cells. *Laboratory Investigation*, **65**, 622–30.

Merz, P. A., Somerville, R. A., Wisniewski, H. M. & Iqbal, K. (1981). Abnormal fibrils from scrapie-infected brain. *Acta Neuropathologica*, **54**, 63–74.

Meyer, R. K., McKinley, M. P., Bowman, K. A., Braunfeld, M. B., Barry, R. A. & Prusiner, S. B. (1986). Separation and properties of cellular and scrapie prion protein. *Proceedings of the National Academy of Sciences, USA*, **83**, 2310–14.

Mobley, W. C., Neve, R. L., Prusiner, S. B. & McKinley, M. P. (1988). Nerve growth factor increases mRNA levels for the prion protein and the beta-amyloid protein precursor in developing hamster brain. *Proceedings of the National Academy of Sciences, USA*, **85**, 9811–15.

Oesch, B., Westaway, D., Walchli, M., McKinley, M. P., Kent, S. B. H., Aebersold, R., Barry, R. A., Tempst, P., Teplow, D. B., Hood, L. E., Prusiner, S. B. & Weissmann, C. (1985). A cellular gene encodes scrapie PrP 27–30 protein. *Cell*, **40**, 735–46.

Pan, K.-M., Baldwin, M., Nguyen, J., Gasset, M., Serban, A., Groth, D., Mehlhorn, I., Huang, Z., Fletterick, R. J., Cohen, F. E. & Prusiner, S. B. (1993). Conversion of alpha-helices into beta-sheets features in the formation of the scrapie prion protein. *Proceedings of the National Academy of Sciences, USA*, **90**, 10962–6.

Petersen, R. B., Parchi, P., Richardson, S. L., Urig, C. B. & Gambetti, P. (1996). Effect of the D178N mutation and the codon 129 polymorphism on the metabolism of the prion protein. *Journal of Biological Chemistry*, **271**, 12661–8.

Pocchiari, M. (1994). Prions and Related Neurological Diseases. *Molecular Aspects in Medicine*, **15**, 195–291.

Pocchiari, M., Schmittinger, S. & Masullo, C. (1987). Amphotericin B delays the incubation period of scrapie in intracerebrally inoculated hamsters. *Journal of General Virology*, **68**, 219–23.

Priola, S. A. & Chesebro, B. (1995). A single hamster amino acid blocks conversion to protease-resistant PrP in scrapie-infected mouse neuroblastoma cells. *Journal of Virology*, **69**, 7754–8.

Priola, S. A., Caughey, B., Race, R. E. & Chesebro, B. (1994). Heterologous PrP molecules interfere with accumulation of protease-resistant PrP in scrapie-infected murine neuroblastoma cells. *Journal of Virology*, **68**, 4873–8.

Priola, S. A., Caughey, B., Wehrly, K. & Chesebro, B. (1995). A 60-kDa prion protein (PrP) with properties of both the normal and scrapie-associated forms of PrP. *Journal of Biological Chemistry*, **270**, 3299–305.

Prusiner, S. B. (1982). Novel proteinaceous infectious particles cause scrapie. *Science*, **216**, 136–44.

Prusiner, S. B. (1991). Molecular biology of prion diseases. *Science*, **252**, 1515–22.

Prusiner, S. B., McKinley, M. P., Bowman, K. A., Bendheim, P. E., Bolton, D. C., Groth, D. F. & Glenner, G. G. (1983). Scrapie prions aggregate to form amyloid-like birefringent rods. *Cell*, **35**, 349–58.

Prusiner, S. B., Scott, M., Foster, D., Pan, K. M., Groth, D., Mirenda, C., Torchia, M., Yang, S. L., Serban, D., Carlson, G. A., Hoppe, P. C., Westaway, D. & DeArmond, S. J. (1990). Transgenetic studies implicate interactions between homologous PrP isoforms in scrapie prion replication. *Cell*, **63**, 673–86.

Race, R. E., Graham, K., Ernst, D., Caughey, B. & Chesebro, B. (1990). Analysis of linkage between scrapie incubation period and the prion protein gene in mice. *Journal of General Virology*, **71**, 493–7.

Race, R. E. & Ernst, D. (1992). Detection of proteinase K-resistant prion protein and infectivity in mouse spleen by 2 weeks after scrapie agent inoculation. *Journal of General Virology*, **73**, 3319–23.

Riek, R., Hornemann, S., Wider, G., Billeter, M., Glockshuber, R. & Wiithrich, K. (1996). NMR Structure of the mouse prion protein domain PrP (121–231). *Nature*, **382**, 180–2.

Robakis, N. K., Sawh, P. R., Wolfe, G. C., Rubenstein, R., Carp, R. I. & Innis, M. A. (1986). Isolation of a cDNA clone encoding the leader peptide of prion protein and expression of the homologous gene in various tissues. *Proceedings of the National Academy of Sciences, USA*, **83**, 6377–81.

Rohwer, R. G. (1991). The scrapie agent: 'a virus by any other name'. *Current Topics in Microbiology and Immunology*, **172**, 195–232.

Rubenstein, R., Kascsak, R. J., Merz, P. A., Papini, M. C., Carp, R. I., Robakis, N. K. & Wisniewski, H. M. (1986). Detection of scrapie-associated fibril (SAF) proteins using anti-SAF antibody in non-purified tissue preparations. *Journal of General Virology*, **67**, 671–81.

Rubenstein, R., Merz, P. A., Kascsak, R. J., Scalici, C. L., Papini, M. C., Carp, R. I. & Kimberlin, R. H. (1991). Scrapie-infected spleens: analysis of infectivity, scrapie-associated fibrils, and protease-resistant proteins. *Journal of Infectious Diseases*, **164**, 29–35.

Safar, J., Wang, W., Padgett, M. P., Ceroni, M., Piccardo, P., Zopf, D., Gajdusek, D. C. & Gibbs, C. J., Jr. (1990). Molecular mass, biochemical composition, and physicochemical behavior of the infectious form of the scrapie precursor protein monomer. *Proceedings of the National Academy of Sciences, USA*, **87**, 6373–7.

Safar, J., Ceroni, M., Gajdusek, D. C. & Gibbs, C. J., Jr. (1991). Differences in the membrane interaction of scrapie amyloid precursor proteins in normal and scrapie- or Creutzfeldt–Jakob disease-infected brains. *Journal of Infectious Diseases*, **163**, 488–94.

Safar, J., Roller, P. P., Gajdusek, D. C. & Gibbs, C. J., Jr. (1993). Conformational transitions, dissociation, and unfolding of scrapie amyloid (prion) protein. *Journal of Biological Chemistry*, **268**, 20276–84.

Sakaguchi, S., Katamine, S., Nishida, N., Moriuchi, R., Shigematsu, K., Sugimoto, T., Nakatani, A., Kataoka, Y., Houtani, T., Shirabe, S., Okada, H., Hasegawa, S., Miyamoto, T. & Noda, T. (1996). Loss of cerebellar Purkinje cells in aged mice homozygous for a disrupted PrP gene. *Nature*, **380**, 528–31.

Scott, M. R., Kohler, R., Foster, D. & Prusiner, S. B. (1992). Chimeric prion protein expression in cultured cells and transgenic mice. *Protein Science*, **1**, 986–97.

Shinagawa, M., Munekata, E., Doi, S., Takahashi, K., Goto, H. & Gato, G. (1986). Immunoreactivity of a synthetic pentadecapeptide corresponding to the N-terminal region of the scrapie prion protein. *Journal of General Virology*, **67**, 1745–50.

Shyng, S. L., Lehmann, S., Moulder, K. L. & Harris, D. A. (1995*a*). Sulfated glycans stimulate endocytosis of the cellular isoform of the prion protein, PrPC, in cultured cells. *Journal of Biological Chemistry*, **270**, 30221–9.

Shyng, S. L., Moulder, K. L., Lesko, A. & Harris, D. A. (1995*b*). The N-terminal domain of a glycolipid-anchored prion protein is essential for its endocytosis via clathrin-coated pits. *Journal of Biological Chemistry*, **270**, 14793–800.

Snow, A. D., Kisilevsky, R., Willmer, J., Prusiner, S. B. & DeArmond, S. J. (1989). Sulfated glycosaminoglycans in amyloid plaques of prion diseases. *Acta Neuropathologica*, **77**, 337–42.

Stahl, N., Borchelt, D. R., Hsiao, K. & Prusiner, S. B. (1987). Scrapie prion protein contains a phosphatidylinositol glycolipid. *Cell*, **51**, 229–40.

Stahl, N., Borchelt, D. R. & Prusiner, S. B. (1990). Differential release of cellular and scrapie prion proteins from cellular membranes by phosphatidylinositol-specific phospholipase C. *Biochemistry*, **29**, 5405–12.

Stahl, N., Baldwin, M. A., Teplow, D. B., Hood, L., Gibson, B. W., Burlingame, A. L. & Prusiner, S. B. (1993). Structural studies of the scrapie prion protein using mass spectrometry and amino acid sequencing. *Biochemistry*, **32**, 1991–2002.

Tagliavini, F., Prelli, F., Porro, M., Salmona, M., Bugiani, O. & Frangione, B. (1992). A soluble form of prion protein in human cerebrospinal fluid: Implications for prion-related encephalopathies. *Biochemical and Biophysical Research Communications*, **184**, 1398–404.

Tagliavini, F., Prelli, F., Porro, M., Rossi, G., Giaccone, G., Farlow, M. R., Dlouhy, S. R., Ghetti, B., Bugiani, O. & Frangione, B. (1994). Amyloid fibrils in Gerstmann-Straussler-Scheinker disease (Indiana and Swedish kindreds) express only PrP peptides encoded by the mutant allele. *Cell*, **79**, 695–703.

Tobler, I., Gaus, S. E., Deboer, T., Achermann, P., Fischer, M., Rulicke, T., Moser, M., Oesch, B., McBride, P. A. & Manson, J. C. (1996). Altered circadian activity rhythms and sleep in mice devoid of prion protein. *Nature*, **380**, 639–42.

Weissman, C. (1996). Molecular biology of transmissible spongiform encephalopathies. *FEBS Letters*, **389**, 3–11.

Wells, G. A. H. & Wilesmith, J. W. (1995). The neuropathology and epidemiology of bovine spongiform encephalopathy. *Brain Pathology*, **5**, 91–103.

Whatley, S. A., Powell, J. F., Politopoulou, G., Campbell, I. C., Brammer, M. J. & Percy, N. S. (1995). Regulation of intracellular free calcium levels by the cellular prion protein. *Neuroreport*, **6**, 2333–7.

Wiley, C. A., Burrola, P. G., Buchmeier, M. J., Wooddell, M. K., Barry, R. A., Prusiner, S. B. & Lampert, P. W. (1987). Immuno-gold localization of prion filaments in scrapie-infected hamster brains. *Laboratory Investigation*, **57**, 646–55.

Will, R. G., Ironside, J. W., Zeidler, M., Cousens, S. N., Estibeiro, K., Alperovitch, A., Poser, S., Pocchiari, M., Hofman, A. & Smith, P. G. (1996). A new variant of Creutzfeldt–Jakob disease in the UK. *Lancet*, **347**, 921–5.

INDEX

α-actinin, 244
$\alpha_v\beta_3$ vitronectin receptor, 29
actin, 98, 238, 259, 312
  bundling, 242
  comet tails, 238
adenovirus, 33
  E1A protein, 115
adenylate cyclase, 48
adhesins, 95, 191
  AfaE, 241
  InvA, 53
Agrobacterium, 265
  Ti (tumour-inducing) plasmid, 266; export, 269
alanine scanning, 333
alfalfa mosaic virus (AMV), 258
amphotericin D, 349
amyloids, 347
antigenic variation, 100, 191
  drift, 171
  shift, 171
Aphthoviruses, 31
arthritis, 74
ATPase, 264
attaching and effacing lesion, 312

17-β-hydroxysteroid-dehydrogenase, 336
±-2 microglobulin, 33
bacterial capsule, 99
bacterial nucleoid, 195
bacteriophage lambda, 198
BAT1 protein, 336
bean dwarf mosaic geminivirus (BDMV), 259
Bordetella pertussis, 97
bovine spongiform encephalopathy(BSE), 341
bundle forming pilus, 312

Cadherins, 95
Calcium-dependency, 46
Candida albicans, 80, 96
Canine parvovirus (CPV), 297
canyon hypothesis, 24, 283
Cardioviruses, 31
cauliflower mosaic virus, 258
CD4, 28, 288
CD46, 29
CD55, 28
CD66 (CEA, carcinoembryonic antigen), 105
CD66a (BGP, biliary glycoprotein), 105
CELICS; cloning by immunocolorimetric screening, 28
cell-mediated immunity, 170
chlamydia
  C. pecorum, 166
  C. psittaci, 153
  C. trachomatis, 153
  attachment to host cells, 160
  genomic sequencing, 167
  growth cycle, 159
  histone protein, 165
  incomplete development, 166
  protective immunity, 169
  taxomomy, 154
coagulase, 67, 81
collagen, 73, 75
  binding protein(s), 67–68, 73
Congo red, 348
cortactin, 245
Coxsackievirus, 30
  A9 receptor (±v±3 vitronectin), 30
  serotypes A13, A15, A18 and A21, 30
Creutzfeldt–Jakob disease (CJD), 341
cryoelectron microscopy, 288
cucumber mosaic virus (CMV), 261
cytokines, 36, 237
cytoskeleton, 322
  reorganization, 239, 248
cytotoxic T-cells, 174

daa (F1845), 199
decay accelerating factor (DAF), 28
desmotubule, 253
diarrhoeal diseases, 237, 311
DNA
  bending, 198
  curved, 196
  gyrase, 201
  invertible, 193
  methylation, 200
  positively supercoiled, 201–202
  supercoiling, 200
  synthesis, 196
  topoisomerases, 200
  vaccine, 179
  viruses, 327
Drosophila, 272
drug design, 134
dsRNA analogues, 227

EBER-1, 214
EBNA2, 114
echovirus, 28
Echovirus 22, 29
EF-hand, 77, 79–80

elastin, 82
elastin-binding protein, 69, 82
Encephalomyocarditis virus (EMCV), 31
endocarditis, 67, 73, 75, 80–81, 87
endocytosis, 162
endothelial cells, 73
enterocyte effacement (LEE) region, 314
E. coli, 192
  enterohemorrhagic, 312
  enteropathogenic (EPEC), 311
  enterotoxigenic, 197
  exponential growth, 198
enteroviruses, 25, 27, 70, 29
environmental signals, 203
epithelial cells, 311
Epstein–Barr virus, 214
extracellular matrix (ECM) components, 102

fae (K88), 199
fan (K99), 199
fatal familial insomnia, 341
fibrin, 82
fibrinogen, 75, 77, 79–82
  binding protein(s), 67, 69, 75, 80–81
fibronectin, 70–73, 80, 86, 98
  binding protein(s), 67, 69–70, 71, 75, 80, 86–87
  receptor ($\alpha 5\beta 1$, FNR), 104
fimbriae, 95
  type 1, 192
fimA gene, 193
fimB gene, 193, 203
FimB protein, 193, 200
fimE gene, 193
FimE protein, 193
foot and mouth disease virus, 31

G-proteins, 247
gamma interferon, 166–167
gangliosides, 104
gene knockout, 343
Gerstmann–Straussler–Scheinker syndrome, 341
globoside, 297
glucocorticoid receptor, 263
guanosine tetraphosphate, 198

H-NS protein, 195–197, 204
Haemophilus influenzae, 100
heat shock proteins
  Hsp60, 175
  Hsp70, 175
hepatitis A, 32
herpes simplex virus, 111
  Vmw63, 120
  Vmw68, 118
  Vmw110, 121
  Vmw175, 116
heterodimer model, 348
himA, 198
himD, 198
hip, 198
hns, 196
hns mutants, 196
human immunodeficiency virus (HIV), 129

ICAM-1, 30, 281, 283, 291
IL-8 secretion, 51
immunoglobulin (Ig) superfamily, 27, 30, 95
  immunoglobulin genes, 202
  immunoglobulin heavy chain class switching, 203
influenza viruses, 328
  genetic engineering, 330
  non-structural proteins (NS1), 327
  nucleoprotein (NP), 327
  nucleoprotein interactor 1 (NPI-1), 331
intasome, 198
integration host factor (IHF), 198
  IHBβ, 198
  ihfA, 198
  ihfB, 198
  IHFα, 198
integrin(s), 70, 75, 77, 79–80, 95, 237
integrin VLA-2, 28
intercellular
  communication, 265
  traffic, 253
  transport, 258
intimin, 313–314
intimin receptor, 315
intracellular
  calcium, 313
  differentiation, 163
  targeting, 319

IP_3, 313
Ipa proteins, 239
IRES, 215
IS*1*, 193

karyopherin α, 263, 332
karyopherin β, 263, 332
kuru, 341

L8 LPS immunotype, 100
lac repressor, 140
LcrD, 54–55
LcrV, 46, 51–52
Legionella pneumophila, 96
Leishmania, 98
leukocyte complement receptor CR3, 96
leuX gene, 203
lipopolysaccharide (LPS), 104, 176
Listeria, 237
long terminal repeat (LTR), 129

low density lipoprotein receptor (LDLR)
  family, 31
LPXTG motif, 67–69, 74–75 81
lysosomal glycoproteins, 321

M cells, 98, 237
mxi-spa locus, 239
macromolecular chelation complexes, 8
macrophages, 46
macropinocytosis, 318
mad cow disease, 341
measles virus P protein, 332
membrane
  leaflets, 242
  ruffling, 318
mesophyll cells, 261
MHC class II
  adherent protein, 69
  analogous protein, 81
microspike structures, 242
MIDAS, 79–80
Mycobacteria avium/intracellulare, 96
myosin, 242, 312

Neisseria meningitidis, 95
NF-κB transcription factor, 263
nitric oxide, 174
novobiocin, 201
NS1-I (NS1-Interacting) protein, 336
nuclear import, 327
  cytoplasmic anchoring, 263
  cytoplasmic receptors, 264
nuclear localization signal (NLS), 262, 332
  bipartite NLS, 262
nuclear pore, 328
  complex (NPC), 262
nucleoplasmin, 333

Olm phenotype, 238
opsonins, 105
opsonization, 73
osteomyelitis, 67, 74

pap (P), 199
papillomavirus E2 protein, 119
parvovirus B19, 297
paxillin, 247
peptoids, 134
pertussis toxin, 97
phase variation, 191
PhoP/PhoQ, 320
picornavirus, 23
  receptors, 25, 283
PilC, 101
pilG, 195
pili, 95
plant
  cell wall-associated protein kinase, 265
  NLS receptors, 273
  transgenic, 270
  viruses, 214
plasmin, 82
plasminogen, 67, 82–83
plasmodesmata, 253
  increase in permeability, 260
  localization signal (PLS), 264
  permeability, 255
  secondary, 265
  size exclusion limit, 255
plastin, 242
poliovirus, 24, 27
  'A'-particles, 292
  receptor (PVR), 27
polyanions, 348
polyene antibiotics, 348–349
polymorphonuclear leukocytes, 46
potato virus X, 261
pp60$^{c\text{-}src}$, 245
prion hypothesis, 342
protein
  association, 10
  contact areas, 1
  contact residues, 5
  cytoskeleton interaction, 259
  disorder to order transitions, 6
  hydrogen bonding, 12
  hydrogen bonds, 6
  hydrophobic interactions, 13
  large conformational change, 7
  nucleic acid complexes, 253
  residue packing, 6
  RNA complex, 258
  stability of protein–protein complexes, 10
protein A, 69–70
protein IHF, 198
protein LRP, 199
protein StpA, 197
protein-only hypothesis, 342
prothrombin, 67, 81
PrP, 343
  biosynthesis and metabolism, 344
  fragment, 344
  protease-resistant, 342
PrP-res
  aggregates, 347
  auto-induction of formation, 346
  conformational analyses, 344
  formation, 344, 347
  homologous PrP-sen PrP-res interactions,
    345
  inhibition of accumulation, 348
  propagation strain specific biochemical
    phenotype, 347
  species specificity in conversion to PrP-sen,
    346
prr exclusion system, 227

pYV plasmid, 46

Rac/Rho family, 247
Ran (GTPase), 263
  interacting protein p10, 263
red clover necrotic mosaic dianthovirus (RCNMV), 258
retroviruses, 214
RGD motif, 96
Rhinoviruses, 30–31
  HRV1A, 293
  HRV14, 285, 291
  HRV16, 294
  major group receptor, 287
ribonucleoprotein (RNP), 328
ribosome
  shunting, 217
  leaky scanning, 217
RNA, 258
  helicase, 336
  phages, 212–213
  polymerase, 196
  polymerase II, 140
  viruses, 327
RNase L, 224
RNaseH, 143
RpoS, 198

S. epidermidis, 67, 72, 86
Salmonella, 237, 311
scrapie associated fibrils, 342
Selectins, 95
sep, 314
septic arthritis, 67
septicaemia, 67
sfa (S), 199
Shigella flexneri, 196
Shigella, 237
Shigella 220 kDa virulence plasmid, 238
sigma factor, 165, 198
  RpoS, 203
signal
  masking, 263
  transduction, 311
Simkania, 158
SKI, 228
SP1 binding sites, 146
Specific Yop Chaperone (Syc), 57
  SycD, 57–58
  SycE, 57–58
  SycH, 57–58
splicing, 119
  block, 336
squash leaf curl virus (SqLCV), 259
SRP1 protein, 331
ssDNA binding protein (SSB), 268
staphylokinase, 67, 82
stationary phase, 196, 198, 203
Streptococcus dysgalactiae, 72
Streptococcus, 72
SV40 large T-antigen, 333

T-helper cells, 173
talin, 244, 312
TAR, 129
  conformational rearrangement, 132
  NMR studies, 132
  RNA 'decoy', 133
Tat, 129
  cellular co-factors, 147
TATA element, 146
thrombospondin, 82
thymidylate synthase, 197
TNF a, 36
tobacco mosaic virus (TMV), 257
  30 kDa protein (P30), 257; kinase activity, 265; microinjection, 261
tobacco rattle virus (TRV), 261
topA gene, 201
topoisomerase I, 200
trachoma, 153, 171
transactivators, 112
  HCMV ie2, 115
transcription
  cell-free systems, 136
  elongation, 134
  elongation complex 140
  factor: E2F, 115; TFIIB, 115; TFIID, 115
  initiation complex, 111
transgenics, 337
translation
  coupling, 220
  frameshifting, 219
  host cell shutoff, 212
  inhibition, 224
  initiation, 219; eIF2 kinase PKR, 223; eIF2 phosphorylation, 222; eIF4E dephosphorylation, 222; eIF4G, 212; eIF4G cleavage, 223
  reinitiation, 218
transmissible spongiform encephalopathies (TSE), 341
  familial diseases, 343
  therapy, 348
trichome cells, 261
tRNA cleavage, 221
tubulin, 259
tyrosine phosphorylation, 245, 313

ubiquitin specific protease, 123

vinculin, 239
Viral movement proteins, 257
virB gene, 196
virD, 266
VirD2, 268

virE, 266
VirE1, 269
VirE2, 268
VirE2 NLSs, 272
VirF, 58
virulence mechanisms, 311
vitronectin, 33, 82
  receptor (αvβ3, VNR), 104

'WIN' compounds, 293

Xenopus, 272

YadA, 53, 98
yeast two-hybrid system, 327
  'bait' protein, 331
  'catch' protein, 331
  'prey' protein, 331
Yersinia enterocolitica, 45
Yersinia pestis, 45
Yersinia pseudotuberculosis, 45
Yersinia, 46, 237
Yersinia outer proteins, 46
YmoA, 58–59
Yop proteins, 46–59, 239
  Feedback inhibition of synthesis, 59
  YopB, 49, 52, 58
  YopD, 47–49, 51–52, 58
  YopE, 46–49, 51–52, 56–58
  YopH, 49, 51–52, 56, 58
  YopM, 49, 51–52, 56
  YopN, 53–55
  YopO, 51, 56
YpkA (YopO), 49, 52
YscC, 55
YscJ, 55
YscN, 55